Advances in
Ergonomics Modeling
and Usability Evaluation

Advances in Human Factors and Ergonomics Series

Series Editors

Gavriel Salvendy
Professor Emeritus
Purdue University
West Lafayette, Indiana

Chair Professor & Head
Tsinghua University
Beijing, People's Republic of China

Waldemar Karwowski
Professor & Chair
University of Central Florida
Orlando, Florida, U.S.A.

Advances in Human Factors and Ergonomics in Healthcare
V. Duffy

Advances in Applied Digital Human Modeling
V. Duffy

Advances in Cross-Cultural Decision Making
D. Schmorrow and D. Nicholson

Advances in Cognitive Ergonomics
D. Kaber and G. Boy

Advances in Occupational, Social, and Organizational Ergonomics
P. Vink and J. Kantola

Advances in Human Factors, Ergonomics, and Safety in Manufacturing
and Service Industries
W. Karwowski and G. Salvendy

Advances in Ergonomics Modeling and Usability Evaluation
H. Khalid, A. Hedge, and T. Ahram

Advances in Understanding Human Performance: Neuroergonomics,
Human Factors Design, and Special Populations
T. Marek, W. Karwowski, and V. Rice

Advances in
Ergonomics Modeling
and Usability Evaluation

Edited by
Halimahtun Khalid
Alan Hedge
Tareq Z. Ahram

CRC Press
Taylor & Francis Group
Boca Raton London New York

CRC Press is an imprint of the
Taylor & Francis Group, an **informa** business

CRC Press
Taylor & Francis Group
6000 Broken Sound Parkway NW, Suite 300
Boca Raton, FL 33487-2742

First issued in paperback 2017

ISBN-13: 978-1-4398-3503-6 (hbk)
ISBN-13: 978-1-138-11806-5 (pbk)

Visit the Taylor & Francis Web site at
http://www.taylorandfrancis.com

and the CRC Press Web site at
http://www.crcpress.com

Table of Contents

Section II. Vision and Visual Displays

Section III. Product Design and User Interfaces

Section IV. Input Devices and Computer Based Systems

Section V. Individual and Environmental Technology Related Issues

Preface

Human Factors and Ergonomics is concerned with the design of products, services and work systems to meet human productive, safe and satisfying use. This book focuses on advances in the use of ergonomics modeling and on the evaluation of usability, which is a critical aspect of any human-technology system. The research described in the book chapters is an outcome of dedicated research by academics and practitioners from around the world, and across borders, to advance progress beyond the state-of-the-art in this dynamic and all-encompassing discipline.

The book is organized into five sections:

I. Models and Methods
II. Vision and Visual Displays
III. Product Design and User Interfaces
IV. Input Devices and Computer Based Systems
V. Individual and Environmental Technology Related Issues

Each section contains chapters that have been reviewed by members of the Editorial Board. Our sincere thanks and appreciation to the Board members as listed below.

This book is a reflection of the international "state of the art," and it is a valuable contribution toward enrichment of knowledge and skills in applied areas of human factors and ergonomics, to enable better designs of products and services for global markets.

April 2010

H. Khalid, A. Hedge, T. Ahram
Miami, Florida
USA

Editors

Modeling Operator Performance Under Stress and Fatigue: What Can a Cognitive Architectural Model Tell Us?

Jong W. Kim, Peter A. Hancock

Department of Psychology
University of Central Florida
Orlando, FL 32816, USA

ABSTRACT

Formalized modeling is one of the most logical and reasonable of all methods to help us refine and advance our understanding of real-world operational effects. Models force us to make our assumptions explicit and to expose the veracity or fallacy of those assumptions as experience is compared to prediction. Here, we examine the use of the ACT-R architecture and the way in which it has been and can be employed to understand the often deleterious influences of stress and fatigue on operator performance. It is well established that both physiological and psychological sources of stress (e.g., heat, cold, workload, time pressure, etc.) as well as the precursors to fatigue (e.g., hours of work, work repetition, demand overload circadian phase, etc) significantly moderate both physical and cognitive performance capacity. In this paper, we survey the state of present understanding as to (a) what stressors have to date been modeled using the ACT-R architecture, and (b) what theories and mechanisms can be identified by this work as moderating performance under such stressed and fatigued conditions. In examining the implications and limitations of this and similar applications we provide a roadmap to advance modeling and simulation of performance under all adverse operational circumstances.

Keywords: Stress, Time-Constrained Performance, Cognitive Architecture, ACT-R

INTRODUCTION

Physiological and psychological sources of stress (e.g., heat, workload, fatigue, etc.) are well known to significantly moderate the human capacity to accomplish physical and cognitive performance (Hancock & Szalma, 2008). However, these stresses are often limited to particular operational environments. In contrast, time pressure is a ubiquitous source of stress. An individual performing under time pressure exhibits a variety of forms of systematic failure. For example, individuals under time stress frequently fall prey to an omission of one step in a procedural sequence even when that sequence is a relatively well learned one. This error, or lapse, can be considered as a form of memory failure (Reason, 1997). Such lapses are ubiquitous across various forms of human procedural tasks in both the civilian and military worlds. While on many occasions the occurrence of such events provides no critical outcome such lapses do represent significant opportunities for tragedy to occur and they have been implicated in any numbers of major disasters.

To help our understanding of this phenomenon, we seek to use a formalized cognitive architecture. This is because a cognitive architecture has been developed to provide *complete processing models* (Newell, 1973). Rather than an isolated or divided subfields in cognitive psychology, a cognitive architecture is an implementation of unified theories of cognition (see Newell, 1990), providing a methodology through which to model the richness of the whole spectrum of human operator performance. The purpose of applied human factors and ergonomics also focuses on human operator performance in an integrated system. Thus, Newell's desire toward complete processing models can serve to support human performance modeling endeavor in the domain of human factors and ergonomics.

One of the most widely used cognitive architectures is ACT-R, *Adaptive Control of Thought—Rational* (see Anderson et al., 2004; Anderson & Lebiere, 1998). ACT-R supports a wide range of empirical phenomena in cognitive psychology including memory, learning, problem-solving, and perception/action response. However, most cognitive architectures including ACT-R have not been developed with theories of stress and fatigue in mind. Thus, it is important to explore how theories of cognition and stress would be unified in this or any other cognitive architecture. Here, we discuss some example research studies concerning stress and fatigue effects in ACT-R (e.g., Gunzelmann, Gross, Gluck, & Dinges, 2009) and techniques to incorporate theories of stress into a cognitive architecture (e.g., overlay, Ritter, Reifers, Klein, & Schoelles, 2007). We will also show how time pressure can be modeled in ACT-R, which we consider one of the most prevalent of all operator stress factors.

THE ACT-R COGNITIVE ARCHITECTURE

Gray (2008) provides a taxonomic analysis of various cognitive architectures and their use in human factors and cognitive engineering. A cognitive architecture as a unified theory of cognition (see Newell, 1990) combines multiple memory systems and embodied cognition to produce human behavior. ACT-R is a hybrid cognitive architecture containing the symbolic and subsymbolic constructs and relies on a modular organization to represent the brain's functional constraints in local regions (Anderson, 2007a, 2007b; Anderson et al., 2004; Anderson & Lebiere, 1998). Anderson (2007a) describes the symbolic level in ACT-R as an abstract characterization of how brain structures encode and process knowledge and the subsymbolic level as an abstract characterization of the role of neural computation in making that knowledge available. ACT-R basically consists of eight modules that are mapped onto brain regions. Figure 1 shows a schematic representation of the ACT-R architecture and the modules that correspond to each brain region.

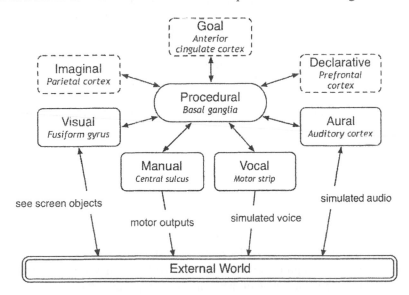

Figure 1. An overview of the ACT-R architecture. Solid single line boxes are modules that interact with the world (external) and dashed line boxes are internal modules. The hypothesized corresponding brain regions for each module are italicized.

The mechanisms representing performance responses in ACT-R include the activation of a declarative memory item that is dependent on how often (frequency) and how recently (recency) a chunk is used. Whenever an item in memory is used, the base-level activation increases and then decreases as a power function of the time since use. Base level activation for a chunk i is represented as the following, where β is a constant, n is the number of presentations for a chunk i, t_j is the time since the jth presentation, and d indicates the decay parameter.

$$B_i = \beta + \ln(\sum_{j=1}^{n} t_j^{-d})$$

Equation 1.

Production compilation is a mechanism of production rule learning in ACT-R (Taatgen & Lee, 2003). ACT-R uses this mechanism to provide performance change by collapsing two productions into a single production. Production compilation combines both proceduralization and composition mechanisms into a single mechanism. ACT-R generates a compiled rule by eliminating the retrieval request in the first rule and the retrieval condition in the second rule. The process to retrieve an item from declarative memory is slow because only one item in memory can be retrieved at one time. Production compilation allows a speed-up process by generating task-specific procedural knowledge where previously there was a string of declarative retrievals. Like the activation mechanism, productions have their own utility values. Based on these utility values, one production can be selected over another. Also, the utilities can be learned from experience. Let $U_i(n-1)$ is the utility of a production i after its $n{-}1^{th}$ application and $R_i(n)$ is the reward the production receives for its nth application. $U_i(n)$ indicates the value of an item i after its nth occurrence and $R_i(n)$ indicates the reinforcement of a reward or a penalty on the nth occurrence. α is the rate of learning, $0 < \alpha < 1$. The utility, $U_i(n)$ after its nth is:

$$U_i(n) = U_i(n-1) + \alpha[R_i(n) - U_i(n-1)]$$

Equation 2.

While the latest version of ACT-R uses the utility value based on Equation 2, the previous version of ACT-R used a simpler equation, $U_i = P_i G - C_i + \varepsilon$, where P_i represents the estimated probability that the goal will be achieved if that production is chosen, G represents the value of the goal, and C_i represents the estimated cost of achieving the goal if that production is chosen. This utility equation is limited to learning from a binary feedback—that is, whether a reward is received or not. This is not sufficient to represent the feedback from the environment and thus the latest version of the ACT-R 6 architecture uses a new utility mechanism as shown in Equation 2, which is similar to the reinforcement learning of Rescorla and Wagner (1972).

As shown in Figure 1, the ACT-R architecture consists of several core modules. The procedural module plays a central role in coordinating productions that interact with other modules. This module specifies productions and matches a production to fire. The declarative module retrieves and stores an item, called a chunk. The goal module produces goal-directed behavior, tracking the current state of a model and holding relevant information for the current task. The only action of the goal module is to create new chunks, and they are placed into the goal buffer. The imaginal module creates new chunks that are the model's internal representation of new information, maintaining context that is relevant to the current task. The other four modules—the visual and aural modules address stimuli from the environment, and the manual and vocal modules produce outputs of hands and voice to the world—provide a way to interact with an environment. With this modular characteristic of ACT-R, one can extend the architecture by adding a module to it, such as a spacing module representing the spacing effect of practice (see Pavlik & Anderson, 2005) and a timing module representing the passage of time (see Taatgen, van Rijn, & Anderson, 2007).

MODELING STRESSORS IN ACT-R

As one way to model stress in a cognitive architecture, a technique called an overlay has been used. This is a technique to include a theory of stress affecting cognition across all models within a cognitive architecture (Ritter, Reifers, Klein, & Schoelles, 2007). An overlay can be described as a set of adjustments of parameters in some operational mechanisms or mechanisms that directly modify parameters to reflect changes due to stress and fatigue.

For example, there have been several attempts to model stress and fatigue in a cognitive architecture. One of the very earliest was that of Jongman (1998) who attempted to model mental fatigue using the spreading activation mechanism in ACT-R. As a more developed work on modeling fatigue, the integration of biomathmatical models into parameters in a cognitive architecture seemed to allow a cognitive model to account how fatigue from sleep deprivation impacts cognition (Gunzelmann, Gross, Gluck, & Dinges, 2009). Other studies have included modeling stress from time pressure (Lerch, Gonzalez, & Lebiere, 1999) and serial subtraction (Ritter, Schoelles, Klein, & Kase, 2007). These studies provide examples of the use of overlay in a cognitive architecture. Table 1 shows a summary of examples concerning stress and fatigue modeling in ACT-R.

Table 1: A summary of previous studies on stress/fatigue modeling in ACT-R

Reference	Stress/Fatigue	Task	Technique
Jongman, 1998	Mental fatigue	Memory task	Overlay
Lerch et al. 1999	Time pressure	Resource management	Overlay
Ritter et al. 2007	Stress	Serial subtraction	Overlay
Gunzelmann et al. 2009	Sleep deprivation	Psychomotor vigilance	Overlay

In ACT-R related prior studies, it has been reported that a parameter that is used in the process of selecting a production rule is related to "arousal" or "motivation" (e.g., Belavkin, 2001; Jongman, 1998). Gunzelmann et al. (2009) also utilized the ACT-R utility mechanism to model decreased alertness. This approach suggests that it is worth exploring the utility mechanism to model time pressure effects on response capacity; a strategy that we seek to pursue. The reason is because the selection process of a production rule is controlled by calculating an expected utility, $U_i(n)$, for each candidate production rule.

However, it is at present not understood nor conceptualized directly how time pressure can be modeled in the ACT-R architecture to represent changes of operators' performance response. For example, time pressure could influence retrieval of a goal from the goal module, and/or motor output in the manual module, and/or memory item retrieval in the declarative module. Thus, here, we explore the ACT-R theory to provide a potential and theoretical understanding of time pressure effects on performance.

TIME STRESS

Time pressure is, perhaps, the most prevalent form of stress in the world of modern operations. For example, individuals in many forms of emergency response and/or extremely high workload conditions, especially life-threatening combat operations, encounter time-stressed decision-making requirements almost continually. Although these are particularly evident instances, time stress is found on many occasions in most forms of human activity. The issue of memorial lapses, already mentioned, is only one form of potential time-induced failure. Wickens and Hollands (2000) have proposed that there are a number of ways human errors might occur under stress. These include interference to the aforesaid working memory capacity but also failure via interference to attention, decision-making processes, as well as long-term memory storage and access. In respect of working memory failures, it is possible that under time stress (i) working memory might be less available for storing and integrating information, (ii) working memory interruptions mean that acquired knowledge may not be successfully retrieved; and/or (iii) working memory may be relocated to the time-related aspects of performance rather than the task per se.

TIME ESTIMATION IN ACT-R

In cognitive modeling, the perception of time is an important ability which can help explain a number of variations in human response. In general, time interval estimation has been grounded on two theories: (a) the *internal clock theory* and (b) the *attentional gate theory* (and see Block, Hancock, & Zakay, in submission). Based on these respective theories, there are a number of formalized models and the two cited models can be differentiated by the presence or absence of the direct attentional gate.

In the ACT-R architecture, the passage of time can be estimated by a timing module that has been recently incorporated in the overall architecture (Taatgen, van Rijn, & Anderson, 2007). This timing module is based on the former, pacemaker-based internal clock model (see Hancock, 1993). The central pacemaker generates pulses at certain frequencies and an accumulator counts these pulses. The basic assumption has traditionally been that the pacemaker generates pulses at a constant rate (although certain forms of stress and neurophysiological disturbance can affect pulse frequency). However, unlike this constant frequency assumption, the ACT-R timing module introduced by Taatgen et al. increases the interval between the pulses as the interval progresses. This timing module can run independently of other cognitive processes. One potential weakness in Taatgen's approach is that the representation of the longer time intervals may not be precise because time between pulses gets longer, although short time intervals would be accurate.

However, Byrne (2006) raised issues regarding the aforementioned timing module by Taatgen, van Rijn, and Anderson (2007). One issue is that timing of multiple overlapping intervals would be necessarily present in modeling time-sensitive complex tasks. The Taatgen's timing module utilizes non-uniform interval rates, making modeling efforts significantly harder than a constant frequency model. The other issue is that people generally underestimate intervals in most conditions (see Hancock & Rausch, 2010). Byrne argues correctly that the Taatgen's timing module would produce overestimation at the start of the time interval. Based on the alternative, attentional gate model (see Zakay & Block, 1997), Byrne has proposed a different timing module. The pacemaker in this model utilizes a fixed mean pulse rate so that a modeler can easily handle overlapping intervals by subtraction. Through the attentional gate model shown in Figure 2, pulses are registered or periodically missed. That is, pulses that are missed produce underestimation of timing. Moreover, this attentional gate models allows prediction of increased underestimation in terms of cognitive load (and see Block, Hancock, & Zakay, in submission). In contrast, Taatgen's module predicts no effect of cognitive load.

MODELING TIME PRESSURE IN ACT-R

We have started to review existing information as to the methods of modeling time pressure in ACT-R. The study by Lerch, Gonzalez, and Lebiere (1999) is perhaps the first identified attempt to model time pressure in a dynamic task. In this study, time pressure was represented by modifying the rate at which the environment changed and compared the ACT-R time to that rate (see Lerch, Gonzalez, & Lebiere, 1999). This paper does not present the actual Lisp code of this ACT-R version. However, we suspect that this model's capability could be limited because no official timing module for ACT-R was available at that time in 1990s. Therefore, our present modeling begins with an effort to represent user performance under time pressure by exploring the predicted performance response change (e.g., accuracy and latency) based on the inter-relationship between the timing module proposed by Byrne (2006) and the W parameter to manipulate cognitive load in working memory (e.g., Lovett, Daily, & Reder, 2000). It is along this line of structure that our present efforts are progressing.

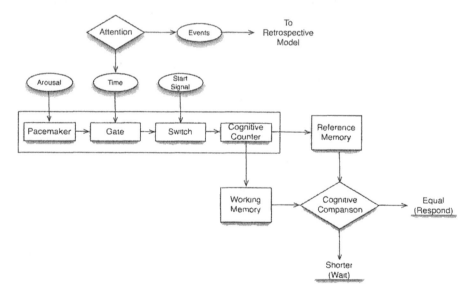

Figure 2. The attentional gate model of prospective time estimation (after Zakay & Block, 1997).

CONCLUSION

We have initially explored the ACT-R architecture to identify what a cognitive architecture can tell us about modeling stress and fatigue. Implementing theories of stress in a cognitive architecture can force the theories to interact with other embodied cognitive mechanisms. This effort can make the theories of stress and cognition more complete and explicit.

It is worth exploring the goals of this roadmap because the current theories of stress are still not sufficiently explicit. The goals specified in the roadmap are ambitious but may serve to provide important outcomes and thus impact relevant operational communities as well as serving to advance our current state of understanding.

1. Find reasonable theories and mechanisms of stress and fatigue.
2. Choose a cognitive architecture.
3. Implement those theories and mechanisms into a cognitive architecture using an extant technique (i.e., overlay).
4. Test human data against the model performance for validation.
5. Generalize the overlay to other stressors.

REFERENCES

Anderson, J.R. (2007a). *How can the human mind occur in the physical universe?* New York, NY: Oxford University Press.

Anderson, J.R. (2007b). Using brain imaging to guide the development of a cognitive architecture. In W. D. Gray (Ed.), *Integrated Models of Cognitive Systems* (pp. 49-62). New York, NY: Oxford University Press.

Anderson, J.R., Bothell, D., Byrne, M.D., Douglass, S., Lebiere, C., & Qin, Y. (2004). An integrated theory of mind. *Psychological Review, 111*(4), 1036-1060.

Anderson, J.R., & Lebiere, C. (1998). *The atomic components of thought.* Mahwah, NJ: Lawrence Erlbaum.

Belavkin, R.V. (2001). Modeling the inverted-U effect with ACT-R. In E.M. Altmann, A. Cleeremans, C.D. Schunn & W.D. Gray (Eds.), *Proceedings of the 4th International Conference on Cognitive Modeling* (pp. 275-276). Mahwah, NJ: Erlbaum.

Block, R.A., Hancock, P.A., & Zakay, D. (in submission). Cognitive load affects duration judgments: A meta-analytic review.

Byrne, M. (2006). An ACT-R timing module based on the attentional gate model. In *ACT-R Workshop Proceedings.* Pittsburgh, PA: CMU.

Gray, W. D. (2008). Cognitive architecture: Choreographing the dance of mental operations with the task environment. *Human Factors, 50*(3), 497-505.

Gunzelmann, G., Gross, J.B., Gluck, K.A., & Dinges, D.F. (2009). Sleep deprivation and sustained attention performance: Integrating mathematical and cognitive modeling. *Cognitive Science, 33*(5), 880-910.

Hancock, P.A. (1993). Body temperature influence on time perception. *Journal of General Psychology, 120*(3), 197-215.

Hancock, P.A., & Rausch, R. (2010). The effects of sex, age, and interval duration on the perception of time. *Acta Psychologica, 133*, 170-179.

Hancock, P.A., & Szalma, J.L. (Eds.). (2008). *Performance under stress*. Aldershot, England: Ashgate Publishing.

Jongman, G.M.G. (1998). How to fatigue ACT-R? In *Proceedings of 2nd European Conference on Cognitive Modeling* (pp. 52-57). Nottingham, UK: Nottingham University Press.

Lerch, F.J., Gonzalez, C., & Lebiere, C. (1999). Learning under high cognitive workload. In *Proceedings of the Twenty-first Conference of the Cognitive Science Society* (pp. 302-307). Mahwah, NJ: Erlbaum.

Lovett, M.C., Daily, L.Z., & Reder, L.M. (2000). A source activation theory of working memory: Cross-talk prediction of performance in ACT-R. *Journal of Cognitive Systems Research, 1*, 99-118.

Newell, A. (1973). You can't play 20 questions with nature and win: Projective comments on the papers of this symposium. In W. G. Chase (Ed.), *Visual information processing* (pp. 283-308). New York: Academic Press.

Newell, A. (1990). *Unified Theories of Cognition*. Cambridge, MA: Harvard University.

Pavlik, P.I., & Anderson, J.R. (2005). Practice and forgetting effects on vocabulary memory: An activation-based model of the spacing effect. *Cognitive Science, 29*, 559-586.

Reason, J. (1997). *Managing the risks of organizational accidents*. Aldershot, UK: Ashgate.

Rescorla, R.A., & Wagner, A.R. (1972). A theory of Pavlovian conditioning: variations in the effectiveness of reinforcement and nonreinforcement. In A. H. Black & W. F. Prokasy (Eds.), *Classical Conditioning II: Current Research and Theory*. New York: Appleton-Century-Crofts.

Ritter, F.E., Reifers, A.L., Klein, L.C., & Schoelles, M.J. (2007). Lessons from defining theories of stress. In W. D. Gray (Ed.), *Integrated models of cognitive systems* (pp. 254-262). New York, NY: Oxford University Press.

Ritter, F.E., Schoelles, M., Klein, L.C., & Kase, S.E. (2007). Modeling the range of performance on the serial subtraction task. In R.L. Lewis, T.A. Polk & J.L. Laird (Eds.), *Proceedings of the 8th International Conference on Cognitive Modeling* (pp. 299-304). Oxford, UK: Taylor & Francis/Psychology Press.

Taatgen, N.A., & Lee, F.J. (2003). Production compilation: A simple mechanism to model complex skill acquisition. *Human Factors, 45*(1), 61-76.

Taatgen, N.A., van Rijn, H., & Anderson, J.R. (2007). An integrated theory of prospective time interval estimation: The role of cognition, attention, and learning. *Psychological Review, 114*(3), 577-598.

Wickens, C.D., & Hollands, J.G. (2000). *Engineering psychology and human performance* (3rd ed.). Upper Saddle River, NJ: Prentice-Hall.

Zakay, D., & Block, R.A. (1997). Temporal cognition. *Current Directions in Psychological Science, 6*(1), 12-16.

Chapter 2

Interactive-Consumer Design & Evaluation (I-CODE): A Method to Investigate Cognitive Structures of User's on Automotive Functionalities

Klaus Bengler[1], Joseph F. Coughlin[2], Bryan Reimer[2],
Bernhard Niedermaier[3]

[1]Technische Universität München

[2]Massachusetts Institute of Technology, USA

[3]BMW Group – Forschung und Technik, Germany

ABSTRACT

Increasing functionality and complexity is entering the automobile. Drivers of all ages must be able on a high level of user experience to easily understand, learn, adopt and use these systems while minimizing distraction and confusion jeopardizing both safety and satisfaction. A new method, I-CODE, is introduced and applied to identify underlying mental models of users and as a technique to engage the consumer as a collaborator in the HMI design process to produce innovations to excite and delight the driver. Results from an exploratory experiment revealing the functionality and design preferences of two age groups 25-35 and 45-55 are presented.

12

Keywords: Human-Machine Interface, HMI, user-centerd-design, collaborative innovation and design, mental model, usability, cognitive engineering.

INTRODUCTION

The number of functionalities available in the modern car is increasing and, for many consumers, overwhelming. It is not only the variety of functions that are growing but also their heterogeneity compared to the classical set of functions that were necessary to operate a car only decades ago. A 'high-end' luxury vehicle in 1957 had only some functionalities. Nearly all the controls were for driving – but an FM radio was already on the center console indicating that more than driving was entering the driver space (Freymann 2006). Introduction of the "new" and the "next" functionality is no longer within the five plus year development cycle of the auto industry, but instead the pace of innovation is now set by the consumer electronics industry often working in "Internet time." Moreover, many of the new functions being introduced to the driver experience are not directly related to operating the vehicle and have their origin in other domains, rather they support connectivity and lifestyle while on the move.

FIGURE 1. Cockpit of a BMW 328 Mille Miglia in 1938 and operation of FM radio in a BMW 700 in 1959.

While the car is changing faster than ever before and new technology-enabled services are being introduced to meet the lifestyle demands of highly mobile consumers, the user is changing as well. Unlike consumer electronics, the first adopters of the high-tech and high-design in the automobile are buyers over 50 years old. The nearly 77 million American Baby Boomers, born between 1946-1964, are now in middle and older age. The Boomer cohort represents a population that grew up with the seamless experience of being mobile and always expecting more functionality in the car. This core automobile consumer group now has between 30 and 45 years driving and car usage experience. The combination of experience and expectations accumulated over four plus decades of driving frames the mental model for auto design and functionality (Coughlin, 2005).

A critical question is, as the vehicle accommodates additional functionality, technology – and complexity – how do designers and engineers integrate, prioritize and place additional systems in the driver workspace, drawing upon the Boomer's expectations, but not constrained by them? This is even of greater importance if we take into account that the layout of the interior and location of functionalty (i.e. displays and controls) is defined by engineers that are often younger than the end user. These young 'creatives' have different technology experiences and driving expectations that shape their own mental models. Designers and engineers routinely follow guidelines and standards (Commission of the European Communities , 2006, The Alliance of Automobile Manufacturers, 2002) to place clusters for information, communication, entertainment and navigation in a way suitable for driving. In this context questions of operability while driving, driver distraction and usability are commonly discussed – however, many of these standards may no longer be adequate to address the volume of innovation finding its way into the car and user experience.

Before the decision of where a display or control cluster is placed, the ultimate design and functionality should seek to meet the user's expectations from the very beginning of the development process. This link or anchor somewhere in the user's mental model of how a function or even the car should work is critical to consumer learning, adoption and ultimately satisfaction. The product property "intuitive to use" in this context means that a product supports instant understanding and perception of its functioning. For this reason, the logic which functionality, is located where, is a crucial decision in the early stages of product design.

DEVELOPMENT OF I-CODE METHODOLOGY TO FACILITATE USER-CENTERED DESIGN & INNOVATION

Is it possible to elicit the mental model of a potential user and to integrate and make explicit those elements in a next generation human machine interface? The question to be investigated is on the location of function-types, their priority relations and the type of information presentation to be used. Typical everyday problems of users are: Where to read time and date in this car? Where and how to adjust temperature here? Where to identify mileage in the cluster instrument among all the other information displayed there?

Current methods to identify what is intuitive or the mental model of the target consumer includes focus groups, product clinics and interviews – each method offers varying degrees of theoretical and analytical rigor. While these methods are useful, they elicit responses from the user based upon what the designer has already created. Is the interface liked or disliked? Does the user understand or not understand the design? Is the device likely to be adopted or rejected? Rather than simply soliciting user opinion, a truly user-centered approach should engage the user as a collaborator in the design process. Engaging the consumer in the design process may provide both insight into underlying mental models and produce an interface that blends both user requirements and innovation.

A method to actively engage the user in the design process – Interactive-Consumer Design & Evaluation or I-CODE was developed. The I-CODE methodology builds upon web design methodologies and card sort techniques used in the social and behavioral sciences (Miller, 1969; Tedesco, et al., 2004) Developed by the MIT AgeLab, I-CODE seeks to reveal user mental models and preferences for in-vehicle HMI (Coughlin, Bengler & Reimer, 2008). I-CODE engages the user in creative play, requiring the user 'to experiment' with their own ideas of design options and function placement. The user is constrained only by the space and range of functionalities and form language provided.

EXPERIMENT

An exploratory I-CODE experiment was conducted to understand generational preferences for automobile dash design and to create a framework for the homogeneous integration of innovative – i.e. never operated before – functionalities into an existing driver interface. If successful, underlying mental models would reveal structure and shorten the learning phase for novice users as well as experts. Researchers were interested in four aspects of the configuration: information dimming; clutter; usability/readability; and style.

15

Materials & Assembly – I-CODE uses three white-magnetic boards, 3'x4' and a variety of more than 70 magnetic gauges, cards and descriptors were arranged on the boards for the experiment. As the information requested is not a design decision but semantic the material was taken from cars that were experienced by subjects during their learning history. The three boards were arranged horizontally as shown in Figure 2. The two outer boards contained the dash gauges. The center board contained an outline of the mock dashboard, area behind the steering wheel and center consol. The dashboard outline is the size of an average driver dash. On the left board, on the top right corner, a set of card sort descriptions and words were lined vertically and covered until after the dashboard configuration.

Participants – Subjects were recruited by e-mail and/or phone. One hundred participants were randomly selected from a subject pool of over 500 individuals in the Generation X and Y age groups of 25 to 35 and younger baby boomers 45 to 55. Twenty-five females and 25 males from each age group were selected. 70% of the sample made $74,000 a year or less.

Experimental Design – A randomization chart was created for the arrangement of the gauges to ensure no bias in item/component selection. Each participant was presented with a unique ordering of the items/gauge/component.

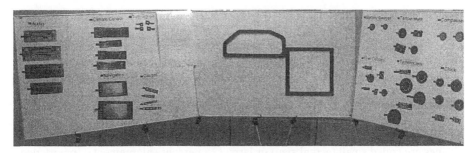

FIGURE 2. I-CODE Experimental platform and setup

Participants were given a choice of 57 gauges. The gauges were broken down into specific categories. Each category was comprised of items of low, medium and high information content determined by the amount of information on each item. Gauges were available in two sizes, small and large. The larger component was 25 percent larger than the smaller.

Procedure – Participants were asked to create their "ideal dashboard." They were required to use four mandatory gauges, as required by the US Department of Transportation, which included two turn signals, a fuel gauge and speedometer. They were told they didn't have to select more than those four gauges and could select as many as they wanted, including duplicates of already selected items.

After participants signed a consent form, they were read instructions. They were given as much time as needed to 'play' or design their ideal dashboard. The researcher also noted any comments made while configuring the dashboard.

The card sort exercise included the names of functions or service concepts not included in the gauge selection or current car models. These included *driver performance, speed, time, navigation, outside climate, safety, seat comfort, engine performance, energy & fuel use, climate control, security, communications*, and *entertainment*. Upon completion of configuring the dash, participants were asked to rank a list of information their car could someday give them. From the ranking, participants chose the most important piece of information and were instructed to place it inside their ideal dashboard.

Coding – A photograph of each completed dash design was taken. An example is shown in Figure 3. The position of each instrument was recorded as a coordinate on a grid. The subject's completed steering wheel and consol area designs, were graphed on a large grid. Coordinates were adjusted for the instruments in the consol. If the instruments were moved after the card sort item was placed inside the dash, the coordinates were rerecorded. In addition to the grid coordinates, the areas were broken into priority zones. The area behind the steering wheel was determined high information priority.

Inside that area, three distinct regions were created. The three areas are as follows: the inner area in direct view of the driver was defined as high priority, the outer areas, represent medium importance; and, the center consol area beyond the steering consol is the low importance area. Dash designs were coded for usability, information dimming, dashboard clutter, and style.

Usability/Readability refers to the accessibility of information based on the size of the gauges on the dashboard. The assumption is larger gauges are easier to read, therefore increasing the usability/readability and accessibility of information. Gauges are coded small (actual size) or large (25 percent enlargement). Importance of usability/readability is determined by using a percentage of large gauges (number of large gauges divided by number of total gauges).

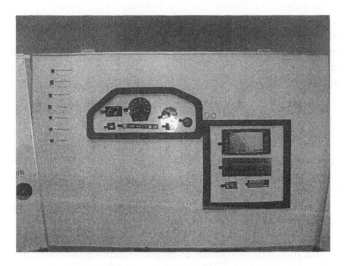

FIGURE 3. Completed I-CODE Dash Design

Information dimming refers to the ranking of important information based on placement of gauges on the dashboard. The assumption is that the number of functions placed outside high-priority regions indicates information of less importance or "dimmed" by the user..

Dashboard clutter is measured by the total number of gauges on the dash. The assumption is that components included on the dashboard will show what drivers want in their cars in addition to the necessary critical gauges.

Style refers to the aesthetic aspect of dashboard configuration. The driver's selection of types of gauges (analog or digital), number (cluttered or sparse) and location (high importance versus low importance) for each component illustrates importance and preferences for specific types of information. The assumption is patterns and themes will emerge between dashboard configuration, its creator and acceptance of new technology in automotive design. The total number of times a gauge is used and the complexity of the gauges selected (e.g., complexity increases with increased information such as number of tick marks, size of numbers, digital versus analog) indicates style preferences of the user.

PRELIMINARY FINDINGS

I-CODE revealed similar and contrasting patterns between age groups. These include the functional and information priorities of users as well as patterns of secondary functions placed beyond the center dash display.

As anticipated, usability and readability features did reveal that older adults were more likely to choose larger and simpler dash gauges than did younger subjects. Surprisingly, there was little variation between the two age groups when prioritizing information. Speed, fuel, engine status and time were highly important to all users. Subjects across age groups and gender also indicated a high priority for displays that provided information on car status and driving conditions. There was little gender difference in desired functions or user design. Although there was an indication that women were far more likely to prioritize time then men across all age groups.

Younger users did choose more functions not necessarily part of the driving task. The majority of both groups did choose entertainment and navigation systems. However, similar to older subjects who chose additional non-driving function features, younger subjects 'dimmed' this information by placing these additional functionalities beyond the high priority region of the dash. Style preferences between the groups indicated some differences. Older subjects were more likely to choose larger and analog displays.

Finally, desire for functions not presented in photographed elements, or for concepts not yet available in the vehicle, did reveal some differences between the groups. Younger cohorts gave priority to new features that would present additional information on time, speed, and energy efficiency. Older participants also identified energy as a priority, but selected safety and comfort as desired values.

CONCLUSION

Experimental results indicate that the I-CODE method is able to detect differences between subjects considering age and gender. There was remarkably little variability for some items in priority and position which might not only be explained by cultural stereotypes of older drivers or individual socialization of experience but from the specific demands "using information while being mobile" for all ages. For example, the majority of subjects indicated the item "time" as high priority and placed either digital or analog time indicators in high priority positions.

This study suggests that I-CODE is a sensitive and suitable method to investigate users' concept in a very early step of development on a semantic level without the necessity to have the final layout available. Locations and priorities as output serve as an important input for information and interaction for designers working on detailed master-layouts and specifications. Moreover the data apply cluster-analytical approaches for further analysis. The I-CODE method shows promise for future research on HMI in the automobile. Further refinement of the method includes improvement in photograph quality of gauge and function elements as well as a reduction in the range of choices, e.g., elimination of oil light indicator. Future modifications to the approach may show productive application to informing the design of other interfaces including consumer electronics, medical devices, industrial applications and packaging.

Acknowledgements & Notes

The authors gratefully acknowledge BMW, the United States Department of Transportation's Region I New England University Transportation Center at the Massachusetts Institute of Technology, and the Santos Family Foundation for support during the preparation of this manuscript.

REFERENCES

Commission of the European Communities (2006), Commission Recommendation of 22 December 2006 on safe and efficient in-vehicle information and communication systems: Update of the European Statement of Principles on human machine interface. Brussels, 22.12.2006.

Coughlin, J. F. (2005), "Not your father's auto industry? Aging, the automobile, and the first of the drive for product innovation." Generations, Winter, 38-44.

Coughlin, Joseph, K. Bengler & B. Reimer. (2008) I-CODE Application of an Interactive- Consumer Design & Evaluation Method to Identify User Mental Models in Automobile Dashboard Interfaces. Technical Report MIT AgeLab

Freymann (2006).HMI: A Fascinating and Challenging Task. IEA - 16th World Congress on Ergonomics. Maastricht.

Miller, G.A. (1969). A Psychological Method to Investigate Verbal Concepts. Journal of Mathematical Psychology, 6, 169-191.

The Alliance of Automobile Manufacturers: Statement of Principles, Criteria and Verification Procedures on Driver Interactions with Advanced In-Vehicle Information and Communication Systems. Detroit 2002.

Tedesco, D., A. Chadwick-Dias, T. Tullis. (2004) Freehand Interactive Design Offline (FIDO): A New Methodology for Participatory Design. Aging by Design 2004 Conference, Bentley University.

Chapter 3

Using TRIZ for Human
Factors Design

Martin G. Helander

School of Mechanical and Aerospace Engineering,
Nanyang Technological University,
639798 Singapore

ABSTRACT

TRIZ is an abbreviation of a Russian phrase meaning "Theory of inventor's problem solving." This theory or framework for inventions was conceived by Altshuller (1946). The purpose of this paper is to analyze how TRIZ design principles can be used for Human Factors Design perspective. The results show that TRIZ can be used to suggest design solutions that save space, reduce task time, increase life length, improve ease-of-use, and improve safety. These are common design criteria used for optimize Human Factors design. TRIZ seems to be well suited to be adopted as a design tool to support Human Factors Design.

Keywords: TRIZ, Human Factors design, Time and space

INTRODUCTION

TRIZ is an abbreviation of a Russian phrase meaning "Theory of inventor's problem solving." This theory or framework for inventions was conceived by Altshuller (1946). During several years in the Gulag of the Soviet Union Genrich Altshuller, a Soviet engineer and researcher analyzed the theory of inventor's problem solving.

TRIZ is a methodology, tool set, knowledge base, and model-based technology for generating innovative ideas and solutions for problem solving (Altshuller, 1984; 1994). It provides tools and methods for use in problem formulation, system analysis, failure analysis, and patterns of system evolution (both 'as-is' and 'could be'). The methodology has been used by several companies. Ford Motor Co. used TRIZ to solve an idle-vibration problem, resulting in several new patents for the company. DaimlerChrysler looked into the future of steering column technology. Johnson & Johnson developed new feminine hygiene products. Saab Aerospace applied TRIZ to solve design problems in submarines (Wallace, 2000).

The design questions and the principles given below in Tables 1-4, were taken from Althuller (1984). The analyses of Design functions and Human Factors Effect are our own contribution.

Table 1. Principles 1-10

Principles for Design	Design Function What does it do	Human Factors effect
Principle 1. Segmentation - Make an object sectional - easy to assemble or disassemble		
Rapid-release fasteners for bicycle saddle etc.	Quick release	Save time
Quick disconnect joints in plumbing systems	Quick release	Save time
Loose-leaf paper in a ring binder	Organize info	Find info
Principle 2. Use only the necessary part or property of an object		
Sound of a barking dog (with no dog) as a burglar alarm	Protect home	Feel safe
Scarecrow (to substitute for a real crow)	Protect harvest	Save harvest
Principle 3. Multifunction tools		
Combined can and bottle opener	Combination use	Save time
Hammer with nail puller	Combination use	Save time
Rubber on the end of a pencil	Combination use	Save time
Night-time adjustment of rear-view mirror	Enhance Vision	See clearly
Principle 4. Asymmetry may include ergonomics features		
Cutaway on a guitar access high notes	Improved reach	Reach easily
Human-shaped seating,	Ergonomics Seat	Sit comfortably
Design for left and right handed users	Ergonomics design	Grasp easily
Finger and thumb grip of objects	Ergonomics design	Grasp easily
Principle 5. Dual Function		
A. Merge similar objects		
Bi-focal lens spectacles	Better vision	See clearly
Window triple glazing	Noise reduction	Hear well
B. Design tasks to be serial in time		
Combine harvester	Combine tasks	Save time
Lawn Grass collector	Combine tasks	Save time
Principle 6. Make an object perform multiple functions or tasks		
Child's car safety seat converts to pushchair	Transform seat Heat/bake food	Save space Save space

Grill and microwave oven	Cut, carve, open	Save space
Swiss Army knife	Change functions	Save space
Cordless drill with screwdriver, sander,etc		
Principle 7a. Nested doll. Place one object inside another		
Paint-brush attached to lid of nail-varnish bottle	Shared space	Save space
Principles for Design	*Design Function* **What does it do**	**Human Factors effect**
Nested tables and stacked chairs	Reduce storage	Save space
Measuring cups/spoons	Reduce storage	Save space
Prnciple 7 b. Store item in itself		
Retracting tape measure	Easy storage	Save space/time
Retracting Seat belt	Easy storage	Save space/time
Principle 8. To compensate for objects weight use aerodynamic, hydrodynamic or other force		
Hydrofoils lift ship	Raise Ship	Travel faster
Magnetic repulsion reduces friction in Maglev train	Raise train	Travel faster Save time
Principle 9. A useful process that is unsafe can be controlled		
Lead apron for X-rays	Filters X-rays	Avoid sickness
Decompression chamber to prevent divers' bend	Maintain low pressure	Avoid sickness
Principle 10. Place objects for convenient use		
Store car jack, wheel brace and spare tire together	Organize task	Save time
Collect tools and material before starting a job	Organize task	Save time

Table 2. Principles 11-20

Design Principle	*Design function* **What does it do**	**What is the effect**
Principle 11. Cushioning for Safety		
Air-bag in car	Soft crash	Improve safety
Motorway Crash Barriers	Soft crash	Improve safety
Principle 12. Reduce lifting		
Spring-loaded parts delivery system in factory	Quick delivery	Save time
Mechanic's pit in a garage	No lifting of car	Save time
Descending cable cars balance weight of ascending cars	Easy pulling	Save energy
Principle 13. Make movable parts fixed, and fixed parts movable		
Moving sidewalk with standing people	No walking	Less effort
Drive through restaurant	No walking	Save time
Turn an assembly upside down to insert fasteners	Good posture	Ergonomics
Principle 14. Use spherical surfaces rather than flat		

Curvature change on multifocal lenses alter light refraction	Ergonomics	Improve vision
Principle 15. Change the object to adapt to the user and situation		
Adjustability of office furniture, e.g. seat	Fits all users	Improve comfort
Bifurcated (Split) bicycle saddle	Softer	Feels better
Design Principle	*Design function* **What does it do**	**What is the effect**
Bendable drinking straw	Drinking easy	Improve comfort
Principle 16. Excessive actions. If you can't achieve the exact effect - then go for more		
Overspray when painting, then remove excess	Even paint	Looks good
Over-fill holes with plaster and then to smoothen	Even holes	Looks good
Principle 17. Add dimensions		
A. Add dimensions – from two to three		
Coiled telephone wire	Less storage	Save space/time
Spiral staircase	Less floor area	Save space
B. Go from single layer to multi-layer		
Book case, CD RACK	Keep books/CDs	Save space
Multi-story car-park	Store cars	Save space
C. Incline an object, lay it on its side		
Cars on road transporter inclined	Store cars	Save space
D. Use the Other Side		
Electronic components on both sides of a circuit board	Small board	Save space
Principle 18. Use vibrating tool		
Electric carving knife with vibrating blades	Cut meat	Less labor
Hammer drill	Drill holes	Less work
Dog-whistle with sound outside human range	Get dog	Save ears
Principle 19. Periodic action		
Siren with pulsed sound	Pulsed sound	Hear better
Pulsed bicycle lights more noticeable to drivers	Pulsed light	Improve visibility
ABS car braking systems	Pulsed break	Brake quicker
Principle 20. Continuity of useful action		
Double-ended Kayak paddle with recovery stroke	Double strokes	Save time
Computer uses idle time for housekeeping tasks	Dual tasking	Save time

Table 3. Principles 21-30

Design Principle	Design function What does it do	What is the effect
Principle 21. Conducting a process at high speed		
High speed dentists drill avoids heating tooth	Quick drilling	Reduce pain
Principle 22. Use harmful factor to achieve positive effect		
Painful Vaccination	Inject vaccine	Avoid cold
Low body temperature slows metabolism during surgery	Slow metabolism	Recover quickly
Principle 23. Introduce feedback to improve a process or action		
Operator performance feedback simplifies process control	Performance feedback	Learn quicker
Feedback in learning improves performance	Learning feedback	Learn quicker
Principle 24. Use an intermediary tool or process		
Play a guitar with a plectrum	Use Plectrum	Play quicker
Joining papers with a paper clip	Organize information	Save time
Principle 25. Self-Service. An object services itself through auxiliary functions		
Self-cleaning oven	Clean oven	Save time
Use pressure difference to help sealing	Seal	Save time
Principle 26. Copying		
A. Replace unavailable, expensive, fragile object with inexpensive copies		
Astroturf	Replace grass	Soft surface
Crash test dummy	Testing Safety	Improve safety
B. Perform task using images		
Virtual reality modeling as design tool	Visualization tool	Design better
Invisible infrared light to detect intruders	Detect thieves	Improve security
Principle 27. Replace expensive object with inexpensive objects		
Disposable nappies paper-cups/cameras	One-time use	Save time
Throw-away cigarette lighters	One-time use	Save time
Principle 28. Enhance human sensing		
Add a bad smell to natural gas to alert users to leaks	Bad smell	Warn people
Voice activated telephone dialing	Voice dialing	Save time
Principle 29. Substitute solid parts for gas and liquid		
Inflatable furniture; mattress	Soft sleeping	Improve Comfort
Gel filled saddle adapts to user	Soft riding	Improve comfort
Principle 30. Protection using flexible membranes		
Bubble-wrap	Protect goods	Avoid damage
Bandages/plasters	Protect wounds	Avoid infection

Table 4. Principles 31- 40

Design Principle	Design function What does it do	What is the effect
Principle 31. Porous materials		
Holes reduce weight.	Weighs less	Ease of use
Porous building blocks	Insulator	Less noise, cold
Principle 32. Change color of an object or its external environment		
Plastic spoon changes color when hot	Change color	Warn high temperature
Camouflage	Difficult to see	Protects user
Principle 33. Homogeneity of objects		
Ice-cubes out of the same fluid as the drink	Same material	Tastes better
Bio-compatible materials in transfusions	Bio-compatible	Reduces risk
Principle 34. An object dissolves after use		
Dissolving capsules for medication	Dissolving capsule	Tastes better
Bio-degradable bags	Bio-degradable	Improve environment
Principle 35. Concentration and dilution		
Transport gas as a liquid to reduce volume	Reduce volume	Save energy
Dilute liquid soap	Clean hands	Ease of use
Principle 36. Use phase transitions from water to ice to steam and ensuing volume change		
Freezing water expands and breaks rock	Break rock with ice	Save energy
Water-to-steam transition in pressure cooker	Use pressure to cook	Save time
Principle 37. Use thermal expansion and contraction of materials.		
Fit a joint by cooling inner part and heating outer part	Contract/expand	Tighten joint
Bi-metallic strips used for thermostats, etc	Temp measurement	Increase life
Principle 38. Accelerated oxidation		
High pressure oxygen kills anaerobic bacteria in wounds	Kill bacteria	Recover quickly
Place asthmatic patients in oxygen tent	Quick recovery	Suffer less
Principle 39. Replace a normal environment with an inert environment		
Vacuum packaging	Long life	Tastes good
Foam starves a fire of oxygen in air	Extinguish fire	Increase safety
Principle 40. Use composite rather than uniform materials		
Composites in golf clubs, tennis rackets, etc.	Durable flexible	Long lasting
Composites in suitcases and briefcases	Durability	Long lasting

ANALYSIS OF HUMAN FACTORS/ERGONOMICS EFFECTS

As shown in the last column of tables 1-4 TRIZ analyses the impact or effect of inventions on the users. Table 5 shows a summary of effects of the design outcomes.

Table 5.Number of Human Factors implications for 85 design outcomes

Save space	21	Increase safety	3
Save time	12	Improve comfort	3
Increase life length	5	Avoid sickness	3
Improve safety	5	Warning	2
Ease of use	4	Travel faster	2
Tastes good	3	Sit comfortably	2
See clearly	3	Recover quickly	2
Save money	3	Looks good	2
Less work	3	Hear better	2
Reduce pain	3	Do quicker	2

The two most frequent implications of the design were: "Save space", which was the outcome in about 25 % and "Save time" with 14% of the design cases. We claim that most of these have Human Factors design implications.

Table 6 presents the cases for saving space and Table 7 documents the principles involved in saving time.

Table 6. Save space

Principles for Design	Design Function What does it do	Human Factors effect
Principle 6. Make an object perform multiple tasks		
Child's car safety seat converts to pushchair	Transform seat	Save space
Grill and microwave oven	Heat/bake food	Save space
Swiss Army knife	Cut, carve, open	Save space
Cordless drill with screwdriver, sander, etc.	Change functions	Save space
Principle 7a. Nested doll. Place one object inside another		
Paint-brush attached to lid of nail-varnish bottle	Shared space	Save space
Nested tables and stacked chairs	Reduce storage	Save space
Measuring cups/spoons	Reduce storage	Save space
Prnciple 7b. Store item in itself		
Retracting tape measure	Easy storage	Save time/space
Retracting Seat belt	Easy storage	Save time/space

Principles for Design	*Design Function* What does it do	Human Factors effect
Principle 17. Add dimensions		
A. Add dimensions – from two to three		
Coiled telephone wire	Less storage	Save space
Spiral staircase	Less floor area	Save space
B. Go from single layer to multi-layer		
Book case, CD RACK	Keep books/CDs	Save space
Multi-story car-park	Store cars	Save space
C. Incline an object, lay it on its side.		
Cars on road transporter inclined	Store cars	Save space
D. Use the Other Side		
Electronic components on both sides of a circuit board	Small board	Save space

Table 7. Save time

Principles for Design	*Design Function* What does it do	Human Factors effect
Principle 1. Segmentation - Make an object sectional - easy to assemble or disassemble		
Rapid-release fasteners for bicycle saddle etc.	Quick release	Save time
Quick disconnect joints in plumbing systems	Quick release	Save time
Principle 3. Multifunction tools		
Combined can and bottle opener	Combination use	Save time
Hammer with nail puller	Combination use	Save time
Rubber on the end of a pencil	Combination use	Save time
Principle 7b. Store item in itself		
Retracting tape measure	Easy storage	Save time
Retracting seat belt	Easy storage	Save time
Principle 8. Use aerodynamic, hydrodynamic force		
Hydrofoils lift ship	Raise Ship	Save time
Magnetic repulsion reduces friction in Maglev train	Raise train	Save time
Principle 10. Place objects for convenient use		
Store car jack, wheel brace and spare tire together	Organize task	Save time
Collect tools and material before starting a job	Organize task	Save time
Principle 12. Reduce lifting		
Spring-loaded parts delivery system in factory	Quick delivery	Save time
Mechanic's pit in a garage	No lifting of car	Save time
Principle 13. Make movable parts fixed, and fixed parts movable		
Drive through restaurant	No walking	Save time
Principle 20. Continuity of useful action		
Double-ended Kayak paddle with recovery stroke	Double strokes	Save time
Computer uses idle time for housekeeping tasks	Dual tasking	Save time

Principles for Design	Design Function What does it do	Human Factors effect
Principle 25. Self-Service. An object services itself through auxiliary functions		
Self-cleaning oven	Clean oven	Save time
Use pressure difference to help sealing	Seal	Save time
Principle 27. Replace expensive object with inexpensive objects		
Disposable nappies paper-cups/cameras	One-time use	Save time
Throw-away cigarette lighters	One-time use	Save time
Principle 28. Enhance human sensing		
Voice activated telephone dialing	Voice dialing	Save time
Principle 36. Use phase transitions from water to ice to steam and ensuing volume change		
Water-to-steam transition in pressure cooker	Pressure to cook	Save time

Most of the cases in Table 6 describe the spatial design or storage facility; how the item itself or the storage space for it can be reduced in size. We note that in daily life storage facilities are more common than we may usually reflect about

Table 7 analyzed Principles for Design that can save time. In this case there were a variety of Design Functions including: Mechanisms for Quick release, Task Organization and Use of Disposable items.

DISCUSSION

The forty principles in TRIZ are well known. What is new are the broad ergonomic and human factors engineering implications of the principles. Some of the principles support human perception and human performance. Some improve safety and health. Most of them have ergonomic implications in the sense that the new designs or systems are simpler and easier to use.

In contrast to techniques such as brainstorming (which is based on random idea generation), TRIZ aims to create an algorithmic approach to the invention of new systems, and the refinement of old systems.

REFERENCES

Altshuller, Genrich (1984). Creativity as an Exact Science. New York, NY: Gordon & Breach. ISBN 0-677-21230-5.
Altshuller, Genrich (1994). And Suddenly the Inventor Appeared. Worcester, MA: Technical Innovation Center. ISBN 0-9640740-1-X.
Wallace, M., 2000. The science of innovation, Available at Website: http://www.salon.com/tech/feature/2000/06/29/altshuller/index.html

Chapter 4

Graphical User Interface Model Based on Human Error Using Analytic Hierarchy Process

Sumie Yamada, Takako Nonaka, Tomohiro Hase

Ryukoku University

ABSTRACT

In this paper, we propose a new model for the design method of graphical user interfaces for an audio visual remote controller based on analytic hierarchy processes. The goal of this model is to reduce the human error in order to modify the graphical user interface of a wireless remote controller to obtain the most suitable interface for every user. This paper proposes a new model with six evaluation criteria; personal attributes, misunderstanding, involuntary failure, lack of ability, lack of knowledge, and neglect. As alternatives, we decided on four design strategies for the user interface; vision assistance, cognition assistance, operation assistance, and memorizing. The proposed method is evaluated by a prototype assuming a real-time OS on an embedded microprocessor. Furthermore, we confirmed this proposal as effective.

Keywords: AHP, Human Error, Remote Controller

INTRODUCTION

These days, remote controllers for audio visual (AV) systems are used by many kinds of user, such as the elderly, these with visual impairments, and those with different levels of ability for AV devices. As the user interface of a conventional remote controller is fixed by the product maker, the design is not a suitable design for every user to operate it.

To solve this problem, we need to think about a graphical user interface (GUI) design strategy which provides the most suitable GUI for every user.

The authors have previously proposed a design strategy model for a GUI which aimed at being easy to use (Yamada et al., 2009). Its goal was an easy to use AV remote controller for every user and proposed GUI design model based on user attributes and AV system environment for the AV remote controller using an analytic hierarchy process (AHP) (Saaty and Vargas, 2001).

However a few human errors still remain. This paper proposes a new GUI design model based on human error for the AV remote controller using an AHP. Also, we evaluate the new and earlier models.

CONCEPTUAL MODEL OF GUI DESIGN USING AHP

OUTLINE OF HUMAN ERROR
Optimizing the human factors is thought to help prevent human errors. There are some well-known models and proposals regarding human factors. One is the SHEL model which consists of software, hardware, environment, and liveware. Another models, called the 5M Factor, which means Man, Machine, Medium, Mission and Management, is used to examine the nature of accidents in the transport industry.

Omission errors, commission errors, extraneous acts, sequential errors and time errors are typical human errors (Komatsubara, 2008).

Furthermore, this paper uses the cause of these human errors as criteria. The next section describes the criteria.

PROPOSED MODEL FOR GUI DESIGN STRATEGY BASED ON AHP
In this section, an outline of the evolution criteria for the conventional model and its problems are given before discussing the new GUI design strategy. Figure 1 shows the conceptual model proposed by the authors.

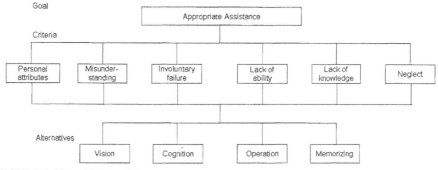

FIGURE 1 New AHP model.

This model consists of three sections, the goal, criteria and alternatives. The goal of this model is to revise the GUI of a remote controller to obtain the most suitable strategy for each user. The following six criteria are defined;

(1)"Personal attributes" criterion, which are the limits of the user's ability.

(2)"Misunderstanding" criterion, mistakes in judgment including misinterpretation, misunderstanding and erroneous ideas.

(3) "Involuntary failure" criterion, ignorance of something to be done and lack of memory of doing something.

(4) "Lack of ability" criterion, lack of ability and/or skill to pursue some operation.

(5) "Lack of knowledge" criterion, lack of knowledge of what to do.

(6) "Neglect" criterion, such as intentional negligence and procrastination.

Finally the following four interfaces were chosen as alternatives vision assistance, cognition assistance, operation assistance, and Memorizing, and evaluated from Eigen value calculations by AHP.

SAMPLE CALCULATION OF PROPOSED MODEL AND PROTOTYPE

SAMPLES CALCULATION OF PROPOSED MODEL

This section describes specific sample calculations using AHP. The evaluation values for the six criteria were given concrete numerical values and fine weights were calculated. Table 1 shows the values of the six criteria. The assumed user and environment are given below. The first user is a child with no visual problems and limited knowledge. The AV device and the equipment are not complicated. Also they are using the AV remote controller for the first time. In addition, there is no particular stress regarding the GUI. The consistency index (C.I.) value is 0.027, which is lower than 0.1, indicating that it is effective.

Table 1 Evaluation Values for the Six Criteria

	PER	MIS	FAI	ABI	KNO	NEG	WEIGHT
PER	1	7	7	9	5	7	0.474
MIS	1/7	1	1	1/5	1/3	1	0.051
FAI	1/7	1	1	1/5	1/3	1	0.051
ABI	1/3	5	5	1	3	5	0.251
KNO	1/5	3	3	1/3	1	3	0.123
NEG	1/7	1	1	1/5	1/3	1	0.051

C.I. = 0.027

Next, the values are set for the four alternative design strategies according to the criteria, as shown in Table 2. This is the case of Physiological attribute criteria.

Table 2 Evaluation Value for Personal attribute

	VIS	COG	OPE	MEM	WEIGHT
VIS	1	1/5	1	1/3	0.095
COG	5	1	5	3	0.560
OPE	1	1/5	1	1/3	0.095
MEM	3	1/3	3	1	0.249

C.I. = 0.014

The C.I. value is 0.014, which is lower than 0.1, indicating that it is effective. The five other criteria were computed in the same way. Based on the results, comprehensive evaluation results were determined using Eq(1).

$$\begin{bmatrix} Vision \\ Cognition \\ Operation \\ Memorizing \end{bmatrix} = \begin{bmatrix} 0.095 & 0.200 & 0.089 & 0.200 & 0.095 & 0.095 \\ 0.560 & 0.522 & 0.209 & 0.522 & 0.560 & 0.249 \\ 0.095 & 0.200 & 0.089 & 0.200 & 0.095 & 0.560 \\ 0.249 & 0.078 & 0.613 & 0.078 & 0.249 & 0.095 \end{bmatrix} \cdot \begin{bmatrix} 0.474 \\ 0.051 \\ 0.051 \\ 0.251 \\ 0.123 \\ 0.051 \end{bmatrix} = \begin{bmatrix} 0.126 \\ 0.515 \\ 0.150 \\ 0.208 \end{bmatrix} \cdots (1)$$

This example concluded that Cognition was the most important factor for the user, followed by Memorizing and Operation, and that Vision was not important.

Next, a case of an elderly person was calculated. This concluded that Vision was the most important factor, while Operation was not important. In addition, in a previous paper, we used user attributes and an AV system environment as criteria. Our results show that Cognition was regarded as most important design strategy for a GUI for children, and it was Vision for the elderly. When we used human error as the criteria in this paper, the same results were obtained. As a result of having two different approaches for one goal, it was found that the new model corresponds with the previous model quantitatively and qualitatively. Therefore, it is thought that both models are effective.

EXAMPLE OF PROTOTYPE OF THE PROPOSED MODEL

This calculation algorithm was used to evaluate the AV remote control. Figure 2 shows a prototype of our new AV remote controller. This assumed that the prototype is a wireless remote controller for household appliances with small CPU resources and a real-time OS. The evaluation confirmed its operability.

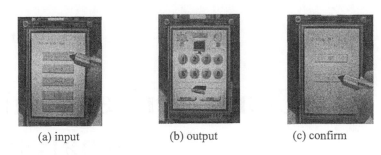

|(a) input|(b) output|(c) confirm|

FIGURE 2 Example of prototype of the suggested model.

CONCLUSIONS

In this paper, we have proposed a new model for the design of graphical user interfaces for an AV remote controller based on AHP. The goal of this model is to reduce the human error in order to modify the graphical user interface of a wireless remote controller to obtain the most suitable interface for every user. This paper proposed a new model with six evaluation criteria: personal attributes, misunderstanding, involuntary failure, lack of ability, lack of knowledge, and neglect. As alternatives, we decided four design strategies for the user interface; vision assistance, cognition assistance, operation assistance, and memorizing. This proposed method was evaluated using a prototype system with a real-time OS in an embedded microprocessor. As a result, this proposal was confirmed as effective. It was found that vision was the most important alternative for an elder user, and cognitive was that for children. This calculation algorithm is easy to implement in the evaluation system. This system could display a GUI suitable for each user, and the proposed model is confirmed as being effective.

REFERENCES

Yamada, S., Uesugi, Y., Nonaka, T., & Hase, T. (2009). "Suitable GUI selection using analytic hierarchy process.", 13th IEEE International Symposium on Consumer Electronics (ISCE2009), 375-377.

Saaty, T. L., & Vargas, L. G. (2001). *Models, methods, concepts & applications of the analytic hierarchy process*, Norwell, MA: Kluwer Academic publishers.

Komatsubara, A. (2008). Human Error 2nd Edition (pp.18-20). Tokyo: Maruzen.

Chapter 5

Computerised Link Analysis

Yu Zhao, Sue Hignett, Neil Mansfield

Department of Ergonomics
Loughborough University, UK

ABSTRACT

Link Analysis (LA) is a useful method to study relationships and arrangements of elements within a system. It has been used more and more in the healthcare domain to create appropriate work conditions for clinical staff and provide patients with effective, efficient and safe services. However LA has limitations, such as the time-consuming manual data recording and results generating processes. Thus, the Computerised Link Analysis (CLA) system was developed and is being tested to provide flexibility by computerising template generation, and data collection and analysis processes. This paper will describe the development and iterative testing (technical, usability and field) of the CLA system.

INTRODUCTION

Link Analysis (LA) is a useful ergonomic method in layout and task design. In Link Analysis, systems are composed of components and links. Components comprise items in the system (for example, chairs and tables in an office), while links are represented by the interactions between components, such as hand, eye or body movements. The purpose of LA is to capture these interactions and attempt to produce an improved layout of workspace by rearranging components to achieve a minimum of overall link distance, in order to increase efficiency. LA has been increasingly employed in the healthcare industry to improve layouts and enhance performance (Hignett & Lu, 2007; Jones, Hignett & Benger, 2008; Ferriera & Hignett, 2004). Currently, the majority of researchers are still using a traditional 'pen and paper' method of LA. This is labour-intensive and time-consuming as researchers have to observe a task carefully while trying to record every interaction by drawing line between components on a link diagram, and then calculate the link frequencies by counting the lines on the diagram. These data (link frequency in link table and interaction map in link diagram) have limited usefulness, as more parameters (e.g. link directions, chronological orders, time durations, etc) are required to take account of complex tasks, especially tasks in the healthcare domain. Due to these limitations, a Computerised Link Analysis (CLA) system has been developed (Zhao, Hignett & Mansfield, 2009) and tested to provide researchers with an enhanced software tool to reduce the workload and save time and resources. This paper will describe the development and iterative testing (technical, usability and field) of the CLA system.

COMPUTERISED LINK ANALYSIS (CLA) SYSTEM

As techniques develop many achievements of simplifying the method of LA have been documented, such as Computer-Aided Link Analysis system (Glass, Zaloom & Gates, 1991), Link Analyzer system (Thorstensson, Axelsson, Morin & Jenvald, 2005), etc. However, none are able to either capture and analyse real-time data effectively and efficiently, or were designed especially for use in the healthcare industry. Thus, the Computerised Link Analysis (CLA) system has been developed and tested to provide analysts with an efficient and effective tool which integrates computerised layout creating, real-time data recording and real-time results generating processes. The system consists of two applications, the Template Generator (TG) and Data Recorder (DR). The TG is used to develop user-specified 2-D layout design, in which components are represented by rectangular boxes. The purpose of the DR is to read in the layout, collect interactions in real-time, and generate LA results instantly. The benefits of CLA are:

- Reduction in time and resources required compared with the traditional LA method, as interactions/links are recorded simply by selecting the rectangular boxes in the layout according to the naturalistic operations;
- Increased data analysed, such as chronological order, time duration and direction of links.

Early versions of CLA were tested for accuracy, usability and real-time ability, etc. in Technical Validation and Usability tests.

TECHNICAL VALIDATION TEST

Technical Validation aimed to ensure the CLA system worked technically and achieved acceptable accuracy. This test was conducted using previous LA studies with a range of different complexity tasks. Comparisons were made between results of CLA and the manual results (obtained from previous reports) to assess its accuracy.

METHOD

This test was conducted in two stages, Stage 1: previous study of assessing the reproduction of a two-sided document using a photocopier (Whitehead, 2008), which consisted of approximately 10 links; Stage 2: ambulance patient compartment (Thorne, 2008) was re-evaluated in CLA system using video clips of clinical staff simulating patient (mannequin) treatment in a ambulance mock-up. Comparisons were made between manual results and CLA results.

RESULTS

Stage 1: As there were four participants in the previous report, four sets of comparisons were made on both link tables and link diagrams. Two sets of CLA results out of four (including link tables and link diagrams) were identical with manual results (Participant A and C). Although results of Participant B & D from two methods were slightly different, results of the CLA system were proven to be correct after Trial B & D was repeated using both traditional pen and paper method and CLA method (tables 1, 2, 3; figure 1).

Table 1 Photocopying activities of Participant D (adopted from Whitehead (2008)).

STEP	ACTION	CODE
1	'Original' document placed in feeder tray	D
2	Presses 'ON/OFF' button (machine powers up)	A
3	Finger hovers over LCD screen	J
4	'Two-sided' button pressed	E
5	Finger hovers over LCD screen	J
6	'2 > 2-Sided' button pressed	F
7	'OK' button pressed	G
8	'START' button pressed (Document in feeder is drawn into the machine, document copied and copy plus original dispensed)	B

Table 2 Link tables for Participant D, manual result (Whitehead, 2008). Numbers in red and underlined are not correct; the correct ones are shown in Table 3 CLA result. (A: On/Off; B: Start; C: Cancel; D: Feeder Tray; E: 'Two-sided'; F: 2 > 2-sided; G: OK; H: Paper Select; I: Special Features; J Hover/Scan/Move Finger Away; K Lid)

	A	B	C	D	E	F	G	H	I	J	K
A											
B											
C											
D	1										
E											
F					1						
G		1				1					
H											
I											
J	1				1	0					
K											

Table 3 Link tables of Participant D, CLA result

	A	B	C	D	E	F	G	H	I	J	K
A		0	0	0	0	0	0	0	0	0	0
B	0		0	0	0	0	0	0	0	0	0
C	0	0		0	0	0	0	0	0	0	0
D	1	0	0		0	0	0	0	0	0	0
E	0	0	0	0		0	0	0	0	0	0
F	0	0	0	0	0		0	0	0	0	0
G	0	1	0	0	0	1		0	0	0	0
H	0	0	0	0	0	0	0		0	0	0
I	0	0	0	0	0	0	0	0		0	0
J	1	0	0	0	2	1	0	0	0		0
K	0	0	0	0	0	0	0	0	0	0	

38

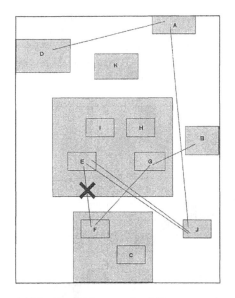 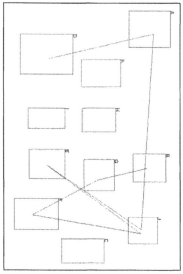

(a) The line with cross should be removed   (b) Correct result

FIGURE 1 Link diagrams of Participant D. (a) manual method (b) CLA method

Stage 2: One simulation scenario with two clinical staff was assessed and two sets of comparison on link tables were obtained. Activities of Staff 1 registered using two methods were the same, however, there were some differences in results for Staff 2 (tables 4 and 5). It was concluded that the link table generated by the CLA system was correct.

Table 4 Link table Staff 2 (CLA result).

		1	2	3	4	5	6	7	8	9	10	11	12	13	14	15	16	17	18
1	Paramedic	6																	
2	patient	16	1																
3	Oxygen tap	2																	
4	Work top																		
5	Defibrillator	15	1			1													
6	Work-top	2																	
7	Clinical waste bins	2																	
8	sharps bin																		
9	High level cabinets																		
10	Worktop	3	1																
11	Bag on floor	3																	
12	Suction unit																		
13	Chair1																		
14	Responder bags	1											1						
15	Chair2																		
16	Chair3																		
17	Another staff	7	1																
18	High level cabinet	2																	

Table 5 Link table Staff 2 (manual result; Thorne, 2008)

	1	2	3	4	5	6	7	8	9	10	11	12	13	14	15	16	17	18	19	20	21	22	23	24	25	26	27	28	29	30	31	32
Paramedic/technician	8*																															
Patient	17	1*																														
Oxygen tap	2																															
Worktop1																																
Defibrillator	15	1		1*																												
Worktop2	2																															
Clinical waste bins	1																															
Sharps bin																																
High level cabinet																																
Stretcher																																
Bulkhead worktop	2	1																														
Responder bag on bulkhead worktop																																
Responder bag on floor	3																															
Responder bag in cabinet	1											1																				
Responder bag on chair																																
Responder bag on attendant chair																																
Responder bag on stretcher																																
item placed on floor																																
item placed on chair																																
item placed on attendant chair																																
Second paramedic/technician	7																															
Third paramedic/technician																																
Drugs safe																																
Drugs bag in cabinet																																
Drugs bag on floor																																
Drugs bag on bulkhead worktop																																
Drip hung up																																
Suction Unit																																
High level cabinet2	1																															
Bulkhead door																																
Left ambulance via back door																																
Phone at bulkhead																																

The Technical Validation test was successful and it was concluded that the CLA system, both TG and DR applications, operated reliably and accurately throughout validation test, and was capable of recording and analysing data in real time.

USABILITY TEST

The Usability Test aimed to identify improvements and bugs of the CLA system to both reduce the probability of failure/deficiency and increase reliability/usability. 12 users with various backgrounds (for example, Ergonomics, Computer Sciences, etc.) were invited to operate the CLA system in a laboratory environment. Virzi (1992) suggests that 65% of usability problems can be uncovered by three subjects, 80% can be found by five subjects, and as high as 95% of problems can be detected by nine. The task was to create layouts (TG) and collect interactions (DR) from four observational sets of data (two real and two video tasks). The researcher (YZ) accompanied each user throughout the test to monitor his/her performance from the PC screen and log bugs/failures of system, confusing steps or errors in time, as users might miss reporting some bugs. Further bugs and improvement ideas were identified through a post-test questionnaire. System enhancements were carried out to develop the CLA system to the next generation.

METHOD

The Usability Test was designed to be carried out in laboratory environment (Healthcare Ergonomics and Patient Safety research Unit (HEPSU) at Loughborough University). 12 people with various research backgrounds (e.g. Ergonomics, Computer Studies, etc.) were recruited as system users. As they were novices, they were provided with CLA system manual, in order to familiarise themselves with the system before data collection. During this practice period, users and their performance were monitored carefully by the researcher (YZ), to identify any bugs and breakdowns of system and confusing steps and unclear

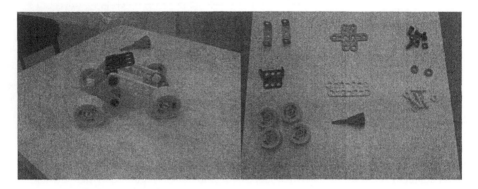

FIGURE 2 Assembly task one.

FIGURE 3 Assembly task two.

statements in user manual; these may take more time than expected to execute certain commands. Two assembly tasks were chosen (figures 2 and 3) and video-recorded prior to the test. Therefore, there were four tasks to be analysed using CLA system (two naturalistic data and two video data). Randomisation was applied with the tasks analysed in different orders by different users. Users were required to do their best to register every link, however, accuracy of results was not examined, as it was assessed in the Technical Validation Test. After the test, suggestions of improvements and bug experiences were obtained using questionnaires.

RESULTS

Suggestions of improvements and bug experiences were collected from monitoring the users' performance as well as questionnaires, such as data buffer bug (the buffers were not emptied before a new data collection, this lead to results adding up). All the improvements and bugs were analysed, system developments and improvements were carried out using this feedback.

BETA TEST

Beta-Testing of software systems or programmes is the last stage before release for public use. The Beta-Test of CLA system aims to review its real-time data recording and analysis abilities and address further improvements. This is being conducted in a real environment – a hospital. This test is being carried out at Emergency Department (ED) of Leicester Royal Infirmary (LRI), UK. The physical movements of clinical staff at ED are shadowed by three system users (final year undergraduate students in Ergonomics, Loughborough University), with one student for doctors, one for nurses and one for healthcare assistants (HCAs)/housekeepers.

METHOD

Three system users follow the staff around the ED except inside patient cubicles when staff are talking or providing treatment to patient. They attempt to register every physical movement of the staff they are following using the CLA system. Each system user needs to shadow 20 staff (3 x 20 = 60 staff), and observations are made for up to 2 hours for each staff member, to collect enough movement data to produce valid link analysis results. The system users will be interviewed after data collection at ED for software improvements. Additionally, the link analysis results generated from the beta-test will be studied to review the layout and storage of equipment and consumables in the ED to increase efficiency and enhance performance.

RESULTS

Four pilot tests at the ED were carried out to obtained a detailed ED layout and familiarise the novice system users with the busy ED environment. Although data are being collected using the CLA system, some ideas/improvements were fed back by system users, and developments are being conducted.

CONCLUSION

This paper has described the CLA system and three iterative tests (technical, usability and field). Technical Validation and Usability Test were satisfactory and the CLA system was proven to be accurate and reliable. The Beta-Test is being carried out, it aims to review real-time data recording and analysis abilities of the CLA system. Further improvements will be made to enhance functionality of CLA system.

REFERENCES

Ferreira, J. & Hignett, S. (2005), *Reviewing ambulance design for clinical efficiency and paramedic safety*. Applied Ergonomics, 36, 97-105.

Glass Jr, J. T., Zaloom, V., & Gates, D. (1991), *Computer-aided link analysis (CALA)*. Computers in Industry, 16(2), 179-187.

Hignett, S. & Lu, J. (2007), *Evaluation of Critical Care Space Requirements for Three Frequent and High-Risk Tasks*. Critical Care Nursing Clinics of North America, 19, 167-173.

Jones, A., Hignett, S. & Benger, J. (2008), *Identifying current and future care activities in ambulances, emergency departments and primary care*. Emergency Medicine Journal, 25 (Suppl):A3

Thorne, E. (2008), *Development of a protocol to evaluate ambulance patient compartments using link analysis and simulation*. Unpublished BSc Report, Department of Human Sciences, Loughborough University, Loughborough, UK.

Thorstensson, M., Axelsson, M., Morin, M., & Jenvald, J. (2001), *Monitoring and analysis of command post communication in rescue operations*. Safety Science, 39(1-2), 51-60.

Virzi, R. A. (1992), *Refining the test phase of usability evaluation: How many subjects is enough?* Human Factors: The Journal of the Human Factors and Ergonomics Society, 34(4), 457-468.

Whitehead, M. (2008), *Task Analysis - An Ergonomic Assessment of an Office Photocopier*. Unpublished Report, Department of Human Sciences, Loughborough University, Loughborough, UK.

Zhao, Y., Hignett, S., & Mansfield, N.J. (2009), *Computerised link analysis*. Proceedings of the 2009 Naturalistic Decision Making London, NDM9 "NDM and Computers", 390-398.

Chapter 6

A Methodology for Evaluating Multiple Aspects of Learnability: Testing an Early Prototype of the New BMW iDrive

Bryan Reimer[1], Bernhard Niedermaier[2], Klaus Bengler[3],
Bruce Mehler[1], Joseph F. Coughlin[1]

[1]MIT Center for Transportation & Logistics
USA

[2]BMW Group – Forschung und Technik
Germany

[3]Technische Universität München
Germany

ABSTRACT

This paper describes a usability evaluation methodology developed and tested at the Massachusetts Institute of Technology to assess the intuitiveness and learnability of the interaction between a control-element and display in an early stage prototype of the new generation BMW iDrive in-vehicle information and telematics system. This was one component of a multi-stage consumer focused design and evaluation process undertaken by BMW during the redesign of the BMW 7 Series human-machine interface. The output generated using this methodology was utilized by the BMW development team to gain insight into the mental models of US based users of differing age, gender, technology experience, and educational backgrounds. The results contributed to the design team's success in further enhancing the intuitiveness and efficiency of interaction in the release version of the product.

Keywords: Usability, User-Centered Design, Heuristic Evaluation, Human Machine Interface (HMI), In-Vehicle Information Systems (IVIS), BMW, iDrive.

INTRODUCTION

The iDrive system, introduced by BMW in 2001 as a novel human-machine interface (HMI), was designed to combat the increasing clutter of automotive dashboards by providing drivers with an expandable interface for current and future telematic engagement. The HMI consists of two major components: a distributed controller and a display. Both components are ergonomically configured, with the controller located within easy reach of the driver's right hand and the display at the top of the dashboard. Operations with the controller can be completed with minimal visual disruption to the forward scene, and items requiring visual attention are centrally located. The overall concept of a distributed automotive HMI has since been adopted by other automotive manufacturers. In 2008 BMW introduced an upgraded version of the iDrive system. Rather than being solely technology driven, the new iDrive design was developed through a multi-stage consumer focused design and evaluation process that sought to maximize the intuitiveness and the efficiency of driver interactions (Niedermaier, Durach, Eckstein & Keinath, 2009).

The development and implementation of effective design is a complex challenge involving many disciplines and requires a balancing of often conflicting requirements (Norman, 2002). The design of in-vehicle systems for the modern automobile is best described in the context of "extreme" usability. Extreme usability exists where basic functions of the system have to be usable with little or no formal training under heterogeneous complex and potentially demanding conditions. Furthermore, operation of the system cannot have any impact on driving performance. To achieve extreme usability, the operation of an in-vehicle device must be intuitive, i.e. there must be little or no disconnect between the operator's model of how the system should work and its actual function. A driver confronting a non-intuitive interaction can become distracted, which is associated with extended visual and/or cognitive demand. Learnability and good fitting mental models are a prerequisite for minimal distraction. Design guidelines such as the Society of Automotive Engineers 15-second rule (*SAE* J2364) and assessment methodologies such as visual occlusion try to help guide designers to minimize the demand of individual tasks. However, these methodologies provide little guidance on how users adapt to and learn the functionality of a new interface.

ASPECTS OF LEARNABILITY

As Bengler (2007) points out, there are different phases or aspects of learnability and it is important to understand these different aspects in both designing and evaluating a system. One useful distinction is between how easy a system is to learn for a novice user versus the extent to which a system is ultimately learnable. In some circumstances, the introduction of novel functionality may require presenting a potential user with an innovative HMI design for which they have little or no prior mental model, or one that requires adopting a mental model that varies in some significant detail from their previous experience. If there is sufficient benefit to be gained from the functionality of the new design, then the effort of learning a new model or adapting an existing model should be well worth the learning investment. As Bengler argues, in such situations it is important to separate out performance during the acquisition phase and the ultimate performance gains after the HMI has been mastered. Advances in technology provide a designer with the ability to incorporate functionality that requires new models of interaction. The iPod® HMI is an example of an interface for which users did not have a mental model; however with some interaction, most users were able to become familiar with this design element rapidly and utilize it successfully.

An idealized system is one that extends a user's mental model for a similar application or requires a minimal degree of new learning to grasp the basic features of the system while also offering the user enhanced functionality over previous designs. The methodology presented in this chapter was developed to assess various aspects of learnability associated with a potential user's early interactions with the system. In brief, the goal was to address the following questions:

1. How intuitive is the basic logic or mental model of the HMI as perceived by the individual before he/she actually interacts with the device?

   This question attempts to address the situation when an individual encounters a system for the first time, as might occur when looking at a system on a store shelf, in a showroom or in a borrowed or rental car. The resulting impression of the degree of intuitiveness may impact the extent to which a potential user may feel attracted to, or uncomfortable with, trying to interact with a system.

2. How much is the user likely to adapt his/her mental model through a brief, unguided exploration?

   This is intended to capture aspects of the experience an individual will have trying to use a system for the first time. This experience will affect initial attractiveness and the likelihood that a user will invest additional time to learn the full feature set or decide that the device is not worth the trouble of learning. This may significantly influence his/her feelings of comfort and confidence in the overall product.

3. To what extent is the underlying logic of the system (and features available) easily learnable (and retained) through a brief guided instruction period?

The introduction of an advanced design sometimes requires that the user develop a new mental model for interaction. To what extent is this new mental model readily learned if a guided introduction is provided?

4. To what extent does a user become aware of and extend a mental model associated with function A to function B or across the system as a whole?

This addresses the question of consistency and transfer of mental models.

CONSIDERING DEMOGRAPHIC INFLUENCES

It is a truism of usability research that subjects should be selected that are representative of the target user population (Proctor & Van Zandt, 2008). However, simply selecting a representative sample and summing their responses can result in the loss of important information. This is especially true in the development of an HMI where demographic factors such as culture, age, gender, technology experience, and education can significantly influence the mental models of the individual interacting with the system.

Demographic and financial trends show the individuals of the "boomer" generation are the core purchasing power that drives high-end automobile purchases; these individuals are likely to have a wider range of experience with new technologies and HMIs than a younger cohort. In addition, women are becoming increasingly influential in automobile purchasing decisions (Coughlin, 2005); yet women appear to be less likely to adopt infotainment technology in the home and it is conceivable that this resistance will extend to the automobile environment. Thus, as the HMI design team continues to innovate as a method of maintaining a competitive advantage, theymust understand how various segments of the target population may differ in their ability to understand and adopt novel technologies. An HMI must relate in some manner to a driver's existing model of how an interface should work, or appear so intuitive that the driver's model is easily modified without confusion.

THE EVALUATION CONTEXT

With the launch of the new 7 series, BMW introduced the next generation iDrive system. In this paper, we describe the methodology of an assessment commissioned by BMW to explore users' interactions with a prototype of the updated iDrive. The assessment was developed to evaluate the intuitiveness of the relationship between the controller (figure 1a) and the display (figure 1b). Figure 1c shows how the controller and display are actually integrated into the vehicle.

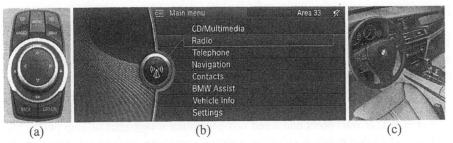

(a)   (b)   (c)

FIGURE 1. (a) iDrive controller with short cut buttons, (b) iDrive main system menu, and (c) the iDrive controller and display as integrated into the 2008 7 Series HMI.

A period of unconstrained interaction (e.g., show room type exposure) with a working prototype and a period of defined interactive training were employed to assess different aspects of system learnability. Comparisons across age, gender, technology experience and educational levels were undertaken to assess the extent to which these demographic characteristics influenced user's mental models and the effective usability of various features. Conducting the study with a US sample was intended to provide the German based design team with awareness on any potential culture based differences in interaction style with the HMI. The overall goal was to provide feedback and insight to the design team that could be utilized to adjust or otherwise enhance the final implementation. In addition, the assessment aimed to identify the frequency with which different characteristics of the HMI were used and how users rated their experience with components of the system.

METHODS

The evaluation included three interactive assessments, two learning periods and three questionnaires for a total of eight stages. The stages were ordered: (1) pre-experience questionnaire; (2) naive static assessment; (3) unguided learning; (4) post-exploratory static assessment; (5) mid-assessment questionnaire; (6) guided learning; (7) knowledge recall assessment; and (8) post-experience questionnaire.

QUESTIONNAIRES

The pre-assessment questionnaire gauged participants prior experience with different technologies, e.g., personal computers, cell phones, hands free systems, personal organizers, and vehicle navigation systems. Also collected were measures of driving experience, previous vehicle ownership and basic demographics, such as education and income. The second questionnaire was used to gauge the degree to which participants understood the interface, including the five different types of movements of the controller, e.g., rotation, right translation, etc. The final questionnaire was used to assess how close participants felt different operations of the controller and features of the system came to their mental model of the system.

STATIC ASSESSMENT

The intuitiveness of the iDrive rotary controller was initially assessed by observing novice users' navigation between items in a non-interactive (static) representation of the iDrive display (Figure 2). Twenty images that represented different combinations of conceptual movements within menus and between sub-menus were created. The images were constructed by overlaying a green box on the image indicating a task start point and a red box indicating a task goal or stop point. The starting and ending points highlighted specific operations such as selecting a radio station. A functional prototype of the controller (see Figure 1a for an image of the production version) allowed users to physically manipulate the device through its range of functional movements: rotation; translation (tilting the controller to one of four quadrants – left, right, forward or back); and select (push the controller down). During the static assessment, movements of the controller were not represented in the display; however, users were asked to manipulate the controller in the manner they thought would produce the desired actions and to verbalize what action they were taking. Having the users talk through the movements was intended to make overt their assumptions about the required actions as well as to facilitate easier coding. Disconnecting movement of the controller from actions on the display screen allowed separation of the user's intuitive assumptions about the interface from the interactive learning that would have taken place if the connection was active. (This same approach could also be used for assessment of intuitive mental models earlier in a design process when functional prototypes might not yet exist or with actual products

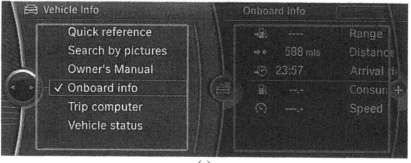

(a)

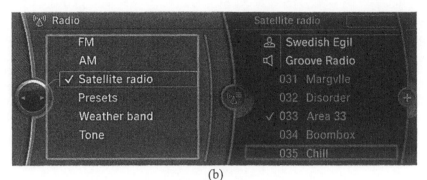

(b)

FIGURE 2. Example of non-interactive representation of the display with green starting position and red ending position for (a) an single translation task, (b) a translation and rotation task. The overlaid green box outlines the task start point and the red the goal stop point.

Figure 2a illustrates a basic one movement task. To complete the task successfully, a participant needed to make one right translation of the controller (tilt right) to move from the vehicle *Info menu* to the *Onboard info* submenu. In the more complex two movement example (figure 2b), participants needed to demonstrate a right translation (tilt right) to move from the radio menu to the station list. A right rotation of the controller was then needed to move the cursor from the default landing position (highlighted with the yellow check station) down to station #035. Participants then needed to press the controller to complete the selection.

During the assessment, participants were instructed that *"In the next task we are investigating the intuitiveness of movement between items in a telematic interface. There are no correct or incorrect answers. We are just interested in seeing how you navigate between items."* A description of the translation, rotation and pressing operations of the controller were provided. Before beginning the assessment, participants were given a period to practice controller movements (without observing the effects on the display).

In the case of translation movements (tilt actions), participants were scored on both the direction and number of movements required to complete a task. In tasks requiring rotation to scroll across lists, only the proper movement direction was required. This simplification was made since the interactive feedback of the controller was not engaged. In the actual working system multiple translation movements (tilt actions) are required to move between menus but scrolling motions can move past multiple items on a list. Feedback is provided in the form of a tactile "click" as each item of the list is passed.

Tasks of three levels of complexity were included in the assessment. Ten easy tasks required a single movement of the controller. Five moderate tasks involved two sequential movements. The remaining, more involved tasks encompassed three or four movements in a specific order. Errors were recorded and then classified using a pre-defined taxonomy of error types to produce a primary outcome measure. Each missing or extra motion of the controller was assigned an error code. For example, a participant who indicated that two right translations would be required to move from a green start point to the red completion point an error would be recorded if only one translation was required. Cases where an error was likely due to a limitation in the assessment methodology were categorized separately.

The first phase of the static assessment provides a measure of how well a user's initial mental model of the device's operation overlaps with that of the designers. Following a brief unguided training (described below), the static assessment was repeated with items presented in a new random order. A change in the total number and distribution of errors provides an indication of how well the intended relationship between the rotary controller and display can be quickly learned.

UNGUIDED LEARNING - FREE EXPLORATION

This learning period was intended to model what a user would experience the first time he/she attempts to understand a new automotive HIMI on their own. Most users in the United States tend to try new HMIs before reading a user manual or reviewing any training materials.

Between the two static assessments, users were given 10 minutes to explore a fully functional prototype of the new iDrive system. The menu was displayed on a laptop screen and controlled through an attached iDrive controller that contained all of the major features that appear in the final production version (figure 1a). Movements of the controller were directly related changes in the display. Force feedback enhanced the feel of the system by amplifying the effect of certain movements. In this format the haptic interface allows user to "feel" movements as they occur. For example, as described earlier, the rotation of the controller embodies tactical clicks to indicate movement past an item in a list as well as "soft stops" (where the controller's resistance increases) at the top and bottom of the list.

During this phase of the protocol, no specific tasks were provided; the user was free to explore the device in whatever manner he/she wished. Instructions told participants to *"explore the advanced interface for an in-vehicle telematic system. This system allows you to control and monitor the Navigation, Telephone, Radio, CD/Multimedia, as well as other settings of your vehicle. Please try to understand as much of the system as possible in the allotted 10 minutes."*

GUIDED LEARNING

This phase was designed to model the experience of being given hands on instructional training on how to use the interface. During this phase of the assessment, users were first provided with a description of a number of additional featured enabled through shortcut buttons that surrounded the rotary controller (static testing did not involve these features). They were then taken step by step through a series of ten tasks similar to those used in the static test. In contrast to the static assessment, the haptic feedback features of the interface were present to help guide user movements as intended by the design team. During each step, participants were required to make the described movements and the prototype interface represented all changes visually. For example, *"we are going to call David Baker by using the address book. Scroll to the address book. Select the address book. Scroll clockwise five clicks to Baker, David. Select Baker, David. From here, you can navigate to David Baker by selecting the Navigate To button or call him by selecting the green phone icon. Select the green phone icon. Select End Call. Now that you have called David Baker, press the MENU to return to the main menu."* The ten training exercises were presented in random order.

KNOWLEDGE RECALL ASSESSMENT

This phase was intended to evaluate the extent to which the user could learn and retain the concepts highlighted as part of the instructional training. Some of the tasks required participants to extend learned concepts. Participants were instructed that for the next series of tasks, *"I will not be providing any instructions. You will be responsible for completing the tasks on your own."* Thirteen tasks such as *"Call Audrey Brown, Change radio to WKTU-FM,"* and *"Navigate to Third Ave, Albany, New York using Enter Address Menu"* were presented. During the knowledge recall testing, a participant's ability to successfully complete each task was recorded. In addition, strategies used to complete the task were coded to evaluate the frequency with which different shortcut menus were used in place of full navigation of the menu system. Cases where the *back, menu* and *option* buttons were used were recorded to assess their usefulness.

RESULTS

Detailed data on the type and frequency of errors at each level of assessment were generated and supplied to the development team along with the impressions and recommendations of the assessment group. The initial static assessment was useful in several ways. In addition to identifying areas where the intuitiveness of the interface might ideally be improved, it provided detailed insight into how users of different ages, genders, income levels, and educational backgrounds learned how to interact with the HMI. In addition, it identified points or concepts that should be highlighted in the user instruction guide and other learning support materials (e.g., video tutorials, etc.). Such targeted highlighting may assist users to quickly grasp the logic behind those features that some individuals do not find intuitive under unguided conditions.

The dynamic assessment with knowledge recall task identified a greater than expected usage of the short cut buttons (i.e. radio, CD, back). Drivers appeared to readily adopt the usage of these features. The assessment supported the design team's belief that the inclusion of haptic feedback and subtle indications of the appropriate controller operations on the representation of the controller in the display were useful components that assisted in learning to operate the system.

The methodologies discussed here assessed the intuitiveness and learnability of the system under non-driving conditions. Thus, they considered base level issues of specific aspects of the HMI design. As noted previously, intuitiveness of interaction is a critical design component since any disconnect between a driver's mental model of how an interaction should take place and what they encounter will result in distraction. Evaluation of usability characteristics under simulated and actual driving conditions clearly are required in addition to this class of testing (Bengler, 2007). In the case of the new iDrive system, device operation with the added demand of driving was assessed independently by BMW (Niedermaier et al., 2009).

While the assessment methodology presented here generated useful output as discussed above, some limitations were identified. Results from the easy and moderate static assessment tasks produced robust data on deviations between the designers' and users' mental models. This assessment methodology, however, did not work as well for representing and testing the more complex three and four movement tasks. In the case of these tasks, it appeared that actual changes in the display were needed to fully represent the actual operation of the HMI. A possible improvement in the methodology to would be to add an additional assessment component with the dynamic features of the HMI activated for examining multi-step processes.

CONCLUSIONS

The evaluation resulted in a number of suggestions for improvement in the model of operation and the location of selected functions in the prototype to further align functions with users' models of the system. The multi-stage assessment methodology supported the evaluation of different aspects of the learnability of the system. The methodology extends the procedure known as a cognitive walkthrough by combining it at various points with interactive prototypes, thus supporting a more in depth analysis. As expected, differences in individuals' initial models of operation, and their ability to learn methods of operation, were associated with age and technology exposure. Results from this evaluation of a design prototype were used by the design team to optimize the user interaction model and feature selection for the production version of the new iDrive system. This level of investment in the design process is important both for maximizing customer satisfaction and eliminating potential sources of driver distraction.

Acknowledgements & Notes

The authors gratefully acknowledge BMW and the United States Department of Transportation's Region I New England University Transportation Center at the Massachusetts Institute of Technology for support during the preparation of this manuscript. Note that support provided by the latter source was for the advancement of research methodologies and the understanding of HMIs and should not be taken as an implied endorsement of a specific product.

REFERENCES

Bengler, K. (2007), "Subject testing for the evaluation of driver information systems and driver assistance systems – learning effects and methodological solutions." In C. Cacciabue (Ed.), *Modeling Driver Behavior in Automotive Environments; Critical Issues in Driver Interactions with Intelligent Transport Systems* (pp. 123-134). London: Springer.

Coughlin, J. F. (2005), "Not your father's auto industry? Aging, the automobile, and the first of the drive for product innovation." *Generations,* Winter, 38-44.

Niedermaier, B., Durach, S., Eckstein, L., & Keinath, A. (2009), "The new BMW iDrive – Applied processes and methods to assure high usability." In V.G. Duffy (Ed.), *Proceedings of the Digital Human Modeling, HCII,* 443–452.

Norman, D.A. (2002), *The Design of Everyday Things*. New York: Basic Books (Perseus).

Proctor, R. W. & Van Zandt, T. (2008). *Human Factors in Simple and Complex Systems. Second Edition.* New York: CRC Press.

Stanton, N.A., Salmon, P.M., Walker, G.H., Baber, C., & Jenkins, D.P. (2005), *Human Factors Methods: A Practical Guide for Engineering and Design.* Hampshire, England: Ashgate Publishing Limited.

CHAPTER 7

The Method for Risk Assessment of Accident Caused by Technological Process, Human Factor and Environment

[1]Miroljub Grozdanovic, [2]Evica Stojiljkovic, [3]Dragutin Grozdanovic

[1,2]Faculty of Occupational Safety
University of Nis, Serbia

[3]Faculty of Medicine
University of Nis, Serbia

ABSTRACT

The method for risk assessment of accident caused by technological process, human factor and environment, aimed for assessment of the level of overall hazard of an accident, however complex, could be redeveloped in a very practical and simple manner. It consists of evaluation of hazards caused by technological processes, human factor and the environment respectively. Parameters that determine each of those basic factors are discussed in detail in this paper.

Keywords: Risk of Accident, Level of Hazard, Human Factor, Environmental Risk.

INTRODUCTION

Analysis of the existing data on industrial accidents testifies that the major threats to a system safety are flows of energy and harmful materials. Basic rules of accident occurrences indicate that: accidental events can be interpreted as relatively rare random events with Poisson law of distribution at limited time intervals and exponential law of distribution of time between their occurrences; the occurrence of each of them most often is not caused by one particular cause, but rather by a chain of events under appropriate prerequisites. Initiators and/or links of such a chain are human errors, technology failures and/or unforeseen external influence (Меньшиков, 2003).

According to data of the International Labor Organization, about 1000 major chemical accidents were recorded in the world up to 1990. Approximately 40% of the overall numbers of accidents occur in production plants, about 35% during transport, and about 25% refers to accidents during storage of chemicals.

According to the data available to the Ministry of Environment of the Republic of Serbia, 20 to 25 small-scale chemical accidents were registered annually in our country before enforcement of international embargo on Serbia in 1992. In most cases, such accidents did not exceed the boundaries of an industrial complex. The basic cause of those accidents was related to human errors (62%), and obsolete technology (20% cases). The amount of chemical accidents in Serbia is below the average registered in industrially developed countries of Europe and North America. According to data published by the Organization of Economic Cooperation and Development (OECD), 30 to 35 chemical accidents of smaller or bigger scale (with human casualties and the environment pollution) happen worldwide every day.

The method for risk assessment of accident caused by technological process, human factor and environment, aimed for assessment of the level of overall hazard of an accident, however complex, could be redeveloped in a very practical and simple manner. It consists of evaluation of hazards caused by technological processes, human factor and the environment respectively. Each of those three basic factors is quantitatively determined by specific parameters.

METHODOLOGY

The method for risk assessment of accident caused by technological process, human factor and the environment, designed to assess the level of overall hazard of accident, is essentially based on some previously published researches (Grozdanovic, 2003., Grozdanovic and Stojiljkovic, 2006., Majer et al, 1998), taking into account certain national regulations, such as "Code of practice on methodology for risk assessment and mitigation of chemical and environmental accidents" (Official Gazette of Republic of Serbia, No. 60/94)

The method itself is composed of the following steps: assessment of hazard caused by the technological process (TP), assessment of hazard caused by the human factor (HF), assessment of hazard caused by impact from the environment (E), assessment of the level of overall hazard of accident (A), recommendations for the assessment of the level of overall hazard of accident.

The level of overall hazard of accident is given in the following expression:

$$A = TP \cdot E - HF \cdot \frac{TP}{30}$$

Hazard caused by a technological process is determined by the expression:

$$TP = CA \cdot ER \cdot PP \cdot PM$$

where:

CA –magnitude of possible consequences of accidents
1 - negligible, 2 - significant, 3 - serious, 4 – very serious, 5 – extreme.
ER –exposure to risk of accidents
1 - small, 2 - medium, 3 - large.
PP –possibilities of prevention
1 – poor, 1.5 – acceptable, 2 – good.
PM – preparedness measures
1 - low level of preparedness, 1.5 – medium level of preparedness, 2 – high level of preparedness.

Hazard caused by a human factor is determined by the expression:

$$HF = QL + RA + ER$$

where:

QL – level of skillfulness of the people to act in case of accident
1 –poor skilled, 1.5 – medium level skilled, 2 – good skilled.
RA – adequacy of response to an accident
1 - inadequate, 1.5 – semi-adequate, 2 - adequate.
ER – evaluation of remediation likelihood
1 – poor, 2 – good, 3 – excellent.

Hazard caused by the influence of environment:

$$E = EC + CPS + OI$$

where:

EC – arrangement of surroundings condition
1 – poor condition, 1.5 - satisfactory, 2 – excellent condition.
CPS – compatibility of people and response systems
1 - poor compatibility, 1.5 – medium compatibility, 2 – good compatibility.
OI –other influences
0.5 - almost negligible, 1 – little influence, 1.5 – great influence.

The level of overall hazard of accidents is in the range of the minimal Amin=2.4 to the maximum Amax=316 (Stojiljkovic, 2007). Table 1 explains recommendations for assessment of the level of overall hazard of an accident.

Table 1 Assessment of the level of overall hazard of accident

Levels of overall hazards		Comment
I	2.4 - 60	Relatively harmless condition. Negative consequences of an accident (if they occur) are limited to an industrial site, and negative effects on wider surrounding are not expected.
II	60 – 120	It is necessary to reduce the hazard. Consequences of the accident oould engulf one portion or entirety of an industrial compound, however without negative consequences on wider surrounding.
III	120 – 180	Significant hazard. Consequences of the accident can be transferred to environment and the consequences are expected to affect a part of territory, or the whole territory of a municipality or a city.
IV	180 – 240	Great hazard. Consequences can spread over the territory of several municipalities.
V	240 – 316	Extremely great hazard. It is a large-scale accident and its consequences threaten to expand beyond national borders.

RESULTS

Described method for risk assessment has been applied at company "Messer Tehnogas"AD (production plant in Nis, Serbia). Production site comprises oxygen, nitrogen and crude argon manufacturing; distribution centers of oxygen and carbon dioxide, and warehouses of these technical gases (Mitic, at al. 2005).

The method is based on the assessment of the following steps:

Assessment of hazard caused by the technological process – observed work process combines a specific manufacturing technology with operating storage facilities and filling steel tanks for broad consumption. Oxygen stored in quantities of more than 200t belongs to category of hazardous materials and the technology includes a series of processing equipment such as containers under pressure. Taking into account all the circumstances it is assessed that accidents can occur as mechanical explosions of equipment and local installations, as explosions with accompanying fires due to contact with organic material and due to gas spill.

In this hazard estimate it was necessary to quantitatively determine the values of possible consequences of potential accidents (CA).

Causes of possible accidents at "Messer Tehnogas"AD production site are: human factor, mechanical failures, disruptions in the transport of products, natural disasters, any possible war situations and sabotage, fires and explosions.

On the basis of performed analysis, expert assessment on the consequences of potential accidents provides the value of 3 (CA=3).

Quantitative determination of exposure to risk is performed on the basis of several relevant conclusions, such as:

- in case of leakage of any of technical gases (that has been assessed as possible scenario), it is probable to expect a minor material damage and possible injuries of the participants in the event (assessment of exposure to risk 1);
- in case of fire on the transport vehicle there is a great hazard for the lives of the people present there and possibility of significant material damage (assessment of exposure to risk 2);
- despite law probability of events with total damage of stored tanks, consequences of such occurrences would be the most disastrous (assessment of exposure to risk 3).

On the basis of performed analysis, expert assessment on the risk exposure due to potential accidents gives the value of 2 (ER=2).

Analysis of some previous accidents caused by errors in technological process implies significance of both preventive measures and emergency procedures taken on the spot of the accident; a great deal of planning is needed in order to prevent possibility or reduce probability of accident occurrence and mitigate it's consequences.

Quantitative expert assessment of possibilities for preventing the accident in given circumstances is described by value of 2 (PP=2).

Preparedness (ability to provide the most adequate response to a possible accident), is achieved trough interaction of competent subjects with available equipment while applying prescribed techniques. The level of preparedness could be increased by planning and designing appropriate safety systems based on hazard assessment and accident probability. In given setup, safety expert assessment reveals value of preparedness of the system as 1.5 (PM=1.5).

On the basis of these quantitative determinations, the hazard caused by a technological process is estimated as:

$$TP = CA \cdot ER \cdot PP \cdot PM$$

$$TP = 3 \cdot 2 \cdot 2 \cdot 1.5$$

$$TP = 18$$

Assessment of hazard due to the human factor refers to:

- skill level of the people who are supposed to act in case of an accident (QL),
- to apply an adequate response to the accident (RA) and
- evaluation of remediation likelihood (ER).

Considering the fact that in observed case basic training has been provided for employees that are supposed to act in unpredictable situations, skill assessment provided the value of 1.5 (QL=1.5).

An adequate response to an accident starts at the moment when the right information about the accidents has been received.

In the company "Messer Tehnogas" AD there is an appropriate scheme of responses to an accident that prescribes information flow among the most responsible workers, managers and services that take part in giving response to an accident. Therefore, quality of response to accident has been assessed as satisfactory (RA=1.5).

Evaluation of remediation capacity and likelihood is greatly hypothetical because to judge remediation measures to compensate for consequences of an accident would mean monitoring the post-accidental situation, restoration and remediation of the living environment, overall recovery, as well as removing the risk of accident recurrence. Therefore, estimated level of remediation capacity and likelihood is described with the value of 1 (ER=1).

On the basis of previous quantitative determinations, hazard due to the human factor is estimated as follows:

$$HF = QL + RA + ER$$

$$HF = 1.5 + 1.5 + 1$$

$$HF = 4$$

Assessment of hazard caused by the environment is based on the analysis of the arrangement of surroundings (EC), compatibility of people skills with available technological system (CPS) and other negative influences (OI).

Analysis of immediate surroundings begins with the survey of disposition of objects in an industrial compound. In the system observed, the impact of the materials used in manufacturing process on structural elements in case of accidental leakage is reduced to a minimum.

On the basis of analysis of the data about the location and reaction of materials in the process of handling the materials of the constructions, the value of expert assessment on arrangement of surroundings condition is found to be 1.5 (EC=1.5).

Assessment of compatibility of people skills and response system features depends on system complexity (structure and interactions between a man and elements of the system).

Using the graph system for potential accidents, it has been assessed that the compatibility level between the people and the system was 1.5 (CPS=1.5).

Assessment of hazard due to other negative impacts is based on assessment of probability that some form of natural or man-made disasters could happen (storm winds, thunder, and impact of flying objects, war situation and sabotage).

It is assessed that there is a slight hazard of external impacts, i.e. 0.5 (OI=0.5).

Based on previous quantitative determinations hazards caused by external factors are estimated as:

$$E = EC + CPS + OI$$

$$E = 1.5 + 1.5 + 0.5$$

$$E = 3.5$$

On the basis of the research done, results of assessment of the level of overall hazard of an accident to occur in "Messer Tehnogas"AD are shown in Table 2.

Table 2 Results of assessment of the level of overall hazard of an accident incidence

CA	ER	PP	PM	TP	EC	CPS	OI	E	QL	RA	ER	HF	Hazard of Accident (A)
3	2	2	1.5	18	1.5	1.5	0.5	3.5	1.5	1.5	1	4	60.6

The results of the assessment of overall hazard of accident (A=60.6) suggest a level of hazard that would not endanger surroundings, over the boundaries of observed industrial compound.

CONCLUSIONS

The most important methodological step for risk management is assessment of probability and consequences of an accident. That is complex procedures that in a way depict all the weight of the problem related to endangered environment and consequences of industrial development. By assessment of probability and consequences of an accident we come to a conclusion whether the risk of hazardous activities on particular place is acceptable.

The process of assessment of probability of an accident can be divided into phases according to various criteria, depending on the scope and complexity of perception of the problem. Each of the phases, with its qualitative characteristics, separately consists of a complex of procedures and activities undertaken with the aim to assess the hazard of accident and serves as the basis for further improvement of knowledge in this field.

The assessment procedure of probability of accident has the aim to identify and quantify the regions where potentially accidents can occur. That is a research process that must be professionally and scientifically based, with a multidisciplinary approach. A well done assessment of probability and consequences of an accident is a precondition for adequate prevention planning, preparations, reactions to an accident and remediation of consequences. It offers enough relevant data for the process of risk management, because, on the one hand, it indicates the status of safety in the observed system, and on the other hand it refers to options for risk treatment and to the necessity for improvement of the systems for protection of working and living environment within the risk control.

ACKNOWLEDGEMENTS

This paper is a part of the results from research project titled "Research and development of expert systems and methods for ergo-ecological risk assessment of accidents in Electric power company of Serbia", TR 21030, 2008-2010., under auspices of Ministry of Science and Technological Development, Republic of Serbia.

REFERENCES

Grozdanovic, M., and Stojiljkovic, E. (2006). Framework for human error quantification. In N. Bozilovic (ed.), Facta Universitatis, Series: Philosophy, Sociology and Psychology, 5(1), 131-144.

Grozdanovic, M. (2003). Information systems for the risk management in working and living environment, The Journal "Economics", XLIX (5-6), Faculty of Economics, Nis, pp.33-54, (in Serbian).

Majer, I., Oravec, M., Sinay, J., and Sloboda, A. (1998). Methods of the Risk Evaluation, 3rd International Conference "Global Safety", Bled, Slovenia.

Меньшиков, В.Б. (2003). The identification professional risks, International Conference "Professional risk assessment - theory and practice", Faculty of Occupational Safety, Nis, pp. 9-12, (in Serbian).

Mitic, D. et al. (2005). Risk assessment and Mitigation of chemical and environmental accidents, measures and prepare measures for elimination of consequences in company „Messer Tehnogas"AD. Faculty of Occupational Safety, Nis, (in Serbian).

Stojiljkovic, E. (2007). Methodological framework for assessing the probability of environmental accidents, Master Thesis, Faculty of Occupational Safety, Nis, (in Serbian).

Code of Practice on Methodology for Risk Assessment and Mitigation of Chemical and Environmental Accidents (Official Gazette of Republic of Serbia, No. 60/94), (in Serbian).

CHAPTER 8

Methodology for Tests Execution in Usability Laboratory Considering Cross-Platform Devices

Claudia de Andrade Tambascia, Robson Eudes Duarte

Fundação CPqD – Centro de Pesquisa e Desenvolvimento em Telecomunicações
Campinas, São Paulo, Brasil

ABSTRACT

Several applications have been widely developed for multi-platform devices such as Digital TV, smartphones, internet, palmtops, etc in present days. In this context, many aspects concerning usability play such import role as essential requirements to the success of those applications. It is necessary, then, a methodology for evaluating the usability of such complex applications and a specific environment for testing the usage of those applications running on the most varied platforms considered in a given project.

Keywords: Usability Laboratory, Human Computer Interaction, Cross-Platform Devices

INTRODUCTION

The idea of a usability laboratory comes with the necessity emerged from the growing amount of digital inclusion projects. In this context a project for services for interactive television is being developed, aiming to include in the information society people who are twice-excluded (social and digitally). The success of this project, named SMTVI (Soluções Multiplataforma para TV Interativa) depends on giving to the users a complete and likeable use of all functionalities of the services provided considering also the use of assistive tools, if it is necessary.

The intention is that this set of tasks, corresponding to suitability and usability factors although articulated in a pleasurable experience to the user, thus providing the appropriation of information technologies and communication technologies (ICTs) by people with little or no expertise with them. Figure 1 shows the main factors to be considered in relation to user's experience.

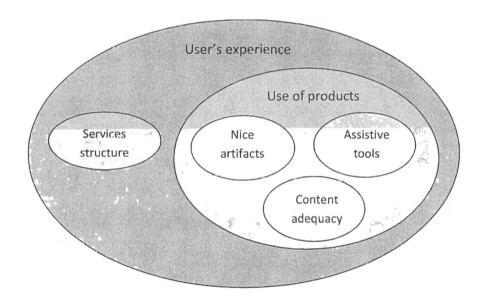

Figure 1. Domains of human systems integration.

The implementation of the project involves an opportunity for the digitally illiterate people to use ICTs. To be able to break through the barriers that prevent the free enjoyment of use of ICT by these people, user's Quality of Experience (QoE) must be strictly observed in the tests of the interfaces of the services that are to be developed under the project. To do so, one of the major challenges is to formulate a testing methodology that contributes to the continuous improvement of prototypes proposed in order to maximize the user QoE.

Thus, the contributing authors worked on the implementation of a testing laboratory, to serve as infrastructure for implementation of a methodology for testing which aims to guarantee a high level of QoE of the user, encompassing aspects of usability, accessibility, intelligibility, pleasantness and QoS, and also consider the evaluation of products and services resulting from the project SMTVI.

It's not an easy task to propose a testing methodology for a project with such expansive proportions as SMTVI, whose audience is particularly diverse and yet so special in terms of their specificities, such as disability (low literacy or sensory disabilities), that are skills developed to compensate for each disability or even those so-called normal capacity. Given these combinations of abilities and disabilities, specific to each of the possible categories of user, the methodology should provide test sequences that are both relevant to the various user groups, without losing its unity method. Moreover, there is little research involving the target audience of this project, resulting in little literature available to follow.

USABILITY LABORATORY

The aim of setting up a usability laboratory is mainly to support an environment for observing, monitoring, analyzing the use of a device in a set of activities and identifying issues in the developed user interfaces, taking into account not only the final application, but also the target user. Problems related to the limitation of the user him/herself before the ICT may be frequently found, and these problems may vary according to the target users and the different technologies considered. In many cases, the users selected may have their practices with the ICT altered due to their accumulated experience with determined device, which modifies the way a same task is executed along in the time.

The project of the laboratory considered took into account the following aspects: needs of the users for applications designed to mobile devices; aspects related to the project of user interface; adequate environment for supporting usability tests and heuristic evaluation; assisted usage of developed applications. The activities and experiments done in the laboratory involved the knowledge acquisition of the needs of the target users and their limitation in the use of ICT solutions; the efficacy and the quality of experience in the usage of the services; response to questions during the project development; application inspection simultaneously to real practices of use; comparison of design alternatives; adherence to standards and patterns.

The usability laboratory is ready to receive external users to the corporation, including people with sensory and motor disabilities and older people. It is composed by two separate rooms, one of them being the tests room and the other the observation room.

The tests room is mainly destined to accommodating users for the usability tests execution, and it contains three desktop computers connected to the Internet and disposed into a strategic layout, allowing sound and video capturing during the realization of the activities. Such captures are done through multimedia resources from the computers themselves, and also through screen readers and softwares that help the usability inspections. Too in this room there is a Braille printer, a multi-functional printer, a LCD TV with 42" a touch screen television and three cameras strategically placed on its ceiling. The whole furniture in the room is according to accessibility standards and patterns, which allows people on wheelchairs or other motor disabilities to richly use the test room.

The observation room is located besides the testing room, separated by a mirror, where some professionals are positioned to watch the tasks execution prescribed in the tests schedule. Also in this room it is located the equipment responsible for controlling the audio and video recording and the ceiling cameras positioning. The observation room contains three laptops connected to the Internet, a server for images and sounds, the control table for observation cameras (the ceiling cameras) and a table for sound mixing, for the audio and video treatment, captured during the tests.

With this infrastructure, the possibility of testing usability is guaranteed to validate the human-computer interaction in relation to applications developed for each of the devices required.

For this purpose, it was established a methodology for conducting these tests. First the users, devices and interfaces to be evaluated are selected as well as the team that will participate in the test conduction and observation process. Second, test scenarios to be executed are defined, and for each of them, the set of criteria and attributes that should be considered and determined. The roles of those professionals involved are then distributed in the testing process and after the users' selection, the test execution starts with all the involved people executing their own roles. It is important to quote that the users participants of the test are aware of all observation and audiovisual capture processes.

At the end of the test data are tabulated and the video and audio captured are analyzed, so that no information is lost and all the problems and recommendations for interface improvement are properly documented. Thus, it is possible to guarantee that the usability criteria are being met and the user satisfaction is being achieved by improving their perception of the application developed.

A QUALITY OF EXPERIENCE (QoE) AS FOCUS METHODOLOGY

According to Tambascia et al. (2007), in the early 90's, when communication was still strongly marked by computer, on the one hand, and information technology, on the other hand, the ITU-T (1994) has defined Quality of Service (QoS) from the user's perspective. While providing a framework of performance concepts, which contributed to the quality of service perceived by the user, the orientation was the underlying metrics of technical parameters.

Factors that directly affected the perception of the user in the level of quality offered - like the traditional delays in establishing a telephone call or the level of noise in voice transmission, mainly dependent on the performance of the network - were monitored through rigorous methods and evaluated objectively in data bases. However, the effect of delays in human perception - such as irritability or dissatisfaction - were not considered under the same methodological rigor, except for one of its consequences: the number of retries to establish a call which, in turn, degraded network performance and amplified further delays, causing an avalanche effect.

From the diffusion of Internet and multimedia services and content, the design quality is to include new performance factors, and further focus on the perception of the end user. This evolution can be noted in Recommendation G. 1010 ITU-T (2001), which provides guidelines for QoS for multimedia services and applications based on Internet protocol (IP), provided by different technology platforms and also different requirements. Although including factors that directly affect the perception of the user and that can be measured both objectively and subjectively at the point of access to the service, instead of emphasizing the causes of anomalies only in the core network, the approach is still focused on technical parameters quantified as delays and loss of bits. Even the process of recovery and observation components of HTML Web pages have their experience evaluated the effects of delays acceptable (not exceeding 10 seconds, for example).

With the advent of convergence and the widespread use of information technologies and communication technologies (ICTs), the boundaries between the worlds of telecommunications, information technology and multimedia terminal industry - from television to mobile devices - will lose its sharpness. Thus, the need for further expansion of the conceptual quality of the experience, beyond the effects that the network performance exerts a user terminal and into the cognitive and cultural needs of users. Quality that should be considered on the one hand encompasses the effects of the combination of sound, image, data and voice bring to fruition new content formats and, secondly, the level of demand and growing confidence in the services and resources characterize the information society.

It is in this context that the concept of QoE emerges, bringing a new approach by increasing methodological and interdisciplinary research, to cover gaps related to the QoS and to develop new models in order to understand correlation of the various aspects that influence the perception of the User.

The question on the concepts of QoS and QoE has been treated by several authors in the field of technological sciences in their research, e.g., McNamara & Kirakowski, 2005, Eriksen, S. et al., 2007; O'Neil, 2007; Khirman & Henriksen, 2007. Such concepts reflect, to some extent, the discussion about what is considered objective versus subjective - and which methods to deal with these issues - not just on the border between the technological sciences and social sciences, but also within the social sciences.

The Muntean (2005) thesis seeks to increase awareness of the End User QoE in different environments. The QoE proposal takes into account many factors that affect it, such as Web components and connecting networks. It uses a new model for perception of performance that takes into account a variety of performance metrics so that much can be learned about the characteristics of the operating environment of Web User, the changes in network connection and the consequence of these changes on users' QoE. This model also considers the views of the User on QoE, increasing their effectiveness.

Approaches that deal with QoE bring focus to the composition of perceptions of the User and, unlike the QoS, assessed subjectively. On the client side, the User Experience is influenced by factors such as type User Interface, models of interaction, language employed, privacy and security, nature of content, utility service and more. In relation to the user consumption (or enjoyment), the experience is shaped by cognitive and emotional equipment of the individual, as well as its cultural context, as undivided and integrated response to all *stimuli* presented in the bid.

The analysis of QoE involves exploratory research and qualitative methods through interviews and careful observation to learn how and in which situations services are used and how they are perceived by users. The QoE looks toward describing how well a service affects the expectations of the User, is a classification of service performance from the perspective of the User.

QoE CONCEPTS AND EVALUATION

Because of being a recent subject of discussion and research, there is still no consensus on the definition of QoE (MAPPING, 2007, Morris & Turner, 2001). Bibliography often suggests the need for new research and new tests to the improvement QoE conception and the way they perform their measurements. Also, there is still a metric model as the basis for the development of tests and QoE, even though there would be a model developed based on research conducted in developed countries, people familiar with the use of ICT, an audience quite different from that considered by the project SMTVI (illiterate, elderly, rural workers with hearing and visual impairment).

As the main goal of the project SMTVI is to provide access to understandable ICT to people who are currently unable to do so due to different types of barriers (Holanda and Dall'Antonia, 2006), we believe that it will be achieved once the user interfaces are satisfactorily usable, taking into account the user's cultural, linguistic, and cognitive, sensory, and physical limitations. The services has also to be accessible in terms of network data quality and the availability of appropriate equipment and adequate infrastructure. In other words, the goal will be achieved as the quality perceived by the User Experience is satisfactory.

Therefore, the concept of QoE employed under this methodology is very comprehensive, once it encompasses the concepts of usability, accessibility, intelligibility, pleasantness and QoS related aspects identified in Figure 1. So, to evaluate the QoE of user it should be considered all the aspects listed in the Figure 2.

Regarding the QoS guarantee, we will follow the recommendation given in Muntean (2005), that the main point to be considered in web-based systems is to minimize the response times perceived by the user, defined as the time between a request and presentation of response by the system. This time interval, as suggested by the ITU G.1010 standard, should be preferably less than 2 seconds, though it is also acceptable an interval lasting less than 4 seconds. Tests of this nature might be done outside the Usability Laboratory, being part of the systemic tests process included in QoE methodology.

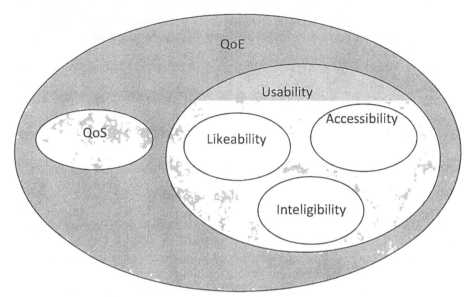

Figure 2: Aspects to be considered in testing process

The methodology for evaluating the usability is described in the next section of this paper. The design of this methodology considers the various factors that impress the User of ICT and that directly affect their experience in using it. These factors include, for example, the interface itself offered to the user, the navigation map in the case of web-based service, the reliability through the support, and the attractiveness of services (META Group White Paper). Focus firstly on the user and then on the above parameters should also be evidenced by the incorporation into the system of an ombudsman for users with the implementation of the polls, as well as collection of spontaneous opinions. This may influence which factors should be reworked to improve the quality of User Experience in the use of ICT (META Group White Paper).

METHODOLOGY DEFINITION TO BE HELD IN THE USABILITY LABORATORY

The methodology looks forward to undertake the scientific research that combines the inferential processes above, looking for a more systematic and consistent assessment with the use of data already known about user interfaces and interaction models, based on the proposition of initial hypotheses relevant in the deduction of testable consequences, in the design of the test sample that is more representative of the target population and, ultimately, in achieving the proposed tests.

In the specific context of a methodology for testing new user interfaces and new models of interaction, in which it is known *a priori* a certain set of data, both concerning the characteristics of the different user profiles and the aspects of the UI (e.g. skills required for the interface) should be considered. The evaluation process can begin by proposing hypotheses concerning the possible implications of these observational data on the supposed abilities or disabilities of the subjects. For each hypothesis, it is necessary to infer consequences with regard to possible problems of that category of interaction with the User interface features that are offered.

It is also part of the design of tests to select the User Profiles representative of the target and measuring the number of subjects required for each category in order to give statistical validity to the results or, alternatively, support them in the accepted heuristics HCI community (Nielsen and Landauer, 1993, Nielsen and Mack, 1994). This allows you to identify a given set of tests to be applied to a given set of subjects to try to validate the original hypotheses. If validated, these assumptions become data that comes to be part of the set of information available, as illustrated in Figure 3, and this may allow proposing new hypotheses. Thus, for example, an initial hypothesis H1 can be derived some consequences (C1.1, C1.2, C1.3) For defining tests (T1.1, T1.2) Appropriate for validating the initial hypothesis. It is then a cyclic process of whether fixing belief or learning

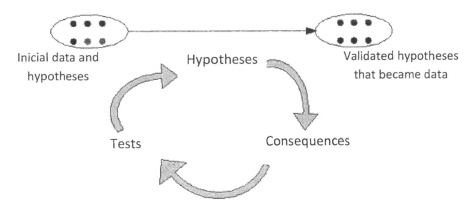

Figure 3 - Process of expanding knowledge

CONCLUSIONS

Human Systems Integration (HSI) is becoming a critical piece of complex systems to help resolve system designs. This proposal has presented a growing body of knowledge for HSI and new technologies that are being developed to capture critical aspects of HSI. The development of a framework for Human Systems Integration with Systems Modeling Language (SysML) will enable teams to collaborate better by providing a common language and process to distribute models and share information. The Human Systems Integration component in systems engineering will be able to recognize the human as an integral element of every system by representing behaviors, constraints, states, and goals through-out the entire life cycle.

The usability tests conducted in the laboratory provide qualitative and quantitative results in which quality will refer to the knowledge of the strategies applied by users during their interactions. They can confirm expected behaviors and, in contrast, reveal unexpected behaviors of the users, expressed by feelings as surprise, disappointment, joy, satisfaction, etc. Moreover, they could show what takes a user to deviate from an ideal planned trajectory, and also how he proceeds to restore normal system way after an incident.

The quantitative results, in turn, classify the frequency and duration of events in terms of effectiveness and efficiency of users during simulations. It can be useful to determine the percentage of users who succeeded in the task, how long they took to understand it and how much of that time the User was involved in productive and unproductive activities (gaps, errors and help).

REFERENCES

Tambascia, C. A.; Costa, R.G.; Hollanda, G.M. 2007. "Uma perspectiva da experiência do usuário em soluções de inclusão digital". Cadernos CPqD.

ITU-T. Recommendation E.800. "Terms and definitions related to quality of service and network performance including dependability". Geneve, 1994.

ITU-T Recommendation G.1010. End-user multimedia QoS categories. Series G: Transmission systems and media, digital systems and networks. 2001.

Mcnaamara, N. & KiraskowskI, J. Defining Usability: Quality Of Use Or Quality Of Experience? Professional Communication Conference, 2005. IPCC 2005. Proceedings. International Volume, Issue, 2005.

Eriksén, S. Et Al. Mapping Service Quality – Comparing Quality Of Experience And Quality Of Service For Internet-Based Map Services. Proceedings Of The 30th Information Systems Research Seminar In Scandinavia IRIS 2007.

O'neil, T. M. Quality of Experience and Quality Service. http://www.h323 forum.org/ papers/polycom/QualityOfExperience&ServiceForIPVideo.pdf, acessado em 03 de setembro de 2007.

Muntean, C.H. Quality of Experience Aware Adaptive Hypermedia System. Submitted for the fulfilment of the requirements for the degree of Doctor in Philosophy (Ph.D.), Dublin City University, Faculty of Engineering and Computing, School of Electronic Engineering, 2005.

Mapping Service Quality – Comparing QoE and QoS for Internet-based map services. Blekinge Institute of Technology, Suécia, 2007.

Morris, M. & Turner, J. (2001). Assessing users' subjective quality of experience with the world wide web: an exploratory examination of temporal changes in technology acceptance. Int. J. Human-Computer Studies, 54, 877-901.

Holanda, G. & Dall'Antonia, J. An Approach for e-inclusion: Bringing Illiterates and Disabled People into Play. J. Technol. Manag. Innov. 2006, Volume I, Issue 3.

Nielsen, J., & Landauer, T.K.: A Mathematical Model Of The Finding Of Usability Problems, Proceedings Of Acm Interchi'93 Conference. Amsterdam, The Netherlands, , Pp. 206-213 (1993).

Nielsen, J. & Mack, R. Usability Inspection Methods. New York, John Wiley & Sons, Inc.,1994.

Chapter 9

The Study of Sizing System with 3D Measurement Data for Preschool Children in Central Taiwan

Yu-Cheng Lin, Jia-Shing Chen

Department of Industrial Engineering and Management
Overseas Chinese University
Taichung City, 407, TAIWAN

ABSTRACT

With the improvement of nutrition and development of life quality in recent years, the growth of body and its proportion for preschool children in Taiwan have changed. Most of the past anthropometric data and sizing system for Taiwanese preschool children do not met present applications anymore. Furthermore, the variation between north Taiwan, central Taiwan and south Taiwan becomes greater and greater since the migration from countries to cities with the demand of work and sociality. The purpose of this study was to collect the 3D digital body data from preschool children aged from 4 to 6 years in central Taiwan with a 3D body scanner and to measure useful dimensions from scans for children's clothes. Totally, fourteen typical clothing dimensions, including 9 dimensions for shirts and jackets, 5 ones for pants and stature, were retrieved from the collected digital body scans. More than 150 preschool children aged from 4 to 6 years were invited and 58 preschool children, 32 boys and 26 girls, were completed the measurements finally. Principle component analysis was conducted to retrieve important shirt dimensions and pants dimensions based on their loadings respectively. With these important dimensions, the shirt sizing system and pants sizing system for preschool children were obtained. The results are able to be applied in designing and manufacturing more suitable clothes for the preschool children in central Taiwan.

Keywords: 3D Scan, anthropometry, sizing system, preschool children

INTRODUCTION

During recent years, the quality of life is increasing with the development of society and economy in Taiwan. The change on body size and shape for preschool children is obviously. Furthermore, the segmental proportion is varied with the improvement in nutrition and living type. Therefore, previous anthropometric data of Taiwanese children becomes out of time. The demand of new children's anthropometric data for designing and manufacturing more convenient children products is urgent. One of the urgent requests is the data for children's sizing system. The tailoring firms in Taiwan usually adopt other country's children sizing system to manufacture domestic children garment because of the incompleteness and outmoded domestic sizing system. Besides, there are various considerations to formulate the domestic children sizing systems for different countries (Chung, 2003). Without exact data, no applicable sizing system could be achieved. Proper anthropometric data should be sufficient, correct and up to date. Moreover, the regional diversity in Taiwan should not be disregarded since the migration from countries to cities with the demand of work and sociality. Most of the past anthropometric surveys were done for the children in northern or southern Taiwan but rare for children in Central Taiwanese. The study of establishing Taiwanese anthropometric database conducted a large-scale survey to collect about 10 thousand people's data (Wang et al., 1999 and Wang et al., 2001). The elementary student sample is also an important part of the study. However, only students aged above 6 were measured. Thus, an up-to-date sizing system to meet the preschool children's body shape and size is important in Central Taiwan.

The optical 3D body scanning system is one of the major measuring instruments now (Tsai, 1997). Adopting optical measurement method can obtain body surface data clouds quickly and exactly within seconds and then construct the 3D digital model. Although the model may be fragmented because of shaded surface and poor reflection quality, the applications of 3D model are more extensive then traditional anthropometric methods. In order to avoid the fragmented scanned images, the standing posture was suggested (Daanen, 1998).

It is not easy to establish an optimal sizing system for children aged from 4 to 6 years since the difficulty to measure dimensions and the high variation in body shape and proportion. The difference between regions is also important (Tsai, 2000). Three indices are commonly used to evaluate a sizing system. The indices are the number of size, the coverage of sizing system, and the proportion of fitness (Chung, 2003). The data provided by a robust sizing system should be the body dimensions in place of real clothes dimensions in order to avoid the problems of applications (Hung, 2001).

The purpose of this study was to collect the 3D digital body data from preschool children aging 4 - 6 in central Taiwan with a 3D body scanner and to measure useful dimensions from scans for children's clothes. A new sizing system was retrieved according these data. The results were provided to the children's garment manufacturers for references.

METHODOLOGY

INSTRUMENT AND SAMPLE

The portable 3D Body Scanner manufactured by LT-tech co. was employed to retrieve the children body surface data. A set of scanner contain three scan modules. The precision is ± 1.0 mm and the resolution is 2.5 mm. About eighty thousand 3D data points were taken within 5 seconds. The scanned images were edited and modified by the software, LTBODYCAM, and anthropometric dimensions were retrieved from the 3D digital images. The scanning process was conducted in a darkroom to avoid the influence of light.

FIGURE 1. One module of Portable 3D Body Scanner.

One hundred and fifty preschool children aged from 4 to 6 years were invited randomly form 6 kindergartens in central Taiwan. The letters of parent's consent and authorization were taken home by invited children before experiment. Only those children who have parent's consent would be measured. Finally, 32 boys and 26 girls were completed the measurements. Total time to complete one subject was 5 minutes approximately. Because these preschool children are vivacious and interested, their teachers had to keep with them during the experiment.

DIMENSIONS

Fourteen typical clothing dimensions, including 9 dimensions for shirts and jackets, 5 ones for pants and stature, were retrieved from the collected digital body scans. These dimensions are important and useful for manufacturing children clothes and establishing sizing system. The 14 dimensions are shown in Table 1 and Figure 2.

Table 1: the 14 typical clothing dimensions

Index	Dimension	Index	Dimension
1	Stature	8	Wrist circumference
2	Head circumference	9	Thigh circumference
3	Neck circumference	10	Shoulder breadth
4	Chest circumference	11	Sleeve length
5	Waist circumference	12	Shoulder to waist length
6	Hip circumference	13	Waist to crotch length
7	Axillary arm circumference	14	Crotch height

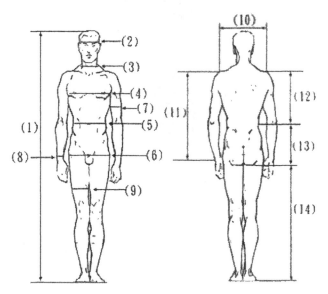

FIGURE 2. The illustration of measured dimensions.

RESULT

ANTHROPOMETRIC DATA FROM 3D SCANNED MODELS

Each scan module of the 3D body scanner took a 3D image simultaneously and all images were combined to a 3D digital model. The 14 anthropometric data then measured from the 3D model for every subject. The average age is 5.17. Table 2 shows the results of anthropometric data. Four out of 14 dimensions show significant differences only. Except stature, the three significant dimensions are all related to children's chunk.

Table 2: the anthropometric data for preschool children aging 4 - 6 in central Taiwan

	Dimension	Boy		Girl		Total		Sig. between gender (α =0.05)
		mean	std.	mean	std	mean	std.	
1	Stature	111.92	5.30	107.96	4.55	110.14	5.35	*
2	Head circumference	50.18	1.92	50.79	2.46	50.45	2.20	
3	Neck circumference	26.91	2.02	27.25	2.18	27.06	2.10	
4	Chest circumference	56.64	2.84	55.75	3.51	56.24	3.18	
5	Waist circumference	57.05	4.46	56.61	4.55	56.85	4.51	
6	Hip circumference	63.19	3.97	63.02	4.03	63.12	4.00	
7	Axillary arm circumference	20.80	2.60	20.42	2.94	20.63	2.76	
8	Wrist circumference	13.83	1.52	13.44	1.08	13.66	1.35	
9	Thigh circumference	32.64	3.72	31.61	2.78	32.17	3.37	
10	Shoulder breadth	26.49	2.22	24.54	2.65	25.61	2.61	*
11	Sleeve length	28.78	2.34	27.58	2.22	28.24	2.36	*
12	Shoulder to waist length	26.35	2.23	24.37	3.19	25.46	2.87	*
13	Waist to crotch length	10.72	2.23	10.73	2.07	10.72	2.16	
14	Crotch height	51.00	3.99	50.15	3.15	50.62	3.67	

THE CLASSIFICATION OF SIZING

The 14 dimensions were separated into two groups. The first group is related with top and blouse and contains 10 dimensions. The second group is related with trousers and pants and contains 6 dimensions. The principle analysis was employed to obtain the major dimensions for sizing classification. By considering the explained variation and loading factor, chest circumference and shoulder breadth were chosen as the major dimensions for tops and blouse. The dimensions for trousers and pants are crotch height and hip circumference.

To implement the sizing classification, the following issues should be noticed.

First, the sizing system should be applied to most of children aging 4 -6. Second, the fitness and growth speed should be both considered. Therefore, maximal union classification method was utilized to implement the sizing system by this study. According to precious sizing system for Taiwanese preschool children, the range of each single size is 6 mm, that is, the upper / lower values for each size equal the central value plus / minus 3 mm. However, almost 12% of sample couldn't be involved in any size. Thus, the range of size was extended to 3.5 mm. The total data were classified into three sizes, S, M and L. The results of size classification for top/blouse and trousers/pants are illustrated in FIGURE 3 and FIGURE 4 respectively. Thus, the sizing system for top/blouse and trousers/pants were achieved and summarized in Table 3.

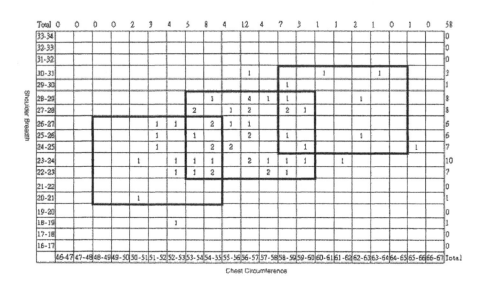

FIGURE 3. The result of size classification for top/blouse.

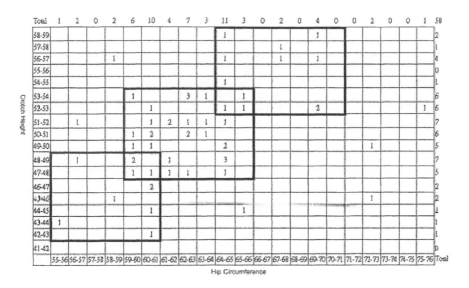

FIGURE 4. The result of size classification for trousers/pants.

Table 3: the sizing system (unit: mm)

Size	top/clouse		trousers/pants	
	Chest circumference	Shoulder breadth	Hip circumference	Crotch height
S	48 - 55	20 - 27	55 – 61	42 – 49
M	53 - 60	22 – 29	59 – 66	47 – 54
L	58 - 65	24 - 31	64 - 71	52 - 59

THE SUGGESTED VALUES FOR CLOTHES

Based on above results of size classification, the mean and standard deviation for each size were suggested. The suggested values were summarized in Table 4 and Table 5 for top/blouse and trousers/pants respectively. The clothes manufacturers could consult the data in produce children clothes in central Taiwan.

Table 4: the suggested value for top/blouse (n=54; unit: mm)

	Dimension	S		M		L	
		mean	std.	mean	std	mean	std.
1	Stature	108.22	4.10	110.43	5.15	115.06	5.54
2	Head circumference	49.66	2.37	50.69	1.99	51.43	2.16
3	Neck circumference	26.93	2.39	26.86	2.10	27.80	2.36
4	Chest circumference	53.08	1.51	56.24	1.80	60.00	2.03
5	Waist circumference	54.24	2.45	56.40	3.35	60.78	5.45
6	Axillary arm circumference	18.77	1.21	20.81	2.80	22.51	3.64
7	Wrist circumference	12.94	0.77	13.61	1.14	14.61	1.91
8	Shoulder breadth	24.17	1.81	25.63	2.22	27.83	2.09
9	Sleeve length	26.79	1.52	28.13	2.03	29.75	2.19
10	Shoulder to waist length	24.23	2.58	25.20	2.54	26.47	3.07

Table 5: the suggested value for trousers/pants (n=52; unit: mm)

	Dimension	S		M		L	
		mean	std.	mean	std	mean	std.
1	Stature	107.79	3.76	110.34	5.06	113.63	5.33
2	Waist circumference	54.41	3.37	56.15	2.60	59.00	4.59
3	Hip circumference	59.21	1.87	62.34	1.99	66.81	2.07
4	Thigh circumference	30.83	3.02	31.75	2.90	32.95	3.46
5	Waist to crotch length	11.89	1.71	10.74	2.11	9.63	2.41
6	Crotch height	46.27	2.16	50.16	2.41	55.14	2.40

CONCLUSIONS

In order to establish an appropriate and up-to-date sizing system for preschool children aged from 4 to 6 years in central Taiwan, this study conducted an anthropometric survey to collect preschool children's 3D digital models with the 3D body scanner. According the anthropometric data, the sizing system was achieved. The sizing system contains only 3 sizes, S, M and L. Simple sizing system is helpful to reduce the cost and enhance the intension to manufacture.

REFERENCES

Chung, M.J. (2003), *Establish sizing system for boys and girls aging 6-18 by using cluster analysis*, Master thesis, Tsing Hua University. (in Chinese)

Daanen, H.A.M., and Jeroen, G. (1998), "Whole body scanners." *Displays*, 19, 111-120.

Hung, K.L. (2001), *Establish sizing system and standard dress form with 3D human data*, Master thesis, Tsing Hua University. (in Chinese)

Tsai, C.C. (2000), *Apply 3D anthropometric data to establish sizing system for elementary and high school students*, Master thesis, Tsing Hua University. (in Chinese)

Tsai, C.Z. (1997), *Apply anthropometric data to establish sizing system for woman's trousers*, Master thesis, Tsing Hua University. (in Chinese)

Wang, E.M.Y., Wang, M.J., Yeh, W.Y., Shih, Y.C. and Lin, Y.C. (1999), "Development of anthropometric work environment for Taiwanese workers." *International Journal of Industrial Ergonomics*, 23, 3-8.

Wang, M.J., Wang, E.M.Y. and Lin, Y.C. (2001), *Anthropometric data book of the Chinese people in Taiwan*, Ergonomics Society of Taiwan. (in Chinese)

Chapter 10

Perception of Pressure on Foot

Asanka S. Rodrigo[1], Ravindra S. Goonetilleke[1], Shuping Xiong[2]

[1]Human Performance Laboratory
Department of Industrial Engineering and Logistics Management
Hong Kong University of Science and Technology
Clear Water Bay, Hong Kong

[2]Department of Industrial Engineering and Management
Shanghai Jiao Tong University
Shanghai, China

ABSTRACT

This study attempted to identify the factors that affect the pressure pain threshold (PPT) on the heel region of the foot. Using dimensional analysis, it was found that probe area (A), indentation speed (V) and their interaction contribute towards PPT. A power form of these factors can be used to model PPT with an exponent of 0.63.

Keywords: Perception, Power Law, Foot, Pressure Pain Threshold, Dimensional Analysis, Modeling.

INTRODUCTION

Human interaction with different types of equipment always involves various types of forces. When the force per unit area commonly known as pressure is excessive, people will experience pain or discomfort (Gonzalez et al., 1999). In this respect, the pressures acting on feet have received considerable attention due to constant contact of the foot with some type of footwear or directly with ground. The pressure acting on the foot when wearing footwear can result in foot deformities, ulcers, corns, callous, bunions, if the normal or shear forces are excessive or repetitive (Dunn et al., 2004). The pressure distribution on the foot can vary with a person's body weight, type of the activity (walking, running, jumping, and so on) performed (Soames, 1985; Rodgers, 1988) and the type of footwear (Yung-Hui and Wei-Hsien, 2005; Stewart et al., 2007) as well. Researchers are still searching for the optimum distribution and the best means to support load acting on any body part so that potential pain and discomfort are minimized (Goonetilleke, 2001).

Hence, it is no surprise that there is a growing trend to investigate the effects of pressure on the foot. Most research has focused on ways to reduce peak plantar pressure either by introducing differing insole materials (O'Leary et al.,2008; Hinz et al.,2008) or by varying the insole shapes (Yung-Hui and Wei-Hsien, 2005; Stewart et al., 2007). However, little is known as to why some designs are more comfortable or less uncomfortable than others. In this study, we explore potential factors that may influence pressure sensations.

The pressure pain threshold (PPT) (Fischer, 1987) has been found to be a reliable measure to quantify pressure sensations. PPT has been found to reduce with increases in stimulus size (Greenspan et al., 1991; Goonetilleke and Eng, 1994; Greenspan et al., 1997; Defrin et al., 2006; Xiong, 2008) and the number of indentations (Fransson-Hall and Kilbom, 1993; Defrin et al., 2008). Furthermore, PPT increases with increasing rate of change of the stimulus (Defrin et al., 2006; Xiong, 2008). However, the theoretical basis for these changes has not been well documented.

Dimensional analysis is a well-known technique to obtain the explicit functional relationship among variables (Barenblatt, 1987). It can be applied in psychophysics as well (Marinov, 2004; Marinov, 2005). Thus, we attempted to derive a preliminary model for pressure perception based on dimensional analysis and test the validity of the model using perception data.

METHODOLOGY

PARTICIPANTS

Twenty-four participants (12 males and 12 females), from the Hong Kong University of Science and Technology, with informed consent were recruited for the study. The descriptive statistics of the participants are given in the Table 1. All the subjects were selected based on their ability to make reasonable judgments of magnitude estimation and none of them had any visible foot abnormalities or foot illnesses.

Table 1 Descriptive statistics of participants. Standard deviations are in parenthesis.

	Mean (SD)				
	Age (yrs)	Weight (kg)	Height (cm)	Foot breath (cm)	Foot length (cm)
Male	23.67 (2.15)	68.03 (7.06)	172.93 (3.97)	9.71 (0.49)	25.17 (0.90)
Female	22.50 (2.55)	53.55 (9.25)	160.06 (5.07)	8.66 (1.45)	23.15 (0.82)
Total	23.14 (2.36)	61.45 (10.83)	167.08 (7.89)	9.23 (1.14)	24.25 (1.33)

EXPERIMENT PROCEDURE

The experimental design was 4 (size of probes) x 2 (indentation speeds) full factorial design with 2 repeated measurements. Cylindrical rods made of aluminum having a silicon-tip were fabricated for the test. The probe areas were 0.25 cm^2, 0.5 cm^2, 1 cm^2 and 2 cm^2 and the two indentation speeds were 1 mm/s and 2 mm/s. PPT and Pressure Discomfort Threshold (PDT) at the heel center of plantar foot were determined using the Automatic Tissue Tester (ATT) designed and developed by us. The ATT can control the probe indentation and records the force and displacement profile of the probe (Xiong, 2008). The results of PDT are not discussed here.

Each subject was asked to stand on a plexi-glass platform. The control unit that the subject held had two buttons to indicate PDT and PPT. The right foot of the participant was aligned so that the centre of heel coincided with the center of probe (Figure 1). When performing the test, subjects were asked to keep equal weight on both feet and to press the PDT and PPT buttons in the hand-held unit as soon as they felt discomfort and thereafter pain. Prior to the actual test, subjects were given an opportunity to familiarize themselves with the controls and the test procedure.

ANALYSIS AND RESULTS

SPSS statistical software was used for the data analysis. Intra-class correlation (ICC) (Shrout and Fleiss, 1979) type (2,1) was used to check the test-retest reliabilities. The intra-class correlations ($p < 0.001$) were 0.959 and 0.966 for the 1 mm/s and 2 mm/s indentation speeds, respectively. Effect of probe area (A) and indentation speed (V) on PPT were statistically tested using a repeated measures ANOVA (Huck and McLean, 1975; Gorden, 1992). The effect of A, V and their interaction was statistically significant ($p < 0.05$). However, gender and all other interactions were not significant. Hence the male and female data were pooled in all subsequent analyses.

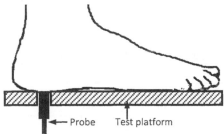

FIGURE 1 Schematic of experiment set-up.

The PPT decreases with increasing probe area and indentation speed (Figure 2).

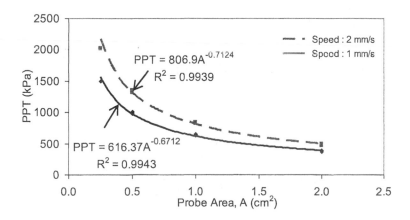

FIGURE 2 PPT variation with probe area and indentation speed.

MODEL OF PPT

Dimensional analysis is based on a comparison of the measurement units of the variables. The probe characteristics can be represented using probe area (A), indentation speed (V) and indentation time (t). Similarly, the tissue characteristics can be represented with tissue deformation (d) and Young's modulus of tissue (E). Since $d=V.t$, only two of these three variables need to be considered in the problem description.

Hence, PPT = f(A, V, t, E) and Table 3 shows the dimensions of these variables.

Table 3 Variables and their corresponding dimensions

Variable	PPT	A	V	t	E
Dimension	$ML^{-1}T^{-2}$	L^2	LT^{-1}	T	$ML^{-1}T^{-2}$
Units	$Pa=Nm^{-2}$	m^2	ms^{-1}	s	Nm^{-2}

According to the Buckingham Π-theorem (Buckingham, 1914), any relationship between n dimensional variables and constants can be reduced to a relation between (n-r) dimensionless variables, where r is the number of independent dimensions. In this study, n=5 and r=3 corresponding to mass (M), length (L), and time (T). Thus, (n-r) = 2 and two dimensionless groups, Π_1 and Π_2, can be formed as follows:

$$\Pi_1 = \frac{PPT}{E} \text{ and } \Pi_2 = \frac{(V.t)^2}{A} ;$$

where $\Pi_1 = \left[\frac{PPT.}{E}\right] = \frac{ML^{-1}T^{-2}}{ML^{-1}T^{-2}} = [1]$ and $\Pi_2 = \left[\frac{(V.t)^2}{A}\right] = \frac{(LT^{-1}T)^2}{L^2} = [1]$;

Hence, $\Pi_1 = f(\Pi_2)$, or $\frac{PPT}{E} = f\left(\frac{(V.t)^2}{A}\right)$;

Thus $PPT = E.f\left(\frac{(V.t)^2}{A}\right)$ or $PPT = E.f\left(\frac{d^2}{A}\right)$.

Since E is a constant at any particular location, the plot of PPT vs $(V.t)^2/A$ could reveal the best fitting function (Figure 3). A power law representation of

$$PPT = c. \left(\frac{(V.t)^{2\beta}}{A^\beta} \right),$$ where c $\approx$ 1424 Nm^{-2}, and β=0.63 seems to fit the data quite well in the heel (R^2=0.993).

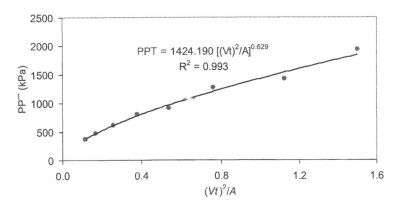

FIGURE 3 Variation of PPT with (Vt)^2/A.

The modeling errors were analyzed using the formula, $$\% \, Error = \frac{Model \ \ Data - Actual \ Data}{Actual \ Data} \times 100\%$$ and are shown the Figure 4.

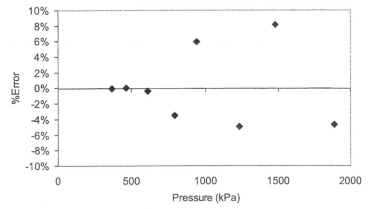

FIGURE 4 Scatter plot of Modeling Error.

DISCUSSION

The aim of this study was to understand the variables related to perceived pressure thresholds. The PPT measurements had very good test-retest reliability (intra class correlation > 0.959) and the variables of probe area, indentation speed and their interaction were found to have a significant ($p < 0.05$) effect on PPT. In general, as found in many previous studies (Greenspan et al., 1997; Goonetilleke and Eng, 1994; and Xiong, 2008.), PPT decreases with increases in area validating the spatial summation theory of pain. It was also seen that PPT increased with increasing indentation speed, consistent with Xiong (2008).

There was no difference in PPT at heel between males and females. Gonzalez et al. (1999) also indicated that there were no gender differences for pressure thresholds on plantar foot even though other studies (Bernnum et al., 1989; Chesterton et al., 2003) have shown differences in gender. The differences among the literature may be related to the site investigated and the tested population.

More importantly, PPT can be mathematically modeled in the form of $PPT = c\left(\dfrac{(V.t)^{2\beta}}{A^{\beta}}\right)$ or $PPT = c\left(\dfrac{d^{2\beta}}{A^{\beta}}\right)$. The coefficient '$c$' can characterize the tissue property and the exponent β is possibly a representation of the spatial summation effect of pressure related pain. The model indicates that PPT increases with the speed, V, as found in the experiment, and time (t) or indentation depth (d).

The scatter plot for modeling error shows that the model has an acceptable fit to the data. Further work is needed to validate the model for other regions of the body.

ACKNOWLEDGMENT

The authors would like to thank the Research Grants Council (RGC) of Hong Kong for funding this study under grant HKUST 613406.

REFERENCES

Barenblatt, G.I., (1987), *Dimensional Analysis*. Gordon and Breach, New York.
Brennum, J., Kjeldsen, M., Jensen, K., and Jensen, T. (1989), "Measurements of human pain-pressure thresholds on fingers and toes." *Pain*, 38(2), 211-217.
Buckingham, E. (1914), "On physically similar systems; illustrations of the use of dimensional equations." *Physical Review*, 4, 345-376.
Chesterton, LS., Barlas, P., Foster, NE, Baxter, GD., Wright, CC. (2003), "Gender

differences in pressure pain threshold in healthy humans." *Pain*, 101, 259-266.

Defrin, R., Givon, R., Raz, N., and Urca, G. (2006), "Spatial summation and spatial discrimination of pain sensation." *Pain*, 126 (1-3), 123-131.

Defrin, R., Pope, G., and Davis, K.D. (2008), "Interactions between spatial summation, 2-point discrimination and habituation of heat pain." *European Journal of Pain*, 12(7), 900-909.

Dunn, J.E., Link , C.L., Felson , D.T., Crincoli , M.G. , Keysor , J.J., and McKinlay, J. B. (2004), "Prevalence of foot and ankle conditions in a multiethnic community sample of older adults." *American Journal of Epidemiology*, 159, 491-498.

Fischer, A.A. (1987), "Tissue compliance meter for objective, quantitative documentation of soft tissue consistency and pathology." *Archives of Physical Medicine and Rehabilitation*, 68(2), 122-125.

Fransson-Hall, C., and Kilbom, A. (1993), "Sensitivity of the hand to surface pressure." *Applied Ergonomics*, 24(3), 181-189.

Gonzalez, J.C., Carcia, A.C., Vivas, M.J., Ferrus, E., Alcantara, E., and Forner, A. (1999), "A New Portable Method for the Measurement of Pressure Discomfort Threshold on the Foot Plant." Fourth Symposium of the Technical Group on Footwear Biomechanics, August 5-7, 1999, Canmore, Canada.

Goonetilleke, R.S., and Eng, T. (1994), "Contact Area Effects on Discomfort." Proceedings of the 38th Human Factors and Ergonomics Society Conference, October 24-28, 1994, Nashville, Tennessee, 688-690.

Goonetilleke, R.S. (2001), "The comfort-discomfort phase change." in: International Encyclopedia of Ergonomics and Human Factors, W. Karwowski (Ed.). Taylor and Francis, pp. 399-402.

Gorden, E.R. (1992), *ANOVA: Repeated Measures*. Calif. Sage Publications, Newbury Park.

Greenspan, J.D., and McGillis S.L.B. (1991), "Stimulus features relevant to the perception of sharpness and mechanically evoked cutaneous pain." *Somatosensory & Motor Research*, 8(2), 137-147.

Greenspan J.D., Thomadaki, M., and McGillis, S.L.B. (1997), "Spatial summation of perceived pressure, sharpness and mechanically evoked cutaneous pain." *Somatosensory and Motor Research*, 14(2): 107-112.

Hinz,P., Henningsen, A., Matthes, G., Jager, B., Ekkernkamp, A., and Rosenbaum, D. (2008), "Analysis of pressure distribution below the metatarsals with different insoles in combat boots of the German army for prevention of march fractures." *Gait & Posture*, 27(3), 535-538.

Huck, S.W, and McLean, R.A. (1975), "Using a repeated measures ANOVA to analyze the data from a pretest-posttest design: A potentially confusing task." *Psychological Bulletin*, 82(4), 511-518.

Marinov, S.A. (2004), "Reversed dimensional analysis in psychophysics." *Perception and Psychophysics*, 66, 23-37.

Marinov, S.A (2005), "Defining the Dimension of a Psychological Variable Using Dimensional Analysis." Proceedings of the Twenty First Annual Meeting of the International Society for Psychophysics, October 19-22, 2005, Traverse City, Michigan, USA, 187-192.

O'Leary, K, Vorpahl, K.A., Heiderscheit, B. (2008), "Effect of cushioned insoles on impact forces during running." *American Podiatric Medical Association*, 98(1), 36-41.

Rodgers, M.M. (1988), "Dynamic biomechanics of the normal foot and ankle during walking and running." *Physical Therapy*, 68(12), 1822-1830.

Shrout, P.E., and Fleiss, J.L. (1979), "Intraclass correlations: use in assessing operator reliability." *Psychological Bulletin*. 86(2), 420-428.

Soames, R.W. (1985). "Foot pressure patterns during gait." *Journal of Biomedical Engineering*, 7(2), 120-126.

Stewart, L., Gibson, J., and Thomson, C. (2007), "In-shoe pressure distribution in "unstable" (MBT) shoes and flat-bottomed training shoes: A comparative study." *Gait & Posture*, 25(4), 648-651.

Xiong, S. (2008), "Pressure perception on the foot and the mechanical properties of foot tissue during constrained standing among Chinese." Ph.D Thesis of Industrial Engineering and Logistic Management, Hong Kong University of Science and Technology.

Yung-Hui, L., and Wei-Hsien, H. (2005), "Effects of shoe inserts and heel height on foot pressure, impact force, and perceived comfort during walking." *Applied Ergonomics*, 36(3), 355-362.

Chapter 11

Development of Methodology to Gather Seated Anthropometry Data in a Microgravity Environment

Karen Young[1], Miranda Mesloh[1], Sudhakar Rajulu[2]

[1]Lockheed Martin
Anthropometry and Biomechanics Facility
NASA, Johnson Space Center
Houston, TX 77058, USA

[2]National Aeronautics Space Association (NASA)
Anthropometry and Biomechanics Facility
NASA, Johnson Space Center
Houston, TX 77058, USA

ABSTRACT

The Constellation Program is designing a new vehicle based on recently developed anthropometric requirements. These requirements specify the need to account for a spinal-elongation factor for anthropometric measurements involving the spine, such as eye height and seated height. However, to date there is no existing data relating spinal elongation to a seated posture. Only data relating spinal elongation to stature have been collected in microgravity. Therefore, it was proposed to collect seated height in microgravity to provide the Constellation designers appropriate data for their analyses. This document will describe the process in which the best method to collect seated height in microgravity was developed.

Keywords: Microgravity, Spinal Elongation, Human System Integration, NASA

INTRODUCTION

The Constellation Program's Crew Exploration Vehicle (CEV) is required to accommodate the full anthropometric range of the future crewmember population according to the requirements stated in the Human-Systems Integration Requirement (HSIR) document (CxP70024, 2007). One critical anthropometric measurement for the CEV is seated height. Seated height allows for the CEV designers to determine the optimum seat configuration in the vehicle. Changes in seated height can have a large impact on the design, on accommodation, and safety of the crewmembers. Historically it is known that the spine changes in microgravity, due to spinal elongation, which will affect crewmembers' seated height.

Spinal elongation is the straightening of the natural curvature of the spine and the expansion of intervertebral disks. This straightening results from fluid shifts in the body and the lack of compressive forces on the spinal vertebrae in microgravity. Previous studies have shown that as the natural curvature of the spine straightens, an increase in overall height of 3% occurs, and this forms the basis of the current HSIR requirements (NASA STD 3000). However, because of variations in the torso/leg ratio, questions arose regarding whether the historical stature data can be applied to seated measurements. Data related to how spinal elongation specifically affects seated measurements are nonexistent.

An experiment was designed to collect spinal-elongation data while subjects were in a seated posture in microgravity. The purpose of this study was to provide quantitative data representing the amount of change that occurs in seated height due to spinal elongation in microgravity environments.

While preparing for the microgravity experiment concerns arose regarding the possibility of crewmembers restraining themselves improperly to the seat pan. To resolve this concern and ensure that the measurements of seated height were accurately collected during the experiment, a simulated microgravity experiment was conducted. A simulated microgravity condition (parabolic flight) was used to conduct three evaluations to test the methodology and procedures of the experiment. During these simulated evaluations, data were collected (a) to ensure that the lap restraint method provided sufficient restraint to eliminate any gap between the subject's gluteal surface and the seat pan, and (b) to document any necessary design and procedural changes needed because of the microgravity environment.

METHOD

One of the major concerns regarding the data collection of seated height in microgravity was the potential separation between the subject and the seat pan. While the crewmember is expected to wear a lap restraint, it was unclear whether the lap restraint's tension, set by the individual crewmember, would be able to restrain the individual to allow for accurate data collection. Three simulated microgravity flights occurred during which seated height, force data, and pressure data was collected.

The test setup included the Shuttle seat and an anthropometer. For Flight 1 and Flight 2, a standard anthropometer was attached to the Shuttle seat (Figure 1-left), and during Flight 3 a prototype of the flight hardware anthropometer was used that attached to the top of the seat back in the same manner as the head rest (Figure 1- right). During the simulated microgravity parabolas, the subject was seated in the Shuttle seat and restrained for the collection of pressure, force, and seated-height data. The subject remained seated during several consecutive parabolas; the number of consecutive parabolas varied between subjects.

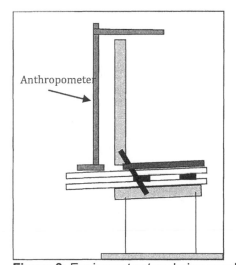
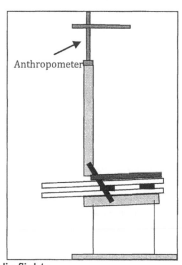

Figure 3: Equipment setup during parabolic flights

Flight 1 was performed to determine if the current hardware on the Shuttle seat was able to ensure proper contact with the seat pan while not affecting the seated- height measurements. This was the proposed method of restraint for the in-flight experiment. The restraint system examined was 3-points of the current 5-point harness on the Shuttle seat. The 3-point harness included the use of the crotch buckle and the two pelvic straps (Figure 2 – right). Data were collected for four subjects using this restraint method.

After learning the results of the first flight, alternative restraint methods were proposed and examined for the second flight. The restraint systems examined were iterations of the restraint method tested during the first flight. However, in discussing the concerns with the stakeholder, an additional restraint method was tested during the flight, adjusting the 3-point harness by wrapping it around the seatback support of the seat (Figure 2 – left). Four subjects participated in Flight 2 in which pressure data and seated-height data were collected.

The third flight, Flight 3, tested the prototype anthropometer assembly and the optimized restraint method from Flight 2, featuring the 3-point harness wrapped around the seatback support bars (Figure 2 – left). Seated-height and pressure data were collected from three subjects during Flight 3.

| 3-point harness nominal (Flight 1 configuration) | 3-point harness after adjustment (Flight 2 configuration) |

Figure 2: Harness Configurations

Seated-height data were collected during all three flights for each restraint system examined. The seated-height data were used to validate which restraint system secured the subject to the seat pan with the smallest amount of variability between the measurements. To ensure minimal variability to the measurement, all measurers followed a standard procedure. Once in microgravity, the measurer ensured the subject was looking straight ahead and collected seated-height measurement, ensuring proper placement, and proper contact of the anthropometer to the top of the subject's head. For Flights 1 and 2, one seated-height measurement was recorded per parabola. During Flight 3, multiple seated-height measurements were recorded per parabola due to time allotment and procedure familiarization of the measurers.

RESULTS

FLIGHT 1

Seated-height measurements were collected during consecutive parabolas for each subject. The total number of data points differed for each subject. The restraint method used during Flight 1 was the 3-point harness of the current hardware on the Shuttle seat. The results showed that the range of seated-height measurements were greater than 1 cm for all values and varied from 1.1 cm to 5.6 cm (Table 1). The variability in the seated-height measurements indicate that the subjects were not tightly restrained and that the subjects' buttocks were not in complete contact with the seat pan. The subjects also reported the lack of restraint; the subjects often felt like their buttocks was floating off of the seat in microgravity even with the lap restraint tightened as far as the restraint system would allow.

Table 1 shows that Subject 3 had the most variability in seated height, which may indicate that this subject was not consistently in contact with the seat pan for all trials. Subject 4 experienced the smallest range of values during microgravity. The smaller the range of seated-height measurements, the higher the repeatability of the measurement. This reduced range is achieved from consistency in tightening the restraint system in the same manner for multiple parabolas, which the participants were unable to do with the current, Flight 1, restraint system.

The results from Flight 1 demonstrate that the current 3-point harness used in its nominal configuration was not sufficient to ensure contact between the subject and the seat pan to collect consistent seated-height measurements. The restraint system must restrain the subject properly so that the seated-height measurements are consistent. Inconsistent measurements will adversely affect the amount of spinal elongation that is reported to the designers, which may impact crew safety and crew fit with the seats, suits, and vehicle after exposure to microgravity.

Table 1. Flight 1 (F1) Seated-height Measurements

	3-point Nominal Harness (cm)			
	Subject 1	**Subject 2**	**Subject 3**	**Subject 4**
Range	1.4	2.3	5.6	1.1
Mean	95.4	90.3	90.5	87.1

FLIGHT 2

As a result of Flight 1, wherein the 3-point harness in its nominal configuration was not sufficient to restrain the subject, alternative methods of restraints were explored during Flight 2, namely:

- Flight 1 3-point configuration
- Flight 1 3-point configuration with a foam insert under the buckle
- A 3-point harness with a loose buckle and a Velcro® leg strap
- Flight 2 3-point harness configuration
- Flight 2 3-point configuration with Velcro® leg strap
- Flight 2 3-point configuration with Velcro® leg and lap straps

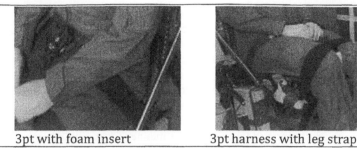

| 3pt with foam insert | 3pt harness with leg strap |

Figure 3: Flight 2 Harness Configurations

Flight 1 configuration represents the attachment point location of the waist belts of the 3-point harness, where the waist belts are attached and used in the normal configuration (Figure 2 – left). Whereas, Flight 2 configuration consists of the waist belts being wrapped around the seatback support bars and then attached to the buckle (Figure 2 – right). The Flight 2 configuration brought the buckle up higher and tighter on the pelvic region. The Flight 2 configuration also pulled the subject back and down into the seat, whereas the Flight 1 configuration only pulled the subject down into the seat and leaving room for the subjects' buttock to rise off of the seat.

According to the seated-height measurements collected during Flight 2, the seated-height range verified that the Flight 2 configuration restrained the subjects more effectively than the Flight 1 configuration. As an example, the greatest range in seated height for Subject 1 was 0.5 cm for Flight 2 (3-point w/leg and lap configuration) compared to 1.4 cm for Flight 1 (Tables 1 and 2). This indicates that the subjects were better restrained in the seat for Flight 2 than for Flight 1. Looking at the three different types of restraints and the ranges of measurement variation, the data indicate that the addition of the leg and lap straps minimize the relative seated-height range for all subjects with the exception of Subject 1. This would signify that the addition of a leg and lap strap may be necessary to restrain the subject to the seat and allow for accurate seated-height measurements. The differences in the average seated heights can be attributed to variation that occurs while collecting anthropometric measurements using traditional methods, i.e., manual measurements using an anthropometer (0.6 cm for seated height). However, when comparing the average seated-height measurements for each configuration, the measurements were consistent between the different configurations (Table 3). Therefore, to reduce the need for transporting and stowing extra hardware, the Flight 2 3-point harness configuration was chosen as the best restraint system.

Table 2. Flight 2 (F2) Seated-height Measurements

	Flight 2							
		3-point (cm)			3-point w/ leg strap (cm)			3-point w/ leg and lap strap (cm)
	S 1	S 2	S 3	S 1	S 2	S3	S 1	S 2
Range	0.1	0.3	1.2	0.4	0.1	0.1	0.5	0.1
Mean	94.0	90.0	94.2	93.7	90.1	94.1	93.6	89.9

Table 3. Average Seated-height Configuration Comparison

Restraint Configuration	S 1 (cm)	S 2 (cm)	S3 (cm)
Flight 2 3-point	94.0	90.0	94.2
Flight 2 3-point w/ Leg Strap	93.7	90.1	94.1
Flight 2 3-point w/ Leg & Lap Straps	93.6	89.9	-
Flight 1 3-point	95.4	90.2	-

FLIGHT 3

Flight 3 tested the Flight 2 3-point harness configuration in addition to the use of the prototype anthropometer to be used during the in-flight experiment. The prototype anthropometer attaches to the seatback in the same manner as the headrest. In addition to testing the 3-point harness, the subjects were also asked to look up, down, and straight forward to determine if the head position does affect the seated height. During Flight 3, it became apparent that the operator and subject must work together to ensure that the restraint system is tight against the subject's pelvis. Several times the restraint was not completely tight, and the subjects responded by stating that they were not cinched down completely and felt as if they were floating. The mean and range of seated-height measurements for the different conditions, head orientations, and restraint tightness, can be seen in Table 4.

Table 4. Flight 3 Seated-height Measurements

	Flight 3									
	Restraint Loose, Looking Forward (cm)			Restraint Tight, Looking Forward (cm)			Looking Up (cm)		Looking Down (cm)	
	S 1	S 2	S 3	S 1	S 2	S 3	S 1	S 2	S 1	S 2
Range	0.9	1.4	0.1	0.6	0.6	0.7	0.5	-	0.6	-
Mean	95.3	91.4	94.7	95.4	91.1	93.7	95.2	90.7	93.3	91.2

In comparison to the loosely restrained subjects in the 3-point harness, the results indicate that the range of seated height decreased for the tightly restrained subjects. For example, Subject 2's measurement range decreased from 1.4 cm to 0.6 cm due to tightening of the restraint system (Table 4). All three subjects seated-height ranges were similar when they were tightly restrained, 0.6 cm, 0.6 cm, and

0.7 cm, indicating that when tightly restrained there is increased consistency in the variation in seated height. The remaining variation observed in the data can be attributed to the effects of microgravity and the variation in collecting manual measurements (as previously described) (Table 4).

Based on the limited amount of data collected during Flight 3, in which the subjects changed their head orientation from looking forward, down, and up, the seated-height measurements did change for subject 1 and subject 2. When comparing looking forward to looking down, there was a greater increase in seated height for subject 1 compared to subject 2. This originated from subject 2 drastically looked down as opposed to subject 1 who did not drastically look down at the floor. This drastic change in head orientation was not seen when the subjects looked up. Therefore, the difference in seated heights when the subjects looked up versus looking forward was very small, 0.2 cm and 0.4 cm respectively for subjects 1 and 2. Another factor that may explain the small variation in seated-height values when looking up was the location on the head where the measurement was collected. The measurements may not have been collected from the most superior location on the head but rather to the middle of the head. From the results of looking down, it can be assumed that if the subjects drastically looked up that the measurements would be collected at the most superior location on the head, affecting the seated-height measurements in the same manner as when the subjects drastically looked downward. Based on the results, it is essential to maintain correct eye position during the microgravity experiment otherwise the values would be erroneous, impacting crew safety, crew selection, and anthropometric requirements.

DISCUSSION

FLIGHT 1

The results obtained from Flight 1 were: (1) the current 3-point harness did not ensure complete contact between the subject and the seat pan, and (2) the set-up, harness and anthropometer, affected the seated-height measurements' consistency. The subjects reported verbally that they often felt as if they were not completely restrained and experienced a sensation of floating between the seat pan and their gluteal surface. These results may be due to the configuration of the restraint system. The configuration of the 3-point restraint system for Flight 1 rested lower on the pelvic region. The crotch buckle was not adjusted by each subject nor was its length extended completely, resulting in the buckle attachment point resting lower on the pelvic region that allowed for a gap to exist between the buckle and the subject. This gap allowed the subject to float unrestrained until the subject was stopped by the restraint buckle (Figure 2 - right).

Two lessons learned while collecting seated-height measurements during Flight 1 were: (1) to make sure the subject is looking straight forward, and (2) make sure the subject is in the center of the seat. Oftentimes, the subject perceived that they were looking forward but they were actually looking slightly downwards. Also, there were times when the anthropometer was not positioned in the middle of the subject's head due to their positioning on the chair. Therefore, the measurer needs to instruct the subject as to how to move their torso/head so that the anthropometer measures to the middle of the top of the head. The two lessons learned can adversely affect the measurement of seated height by providing inaccurate data.

FLIGHT 2

The findings from Flight 2 indicated: (1) additional straps across the waist and lower thighs decrease the range in seated-height measurements, and (2) that the waist restraint should be rerouted through the back of the seatback, this increased the consistency in measurements and the contact between the subject and the seat. The subjects felt greater restrained with this configuration than the configuration of Flight 1 (Figure 3).

The average seated heights resulting from this flight demonstrated that additional straps were not necessary. Therefore, additional hardware was not needed for the on-orbit activity. The decision was made to proceed with the Flight 2 3-point harness configuration for on-orbit activities; wrapping the restraint belts around the seatback joint.

FLIGHT 3

Flight 3 only tested the Flight 2 3-point harness configuration using the prototype hardware anthropometer. The prototype hardware was attached to the seat in the same manner as the headrest. Using the prototype anthropometer during Flight 3 resulted in little variability during parabolas and decreased the inconsistencies of using a standard anthropometer taped to the load cell plate. Also, using the Flight 2 3-point harness configuration decreased the range for seated height due to the tightening of the restraint. During Flight 3, several seated-height measurements were collected when the restraint was not completely tight, resulting in a greater range than when the restraint was pulled tight around the subject's pelvic/waist area (Table 4). The lack of tightness led to inconsistencies within the seated-height measurements. As a result, a lesson learned from Flight 3 is that after the subject gets in the seat and is buckled the operator must pull the side straps as tight as he/she can to ensure the subject is in contact with the seat pan.

Comparing the average seated height for the 3-point harness-restraint system for all three flights, it was determined that the Flight 3 configuration improved the variability compared to Flight 1 (Table 5). Flight 2 had the smallest seated-height range but had the fewest seated-height measurements for each subject; Subject 1 had two measurements and Subject 2 had three measurements.

Table 5. Comparison of Seated Height 3pt Harness Configuration

	Subject 1		Subject 2		Subject 3	
	Average (cm)	Range (cm)	Average (cm)	Range (cm)	Average (cm)	Range (cm)
Flight 1	95.4	1.4	90.3	2.3	90.5	5.6
Flight 2	94.0	0.1	90.0	0.3	-	-
Flight 3	95.4	0.6	91.1	0.6	93.7	0.7

CONCLUSION

In conclusion, the microgravity flights were pertinent to the spinal elongation microgravity experiment. The three microgravity flights allowed the principal investigation team to explore the best methodology for collecting seated-height data. The seated-height data collected during the microgravity flights assisted with crew procedures, data collection methodology, and hardware design. The lessons learned from the microgravity flights will aide in successfully and accurately collecting seated-height data that can be used by the CEV designers to accurately design for the amount spinal elongation due to microgravity. The spinal elongation that crewmembers may experience will effect crew safety, crew selection, and design requirements. If the designers do not have the correct data for spinal elongation, then a crewmember may not have enough clearance upon re-entry to allow for the appropriate vibration and stroke volume or may not properly fit into their re-entry suit; therefore, affecting their safety and the safety of the other crewmembers.

REFERENCES

Anthropometric Source Book, Vol. I: Anthropometry for Designers, (NASA 1024). Edited by Staff of Anthropology Research Project, Webb Associates: Yellow Springs, OH, 1978

Man-Systems Integration Standards, NASA-STD-3000 (1985, Rev B). NASA Johnson Space Center, Houston, TX.

Human-Systems Integration Requirements (HSIR) (C000114) (CxP 70024). NASA Johnson Space Center, Houston, TX, 2007

Chapter 12

Human Factors Engineering Availability and Suitability Verification for Human-System Interface in Nuclear Power Plants

Jung Chang Ra

Korea Power Engineering Company
360-9, Mabuk-dong, Giheung-gu, Yongin-si
Gyeonggi-do, Korea
E-mail: jcn@kopec.co.kr

ABSTRACT

Availability Verification (AV) and Suitability Verification (SV) have been performed to assure that the Human-System Interface (HSI) features accommodate the various issues such as the Advanced Control Room (ACR) issues, Human Engineering Discrepancies (HEDs) from the Research & Development (R&D) stages, and questions or comments by a regulatory body. Human Factors Engineering (HFE) evaluations were well integrated into the overall design process and the associated design activities were completed before carrying out the verification activities. Several HFE evaluations for the HSI design have been iteratively performed along with design activities and the performance-based evaluations (i.e., situation awareness, workload, team interaction, etc.) were performed to verify that the HSI design fully supports safe operation of the plant. This approach allows us to reduce the project risk of new design features by ensuring no serious problems encountered in the early design stages and permitting enough time for improving the HSI design.

Keywords: HFE, Verification, ACR, HSI

INTRODUCTION

Evaluation of HSI designs for Advanced Power Reactor 1400 (APR1400) was conducted throughout the HSI development process. Although the types of evaluations performed vary depending on the specific design process, the methodology used was determined using the appropriate criteria. Aspects of human performance that are important to operational tasks were carefully defined so that the differential effects of design options on human performance could be adequately considered in the evaluation of the HSI design. A number of HFE evaluations for the HSI design have been iteratively performed along with design activities from the R&D stage to the construction of the Nuclear Power Plants (NPPs) to ensure that the HSI is effective in supporting the performance of operator tasks. Human-system interface issues and design discrepancies related to the computerized procedure system, information displays, large display panel, soft controls, alarms, and operation consoles were reviewed during this evaluation. This paper focuses on the AV and SV among various HFE activities and introduces a practical methods based on our experiences, rather than theoretical methods.

DESCRIPTION OF THE ACTUAL WORK

AV and SV applied to the APR1400 in Korea were performed in compliance with Korean regulatory requirements and commitment outlined in Safety Analysis Report (SAR) to assure that key HFE elements were not inadvertently overlooked in the HSI design. At planning stage, detailed plan was prepared to govern the conduct of the verifications. This plan included methodology, procedure, schedule, scope and required resources for the successful verifications. Prior to carrying out the activities, the associated design activities were completed, HFE guideline or other criteria was available, and the related function and task analysis were completed. It was demonstrated, during the evaluation that:

- All required control capabilities and displayed quantities were provided and all parts of the interface were configured as intended and required by human factors guidelines and standardized practices. All conflicts between the various requirements and the interface design were addressed and resolved.
- All applicable interfaces were shown to perform all the intended tasks to be carried out effectively. That is, the interfaces were proven to function as intended.

As a principal material for the verifications, a dynamic mock-up using full scale simulator of referenced nuclear power plants was developed to support the effective evaluation of the HSI design. For the evaluation of the various issues and design discrepancies, dynamic mockup was a facsimile of the dynamic characteristics of the NPPs.

TEAM ORGANIZATION AND COMPOSITION

Figure 1 shows the Korea Power Engineering Company (KOPEC) A/E I&C engineering organization which takes a leading role for the HFE Verification and also shows the other engineering groups such as Electrical Engineering, Mechanical Engineering, Nuclear Engineering, Architect Engineering, etc., which are involved in designing or reviewing the HSI based on their expertise. Within the I&C discipline there are five groups (two system groups, one physical group, the HFE group, and HSI design team). The HSI design team (J5) performs the development of the various HSI design documents, drawings and guidance. The two system groups (J1 &J2) perform I&C system and HSI design as required by the APR1400 project. The HFE design team (J4) has been integrated into the I&C discipline since the beginning of the project and the team leader has guided all participants in the conduct of the Verification to assess plant safety and personnel performances:

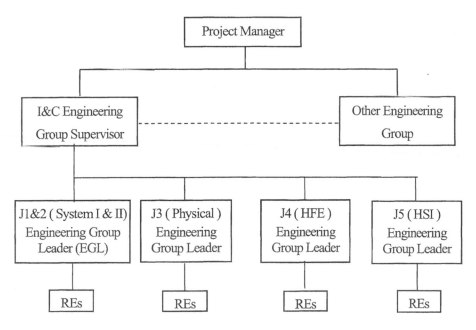

Figure 1.   KOPEC A/E Organization

In addition to the KOPEC engineers, experts including HFE specialists, operating experts, and active operating crews working at the operating nuclear power plants participated in these activities.

AVAILBABILITY VERIFICATION

The first step in the verification process is the AV to confirm the HSI inventory and characterization. The objective of the AV is to verify that the HSI inventory and characterization accurately describes all HSI displays, controls, and related equipment that are within the defined scope of the HSI design. The inventory is based on the best available information sources such as equipment lists, licensing documents, Instruments and Control Requirement (ICR) from task analysis, system design drawings, etc.

Database that provides an accurate and complete description of all HSI components was prepared after the conceptual design was complete. It underwent revision as the design evolved, but it was completed before the HSI design itself was finalized and compared with the HSI design to check whether all HSI components with the appropriate characteristics associated with personnel tasks were included in the final design. The AV was conducted according to the procedure shown in Figure 2 and the findings, the result of the AV, were resolved and incorporated into the final HSI design.

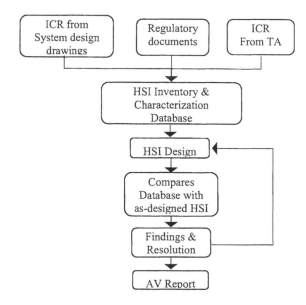

Figure 2. AV Procedure

SUITABILITY VERIFICATION

The second step in the verification process is the SV to evaluate whether the design and characteristics of the HSI conform to HFE guidelines and are acceptable based on the experts' knowledge. The SV was carried out in two steps: Bottom-Up Suitability Verification (BUSV) and Top-Down Suitability Verification (TDSV). These suitability verifications were conducted according to the procedure shown in Figure 3.

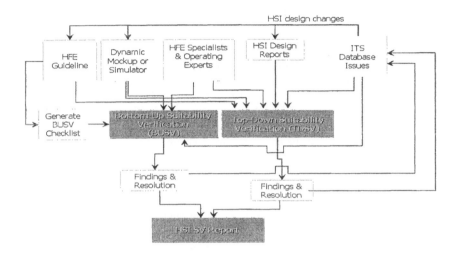

Figure 3. SV Procedure

BOTTOM-UP SUITABILITY VERIFICATION

BUSV was performed to ensure that the HSI was designed to accommodate human capabilities and limitations as reflected in the HFE guidelines such as those provided in NUREG-0700. Design-specific HFE guidelines were developed by the HFE specialist and the checklist was prepared using these guidelines. The characteristics of the HSI components were checked against the HFE guidelines that are applicable to different aspects of the design. The areas covered by the checklist are:

- Information displays including formats, diagrams, mimics, scales, symbols, and codings, etc.
- Interaction including navigation, managing displays, message, response time, and security, etc.
- Soft controls including selection displays, input formats, error detection and correction, etc.

- Computerized procedure system including step informations, procedure managements, etc.
- Alarms including alert functions, priorities, alarm controls, etc.
- Workstation and workplace design
- Safety console including inventories of the controls and displays

If any noncompliance against the HFE guideline was found, it was identified as HED and it was documented by a project team.

TOP-DOWN SUITABILITY VERIFICATION

TDSV was performed by a team consisting of HFE specialists, operation and design experts. The verification evaluated the various aspects of as-designed HSIs in terms of operation tasks to be performed using task scenarios. The scenarios were created by the operation experts with extensive operating experience. Two task scenarios were carefully selected considering that (1) the selection includes conditions that are representative of the range of events that could be encountered during operation of the plant, (2) the selection reflects the characteristics that would be expected to contribute to system performance variation, and (3) the selection considers the safety significance of HSI components. A set of scenarios to guide subsequent analyses and many of the characteristics identified by the operational events were divided into two basic segments according to the time interval. The first segment of each scenario contained a normal operational events and a malfunction was introduced in the second segment. During the TDSV, the evaluators tried to clarify critical HFE issues through the observation of the plant operator's tasks following each scenario. The following plant and human performance data was collected for the assessment of the HFE issues.

- System performance data through data logger
- Situation awareness and workload
- Team communications and coordination
- Response to subjective questionnaires

The HEDs were identified when HSI components and control capabilities needed for task performance were not available in the design and when any characteristics of the displays, controls, and other control room features are not suitable to task and function accomplishments of the operators.

RESULTS

The results of the SV are summarized in Table 1. These HEDs would adversely affect the operator performance and degrade the safe operation of the NPPs. After finishing each scenario or checking the conformance to HFE guidelines, an assessment meeting was held to collect all issues raised by participants. With clear understanding of the issues and comments, the evaluators discussed the participants' opinions and determined whether any design problems raised may adversely affect human performance. Once determined pertinent, recommendation was provided to the designers to correct the problems and required corrective actions were scheduled for implementation in a timely and efficient manner. The results of the SV revealed that the HSI design incorporating the HEDs is quite valuable for improving operators' task performance and reducing human errors. We believe that the HFE evaluation will significantly contribute to enhance the plant safety and availability.

Table 1. Number of HEDs raised during the SV

HSI Resources	Number of HEDs		Remark
	BUSV	TDSV	
LDP & Information FPD	9	0	- Improper Engineering unit - Unclear mimic line - Inconsistent abbreviations
Alarm	3	1	- Lack of alarm awareness - Difficulty in discriminating alarm priorities
Soft Control	0	1	- Difficulty of group controls
Total	12	2	

CONCLUSION

As one of HFE verification activities, the AV and SV were performed to assure that the HSI features support safe operation of the plant and the various issues are resolved. This paper is to introduce the verification activities and results that the KOPEC has performed. A number of HEDs during the verification process were found to improve the HSI design. The results reveal that the HSI design incorporating the HEDs is suitable in terms of the improved operators' task performance and reduced human error. We believe that the HFE will contribute to enhance the plant safety and availability.

REFERENCES

Endsley, M. R., Design and Evaluation for Situation Awareness Enhancement, Proceedings of the Human Factors Society 32nd Annual Meeting, pp. 97-101(1988)

Hallbert, B. P., Sebok, A. L., Haugset, K., Morisseau, D. S., and Persensky, J. J., Interim Results of the Study of Control Room Crew Staffing for Advanced Passive Reactor Plants, Proceedings of the 23rd Water Reactor Safety Meeting, U.S. NRC (2000)

Nygren, T. E., Psychometric Properties of Subjective Workload Measurement Techniques: Implication for Their Use in the Assessment of Perceived Mental Workload, Human Factors, Vol. 33, pp. 17-33 (1991)

O'Hara, J. M. and Hall, R. E., Advanced Control Rooms and Crew Performance Issues: Implications for Human Reliability, IEEE Transactions on Nuclear Science, Vol. 39, pp. 919-923 (1992)

Roth, E. M., Munmaw, R. J. and Stubler, W. F., Human Factors Evaluation Issues for Advanced Control Rooms: A research agenda, IEEE Conference Proceedings, pp. 254-265 (1993)

Sebok, A., Team Performance in Process Control: Influences of Interface Design and Staffing Level, Ergonomics, Vol. 43, pp. 1210-1236 (2000)

U.S. NRC, Human Factors Engineering Program Review Model, NUREG-0711 (2002)

Shin, Y. C., Moon, H.K., Kim, J. H, Human Factors Engineering Verification and Validation for APR1400 Computerized Control Room, ANS (2006)

CHAPTER 13

Using Performance Measures to Assess the Effect of Visual Aesthetics on Usability

Ahamed Altaboli[1,2], Yingzi Lin[1,*], Mohammed Ali[2], Yassin Alterhony[2]

[1] Mechanical and Industrial Engineering Department
Northeastern University, Boston, MA, USA

[2] Industrial and Manufacturing Systems Engineering Department
University of Garyounis, Benghazi, Libya

[*] Corresponding author: Tel: 1.617.373.8610, Fax: 1.617.373.2921,
E-mail: yilin@coe.neu.edu

ABSTRACT

This study is a first exploratory step in a larger in progress research project aiming at developing objective quantitative measures of visual aesthetic of computer interfaces. The main objective of this study was to investigate the possibility of finding high correlation between visual aesthetic and usability when performance measures are used in evaluating usability. An experiment was designed and conducted to test this hypothesis. Results showed relatively high correlations between visual aesthetic and usability measures.

Keywords: Visual Aesthetic, Visual Appearance, Usability, User Interface Design, User Performance and Satisfaction.

INTRODUCTION

Over the past three decades, most of the established user interface design guidelines and principles emphasize that functionality and usability of a system should be the prime objective and that it should be given priority over aesthetic and visual appealing of the interface. However, in the late 1990s, a new research trend in interface design emerged suggesting that aesthetic, visual and physical aspects are as much important to users as usability. This trend was also noticeable in other fields of engineering and design. For example, (Lin and Zhang, 2006) has proposed an integrated design of function, usability and aesthetics for automobile interiors.

Several researches have been conducted to support the idea of that aesthetic and visual appealing of an interface can have a significant impact on system acceptability. Many of them managed to proof that not only aesthetic is an important aspects for user satisfaction and system acceptability, but it also has a significant impact on users' prior perception of usability of a system (Kurosu & Kashimura, 1995; Tractinsky, 1997). Although, the concept of the effect of aesthetic in user experience became more evidence with more research in the area (Hoffmann, R., & Krauss, 2004; Cawthon & Moere, 2007), new questions have arisen: how to evaluate aesthetic of an interface? What types of measures can be used for that? And, is it possible to develop these measures based on performance measures of usability?

Most of researches conducted on aesthetic had used subjective measures to evaluate aesthetic based on users' perception. Few attempts have been conducted in the past few years to develop objective quantitative measures for aesthetic (Bauerly, M. & Liu, Y., 2006; Zain, Tey, & Goh, 2008; Mirdehghani & Monadjemi, 2009). These objective quantitative measures were developed based on visual characteristics and layout elements of the tested interfaces. Examples of these elements include: symmetry, balance (equilibrium), number of visual elements (density), and order. They were evaluated and validated based on users' perceptions and opinions rather than using objective performance measures of usability. The few cases in which objective performance measures were utilized, either used only performance time as a measure of usability (Schmidt, K.E., Bauerly, M., Liu, Y., & Sridharan, S., 2003) or were conducted in uncontrolled environments (Cawthon, N., & Vande Moere, A., 2007) which might make their results less reliable. More investigation is needed to examine the possibility of using performance measures of usability to evaluate and validate aesthetic measures.

This study is a part of a larger research effort aiming at developing objective quantitative measures of visual aesthetic of computer interfaces. It is a preliminary investigation to explore the possibility of using performance measures of usability to evaluate visual aesthetic of an interface.

The main purpose of this study is to investigate the effect of visual aesthetic of an interface on usability. Although it is expected that usability of an interface would be related to its visual appealing; results of previous studies showed low correlation between visual aesthetic and usability of an interface (Kurosu & Kashimura, 1995;

Tractinsky, 1997). One possible reason could be that usability in these studies was measured using selected factors proposed by interface designers rather than using more objective performance measures.

In this study, an experiment was designed and conducted to examine the case of using performance measures to evaluate usability; using performance measures of users performing various tasks with the interface might lead to different results supporting the hypothesis of high correlation between visual aesthetic and usability. To test this possibility, the experiment was designed to answer two questions. First, does visual aesthetic of an interface have any effect on its usability? Second, will a higher correlation between visual aesthetic of an interface and its usability be found?

METHOD

DESIGN OF THE EXPERIMENT

A weekly lectures timetable in an engineering college was used as an interface to carry out the experiment. It was chosen because it has relatively few numbers of visual elements that can be easily manipulated to fit experimental requirements.

Four designs of the table were prepared with different visual appearances. The first design was a plain table with an assumable poor visual appealing. It was prepared to be used as a control group in the experiment. The other three designs were prepared by adding selected visual elements to the plain table that would supposedly enhance its visually appealing. These elements include: using bold font type to increase contrast of the table and the use of shading to highlight selected parts of the table.

A single factor (visual aesthetic) between-subjects design was utilized in the experiment, with the four designs of the table as its treatments (independent variables). Seven participants were randomly assigned to each design (twenty-eight in total). User performance was measured using performance time and number of errors, and satisfaction level was measured using a questionnaire. The two performance variables and questionnaire scores are the dependent variables.

PARTICIPANTS

Twenty-eight students from the final two semesters in an engineering college were selected and distributed randomly among the four designs of the table. All participants were volunteers. Each group has four males and three females (seven subjects per group). Their mean age is 21.82 years with a standard deviation of 1.25 year.

They were selected from the final semesters to ensure that all of them have the same experience with the used tables

THE DIFFERENT DESIGNS OF THE TABLES

Four different designs were used in this experiment (shown in figures 1 to 4). The first one has a plain format and was used as a control group. It was deliberately designed to have an assumable poor visual aesthetic. None of the visual effects that supposed to increase visual aesthetic were used in this design.

The second design is a mixed design with an assumable enhanced visual appealing. It was prepared by adding a mixture of selected visual elements to the plain design. The added elements include: adding a light grey shading to both rows and columns headings; using a bold font style in all text in the table; and using darker and wider separation lines in the table. The intention of adding these elements is that they would add more contrast to the table and highlight its different parts and make them more visible.

The final two designs were prepared by modifying selected elements in the mixed design in each case. In the first of these two designs, only the font style of the "Room no." column was modified from bold style to regular style. This change was made in order to give more contrast to the table. In the second one, light gray shading was used in the "Room no." column. This change was also made to give more contrast to the table by alternate highlighting of the columns.

In all of the four designs the alphanumeric course codes were made different from the ones currently being used. The alphabetic part of the code was converted from IE, ME, EE and ES to UG, HG, TG and CG respectively. The numeric part was modified by adding "400" to it. These changes were made to reduce the effect of participants' past experience.

All the designed tables were presented to participants on an A4-paper size with a landscape orientation.

For simplicity, the designs (groups) will be named; "Plain", "Mixed", "Bold", and "Shading".

PROCEDURE

Each participant was instructed to fill out the lectures timetable for six selected courses (one from each semester). A paper form was used to complete this task. A stopwatch was used to record task completion times. Number of errors was obtained from analyzing the forms that participants had filled out. The number of errors were recorded for each participant in the following manner; one error count was recorded whenever the participant commits one of the following actions: writing the code of a course with a spelling mistake; missing to write one of the lectures times in its suitable cell; mixing- up and writing a course's time in a wrong cell.

After completing the tasks, the questionnaire was presented to the participant. The questionnaire consists of two questions about the participant's level of satisfaction with visual appearance of the table; the two questions measure the same factor (satisfaction level). The reason why two questions were used is to be able to measure the reliability of the questionnaire, which was found to be above 0.7 in most cases. For each question, the participant had to choose a numeric indicator from a five- category scale: excellent (9-10), very good (7-8), good (5-6) acceptable (3-4), and poor (1-2).

	TIME	8.30 - 9.30		9.30 - 10:30		10.30 - 11.30		11.30 - 12.30		2:00 - 3:00		3:00 - 4:00		4:00 - 5:00	
DAYS	SEMSTE & NO	COURSE	ROOM NO	COURSE	ROOM NO	COURSE	ROOM NO	COURSE	ROOM NO	COURSE	ROOM NO	COURSE	ROOM NO	COURSE	ROOM NO
SAT	3	CG 61	16 CG	CG 6	16 CG	CG 60	16 CG	CG 60	16 CG						
	4	UG 608	108 HG	UG 608	108 HG	HG 622	1 JG	HG 622	1 UG	HG 60	LAB	HG 50	LAB	HG 60	1 AB
	5					JG 07	9 HG	UG 07	9 HG	TG 01	108 HG	TG 01	108 HG		
	6	UG 10	8 UG	UG 10	8 UG	UG 06	5 UG	UG 06	5 UG						
	7	UG 804	11 UG	UG 804	11 UG			UG 805	8 UG	HG 12	8 UG	HG 12	8 UG		
	8	UG 804	5 UG	UG 804	5 UG										
SUN	3	HG 60	11 UG	HG 60	11 UG	UG 603		UG 601							
	4	CG 606	16 CG	CG 206	16 CG	UG 604		UG 604							
	5					HG 613	5 UG	HG 613	5 UG						
	6	UG 08	108 HG	UG 08	108 HG	UG 04	11 UG	JG 04	4 UG						
	7									UG 83	LAB	UG 81	LAB	UG 8	LAB
	8	UG 806	8 UG	UG 806	8 UG	UG 802	8 UG	UG 802	8 UG	UG808					
MON	3	CG 61	5 UG			UG 601		UG 601		HG 60	12 HG				
	4	CG 606	16 UG	CG 606	16 UG	HG 622		UG 608	108 HG			HG 602			
	5	UG 03	8 UG	UG 03	8 UG	UG 01	8 UG	UG 01	8 UG			HG 02		HG 01	UG 8
	6	TG 02	108 UG	TG 02	108 HG	UG 02	UG	UG 02	UG						
	7			UG802	UG	UG 802	11 UG	UG824	UG	HG 13	UG				
	8					UG824	UG	UG824	UG						
TUE	3	CG 664	16 CG	CG 664	16 CG	UG 62		UG 62		UG 62		UG 6			
	4	CG 616		CG 616	16 CG	HG 604	5 UG	HG 604	UG						
	5					UG 01	8 UG	UG 07	HG97	HG 25		HG 25		HG 15	LAB
	6	UG 02	8 UG	TG 02	8 UG			UG 06	8 UG						
	7	UG 809	11 UG	UG 80	11 UG	UG 03	11 UG	UG 03	11 UG	UG 802	8 UG	UG 805	8 UG		
	8	UG 822	5 UG	UG 822	5 UG	UG804									
WED	3					CG601	CG	CG 60	CG	HG60	LAB	HG60	LAB	FG60	LAB
	4	CG 616	5 UG	CG 616		UG602	5 UG	UG602	5 UG						
	5	HG613	5 UG			TG701	108 HG	UG 709	8 UG	HG701	8 UG	HG701	8 UG		
	6	UG 710	8 UG	UG 704	8 UG	UG 708	97 HG	UG 708							
	7					UG 817	LAB	UG817 HG809							
	8	UG 806	11 UG	UG 802	11 UG	HG808 UG822		HG808 UG822							

Figure 1 The "plain" (control) design.

TIME		8:30-9:30		9:30-10:30		10:30-11:30		11:30-12:30		2:00-3:00		3:00-4:00		4:00-5:00	
DAYS	SEMSTER NO.	COURSE	ROOM NO.	COURSE	ROOM NO.	COURSE	ROOM NO.	COURSE	ROOM NO.	COURSE	ROOM NO.	COURSE	ROOM NO.	COURSE	ROOM NO.
SAT	3	CG 617	16 CG	CG 617	16 CG	CG 601	16 CG	CG 601	16 CG						
	4	UG 608	108 HG	UG 608	108 HG	HG 622	11 UG	HG 622	11 UG	HG 605	LAB	HG 605	LAB	HG 605	LAB
	5					UG 707	97 HG	UG 707	97 HG	TG 701	108 HG	TG 701	108 HG		
	6	UG 710	8 UG	UG 710	8 UG	UG 706	6 UG	UG 706	6 UG						
	7	UG 809	11 UG	UG 809	11 UG			UG 805	8 UG	HG 712	8 UG	HG 712	8 UG		
	8	UG 804	5 UG	UG 804	5 UG										
SUN	3	HG 607	11 UG	HG 607	11 UG	UG 601		UG 601							
	4	CG 606	16 CG	CG 206	16 CG	UG 604		UG 604							
	5					HG 613	5 UG	HG 613	5 UG						
	6	UG 708	108 HG	UG 708	108 HG	UG 704	11 UG	UG 704	11 UG						
	7									UG 817	LAB	UG 817	LAB	UG 817	LAB
	8	UG 806	8 UG	UG 806	8 UG	UG 802	8 UG	UG 802	8 UG						
MON	3	CG 617	5 UG			UG 601		UG 601		HG 607	22HG				
	4	CG 606	16 UG	CG 606	16 UG	HG 622		UG 608	108 HG			HG 604			
	5	UG 703	8 UG	UG 703	8 UG	UG 701	8 UG	UG 701	8 LG			HG 701		HG 701	UG 8
	6	TG 702	108 UG	TG 702	108 HG	UG702	8 UG	UG702	8 UG						
	7			UG807	11 UG	UG 807	11UG			HG713	5 UG				
	8							UG824	7UG	UG824	7 UG				
TUE	3	CG 661	16 CG	CG 661	16 CG	UG 627		UG 627		UG 627		UG 627			
	4	CG 616		CG 616	10 CG	HG 604	8 UG	HG 604	8 UG						
	5					UG701	8UG	UG707	HG97	HG 725		HG 725		HG 725	LAB
	6	UG 702	8 UG	TG 702	8 UG			UG706	8UG						
	7	UG 809	11 UG	UG 807	11 UG	UG703	11 UG	UG 703	11 UG	UG 805	8TG	UG 805	8TG		
	8	UG 822	5 UG	UG 822	5 UG	UG804									
WED	3					CG601	CG	CG 601	CG	HG607	LAB	HG607	LAB	HG607	LAB
	4	CG 616	5 UG	CG 616		UG602	5 UG	UG602	5 UG						
	5	HG613	5 UG			TG701	108 HG	UG 703	8 UG	HG701	8UG	HG701	8UG		
	6	UG 710	8 UG	UG 704	8 UG	UG 708	97HG	UG 708							
	7					UG 817	LAB	UG817-UG803							
	8	UG 806	11 UG	UG 802	11 UG	HG808-UG822		HG808-UG822							

Figure 2 The "Mixed" design.

TIME		8:30-9:30		9:30-10:30		10:30-11:30		11:30-12:30		2:00-3:00		3:00-4:00		4:00-5:00	
DAYS	SEMSTER NO.	COURSE	ROOM NO.	COURSE	ROOM NO.	COURSE	ROOM NO.	COURSE	ROOM NO.	COURSE	ROOM NO.	COURSE	ROOM NO.	COURSE	ROOM NO.
SAT	3	CG 617	16 CG	CG 617	16 CG	CG 601	16 CG	CG 601	16 CG						
	4	UG 608	108 HG	UG 608	108 HG	HG 622	11 UG	HG 622	1 UG	HG 605	LAB	HG 605	LAB	HG 605	1 B
	5					UG 707	97 HG	UG 707	97 HG	TG 701	108 HG	TG 701	08 HG		
	6	UG 710	8 UG	UG 710	8 LG	UG 706	6 UG	LG 706	6 UG						
	7	UG 809	11 UG	UG 809	11 UG			UG 805	8 UG	HG 712	8 UG	HG 712	8 UG		
	8	LG 804	5 UG	UG 804	5 UG										
SUN	3	HG 607	11 UG	HG 607	11 UG	UG 601		UG 601							
	4	CG 606	16 CG	CG 206	16 CG	UG 604		UG 604							
	5					HG 613	5 LG	HG 613	5 LG						
	6	UG 708	108 HG	UG 708	108 HG	UG 704	11 UG	UG 704	11 UG						
	7									TG 817	LAB	UG 817	LAB	UG 817	LAB
	8	UG 806	8 UG	UG 806	8 UG	UG 802	8 UG	UG 802	8 UG	HG808					
MON	3	CG 617	4 UG			UG 601		UG 601		HG 607	22HG				
	4	CG 606	16 UG	CG 606	16 UG	HG 622		UG 608	108 HG			HG 604			
	5	UG 703	8 UG	UG 703	8 UG	UG 701	8 UG	UG 701	8 UG			HG 701		HG 701	UG S
	6	TG 702	108 CG	TG 702	108 HG	UG702	8 UG	UG702	8 UG						
	7			UG807	11 UG	UG 807	11UG			HG713	5 UG				
	8							UG824	7UG	UG824	7 UG				
TUE	3	CG 661	16 CG	CG 661	16 CG	UG 627		UG 627		UG 627		UG 627			
	4	CG 616		CG 616	10 CG	HG 604	8 UG	HG 604	8 UG						
	5					UG701	8UG	UG707	HG97	HG 725		HG 725		HG 725	LAB
	6	UG 702	8 UG	TG 702	8 UG			UG706	8UG						
	7	UG 809	11 UG	UG 807	11 UG	UG703	11 UG	UG 703	11 UG	UG 805	8TG	UG 805	8UG		
	8	UG 822	5 UG	UG 822	5 UG	UG804									
WED	3					CG601	CG	CG 601	CG	HG607	LAB	HG607	LAB	HG607	LAB
	4	CG 616	5 UG	CG 616		UG602	5 UG	UG602	5 UG						
	5	HG613	5 UG			TG701	108 HG	UG 703	8 UG	HG701	8UG	HG701	8UG		
	6	UG 710	8 UG	UG 704	8 UG	UG 708	97HG	UG 708							
	7					UG 817	LAB	UG817-UG803							
	8	UG 806	11 UG	UG 802	11 UG	HG808-UG822		HG808-UG822							

Figure 3 The "Bold" design.

DAYS	SEMSTER NO.	8:30-9:30 COURSE	ROO M.NO.	9:30-10:30 COURSE	ROOM NO.	10:30-11:30 COURSE	ROOM NO.	11:30-12:30 COURSE	ROOM NO.	2:00-3:00 COURSE	ROOM NO.	3:00-4:00 COURSE	ROOM NO.	4:00-5:00 COURSE	ROOM NO.
SAT	3	CG 617	16 CG	CG 617	16 CC	CG 601	16 CC	CG 601	16 CC						
	4	UG 608	108 HG	UG 608	108 HG	HG 622	11 UG	HG 622	11 UG	HG 605	LAB	HG 605	LAB	HG 605	LAB
	5					UG 707	97 HG	UG 707	97 HG	TG 701	108 HG	TG 701	108 HG		
	6	UG 710	8 UG	UG 710	8 UG	UG 706	8 CG	UG 706	8 CG						
	7	UG 809	11 UG	UG 809	11 UG			UG 805	8 UG	HG 712	8 UG	HG 712	8 UG		
	8	UG 804	8 UG	UG 804	8 UG										
SUN	3	HG 607	11 UG	HG 607	11 UG	UG 601		UG 601							
	4	CC 606	16 CC	CC 206	16 CC	UG 604		UG 604							
	5					HG 613	8 UG	HG 613	8 UG						
	6	UG 708	108 HG	UG 708	108 HG	LG 704	11 UG	LG 704	11 UG						
	7									UG 817	LAB	UG 817	LAB	UG 817	LAB
	8	UG 806	8 UG	UG 806	8 UG	UG 801	8 UG	UG 802	8 UG	HG 808					
MON	3	CG 517	8 UG			UG 601		UG 601		HG 607	22 HG				
	4	CG 606	16 UG	CG 606	16 UG	HG 622		UG 608	108 HG			HG 604			
	5	UG 703	8 UG	UG 703	8 CG	UG 701	8 CG	UG 701	8 CG			HG 701		HG 701	UG 8
	6	TG 702	108 UG	TG 702	108 HG	UG 702	8 UG	TG 702	8 UG						
	7			UG 807	11 UG	UG 807	11 UG	UG 824	7 UG	HG 713	8 UG				
	8					UG 824		UG 824	7 UG						
TUE	3	CG 661	16 CG	CG 661	16 CG	UG 627		UG 627		UG 627		UG 627			
	4	CG 616		CG 616	20 CG	HG 604	8 UG	HG 604	8 CG						
	5					UG 701	8 UG	UG 707	HC97	HG 725		HG 725		HG 725	LAB
	6	UG 702	8 UG	TG 702	8 UG			UG 706	8 UG						
	7	UG 809	11 UG	UG 807	11 UG	UG 703	11 UG	UG 703	11 UG	UG 805	8 UG	UG 805	8 UG		
	8	UG 822	8 CG	UG 822	8 UG	UG 804									
WED	3					CG 601	CG	CC 601	CG	HG 607	LAB	HG 607	LAB	HG 607	LAB
	4	CG 616	8 UG	CG 616		UG 602	8 CG	UG 602	8 CG						
	5	HG 613	8 UG			TG 701	108 HG	UG 703	8 UG	HG 701	8 UG	HG 701	8 UG		
	6	UG 710	8 UG	UG 704	8 UG	UG 708	97 HG	UG 708							
	7					UG 817	LAB	UG817 UG803							
	8	LG 806	11 UG	LG 802	11 UG	HG808 UG822		HG808 UG822							

Figure 4 The "Shading" design.

RESULTS

To answer the first question concerning the effect of visual aesthetic on usability, analysis of variance for each of the measures was conducted.

Table 1 is a summary of descriptive statistics and analysis of variance for the two measures of performance (completion time and number of errors) and the satisfaction scores, for all of the four designs.

Descriptive statistics in Table 1 show, with all the measures, that the poorest performance and the lowest levels of satisfaction were recorded in the plain (control) design. The supposedly more visually aesthetic designs have enhanced users' performance and satisfaction. However, none of the reductions in performance times and number of errors was found significant as results of analysis of variance show. Only the increase in satisfaction level was found significant. Users' were extremely dissatisfied with the plain design as their satisfaction scores indicate. This was also reflected in their performance, although not significant.

Table 1 Summary of descriptive statistics and analysis of variance (time in minutes).

Measure	Group	Descriptive statistics			Analysis of variance	
		Mean	S. D.	Range	P-value	Significant (α = 0.05)
Time	Plain	16.56	6.64	17.17	0.274	No
	Mixed	11.97	2.72	8.08		
	Bold	14.92	4.09	11.96		
	Shading	14.92	2.52	7.79		
Errors	Plain	6.00	4.00	12.00	0.487	No
	Mixed	4.14	3.02	7.00		
	Bold	5.14	2.26	7.00		
	Shading	3.71	2.21	7.00		
Satisfaction	Plain	8.40	4.30	10.00	0.0003	Yes (Plain -all the others)
	Mixed	15.30	2.60	7.00		
	Bold	15.40	2.40	6.00		
	Shading	14.10	1.90	4.00		

These results indicate that no statistically significant effect of visual aesthetic on the two performance measures of usability was found. However, a relatively better (but not statistically significant) performance was recorded with the visually enhanced designs. With satisfaction level, a highly statistically significant effect was found. Users were more satisfied with the enhanced (more visual appealing) designs.

The second question concerning correlation between visual aesthetic and usability was answered by calculating correlations coefficients between nominal values of the four designs (visual aesthetic) and means of the two performance measures and satisfaction scores for each design (usability). A nominal value of "1" was assigned to the plain design and a higher nominal value of "2" was given to each of the other three deigns.

The correlation matrix is given in Table 2. It shows that relatively high correlations were found in most cases. Negative correlations were found between both of the performance measures and visual aesthetic, indicating that performance increases (lower errors and times) with the more visually aesthetic designs (higher nominal value). Also positive high correlation was found between satisfaction score and visual aesthetic, showing that higher satisfaction scores were given to the more visually aesthetic designs.

Table 2 Correlation coefficients between visual aesthetic, and performance measures and satisfaction scores (usability).

	Visual Aesthetic	Error	Time
Error	-0.812		
Time	-0.686	0.653	
Satisfaction	0.984	-0.901	-0.732

Although the procedure used to calculate the correlation coefficients is somehow subjective, but it gives sufficient evidence to support the hypothesis that strong correlation between usability and visual aesthetic of an interface can be found if performance measures were used to measure usability.

CONCLUSIONS

This study was conduct as an exploratory step in a larger in progress research project aiming at developing objective quantitative measures of visual aesthetic of computer interfaces. The main objective of the study was to test the hypothesis of finding high correlation between visual aesthetic and usability when performance measures were used in evaluating usability.

An experiment was designed and conducted to test this hypothesis. Based on the experiment outcomes, relatively high correlations were found between visual aesthetic and usability measures. The highest correlation was found between users' satisfaction with the visual appearance of the interface and its visual aesthetic. This conforms to results of previous researches, which found high correlation between visual aesthetic of an interface and users preferences (Kurosu & Kashimura, 1995; Tractinsky, 1997).

Since a subjective procedure was used in this study to calculate correlation coefficients, the next step should be to design and conduct a more rigors experiment to conform these results. In addition, a computer interface should be used in future experimentation instead of the paper interface used in this preliminary study.

REFERENCES

Bauerly, M. and Liu, Y. (2006). "Computational modeling and experimental investigation of effects of compositional elements on interface and design aesthetics". Int. J. Human-Computer Studies 64, 670–682.

Cawthon, N., and Vande Moere, A. (2007). "The Effect of Aesthetic on the Usability of Data Visualization" 11th International Conference Information Visualization (IV'07).

Hoffmann, R. and Krauss, K. (2004). "A critical evaluation of literature on visual aesthetics for the web". SAICSIT '04: Proceedings of the 2004 annual research conference of the South African institute of computer scientists and information technologists on IT research in developing countries, Western Cape, South Africa. 205-209.

Kurosu, M. and Kashimura, K. (1995). "Apparent usability vs. inherent usability: experimental analysis on the determinants of the apparent usability". CHI '95: Conference companion on Human factors in computing systems, Denver, Colorado, United States. 292-293.

Lin, Y. and W. J. Zhang, (2006) Integrated Design of Function, Usability, and Aesthetics for Automobile Interiors: State-of-the-Art, Challenges, and Solutions, Proc. IMechE Vol. 220 Part I: J. Systems and Control Engineering, 220(I8), 697-708.

Mirdehghani, M. and Monadjemi, S. A. (2009). "Web Pages Aesthetic Evaluation Using Low-Level Visual Features". World Academy of Science, Engineering and Technology 49 2009.

Schmidt, K.E. Bauerly, M. Liu, Y., and Sridharan, S. (2003). "Web Page Aesthetics and Performance: A Survey and an Experimental Study". In Proceedings of the 8th Annual International Conference on Industrial Engineering – Theory, Applications and Practice, Las Vegas, Nevada, USA.

Tractinsky, N. (1997). "Aesthetics and apparent usability: empirically assessing cultural and methodological issues". CHI '97: Proceedings of the SIGCHI conference on Human factors in computing systems. Atlanta, Georgia, United States. 115-122.

Zain, J. Mengkar Tey, M., and and Goh, Y. (2008) "Probing a Self-Developed Aesthetics Measurement Application (SDA) in Measuring Aesthetics of Mandarin Learning Web Page Interfaces". IJCSNS International Journal of Computer Science and Network Security, Vol. 8 No. 1, January 2008.

Chapter 14

Development of an Objective Evaluation System for Advanced Driver Assistance Systems Based on Subjective Criteria - Demonstrated by the Example of ACC

Benedikt Strasser[1], Heiner Bubb[1], Karl-Heinz Siedersberger[2], Markus Maurer[3]

[1]Lehrstuhl für Ergonomie, Technische Universität München
Garching b. München

[2]AUDI AG, Ingolstadt

[3]Institut für Regelungstechnik
Technische Universität Braunschweig
Braunschweig

ABSTRACT

In order to ensure that the complexity in car construction remains manageable, simulation tools are to be enhanced and ideally coordinated with the product development process.

The psychophysical study described below is part of a project which aims at including the simulation more into the functional development process of driver assistance systems. Therefore it will be presented and demonstrated by the example of ACC, how system properties of advanced driver assistance systems can be transformed to objective measurements.

Keywords: product development process of advanced driver assistance systems; psychophysical study; objectification of system properties

MOTIVATION

More and more driver assistance functions are to intervene even without explicit actions of the driver in the dynamics of a vehicle (Maurer 2006). As a consequence of this the reliability of individual systems and a safe interaction with existing vehicle systems as well as with the driver must be ensured (Buld 2002). This growing complexity in the development process of driver assistance systems changes the requirements for the test and simulation tools (Ehmanns 2000). The availability and quality of simulation methods and the proper process integration will become a necessary precondition and a contributing competitive factor (Bock 2008).

This, however, force engineers to face a new challenge: How should measurement results from simulated tests for comfort-oriented advanced driver assistance systems be interpreted?

The state of the art in the application of these systems is the subjective assessment of the development engineers. With the consistent use of virtual development tools, the possibility of experiencing a function exists no longer. The results of virtual "rides" are exclusively measured variables.

Therefore the psychophysical study described below shows how system properties of driver assistance systems are transformed to objective measurements respectively parameters (demonstrated by the example of ACC).

METHODOLOGY

Figure 1 shows the approach to the development of an objective evaluation process for driver assistance systems based on subjective criteria.

The procedure is geared to the classic vehicle dynamics evaluation process (Heißing 2002). By using different maneuvers with variable applications, the target system-behaviour will be found through a psychophysical study. The goal shall be to generate objective parameters and assessment functions.

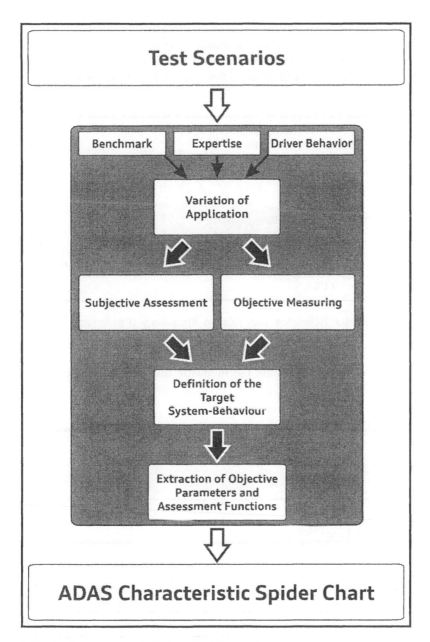

Figure 1: Evaluation Process for Driver Assistance Systems

TEST SCENARIOS AND APPLICATIONS

The selection of suitable test scenarios (maneuvers) is of vital importance to the quality of the assessment. To filter out the relevant maneuvers from the multitude of potential road situations the study of Freyer was analyzed again (52 subjects drive for about one hour without driver assistance systems on highways; (Freyer 2008)). The aim was to find out which ACC relevant situations can be found in general and with which frequency they occur on highways. Figure 2 provides an overview of the results.

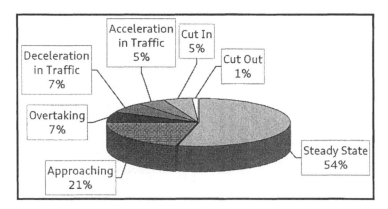

Figure 2: Potential Scenarios with relative Frequency (Percentage of total Travel: 26%)

Because of this investigation the following four maneuvers were selected for further examination:

o   Approaching (21%)

o   Overtaking (7%)

o   Cut In (5%)

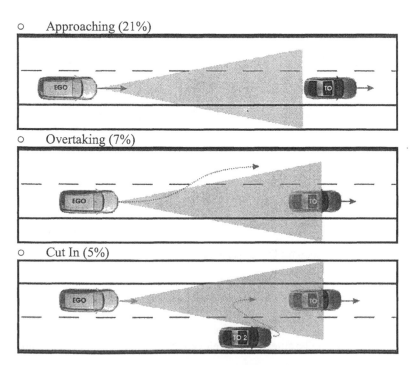

o   Cut Out (1%)

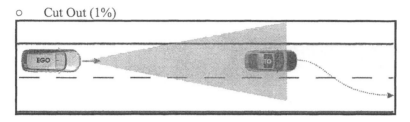

Figure 3: Analyzed Maneuvers

The maneuver "Steady State" (54%) can be neglected because the objective measurement is a maximum allowed scattering around the nominal variables like distance or speed. This does not need to be analyzed in a study.

Based on the factors of the ACC benchmark, engineering expertise and driver behavior three different forms of applications were realized for each of the four maneuvers in a single vehicle. Table 1 shows an overview of the selected system behaviors. The column "Reaction" gives an indication of how soon the system responds to the changing situation (e.g. transition from free traveling to following by the maneuver "Approaching"). Column "Dynamic" indicates the forcefulness of the reaction.

Table 1: Classification of the Application

Approaching	Reaction	Dynamic
Application "moderate"	moderate	moderate
Application "anticipatory"	early	soft
Application "dynamic"	late	strong

Overtaking	Reaction	Dynamic
Application "moderate"	moderate	moderate
Application "dull"	late	moderate
Application "dynamic"	early	strong

Cut In	Reaction	Dynamic
Application "moderate"	moderate	moderate
Application "safe"	early	moderate
Application "anticipatory"	early	soft

Cut Out	Reaction	Dynamic
Application "dull"	late	moderate
Application "dynamic"	early	strong
Application "anticipatory"	early	moderate

METHOD OF MEASUREMENT

For the recording of objective physical parameters a reference system, which consists of a high-accuracy DGPS inertial navigation system was used. By equipping each vehicle involved with one of these systems, measurements of the ego dynamics and environment, such as exact distance, relative velocity and -acceleration, can be recorded (Strasser 2010).

Muriel has discovered in her investigation of the comfort of ACC, that only a comparative study of different system performances will lead to the definition of subjective characteristics (Muriel 2006). Therefore, the three applications which were realized for the study were analyzed through paired comparison. Thus, three paired comparisons had to be done for each maneuver. The interpretation of the data is based on the "Law of Comparative Judgment" (Bortz, 2005), because it allows a scaling of the investigated objects (applications) of one feature (characteristic) by the allocation of the subjects' alternative judgments (Neibecker 2001).

The characteristics to be studied are:
o   safety
o   comfort
o   agility
o   driver's favorite

SUBJECTS

N=36 drivers (18 women and 18 men) participated in the study. The drivers were equally distributed among three age groups: 18 to 29 years ("Younger"), 30 to 49 years ("Medium") and over 50 years ("Older"). The average age overall was 40 years (Younger: 24 years, Medium: 40 years and Older: 55 years). Apart from the prerequisite that each of the subjects had to have a driving experience of at least 10.000 km, there were no further requirements.

RESULTS

For the analysis of the data under the "Law of Comparative Judgment" the frequencies of the preferred judgments are ascertained against the comparison order ("moderate" vs. "dull", "moderate" vs. "dynamic", "dull" vs. "moderate", "dull" vs. "dynamic", "dynamic" vs. "moderate" or "dynamic" vs. "dull"). Via the relative frequency and the z-values, the mean (of the z-values) of each column respectively application can be calculated. In order to carry out the superordinated evaluation the scale values have to be normalized over all characteristics, all applications and all maneuvers. For this purpose, the absolute value of the maximum negative z-value was defined as the origin. Therefore the maneuver "Overtaking" in the application "dull" of the feature "agility" was used to establish the baseline. From expert's point of view this application can be regarded as a total lack of agility, which is why this approach is sensible for finding the origin.

The detected data and resulting conclusions regarding the objective evaluation system for advanced driver assistance systems will now briefly be discussed using the maneuver "Overtaking".

Figure 4 shows the calculated metric scales, which now can be used to accomplish the objectification of each system property.

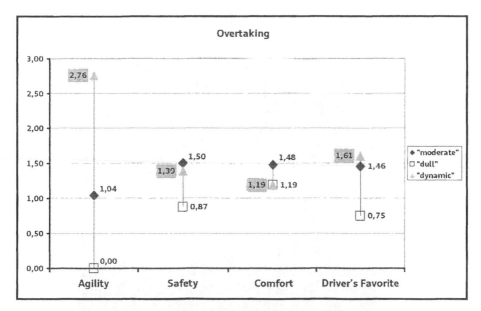

Figure 4: Paired Comparison - Ranking of the Maneuver "Overtaking"

If we now examine the characteristic "agility", a clear gradation between the three applications can be seen. Thus, "dynamic" is perceived to have a high grade of agility, "dull" is whereas seen to be not at all agile and with "moderate" being between the two extremes.

For "safety" there is no distinction between "moderate" and "dynamic", which are therefore both classified as very safe. "Dull", however, is seen to be less safe.

In the characteristic "comfort" no (real) distinction is made between all three applications. Therefore the area in which the system behavior is perceived as comfortable seems to be quite large (in this maneuver).

The final criterion "driver's favorite" shows that "dynamic" and "moderate" are preferred. But some users would also choose the setting "dull".

The aim of this investigation was to link selected, objective and measurable parameters with the subjective evaluation and define graduated areas where the characteristics of interesting are achieved. Therefore the subjects were asked why the chosen application corresponds to their driving style. This should show on the basis of which parameters users evaluate advanced driver assistant systems. For the maneuver "Overtaking" these are at 37% the time until the system reacts to the whish for overtaking and at 45% the strength of the acceleration. At 4% the judgments could not be substantiated and 13% of the answers could not be allocated to measurable parameters.

Figure 5 shows the results graphically by linking the two parameters and the ranking. The color coding of all dark to white stands for "very" to "not at all" in the current characteristic. The boundaries of the ellipses are not as sharp as illustrated and the black area outside the ellipses must be accepted as unknown at the current state of research.

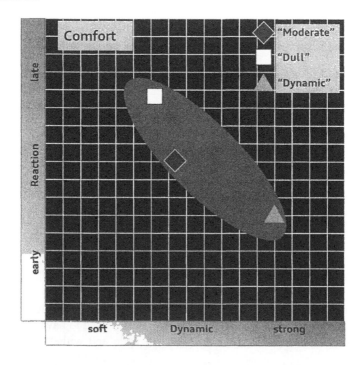

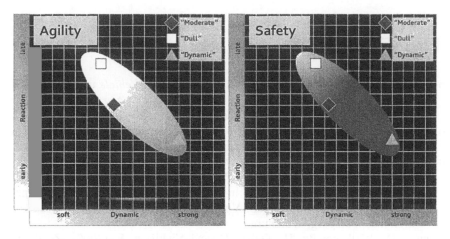

Figure 5: Objective Evaluation for the Maneuver "Overtaking"

CONCLUSION AND OUTLOOK

The psychophysical study presented here shows, that there are significant differences in the evaluation of advanced driver assistance systems of selected characteristics. In addition, physical parameters can be assigned to these characteristics that allow an objective assessment of those systems.

Through this process, there will be two major advantages: First, different system variants can be objectively compared. Second, this method enables the use of simulation tools such as Software-, Hardware- or Vehicle in the Loop (Bock 2005), (Bock 2007) in the development of advanced driver assistant systems as required in the motivation.

For further examinations the unknown areas should be considered by testing different applications that are distributed like a matrix in the Reaction-Dynamic-Layer. Additionally the dependency of the evaluation on the vehicle category should be analyzed.

126

REFERENCES

Bock, T.; Maurer, M.; van Meel, F. and Müller, T. (2008): *Vehicle in the Loop - Ein innovativer Ansatz zur Kopplung virtueller mit realer Erprobung*. In: Automobiltechnische Zeitschrift (ATZ), Jg. 110, H. 01.

Bortz, J. and Döring, N. (2005): *Forschungsmethoden und Evaluation für Sozialwissenschaftler*. 3.Aufl. Heidelberg: Springer Verlag.

Buld, S. and Krüger, H.-P. (2002): *Wirkung von Assistenz und Automation auf Fahrerzustand und Fahrsicherheit*. Projekt: EMPHASIS. Effort-Management und Performance-Handling in sicherheitsrelevanten Situationen. Interdisziplinäres Zentrum für Verkehrswissenschaften (IZVW) an der Universität Würzburg, Würzburg.

Ehmanns, D.; Wallentowitz, H.; Gelau, C. and Nicklisch, F. (2000): *Zukünftige Entwicklungen von Fahrerassistenzsystemen und Methoden zu deren Bewertung*. In: Wallentowitz, H.; Pischinger, S.: 9. Aachener Kolloquium Fahrzeug- und Motorentechnik. Aachen.

Freyer, J. (2008): *Vernetzung von Fahrerassistenzsystemen zur Verbesserung des Spurwechselverhaltens von ACC*. 1. Aufl. Göttingen: Cuvillier Verlag (Audi Dissertationsreihe 9).

Heißing, B. and Brandl, H. J. (2002): *Subjektive Beurteilung des Fahrverhaltens*. 1. Aufl. Würzburg.

Maurer, M. (2006): *Fahrerassistenzsysteme im Kraftfahrzeug*. Materialien zur Vorlesung, Technische Universität München, Garching b. München.

Muriel, D. (2006): *Ein Verfahren zur Messung des Komforts von Abstandsregelsystemen (ACC-Systemen)*. Bericht aus dem Institut für Arbeitswissenschaft der Technischen Universität Darmstadt. Stuttgart: Ergonomia Verlag.

Neibecker, B. (2001): *Hypothetische Konstrukte, intervenierende Variable, Law of Comparative Judgement, Messung, Operationalisierung, Polaritätsprofil, Reliabilität, semantisches Differential, Skalenniveau, Skalentransformation, Skalierungstechnik, theoretische Konstrukte, Validität*. In: Diller, H.: Vahlens Großes Marketing Lexikon. 2. Aufl. München.

Strasser, B.; Siegel, A.; Siedersberger, K.-H.; Maurer, M. and Bubb, H. (2010): *Vernetzung von Test- und Simulationsmethoden für die Entwicklung von Fahrerassistenzsystemen (FAS)*. In: TÜV SÜD Akademie: 4. Tagung Sicherheit durch Fahrerassistenz. München.

CHAPTER 15

Modeling of Environmental Factors Towards Workers' Productivity For Automotive Assembly Line

*[1]Ahmad Rasdan Ismail, [2]Mohd Hanifiah Mohd Haniff, [2]Baba Md. Deros,
[1] Nor Kamilah Makhta, [3]Zafir Khan Mohd Makhbul*

[1]Faculty of Mechanical Engineering, Universiti Malaysia Pahang
26300 UMP, Kuantan
Pahang, Malaysia

[2]Department of Mechanical and Material Engineering
Faculty of Engineering and Build Enviroment
Universiti Kebangsaan Malaysia, 43600 Bangi
Selangor Malaysia

[3]Business Management Programme, Faculty of Business and Economics
Universiti Kebangsaan Malaysia, 43600 Bangi
Selangor, Malaysia

ABSTRACT

The objective of this study is to determine the effects of humidity and of air temperature on the operators' productivity and performance in the Malaysian automotive industry. One automotive components assembly factory was chosen as the sources of subjects of the study. The subjects were the workers in the assembly section of the factory. The examined parameters were the relative humidity (%) and wet-bulb globe temperature (WBGT) of the surrounding workstation area. Two sets of representative data consisting of the relative humidity (%), WBGT and production rate were collected during the study. The production rate data were collected through observations and survey questionnaires, while the relative humidity (%) and WBGT was measured using thermal comfort multi-station (TCM) equipment. Linear regression analysis was performed to obtain the relationship between the effects of relative humidity (%) and of temperature (WBGT) on worker productivity and performance. The linear regression analysis further revealed a linear model with a positive slope between relative humidity (%) and worker productivity for the assembly section involved. The obtained relationship was $Y = 2.79X - 46.1$. For WBGT, the linear regression analysis revealed a linear model with a negative slope between temperature (WBGT) and worker performance for the assembly section involved. The obtained relationship was $Y = -13.3X + 425$.

Keywords: Productivity, Performance, WBGT, Relative Humidity, Relationship.

INTRODUCTION

The automotive industry in Malaysia, which encompasses other similar activities, is a growing industry. Car production in Malaysia has recently also been increasing, with less than 254,000 cars produced in 1999 to double that number (442,000 cars) and employing 47,000 workers in 2007 (OICA, 2007). According to Fisk et al.(1997), productivity is one of the most important factors affecting the overall performance to any organization, from small enterprises to entire nations. Since the 1990s, there has been increasing amounts of attention on the work environment and productivity. Laboratory and field studies have shown that the physical and chemical factors in the work environment could have a notable impact on the health and performance of its occupants and consequently on the productivity. Workplace environmental conditions, such as humidity, indoor air quality, and acoustics, have significant correlation with workers' satisfaction and performance (Tarcan et al. 2004; Marshall et al. 2002; Fisk, 2000). Indoor air quality can have a direct impact on health problems and can lead to uncomfortable workplace environments (Juslen & Tenner, 2005; Fisk & Rosenfeld, 1997; Marshall et al. 2002). In addition, Shikdar and Sawaqed (2003) noted a high correlation between performance indicators and health, facilities, and environmental attributes. In other words, companies with larger health, facilities, and environmental problems could face more performance-related problems such as low productivity and high absenteeism.

Numerous studies have found that human performance and productivity depend on the thermal environment (Srinavin and Mohamed, 2002). Hummelgaard et al. (2007) indicated a higher degree of satisfaction with the indoor environment and a lower prevalence or intensity of symptoms among the occupants in naturally ventilated buildings. A previous study done by Li et al. (2008) shows that motivated people can maintain high performance for a short time under adverse (hot or cold) environmental conditions. However, the room temperature affected task performance differentially, depending on the type of tasks. Previous research done by Gillberg et al. (2003) showed that the performance of experienced operators may not deteriorate during night shift of modern control room work, likely due to a lower workload during the night, lack of monotony and the processes being relatively inert and forgiving to minor operators errors. According to Oesman and Arifin (2007), the optimal condition for using fuzzy logic method in a working environment in the small garment industry are a temperature of 24°C, a noise level of 56 dB, illumination of 300 lux and vibration of no more than 5 m/s².

'Relative humidity' is a term used to describe the water vapor pressure of the air at a given temperature (Bridger, 1995). If the relative humidity is high, the latent heat dissipation ability of the body is decreased due to the decrease in vapor pressure and the increase of sweat remaining on the body (Atmaca and Yigit, 2006). Tsutsumi et al. (2007) found workers' performance to be constant at all conditions but to grow more tired after a step change in relative humidity to 70%, accompanied by a lower evaporation rate of sweat from the human body. Previous research by Gavhed and Klasson (2005) showed that a low relative humidity resulted in more discomfort and more frequent symptoms related to facial skin and the mucous membranes such as dryness of the mouth, throat and facial skin, nasal drip, and more frequent symptoms of the eyes and lips. In addition, Wolkoff and Kjaergaard (2007) pointed out that low relative humidity plays a role in increases of reports of eye irritation symptoms and of cases of alteration of the precorneal tear film. A study on the effects of humidity on the operators' productivity in the Malaysian electronic industry by Ismail et al. (2007) indicated a linear relationship between the relative humidity (%) and productivity of the workers. In addition, Dawal and Zaha (2006) pointed out that the environmental condition especially temperature, humidity, noise and lighting, can affect job satisfaction in the automotive industry.

METHODOLOGY

Selection of Location and Subjects

One automotive manufacturing company was selected as the location for the study. A line producing a product over a period of time and under the effects of certain relative humidity and air temperature was chosen. This criterion is essential to quantify the effects of the relative humidity and of WBGT on the worker productivity based on output of assemblies among operators. The production line consisted of 10 woman operators whose task was to assemble an automotive part (a door check). Figure 1 shows the production line layout, while Figure 2 shows the flow chart of the work sequence on the production line. The standard production rate determined by the previous feasibility study to assemble a complete door check was 240 units for every hour of production.

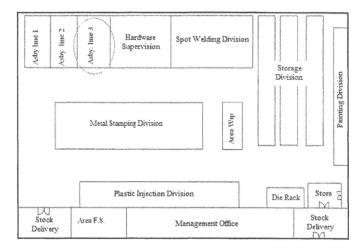

Figure 1. Door Check Production Line

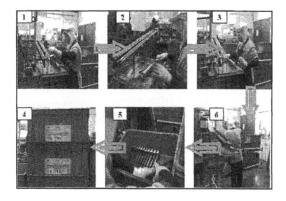

Figure 2. Production sequence for the complete assembly of door checks

Data Gathering and Analysis

The inferential statistics (i.e., production rate, relative humidity and WBGT) were computed to generalize the relationships of production rate to relative humidity and to WBGT. Further regression analyses were performed to obtain the relationship and thus test the hypotheses. The variables in this study were production rate, relative humidity and WBGT. A simple regression analysis was conducted to obtain the mathematical equation to present the effects of relative humidity and of WBGT on the production rate at that particular production line. The sample included 10 female operators whose ages were in the range of 20 – 30 years, comprised mostly of local citizens and non-degree holders who had been working with the organizations for less than 5 years. The majority of the respondents reported that they work for more than 49 hours per week. The relative humidity (%) was measured using TCM environmental equipment. The workers' performance level was represented by the production rate. The quantities of assembled products were recorded every 30 minutes, and data was compared to the relative humidity level and WBGT measurement.

RESULTS FOR RELATIVE HUMIDITY

The relative humidity level were taken to identify the effect of relative humidity on the worker performances. Table 1 shows the data of production rate, relative humidity (%) and the time taken for every 30 minutes. A graph was plotted to show the relationship between the production rate and the illuminance level. Figure 3 shows the graph to describe the relationship between production rate versus illuminance level. Based on the graph in Figure 3, we can note that the production rate were increases as we increase the relative humidity. The coefficient of determination, R^2, of 0.708 indicates that 70.8% of the production rate variation was due to relative humidity (%) variation. The results for regression and ANOVA analysis were presented in Table 2. The hypothesis were as follows:

H_o: $\beta = 0$(The relationship between relative humidity (%) and production rate is not significant)

H_a: $\beta \neq 0$(The relationship between relative humidity (%) and production rate is significant)

Table 1: Relative Humidity, Production Rate and Time Data

Time	Production Target (units)	Production Rate (units)	Relative Humidity (%)
9.05 – 9.35	120	119	60.76
9.35 – 10.05	120	123	59.85
10.05 – 10.35	120	121	59.21
10.35 – 11.05	120	115	59.55
11.05 – 11.35	120	121	59.29
12.05 – 12.35	120	124	59.61
12.35 – 1.05	120	108	55.98
2.10 – 2.40	120	112	55.84
2.40 – 3.10	120	106	55.45

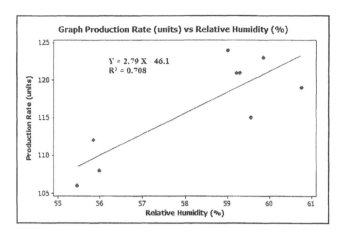

Figure 3. Graph of Production Rate versus Relative Humidity (%)

Table 2 : Regression and ANOVA Analysis

Regression					
Multiple R	0.842				
R Square	0.708				
R Square Adj	0.667				
Standard Error	3.820				
ANOVA					
Model	Sum of Squares	df	Mean Square	F	Sig.
Regression	248.061	1	248.061	16.997	0.004
Residual	102.161	7	14.594		
Total	350.222	8			
Coefficients					

* $p < 0.05$

The F value from the ANOVA is 16.997. The value of the significance level was selected to be 0.05 ($\alpha = 0.05$). Because the P value is 0.004, we can reject H_o: $\beta = 0$ in favor of H_a: $\beta \neq 0$ at the 0.05 significance level. This strongly suggests a significant relationship between the relative humidity and the production rate. Thus, there is a strong evidence that the simple linear model relating production rate and relative humidity (%) is significant.

RESULTS FOR WBGT

The value of WBGT were taken to identify the effect of temperature on the worker performances. The equation (1) is used to compute the value of wet bulb globe. (Wings, 1965)

$$WBGT = 0.7w_b + 0.2 T_g + 0.1 T_d \qquad (1)$$

Where;

w_b = wet-bulb temperature
T_g = globe temperature
T_{db} = dry bulb temperature

Table 3 shows the production rate, WBGT and time data measured every 30 minutes. Figure 4 shows the graph of the relationship between production rate and WBGT. Production rate were decreased when the temperature were increase.

Table 3 : WBGT, Production Rate and Time Data

Time	Production Target (units)	Production Rate (units)	WBGT (°C)
9.05 – 9.35	120	119	22.98
9.35 – 10.05	120	123	23.19
10.05 – 10.35	120	121	23.13
10.35 – 11.05	120	115	23.25
11.05 – 11.35	120	121	23.15
12.05 – 12.35	120	124	22.86
12.35 – 1.05	120	108	23.25
2.10 – 2.40	120	112	23.82
2.40 – 3.10	120	106	24.06

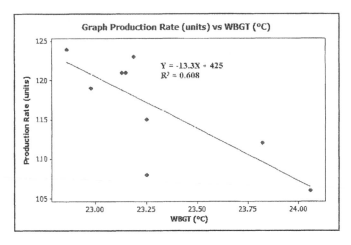

Figure 4. Graph of Production Rate versus WBGT

The coefficient of determination, R^2, of 0.608 indicates that 60.8% of the production rate variation was due to WBGT variation. The results for regression and ANOVA analysis were presented in Table 4. The hypothesis were as follows:

H_o: β= 0(The relationship between WBGT and production rate is not significant)
H_a: $\beta \neq$ 0(The relationship between WBGT and production rate is significant)

Because the P value is 0.013, we can reject H_o: $\beta = 0$ in favor of H_a: $\beta \neq 0$ at the 0.05 significance level. This strongly indicates that the regression relationship between WBGT and production rate is significant.

Table 4 : Regression and ANOVA Analysis

Regression					
Multiple R	0.780				
R Square	0.608				
R Square Adj	0.553				
Standard Error	4.426				
ANOVA					
Model	Sum of Squares	df	Mean Square	F	Sig.
Regression	213.105	1	213.105	10.879	0.013
Residual	137.117	7	19.588		
Total	350.222	8			

* $p < 0.05$

DISCUSSION

In the literature, there are only a small number of studies that have been conducted in the area to establish a mathematical model relating environmental effects to productivity. The authors believe that the study has achieved the objective of obtaining a mathematical model to quantitatively relate relative humidity (%) and WBGT to production rate by inferential statistical analysis. The findings on the effect of relative humidity on productivity are in line with the finding by Tsutsumi et al. (2007), who found that the subjective performance was equal under four different levels of relative humidity. However, Tsutsumi et al. (2007) also reported that their subjects were more tired after a step change in relative humidity to 70%. This finding was also was supported by Ismail et al. (2007), who observed that a linear relationship of relative humidity with productivity. A previous study by Fisk and Rosenfeld (1997) revealed the optimum WBGT for an office worker to achieve thermal equilibrium during work was 23°C to 25°C. Therefore, from this study, the authors expect that in each 30-minutes work cycle, an optimum production rate of 121 units can be achieved. The equation model will be useful for engineers as a guideline during the feasibilities study of production lines to achieve optimum output. The equation model is also useful for engineers in designing proper AHU units or air conditioning systems to minimize the use of power and control while considering the productivity of workers. The obtained mathematical model is only applicable to the current conditions for the selected assembly workstation in the Malaysian automotive industry. From these results, it can be concluded that there is a significant effect of relative humidity (%) and WBGT on production rate. Further tests proved that the model could be used to strongly predict the production rate based on a certain relative humidity (%) and WBGT provided by air conditioning or ventilation systems in a particular environment.

CONCLUSIONS

The results of the study indicate a significant relationship between humidity, wet-bulb globe temperature and workers' productivity. However, to date, research on the relationship between workplace environmental factors and productivity or performance has been very limited and characterized by a short time perspective or by emphasis on survey methods, statistical analysis, satisfaction preferences measurements. This study was done to empirically confirm the previous perception studies based on the role of environmental factors in productivity. This study is expected to be beneficial to the automotive manufacturing industry in Malaysia. The research findings are restricted to the Malaysian workplace environment, where the awareness of productivity among workers is still low. The results may vary for tests carried out with different sample sizes or in different types of industries and countries. The study could be more extensive if the fraction of defects for the product were included in the analysis. Nevertheless, the authors believed the modeling of production rate as time series data is more than adequate to understand the affect of environmental factors on productivity.

ACKNOWLEDGEMENT

The authors would like to thanks National University of Malaysia and Ministry of Higher Education Malaysia for their support in providing a research grant for a project Modeling Relationship of Thermal Comfort and Productivity in Malaysia Energy Intensive Industries, UKM-GUP-TK-08-16-059.

REFERENCE

Bridger, R.S. (1995). Introduction to Ergonomics. International Edition, McGraw-Hill,Inc. Singapore. pp. 1-18, 227-228, 264-298.

Fisk W.J., Rosenfeld A.H. (1997). Estimates of Improved Productivity and Health from Better Indoor Environments. Indoor Air. 7: 158-172.

Fisk, W. J. (2000). Health and Productivity Gains From Better Indoor Environments and Their Relationship with Building Energy Efficiency. Annual Review of Energy & The Environment. Volume 25 No. 2. pp. 537-566.

Gavhed, D., Klasson, L. (2005). Perceived Problems and Discomfort at Low Air Humidity among Office Workers. Volume 3. pp. 225-230.

Gillberg, M., Kecklund, G., Goransson, B., Akerstedt. T. (2003). Operator Performance and Signs of Sleepiness during Day and Night Work in a Simulated Thermal Power Plant. International Journal of Industrial Ergonomics. Volume 31. pp. 101-109.

Hummelgaard, J., Juhl, P., Sæbjörnsson, K.O., Clausen, G., Toftum, J., Langkilde, G. (2007). Indoor Air Quality and Occupant Satisfaction in Five Mechanically and Four Naturally Ventilated Open-Plan Office Buildings. Building and Environment. Volume 42 No. 12. pp. 4051-4058.

International Organization of Motor Vehicle Manufacturers. (2007). Production Statistics. Retrieved 2 November, 2008 from http://oica.net/category/production-statistics

Ismail, A.R., Rani, M.R.A., Makhbul, Z.K.M., Deros, B.M. (2008). Relationship of Relative Humidity to Productivity at A Malaysian Electronics Industry. Journal of Mechanical Engineering. 2008. Volume 5 No. 2. pp. 63-72. ISSN : 1823-5514.

Juslen, H., Tenner, A. (2005). "Mechanisms Involved in Enhancing Human Performance by Changing The Lighting in The Industrial Workplace" International Journal of Industrial Ergonomic. Volume 35 No. 9. pp. 843-855.

Li, L., Lian, Z.W., Li, P., Qian, Y. (2008). Neurobehavioral Approach For Evaluation Of Office Workers' Productivity. The Effects Of Room Temperature. Building and Environment. Volume 44 No. 8. pp. 1578-1588.

Marshall, L., Erica, W., Alan, A., Sanborn, M. D. (2002). Identifying and Managing Adverse Environmental Health Effects: 1. Taking an Exposure History Canadian Medical Association Journal. Volume 166 No. 8. pp. 1049-1055.

Oesman, T. I., Arifin, M. (2007). Temperature, Noise, Illumination and Vibration Analysis in Production Department at Small Garment Industry. Proceeding of International Conference on Ergonomics, Kuala Lumpur. pp. 219-223. ISBN: 978-983-2085-91-1.

Shikdar, A. A., Sawaqed, N. M. (2003). "Worker Productivity, And Occupational Health And Safety Issues In Selected Industries" , Computers and Industrial Engineering. Volume 45 No. 4. pp. 563-572.

Simonson, C. J., Salonvaara, M., Ojanen, T. (2002). The Effect of Structures on Indoor Humidity - Possibility to Improve Comfort and Perceived Air Quality. Indoor Air. Volume 12 No. 4. pp. 243-251.

Dawal, S.Z., Taha, Z. (2006). Factors Affecting Job Satisfaction in Two Automotive Industries in Malaysia. Jurnal Teknologi. Volume 44 (A). pp. 65-80.

Tarcan, E. Varol, E.S., Ates, M. A. (2004). Qualitative Study of Facilities and Their Enviromental Performance Management of Environmental Quality: An International Journal. Volume 15 No. 2. pp. 154-173.

Tsutsumi, H., Tanabea, S.I., Harigayaa, J. I, Guchib, Y., Nakamura, G. (2007). Effect of Humidity on Human Comfort and Productivity after Step Changes from Warm and Humid Environment. Journal of Building and Environment. Volume 42. pp. 4034–4042.

Wolkoff, P., Kjaergaard, S.K. (2007). The Dichotomy of Relative Humidity on Indoor Air Quality. Environmental International. Volume 33. pp. 850-857.

Chapter 16

Toward A Unified Social Robot Taxonomy Based on Human Perceptions

Benedict Tay Tiong Chee, Park Taezoon

School of Mechanical and Aerospace Engineering
Nanyang Technological University Singapore
50 Nanyang Avenue, Singapore 639798

ABSTRACT

Taxonomy is essential to sustainable development in any field research for understanding details. Though some previous works have been done regarding the robot taxonomy, there is still lacking in a scientific and proven classification for social robots about how people perceive them. The aim of this study is to propose a unified view of social robots taxonomy based on four users' perceptions, including perceived personality, animacy, anthropomorphism and threat. Subjective questionnaire consisting of 33 items is used to assess the four subjects' perceptions. In the experiment, a picture of robot is shown to the participants, and each participant is required to complete a set of questionnaire to rate their perceptions towards each of the illustrated robot. The process repeats with subsequent picture until all the 100 pictures of robot have been shown to the participants. A pilot study was conducted to improve the quality and efficiency of our subjective questionnaire. Some deficiencies are revealed and modifications of questionnaire items were done according to the found deficiencies. After the data collection, all the pictures of robot will be classified according to the similar attributes of subjects' perceptions. Subsequently, the relationship between people's perception of a robot and its physical attributes is to be examined for drawing relationship between them.

Keywords: Social Robot, Taxonomy, Classification, Perceived Personality, Perceived Animacy, Perceived Anthropomorphism, Perceived Threat, Subjective Questionnaire.

INTRODUCTION

Good taxonomy is essential to a sustainable development of social robotics research as it helps in understanding of usage of social robots and their relationships with users and the context of use. Although a number of social robots for different purposes including reception and home care, has been developed and tested, but how humans perceive and classify robot is not clearly identified. The aim of this study is to propose a taxonomy which can be used as a common framework across the researchers for design, description and development of social robots. Besides, understanding the underlying factors affecting human classification towards robots may help researchers in getting familiar with the factors of human acceptance towards social robots in the future life.

The specific research intends to identify the underlying factors that affect human judgment on the classification of social robots. The result of this research is proposing taxonomy of social robots based on users' perceptions. The perceptions that will be studied in this research include perceived personality, perceived anthropomorphism, perceived animacy and perceived threat.

BACKGROUNDS

Some previous works have been done regarding robot taxonomy (Ruspini and Khatib 2001; Daughtry 2008), however, most of them are primarily based on subjective classification of the author(s). Though some of the works (Dudek, Jenkin et al. 1996; Gonzalez-Gomez, Zhang et al. 2006) proposed scientific methods to classify robots, the focus is yet mainly on the industrial robots. One thing worth noting is that, unlike industrial robots, the success of social robots will not solely be evaluated by the functional performance of robot, such as quality standard, speed and reliability. The performance of social robots depends highly on the users' acceptance and satisfaction. Such an importance has been highlighted by previous researchers (Bartneck, Croft et al. 2008), which implies that human's perception of robot as a whole is more important for the success of social robots. A proposed taxonomy of social robots based on users' perception will be presented in this paper. Bartneck et al. (2008) did a good summary of past literatures regarding human perceptions of animacy, anthropomorphism, intelligence, safety and likeability towards social robots They studied across various literatures investigating the four human perceptions of social robots and selected one highly recognizable questionnaire with high Cronbach's Alpha value reported for each perception. Then, the authors unified the format of the items by transforming different measuring scales into a unified semantic differential scale.

In our study, four different human perceptions towards social robots, including perceived personality, perceived animacy, perceived anthropomorphism, and perceived threat will be used as the platform of establishing common ground of robot taxonomy.

Perceived Personality

Big Five Personality Model is used as the platform of this study to study subjects' perceived personality. The Big Five model assumes five basic personality factors which are extraversion, agreeableness, conscientiousness, neuroticism, and openness. The Big Five makers reported in the article by Goldberg (1992) have proposed 100 unipolar items which were proven to be highly robust across quite diverse samples of self and peer descriptions. In the research field of personality and social psychology, several studies tried to investigate how raters judged subjects' personality at zero acquaintance, meaning that one person observes another, but the two have never engaged in social interaction. Past attractiveness research (Brunswik 1956; Dion, Berscheid et al. 1972) had shown that consensus at zero acquaintance may be determined by the subject's physical attractiveness. It is therefore believed that physical appearance information such as gender, race, physical attractiveness, and dress style are readily apparent and may accurately or erroneously drive initial perceptions (Albright, Kenny et al. 1988). Therefore, there are sufficient supports for us to expect consensus in personality judgment. Kenny, Horner et al. (1992) studied the consensus at zero acquaintance with the five personality traits ; extraversion, agreeableness, conscientiousness, emotional stability and Culture (Norman 1963). The results showed that consensus is particularly high in ratings of Extraversion, complying with the finding of Funder and Dobroth (1987) ,and also Conscientiousness. Recent research investigated the perceived personality of social robots, its correlation with other factors and users' preferred perceived personality. Sydral D. S. et al. (2007a) studied the perceived personality of robots using the Big Five Model and suggested that participants clearly differentiated between the different robots on the dimensions of extraversion, agreeableness and intelligence, but did not differentiate strongly between them on the emotional stability. The study is expanded by Walters et al. (2008) and reported that participants tended to prefer robots with more human-like appearance and attributes. Also, introverts and participants with lower emotional stability tended to prefer the mechanical looking appearance to greater degree than other participants. The result highlighted that perceived personality of ratees could be affected by the raters' personality.

Perceived Animacy

One of the goals of robotic researches is to produce a robot as lifelike as possible. Animacy, according to Oxford dictionary is "having life, lifely". Previous studies (Heider and Simmel 1944; Michotte 1963) showed certain simple visual displays consisting of 2D geometry shapes is able to give rise to percepts with high-level properties such as causality and animacy. This implies that the interpretations of such phenomena are largely perceptual in nature. Some works (Bartneck, Kanda et al. 2009) has been done to investigate how the design of a robot can influence its perceived animacy and intelligence. The study measured perceived animacy of two robots, Robovie and iCat, with both objective and subjective measurement. The results show that increased realism of robots does not have tremendous effect on children's conceptual beliefs [of animacy]. Therefore, it is believed that it is even less likely that it will have an effect on adults, who are much further developed in their conceptualization of animacy. Also, in the paper, a significant and positive correlation was found between the perceived intelligence and animacy.

Perceived Anthropomorphism

Anthropomorphism is, at its most general, the assignment of human characteristics to objects, events, or non-human animals (Horowitz A. 2007). It can be explained as human-likeness in its simplest form. Mori (1970) proposed an impactful theory, the Theory of Uncanny Valley regarding anthropomorphism in robotics design. It hypothesizes that the more human-likes robots become in motion and appearance, human emotional reaction towards them will become more positive. However, upon reaching certain point, the trends declines and the emotional response quickly become negative. Besides considering user satisfaction and emotion, it is also important to match anthropomorphism of robots with their functions. Duffy (2003) claimed that successful design in both software and robots in Human Computer Interaction (HCI) needs to involve a balance of illusion that leads the user to believe in the sophistication of the system in areas where the user will not encounter its failing. Making social robots too human-like may also defeat the purpose of robots in society.

Perceived Threat

Possibly affected by Sci-fi movies (e.g. Rossum's Universal Robots 1921) or due to unfamiliarity, human may unintentionally carry negative perceptions towards robots. Robots can be a perceived threat to some users who are worry that they will be hurt or their life will be taken away by the robots. Hiror and Ito (2008) investigated the relationship between robot size and psychological threat and reported that out of the three robots of 0.6m, 1.2m and 1.8m tall, 1.2m robot is the most ideal robot size of which the subjects expressed the least anxious and shortest subjective acceptable distance. There are two ways to assess the robot perceived safety; physiological and subjective rating. Rani, Sarkar et al. (2004) used cardiac response, electrodermal response, and electromyographic response to measure the level of subjects' anxiety. Subjective rating questionnaires were used to capture the subjective perception of safety towards robots in some studies. Together with heart rate measurement, Nonaka, Inoue et al. (2004) used 4 items (i.e. surprise, fear, disgust, unpleasantness) of 6 level scales (from 1= never, to 6=very much) to evaluate emotion of human subjects. They also did a comparison study of human psychological evaluation towards real and virtual mobile manipulators (Inoue, Nonaka et al. 2005). 15 semantic differential items were used to evaluate subjects' psychological state, including "Secure/Anxious", "Restless/Calm", "Unreliable/Reliable", "Interesting/Tedious", "Comfortable/Unpleasant", "Unapproachable/Accessible", "Tense/Relaxed" and etc, but no value of Cronbach's Alpha was reported in the study. Kulic and Croft (2007) measured both the physiological measurements, including HeartRate, HRAccel, SCR, dSCR, and

CorrugEMG, and subjective ratings, including anxiety, calm and surprised to access subjects' perceived threat of robot motions. The five extracted features (HeartRate, HRAccel, SCR, dSCR, and CorrugEMG) were fuzzified using simple trapezoidal input membership functions. The outputs of the fuzzy engine were the estimated arousal and valence. The authors did a correlation analysis between estimated arousal and reported subjective responses and reported that estimated arousal is positively correlated with the reported anxiety and surprise, and negatively correlated with reported calm. Hence, in this study, the subjective measurement of perceived threat is validated with the physiological measurement of estimated arousal. In the later research of Bartneck, Croft et al. (2008), the items were transformed into three semantic differential items: "Anxious/Relaxed", "Agitated/Calm", and "Quiescent/Surprised".

METHOD

Subjective questionnaire is designed to capture participants' perception towards different social robots that are shown to the participant in pictures. Albright, Kenny et al. (1988) commented on the pros and cons using photographs stimuli in studying consensus of person perceptions at zero acquaintance. Photographs used in experiments require careful flittering and selection to prevent biases in subjects' ratings. For example, only images which are able to display the clear physical appearance of illustrated robots and of high resolutions will be selected for the use of experiment. Also, we are selecting images from known and reliable resources such as past literatures, existing products and movies. However, robots with strong recognizable backgrounds are excluded from our selection to prevent biases due to the familiarity. Details regarding selection of robot images for the experiment will be illustrated in the later part of this paper. Inoue, Nonaka et al. (2005) conducted a research to investigate subjects' psychological states towards real and virtual mobile manipulates. However, no significant difference was found in the subjects' ratings in the study.

Measurement

In the beginning of the interview, demographic data of the subjects will be collected and their attitudes towards technology will be evaluated as the first part of the study. Then, pre-collected pictures of robots will be shown to each participant one by one. For each picture shown, participants are required to complete a set of questionnaire based on their feedback of perceptions towards the illustrated picture of robot. The identical questionnaire will be used for every subsequent picture shown to participants. Four perceptions, including perceived personality, perceived animacy, perceived anthropomorphism and perceived threat will be assessed in this questionnaire.

Measuring Perceived Personality

Sample questionnaire of 100 items (Goldberg 1992) extracted from International Personality Item Pool (IPIP) is used as the platform to study human perceived personality towards social robots in this study. The original questionnaire is modified according to the context of robot's personality. Selected appropriate items are rewrote and included in the questionnaire. Sample items from the big five domain scale are listed in Table 1.

Table 1 Sample Items from the Big Five domain scale

Item	Questionnaires
Extraversion	This robot looks quiet around strangers This Robot doesn't seems like drawing attention This Robot seemingly feels at ease with people
Agreeableness	This Robot is seemingly interested in people This robot seemingly sympathize with others' feeling This Robot seemingly feels at ease with people
Conscientiousness	This robot seemingly pays attention to details This robot seems always prepared This robot seemingly makes a mess of things
Neuroticism	This robot seems relaxed most of the time This robot seems easily disturbed This robot seemingly gets upset easily
Openness	The robot seems to be full of ideas The robot seemingly will not probe deeply into a subject The robot seemingly can handle a lot of information

Measuring Perceived Animacy

Four items, lifelike, machinelike, interactive and responsive, were proposed by Lee, Min et al. (2005) to measure subjects' perceived animacy. The items were transformed in the later study (Bartneck, Croft et al. 2008) into semantic differential scales with Cronbach's Alpha of 0.702 reported. Six semantic differentials include "Dead/Alive", "Stagnant/Lively", "Mechanical/Organic", "Artificial/Lifelike", "Inert/Interactive", and "Apathetic/Responsive".

Measuring Perceived Anthropomorphism

Bartneck, Kanda et al. (2007) revised six items proposed by Powers and Kiesler (2006) of measuring subjects' perceived human-likeness. The revised semantic differentials, including "Fake/Natural", "Machinelike/Humanlike", "Unconscious/Conscious", "Artificial/Lifelike", and "Moving rigidly/Moving elegantly" are included in the present research to access perceived anthropomorphism.

Measuring Perceived Threat

Kulic and Croft (2007) proposed subjective ratings, including anxiety, calm and surprised, to evaluate subjects physiological measurements. Bartneck, Croft et al. (2008) transformed the subjective ratings into three semantic differential items: "Anxious/Relaxed", "Agitated/Calm", and "Quiescent/Surprised". We include these three items to access subjects' perceived threat.

Participants

Total number of 20 subjects, including 10 males and 10 females, will be recruited from various faculties in the university.

Selection of Pictures

Pictures to be used for the experiment is collected from several sources, including research paper, technical report, newspaper, internet, illustration in literatures, and movies. Among the pictures collected, famous robot images from movies (e.g. R2D2, C3PO from StarWars) are excluded because the familiarity to the original film or story can give strong contextual information which may influence the evaluation of the participant. Only images which are able to show clear and unambiguous physical appearance of a robot will be selected. Additionally, a question asking whether the participant has seen the picture before is added in order to record the previous exposure. Screened pictures are normalized by the size and quality of the picture in order to make them look as uniform as possible.

Procedures

Before the interview, participants are required to complete the pre-experiment questionnaire online and their results will be collected for further analysis.
Next, participants are required to sign the consent form indicating their voluntariness of participation in the research. In the first part of the interview, participant will complete their demographic data, including their email addresses and phone numbers for further collaboration.
In the 2nd part of the interview, which is the main body of the interview, different pictures of social robots will be shown to the participants one by one. For each of the pictures, participants are required to reflect their perceptions towards the particular robot shown in the picture. Each questionnaire consists of total 33 questions to access subjects' perceptions of each illustrated robot. The first 20 questions investigate subjects' perceived personality towards robots. For the rest of 13 questions, 6 of them are measuring perceived animacy, 4 questions are used to access perceived anthropomorphism and 3 questions are measuring perceived threat.

Each questionnaire of a robot will take about 5 minutes and the whole interview consists of 100 pictures of robot. Due to the large number of pictures, the interview session will be segmented into 3 sessions. Each session will take not longer than 2 hours. In total, there will be 3 visits of participants. Each of the subsequent visits will be at least 2 days after the previous visit.

Incentives will be awarded to the participants who take part in the study. Extra incentives will be given to the participants who complete all the 3 sessions.

PILOT STUDY

Prior to the experiment, we had done a pilot study with to improve the quality of our questionnaire and increase the efficiency of the experiment. The pilot study was done with five participants and they were requested to rate their perceptions of the illustrated robot: *Barricade*, a robot in the *"Transformer"*, the movie. The overview result of rating *Barricade* showed that *Barricade* is perceived as one with high extraversion, low agreeableness and neuroticism. Beside the aggressive physical appearance, such an observation can possibly be explained by the biases induced by the ruthless and wicked character of *Barricade* in the movie *"Transformer"*. This shows that subjects' perceptions can be affected by contextual information of the robots. Furthermore, it suggests that, contextual information, such as name of the robot and labeling found on the robot, should be removed from the photo selected for the use of this experiment in order to minimize the possible biases of rating.

One participant, who did not have any background knowledge of *Barricade,* had expressed difficulty in rating the statement "The robot seems always prepared". Other participants, when asked about their understanding of this statement, interpreted the statement as "The robot seems always prepared for battle or initiating a fight." This suggests that, participant may have difficulty in answering this question when they have no prior contextual knowledge of this robot.

Participants took averagely four minutes to complete the set of questionnaire including filling of their demographic information. Some participants were having difficulties in recalling the last time they went to a Personal Computer (PC) fair. This question is included to investigate participants' attitude towards technology. Overall, participants expressed not much difficulty in rating all the items in the questionnaire.

From the pilot study, some deficiencies are revealed and modifications of questionnaire items were done according to the found deficiencies.

CONCLUSION & FUTURE WORKS

The aim of this study is to propose a unified view of taxonomy of social robot to serve as a common framework across the researchers for design, description and development of social robots. The proposed taxonomy is based on four human perceptions, which are perceived personality, perceived anthropomorphism, perceived animacy and perceived threat, of robots. Subjective questionnaire is used as the main tool to access these human perceptions in this study. We adopted items of sufficient validity and reliability from the different literatures measuring each perception and combined them under a set of questionnaire. After the data collection and analysis, robots sharing the similar attributes will be grouped into a classification. The work after data collection will be deciding the appropriate parameters to classify the robots into groups. A cluster analysis will be subsequently conducted after the grouping to validate the proposed taxonomy.

REFERENCES

Albright, L., D. A. Kenny, et al. (1988). "Consensus in Personality Judgments at Zero Acquaintance." Journal of Personality and Social Psychology **55**(3): 387-395.

Bartneck, C., E. Croft, et al. (2008). Measuring the anthropomorphism, animacy, likeability, perceived intelligence and perceived safety of robots. Metrics for Human-Robot Interaction Workshop in affiliation with the 3rd ACM/IEEE International Conference on Human-Robot Interaction (HRI 2008), Technical Report 471, Amsterdam, University of Hertfordshire.

Bartneck, C., T. Kanda, et al. (2007). Is the Uncanny Valley an Uncanny Cliff? Proceedings of the 16 th IEEE International Symposium on Robot and Human Interactive Communication, RO-MAN 2007, IEEE.

Bartneck, C., T. Kanda, et al. (2009). "Does the Design of a Robot Influence Its Animacy and Perceived Intelligence?" International Journal of Social Robotics **1**(2): 195-204.

Brunswik, E. (1956). Perception and the representative design of psychological experiments (2d ed.). Berkeley, CA US, University of California Press.

Daughtry, E. K. (2008). Robots: A Practical Taxonomy. Fine Arts California College of the Arts. **Master:** 94.

Dion, K., E. Berscheid, et al. (1972). "What is beautiful is good." Journal of Personality and Social Psychology **24**(3): 285-290.

Dudek, G., M. R. M. Jenkin, et al. (1996). "Taxonomy for multi-agent robotics." Autonomous Robots **3**(4): 375-397.

Duffy, B. R. (2003). "Anthropomorphism and the social robot." Robotics and Autonomous Systems **42**(3-4): 177-190.

Funder, D. C. and K. M. Dobroth (1987). "Differences between traits: Properties associated with interjudge agreement." Journal of Personality and Social Psychology **52**(2): 409-418.

Goldberg, L. R. (1992). "The development of markers for the Big-Five factor structure." Psychological Assessment **4**(1): 26-42.

Gonzalez-Gomez, J., H. Zhang, et al. (2006). Locomotion Capabilities of a Modular Robot with Eight Pitch-Yaw-Connecting Modules. 9th International Conference on Climbing and Walking Robots. Brussels.

Heider, F. and M. Simmel (1944). "An Experimental Study of Apparent Behavior." The American Journal of Psychology **57**(2): 243-259.

Hiroi, Y. and A. Ito (2008). Are bigger robots scary? - The Relationship Between Robot Size and Psychological Threat-. Advanced Intelligent Mechatronics, 2008. AIM 2008. IEEE/ASME International Conference on.

Inoue, K., S. Nonaka, et al. (2005). Comparison of human psychology for real and virtual mobile manipulators. Robot and Human Interactive Communication, 2005. ROMAN 2005. IEEE International Workshop on.

Kenny, D. A., C. Horner, et al. (1992). "Consensus at zero acquaintance: Replication, behavioral cues, and stability." Journal of Personality and Social Psychology 62(1): 88-97.

Kulic, D. and E. Croft (2007). "Physiological and subjective responses to articulated robot motion." Robotica(Copyright 2007, The Institution of Engineering and Technology): 13-27.

Lee, K. Min, et al. (2005). "Can a Robot Be Perceived as a Developing Creature?" Human Communication Research 31(4): 538-563.

Michotte, A. (1963). The perception of causality / A. Michotte. New York :, Basic Books.

Mori, M. (1970). "The Uncanny Valley." Enery 7: 33-35.

Nonaka, S., K. Inoue, et al. (2004). Evaluation of human sense of security for coexisting robots using virtual reality. 1st report: evaluation of pick and place motion of humanoid robots. 2004 IEEE International Conference on Robotics and Automation, 26 April-1 May 2004, Piscataway, NJ, USA, IEEE.

Norman, W. T. (1963). "Toward an adequate taxonomy of personality attributes: Replicated factor structure in peer nomination personality ratings." The Journal of Abnormal and Social Psychology 66(6): 574-583.

Powers, A. and S. Kiesler (2006). The advisor robot: tracing people's mental model from a robot's physical attributes. Proceedings of the 1st ACM SIGCHI/SIGART conference on Human-robot interaction. Salt Lake City, Utah, USA, ACM: 218-225.

Rani, P., N. Sarkar, et al. (2004). "Anxiety detecting robotic system - towards implicit human-robot collaboration." Robotica 22(1): 85-95.

Ruspini, D. and O. Khatib (2001). "Humanoids and personal robots: Design and experiments." Journal of Robotic Systems 18(12): 673-690.

Walters, M., D. Syrdal, et al. (2008). "Avoiding the uncanny valley: robot appearance, personality and consistency of behavior in an attention-seeking home scenario for a robot companion." Autonomous Robots 24(2): 159-178.

Chapter 17

Automated Analysis of Eye-Tracking Data for the Evaluation of Driver Information Systems According to EN ISO 15007-1 and ISO/TS 15007-2

[1]Christian Lange, [1]Roland Spies, [2]Heiner Bubb, [2]Klaus Bengler

[1]Ergoneers GmbH,
85077 Manching, Mozartstraße 8 ½
Germany

[2]Institute of Ergonomics
Technical University of Munich
Boltzmannstraße 15, 85748 Garching
Germany

ABSTRACT

The present work shows how synchronous recording and analysis of eye-tracking data, data of several video channels and device-specific data can be done with the Dikablis & D-Lab soft- and hardware suite according to EN ISO 15007-1 and ISO/TS 15007-2. Thereby, the experimental process is supported from planning over performing the experiment until analyzing and presenting the data.

Keywords: Eye-Tracking, Experiments, Data Analysis, Driver Information Systems, EN ISO 15007-1, ISO/TS 15007-2

INTRODUCTION

The synchronous recording and analysis of data is always a challenge in experiments. This is especially the case when eye-tracking data and data of several cameras have to be recorded together with device specific data. How this challenge is solved with the Dikablis & D-Lab soft- and hardware suite will be shown in the following.

PLANNING AN EXPERIMENT

The compliancy to test partitioning presented by ISO/TS 15007-2:2001 into „experimental condition", „task", and „subtask" can be defined into one test plan with D-Lab. This way a test can be represented in the form of intertwined and nested intervals in which:

- „experimental condition" an entire Experiment is unfolded (Ex. driving on country roads);
- „task" defines the interaction between a determined system presented within the Experiment (ex. Operation of Navigation Systems);
- A specification of a „task" as „subtask" will be considered. (Ex. Entering the town, streetname and house number).

D-Lab offers the possibility to additionally define „subsubtasks" as the fourth layer, ex. in order to mark the appearance of critical events presented within the experiment (ex. sharp breaking situations) or automatic analysis of the display screen (ex. an individual input screen within the navigation system, for example inputting a destination address).

PERFORMING AN EXPERIMENT

The core when performing an experiment is D-Lab Control. With D-Lab Control, the defined experimental tree can be opened, whereby the experimental conditions, tasks and subtasks are getting represented as clickable buttons. These buttons for marking beginning and end points of each trial interval can be clicked with the mouse as well as by a network event from a different application. Furthermore D-Lab Control is synchronizing itself with the Dikablis recording software which can be controlled remote via the D-Lab Control GUI. Furthermore, D-Lab control allows synchronous recording of several external sources. This can be up to four video streams as well as any kind of device-specific data via network. D-Lab

Control offers a generic network interface for device-specific data like for example driving dynamics data from a driving simulator or the metrics from a car from the car-bus system. Via this network interface this data is received and is saved synchronously to all other measured data. The GUI of D-Lab Control and the Dikablis recording software is shown in figure 1.

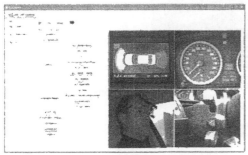 

FIGURE 1 Left: GUI of D-Lab Control with visualized experimental tree and displaying of the four video channels; Right: GUI of the Dikablis recording software with realtime view of the gaze behavior

DATA ANALYSIS

All data is analyzed in D-Lab. With this software, all data of the whole project can be imported and can be analyzed synchronously in one software.

For the analysis of the eye-tracking data, D-Lab offers the possibility for the free definition of Areas of interest AOIs, for which the glance durations are calculated automatically and conforming the eye-tracking standard ISO/TS 15007-2 (see Lange et al. 2009a und Lange et al. 2009b). When the calculation is finished, the glance durations are shown as timeline bars synchronously to the progress bar of the gaze movie respectively to the progress bar of the player for the four recorded video channels (see figure 1). The vertical line under the gaze movie player shows exactly the progress of the gaze movie or the four external videos in the timeline bars of the glance durations to the AOIs.

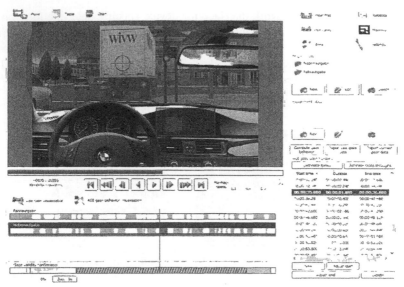

FIGURE 2 Left: Visualization of the glance durations to the defined Areas of Interest "driving task" and "secondary task) as timeline bars below the gaze video player

With the help of the integrated statistics functionality can be defined, which glance metrics shall be calculated for which AOIs during which task. An example for such a calculation would be: Calculate the glance metrics "total glance time", "total glance time as a percentage" and "number of glances" to the AOIs driving task and secondary while entering the navigation destiny. D-Lab calculates these metrics automatically and visualizes the result in a table for all subjects of an experiment. This automated calculation can be done for all defined AOIs, all experimental conditions, tasks and subtasks as well as for all glance metrics. The calculated result which is visualized in a table can be exported to a .csv file which can be imported to SPSS or MS Excel.

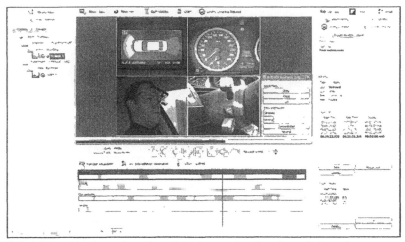

FIGURE 3 Visualization of analyzed tasks as timeline bars below the gaze movie player

For a precise analysis of subject behavior, D-Lab offers an integrated "acitivity analysis". The activity analysis allows the definition of "activity groups" and "activities". Based on the definition, a "trigger GUI" is created which represents all defined activities as clickable buttons. For classifying the activities, one can replay the gaze movie and the four video channels synchronously and press the buttons on the "trigger GUI" for marking specific activities. The marked activities are displayed as timeline bars synchronously to the progress bar of the gaze movie respectively the four videos. Again, a vertical line shows exactly the progress of the gaze movie or the four external videos in the timeline bars of the activity analysis (see figure 3).

As well as for the gaze data there is also a statistics functionality for the activity analysis. This statistics functionality allows for example to calculate the frequency or duration of specific activities during certain tasks. This calculation is also visualized in a table and can be exported to an SPSS and MS Excel compatible .csv file.

Furthermore, the device-specific data is getting visualized as timeline plots synchronously to the gaze movie and the four video channels in an own window (see figure 4). A vertical red line, which is synchronized to the progress bar of the gaze movie player and the four video channels as well as to the timeline bars of the glance durations and the timeline bars of the activity analysis, shows exactly the progress in the device-specific data. As well as for the four video streams an activity analysis can be performed for the timeline plots of the device-specific data. Thereby can be analyzed for example, how often the driver pressed the brake pedal during a specific task,

There is also a statistics functionality which allows calculation of descriptive statistics like for example mean, standard deviation or sum during certain task intervals for the recorded device-specific data. This calculation is also visualized in a table and can be exported to an SPSS and MS Excel compatible .csv file.

152

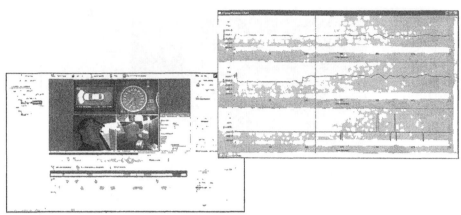

FIGURE 3 Synchronous visualization of recorded device-specific data(top right) and four recorded video channels

In summary, the Dikablis & D-Lab soft- and hardwaresuite is a tool which assists the user in planning and experiment, recording many data synchronously and analyzing this data with lots of useful functionality.

REFERENCES

ISO/TS 15007-2:2001: Road vehicles - Measurement of driver visual behaviour with respect to transport information and control systems - Part 2: Equipment and procedures

Lange C., Wohlfarter M., Bubb H. (2009a) Automated Analysis of Eye-Tracking data for the Evaluation of Driver Information Systems according to ISO/TS 15007-2:2001; In: Proceedings of the 13th International Conference on Human-Computer Interaction, San Diego 2009

Lange C., Bubb H. (2009a) Normgerechte Durchführung von Blickerfassungsexperimenten nach ISO/TS 15007-2:2001, In: Proceedings GfA Gesellschaft für Arbeitswissenschaften Kongress, Dortmund 2009

CHAPTER **18**

Analysis of Glance Movements in Critical Intersection Scenarios

Marina Plavšić, Klaus Bengler, Heiner Bubb

Lehrstuhl für Ergonomie
Technische Universität München
85747 Garching b. München, Germany

ABSTRACT

For designing effective and ergonomic assistance systems for road intersections it is highly beneficial to gain an understanding of the causes of driver's errors. At intersections errors depend mainly on the applied visual strategies and perceived information. This paper reports on a study conducted in the fixed-base driving simulator, with an objective to compare driver's visual strategies among three left-turn intersection scenarios, which can become critical with respect to safety.

The comparison of applied visual strategies showed that in the more complex situation drivers delegate more attention to the task and perform a less risky behavior. Nevertheless, in complex scenarios the driver's visual behavior indicates a prioritization problem regarding the primary directions to scan. A system that in an appropriate moment visualizes the priority roads or even ideal eye-movement sequence could decrease both driver's committed errors and mental workload.

Keywords: Road intersections, Eye movements, Erroneous behavior

INTRODUCTION

Intersection areas present one of the most difficult traffic situations where a lot of accidents happen. In Europe, depending on the country, intersections account for 30 to 60% of all injury-related accidents (Intersafe2, 2009).

In more than 95% of traffic accidents the causes lay in the driver's erroneous behavior. The errorless performance is especially important at intersections, as at an intersection there is a short time span available for the error corrections. In order to safely perform the task, the driver should perceive all relevant information, process it in a correct way, decide on the driving action and then flawlessly execute it. The research results of Gruendl (2005) showed that the main causes for driver's erroneous behavior are in the objectively or subjectively missing information. The wrong decisions and evaluations based on properly seen information are rarely the origin of accidents. This means that the safe performance depends mainly on the driver's ability to perceive important information in optimal time.

The information at intersection is to more then 95% perceived by the visual channel. This means that the information perception depends highly on the driver's eye movements and underlying visual strategies. In this case, the eye movements can be regarded as a result of preceding cognitive processes, which are hard to measure and evaluate. However, measuring and analyzing driver's visual behavior can lead not just to understanding of information perception and prioritization strategies but also to a better understanding of underlying cognitive mechanisms.

Within this work, the applied visual strategies and eye movements at selected intersection scenarios are analyzed in experimental conditions. The aim is to analyze how the visual strategies with resulting erroneous performance of the driver differ from the visual strategies of the drivers who perform the maneuvers safely and whether and how the applied visual strategies depend upon the maneuver. In order to provoke erroneous behavior, the left-turn maneuver is chosen as the most difficult task. Each of scenarios contains a hazard object and it is designed in a way that if the driver does not pay attention to the road with hazardous object, an accident is likely to be caused. The normative driver's behavior does not result in an accident. It is important hereby that hazardous objects have right of way and behave rule-compliant.

In the following, the advanced research with respect to the visual strategies at intersections is presented.

VISUAL STRATEGIES AT INTERSECTION

The basics of visual perception at intersection are similar to the basics of the perception of any arbitrary dynamic scene. The perception process can be divided into several phases. The first phase is so called pre-attentive processing in which the first few fixations give an essential picture of the scene. Schweigert (2003) conducted a field experiment in which he found that the speed of the objects scanning at intersections is between 0.8 to 5 fixations per second. In the second phase, the fixations serve to fill in details and to identify the separate objects. In the same experiment Schweigert found that in the second phase the modal values of fixation durations are between 0.3-0.4s in the urban scenes. After being fixated, the objects are consciously identified and can be kept around 3 to 4s in the short-term memory. Still, the object can be lost even earlier if the new content suppresses the old, which is the case in visually loaded scenes such are intersections.

The eye movements while driving can be described by two mechanisms: bottom up (data-driven) and top-down (knowledge-driven) mechanism. Bottom-up mechanism is controlled by occurrences of objects in the environment and the top-down mechanism indicates controlled strategies that are the result of information processing. For each traffic situation, driver chooses corresponding mental model and applies the visual strategies that correspond to the selected mental model. At intersections this can be considered as the consciously driven rule-based behavior.

The ideal visual behavior at intersections is the successively scanning of all the directions from where the hazards can come from. With experience drivers develop strategies that serve to predict the position of the crucial information. In such a way the relatively rarely seen objects like bicycles and motorcycles can be overseen (Summala, 1996). Langham (2006) found that duration spent looking for hazards at intersection is not higher than 0.5s. This time corresponds to checking just one or two directions which driver considers as relevant before crossing.

When it is necessary to perform several tasks simultaneously, the drivers have to choose the right order of the tasks and sometimes to omit the ones of the less value to them. The information prioritization that drivers do is resembled in the fact that with the increasing difficulty of the driving task, the fixation durations are decreasing, especially with respect to the irrelevant traffic objects and scanning fixations. Schweigert (2003) found that when having to perform additional task, the drivers omit checking for irrelevant traffic objects. At intersections, even 68% of drivers retract from direct scanning for pedestrians. When having to perform these multiple tasks at once it may happen that the driver does not just fail to look in one of the relevant directions but also that the object that is fixated is not perceived.

How the tasks are prioritized is kept in the driver's mental model of the situation. It is of interest to find out how the drivers do prioritize information and how it does depend on the performed task. This can be answered by analyzing the distribution of the driver's attention. Additionally, the analyzed difference between erroneous and safe strategies may indicate the possible assistance support. Therefore, the presented approach focuses on the analysis of the distribution of attention and the difference between erroneous and safe visual strategies.

EXPERIMENTAL STUDY

APPARATUS

The study was conducted in the fixed-base driving simulator located at the Institute of Ergonomics, Technische Universität München. The driving simulator is shown in Figure 1(a). The applied software is the commercial software SILAB (SILAB, 2009). This software enables a very realistic simulation of the traffic situations as well as the full control over all road users. In this way it is possible to create the arbitrary intersection scenarios and to exchange a "look and feel" of the simulation.

The glance behavior was recorded by using eye-tracking system Dikablis (Ergoneers). The system consists of two cameras, one recording the user's eye and the other recording the field of view (see Figure 1(b)).

 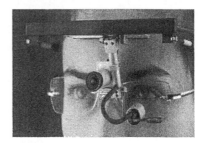

Figure 1. Apparatus applied in the study (a) fixed-base driving simulator of Institute of Ergonomics, (b) Eye tracking system Dikablis (Ergoneers)

EXPERIMENTAL SITUATIONS

Four experimental scenarios are chosen for the investigation. In this paper, the analysis of three scenarios is presented. The fourth scenario was triggered just in some test runs and therefore cannot be appropriately evaluated.

In order to provoke the erroneous behavior of the drivers, the left turn as the most difficult maneuver is chosen for the analysis. Each of three scenarios involved a potentially dangerous vehicle. This vehicle becomes visible in the phase of turn execution and apart from being occluded it does not perform any dangerous or unusual action. All vehicles hereby comply to traffic rules. Also, in all scenarios, the hazardous vehicles have right of way.

The first scenario is one-carriageway intersection with one lane for each direction, regulated by traffic lights. The sketch of the scenario is presented in Figure 2. The green car depicts the own vehicle and it arrives at intersection during the red-light phase. Blue cars in Figure 2 present the vehicles that are not directly dangerous for the own vehicle. During the red phase, the vehicle on the opposite lane that turns left is also visible to the driver. Behind this vehicle, there is an occluded potentially hazardous car (marked with H in Figure 2) that drives straight after the light changes to green.

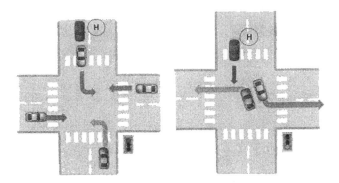

FIGURE 2. Scenario 1 – Left turn at intersection regulated by traffic light. The potentially hazardous vehicle H is occluded for the own vehicle while waiting for the green lights: (a) approaching intersection, (b) turning at intersection

The second scenario is an unordered T-intersection, presented in Figure 3. The own vehicle reaches an intersection at approximately the same time as the vehicle from the left that stops to yield the own vehicle. In that moment the vehicle from the right (depicted by the red car in Figure 3) reaches the intersection and the own vehicle is supposed to yield right of way.

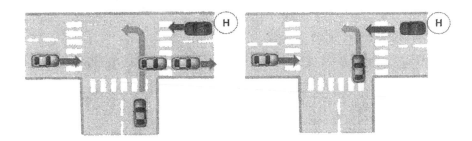

Figure 3. Scenario 2 – an unordered T-intersection with a potentially hazardous vehicle coming from right (a) approaching intersection, (b) turning at intersection

The third intersection consists of two broad lanes in each direction. This intersection is regulated by traffic lights that are in a continuous yellow phase. This is a complex scenario with six groups of independently moving objects, which is on the limit of the driver's processing capacity (Baddeley, 2009). The subject vehicle approaches from the minor street and has to give the right of way the crossing vehicles both from left and right as well as the oncoming vehicle. All three potentially hazardous vehicles reach the conflict area of intersection at the same time with the own vehicle (see Figure 4).

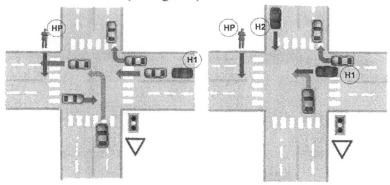

Figure 4. Scenario 3 – Two lanes intersection with three potentially hazardous vehicles, and one vulnerable road user (a) approaching intersection, (b) turning at intersection

TEST SAMPLE AND EXPERIMENTAL PROCEDURE

Thirty drivers took part in the experiment but one third did not finish the test because of the occurring simulator sickness. The presented analysis refers to twenty test subjects that completed the test. The participants possessed the driver's license of the European B category for 8.6 years on average. The group is age-homogeneous with the mean age of 25.8 years (SD = 7.27). The test sample consisted mainly from the students of the Mechanical Engineering department. The gender distribution of 19 males and 1 female reflects the distribution of the student population.

The aim of the study was not revealed to the participants before they finished the trial. The participants got a short introduction about the driving simulator and the eye tracking system and were told that the goal of the study is the evaluation of the simulator. First, the participants drove the practice course to get used to the simulation environment and only afterwards the full course equipped with the eye tracking system. Before starting the test, participants filled-in the demographic questionnaires. After the trial, the participants were confronted with their own drive and eye movements and for each of the three intersections, they were administred a situation-specific questionnaire. The trial lasted about one hour.

RESULTS

The performed analysis includes subjective and objective data. Subjective data consists of the subjective evaluations of the performance, perceived risk and mental workload. The objective analysis is the task and error analysis as well as the analysis of the attention distribution and various glance parameters. The results of the latter two are presented in this paper. The focus hereby is the comparison between the test persons performing an erroneous behavior and causing an accident and test persons who safely performed the maneuver.

The number of participants who performed errors and caused an accident is relatively high: 14, 8 and 10 for the first, second and third scenario, respectively (see Table 1). The reason for erroneous behavior in the first two scenarios is mainly in not searching for and not expecting the hazards. Participants were simply performing the turn without reassuring that no other vehicles come from the oncoming lane and the right street. These directions have been checked but the glances happened too early to see the hazards, already during the segment of approaching intersection. The necessary glance of checking again for the possible hazardous vehicle just before executing the turn was not performed by around 50% of test persons. The reason that such accidents do not happen so often in reality is that alike scenarios are relatively rare and the other road users usually compensate for the driver's errors. Still, this study shows that the drivers' usual visual strategies have deficiencies and offer a high potential for the support.

Table 1. Number of participants causing an accident and performing safely in each scenario

	Scenario 1	Scenario 2	Scenario 3
Erroneous behavior	14	8	10
Safe performance	6	12	10

Closer comparison between subjects causing an accident and subjects who safely performed the maneuver, shows the difference in the distribution fixations. The Figure 5 presents the difference between these two groups for Scenario 1 and Scenario 3. It can be seen that subjects performing safely have on average higher number of shorter fixations, whereby subjects causing an accident had less fixations that lasted longer. Such subjects were scanning the scene relatively slow. The slower scanning may indicate higher processing time or inattentive cognitive state. In that way the mental model of the scenario is not refreshed often enough. This can be an explanation of why elderly drivers have problems at intersections.

The Figure 5(b) shows that in the third scenario both groups showed a more narrow distribution of fixation durations, which is the consequence of the high number of objects in the scene. Accordingly, the number of fixations per second increased from 1,9 in the first to 2,3 in the third scenario, showing higher degree of attention. Still, this attention was often improperly distributed. Even though not visible from the Figure 5 (b), in a contrast to the first and second scenario, in the third scenario two groups of behavior that precedes the collision are distinguished: subjects scanning the scene relatively slow and the ones that scan the scene ineffectively fast. The latter group shows about 50% higher saccade amplitudes and changes of direction of glances than the group not being involved in accident. Such behavior reflects inefficient search strategy where all directions are scanned very fast, probably expecting that in that way more of relevant objects can be perceived. In three of such cases the accident cause was in a "look-but-not-see" error. Hazardous vehicles were fixated for longer than 80ms but the drivers did not

160

perceive them.

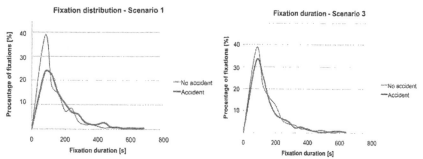

Figure 5. Distribution of fixation durations of test subjects causing an accident and those not causing an accident in (a) Scenario 1 and (b) Scenario 3.

Figure 6 presents the distribution of attention in Scenario 1. As it can be seen the ratio of irrelevant glances is very high. The reason is that glances during the red light phase are also included in the analysis and they mainly contribute to the irrelevant glances. In other two scenarios, the ratio of irrelevant glances is, even though lower, still relatively high: 25% on average. From Figure 6 can be also seen that high number of glances belongs to the planed trajectory. These glances belong mainly to the currently driven segment (up to 1s in front of the vehicle). It can also be seen that drivers involved in an accident had slightly higher attention to the relevant objects than the other group. The similar behavior is observed in other two scenarios as well (see Figure 7 and 8). In general, the highest percentage of irrelevant glances and by that the least attention is present in the phase of leaving the intersection. Hereby, the drivers expect to peripherally capture the vulnerable traffic participants and irregular behavior of other road users. Several drivers endangered the crossing pedestrian in the last segment in the third scenario because of not paying attention to this segment. Conclusively, the main difference between the group of drivers causing an accident in the first scenario and those not causing one is in the number of glances in the direction of the hazardous vehicle. The first group performed one additional glance in all directions just before committing the turn whereby the other group left this glance out. Such visual behavior is observed in other scenarios as well and presents a very dangerous strategy that can be counteracted by an assistance system.

The distribution of attention of both groups of participants, those who committed an erroneous behavior and those who performed the maneuver safely, in Scenario 2 is given in Figure 7. Again, it can be seen that the group of drivers committed an accident had relatively higher ratio of irrelevant glances and glances to the driving path. Also, fewer glances have been given to the crossing lanes both left and right. It can be seen that in both scenarios very low ratio of glances is in the direction of vehicles not having right of way. In the first scenario just 7% (SD=2) and in the second just 15% (SD=3) of glances is in the direction of traffic participants not having right of way. Subjects rely to the high extent on the rule-compliant behavior of others. Increment in the second scenario may indicate uncertainty of regulation rules at an unordered intersection.

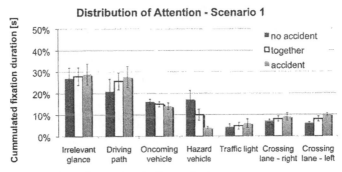

Figure 6. Distribution of attention in Scenario 1.

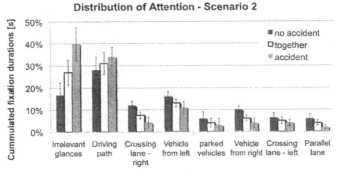

Figure 7. Distribution of attention in Scenario 2.

The distribution of attention in Scenario 3 is presented in Figure 8. The number of glances to the further driving path decreased to 18% (SD=3) and the attention is equally shared among roads in all directions. As already mentioned, the main reason for accidents was either the very fast scanning characterized by short glances or the very slow scanning of the situation with just one glance in each direction. In such a strategy the glances just before the turn were missing. Even though participants in general gave more attention to the road and irrelevant glances decreased for 10%, the drivers had a problem to prioritize the information and very often the sequence of glances was inappropriate.

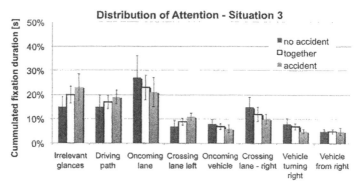

Figure 8. Distribution of attention in Scenario 3.

CONCLUSIONS

The reported study analyzed driver's visual behavior in three left-turn intersection scenarios. The visual behavior showed a specific pattern for each intersection and the applied visual strategies offer a high potential for the support. In the 'simple' scenarios the most problematic are ineffective or insufficient searching strategies. In the third scenario drivers have additional problem to prioritize the directions to scan. To keep in mind the regulations rules continuously and to search simultaneously for numerous potential hazards seem to occupy driver's cognitive capacities to a very high extent. This fact is also confirmed by analysis of subjective data. A solution could be an intersection assistance that is visually informing the driver in critical moment in which direction there are vehicles with right of way.

REFERENCES

Baddeley, A., Eysenck, M., and Anderson, M .(2009), Memory. Psychology Press, Taylor and Francis Group

Ergoneers. *Dikablis - User Manual*. Technical report, Lehrstuhl für Ergonomie, Technische Universität München

Intersafe 2 (2009), User needs and operational requirements for a cooperative intersection safety system, European Union project

Langham, M. (2006) "What do drivers do at junctions?" In 71st Road Safety Congress.

Plavsic, M. Klinker, G. and Bubb, H. (2009). S*ituation Awareness Assessment in Critical Driving Situations at Intersections by Task and Human Error Analysis*. Human Factors and Ergonomics in Manufacturing (in publishing)

Schweigert, M. (2003) Fahrerblickverhalten und Nebenaufgaben. Dissertation, Lehrstuhl für Ergnomie, Technische Universität München

Summala, H., Pasanen, E., Räsänen, M.and Sievänen, J.(1996) Bicycle accidents and drivers' visual search at left and right turns. Accident Analysis & Prevention, 28(2):147 – 153,

SILAB. http://www.wivw.de/ProdukteDienstleistungen/SILAB/index.php.de.

Chapter 19

Contact Analog Information in the Head-Up Display – How Much Information Supports the Driver?

Boris Israel,[1] Maria Seitz,[1] Heiner Bubb,[1] Bernhard Senner[2]

[1]Technische Universität München
Faculty of Mechanical Engineering
Boltzmannstraße 15, 85747 Garching
Germany

[2]AUDI AG
Ingolstadt
Germany

ABSTRACT

New generation cars have an amount of Advanced Driver Assistance Systems (ADAS) which disburden the driver from routine tasks but need to be supervision. To avoid gaze movement while supervisioning these systems more and more car manufactures offer Head-up Displays. The next step is a contact analog Head-up Displays that enable augmented reality displays for the driver and have a great potential for better situation and system awareness. On the other hand there is a major problem of drivers' distraction with such displays in his field of view. This paper introduces an experiment which evaluates augmented reality display concepts for the ACC as an example for ADAS. The experiment shows, that these displays have no impact on the participant's driving performance and lower the participant's workload.

INTRODUCTION

The increasing complexity of traffic requires an elevated driver's attention. Over the last years, this progress led to the development of a broad range of ADAS to disburden the driver from routine tasks. Moreover, these systems aim to increase the drivers' attention to traffic while making routine tasks most convenient. Otherwise it seems that this development enlarges the drivers' duties from the actual operation of the car to the concentration on and evaluation of the traffic and the assistance systems. However, the handling and control of ADAS requires an appreciable amount of the drivers' attention and will shift this from the street to the displays in the cockpit.

To present the assistance system's information to the driver without distraction, car manufactures started to offer Head-up Displays (HUD). With a HUD, information is mirrored on the windshield into the driver's field of view. The driver can recognize the displayed information with a minimum of visual demand and accommodation. HUD technology is not yet fully developed, since so far, as with the instrument cluster, it is only possible to display static information. Therefore, a contact analog HUD was developed (Schneid 2008, Bubb 1986), which displays the HUD's virtual image on the lane. This technology displays virtual information in combination with the real scenery, also known as Augmented Reality. This allows to minimize the visual demand and to reduce the cognitive load to process this information. Also the drivers' attention can guide to the indications overlaid with the reality to keep him in the loop of supervision.

THE CONTACTANALOG HEAD-UP DISPLAY

The contact analog HUD used in this experiment is based on the design of Schneid (2008). It is implemented in an experimental vehicle and allows to display the information in correlation with the street surface in front of the vehicle. This effect is implemented with a tilted image source in the HUD. With such an image source the virtual image of the HUD is tilted also. So the virtual image can be projected horizontally. A translation of the image source moves the virtual image onto the street surface. The design of Schneid has a curved virtual image to overlap the view on the street up to the horizon (FIGURE 1).

But there are also other ways to develop a contact analog HUD. Bergmeier (2008) designs a contact analog HUD by moving a vertical virtual image up to 50m. Thereby he minimized the binocular cues. The overlaid virtual image and the monocular cues enable to display the image into reality. Nakamura (2004) shows a contact analog HUD with stereo vision. It includes two HUDs side-by-side that create a separate image for each eye. Thus, this design uses only binocular cues and an overlaid virtual image.

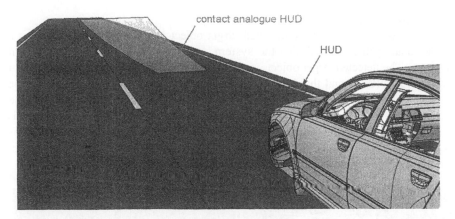

FIGURE 1: Virtual image plane of a Head-up Display and the contact analog Head-up Display

This technology enables an entirely new possibility to display functions of navigation systems, safety distance and current and future Advanced Driver Assistance Systems. This includes the opportunity to mark objects, cars and pedestrians.

DISPLAYCONCEPTS

Contact analog displays in the primary field of view supposedly reduce the driver's attention and lead to tunnel vision and cognitive capture. Distraction may be caused by a moving image in the field of view. However, this moving image can also stimulate drivers' attention during monotony supervision tasks with interesting display content. This stimulation can create a joy of use and can hereby keep the driver in the loop of supervision.

To find the optimal equilibrium between distraction and stimulation in the display, a field test was set up. Two different ACC display concepts for the contact analog HUD was developed and implemented. The first concept (ACC-low) aims, with a reduced and simple display, to reduce the drivers' distraction. In contrast, the second concept (ACC-high) strives to stimulate the driver with interesting display content to increase his/her attention for the supervision of the ACC-system.

In a pretest, participants with ACC experience were asked, how they are using the ACC icons in the instrument cluster. The test should give an indication of the minimum display requirements for the ACC-low concept. The most important were:
- set speed
- preceding vehicle detected
- stand by
- take over request

The more unimportant indications were:

- following distance
- criticality indicator

Therefore ACC-low shows the detected target object and the takeover request as a contact analog display, and also the system activity and desired speed as digital indicator. The detected target object was marked with a green brace at the bottom of the car (FIGURE 2) and this brace turned into a red bar at take over request. The desired speed is shown digital with the brace symbol to avoid confusion with the speedometer. The only dynamic symbol in this display concept is the brace, so it is less distracting but with benefits of better situation awareness, as the detected car is clearly marked. Hence, with this setup misdetection can be identified, system performance can be better understood and system awareness can be improved.

FIGURE 2: ACC-low (left) and ACC-high (right)

The ACC-high concept is meant to stimulate the driver with interesting indication, animations and information about the condition of the ACC. With this stimulation the driver gets joy of watching the display and hereby supervises the assistance system. The animations of this concept are linked to the kinetic perception and provide an optical feedback. Additionally to the green brace, this concept shows the criticality of the ACC-condition, illustrated by a red-green gradient. The driver receives the feedback that the car initiates the braking process as soon as the first part of the area turns red. The desired speed, shown as digital indicator, is animated to a bigger font size when the car changes from follow to free mode and accelerates to the desired speed. With this animation the driver recalls the initial set speed. In this use case the driver's system awareness is increased.

EXPERIMENT

The evaluation of the two display concepts and, as a reference, the display in the instrument cluster is conducted in a field test. The car was an Audi A8 with the described contact analog HUD. The subjects were not impaired by the installed equipment. All 32 participants (average age 37; range 23-57; ♀4, ♂28) have ACC experience, meaning that they have an ACC in their private or company car and use it frequently. 44% of the participants also have HUD experience, meaning they have used it at least once before. Every participant used each display type (i.e. instrument cluster vs. ACC-low vs. ACC-high).

To provoke a higher number of system failure situations and take over requests in the short testing time, the ACC sensors in the experimental vehicle were misaligned by the experimenter. This was told to the participants at the beginning and they had a 10 min drive to adapt. Moreover, the participants had to follow a preceding car to guarantee an ACC target object for the entire experiment.

The experiment took place on a highway part and a country road part and took about two hours. The highway part was selected to show the system and display behavior during high traffic conditions. The misaligned sensor caused misdetections, which normally happen infrequently. The country road part, with low traffic, was used to provoke the Take Over Requests (TOR) through determined braking maneuvers by the preceding car. These exceptional situations should show the different driver behaviors when they were distracted or stimulated by the display. To reduce danger, the experimenter had a braking pedal on the passenger side, but it had never been used during this experiment.

After each display concept, the participants evaluated the displays by means of the NASA TLX questionnaire for workload and with the AttracDiff2 (Hassenzahl 2003) questionnaire for pragmatic quality, hedonistic quality (stimulation, identity), and attractiveness. They also ranked the concepts for distraction, sense of safety, and availability of information. The objective data was the steering wheel standard deviation as driving performance. The ACC-criticality and the braking pressure at drivers take over are being used as a criterion to measure driver's behavior. The experiment is meant to verify if participants are affected by different contact analog displays and if more or less information can influence them (ACC-high vs. ACC-low). The ACC-criticality is a system variable of the ACC-controller. It is computed by the ACC-controller from distance, speed and acceleration. The variable of ACC-criticality is form 0-1024 (1024 is Take Over Requests).

RESULTS

For the statistical analysis a significance level of $\alpha=0.05$ is defined. In the illustrations, the significance is marked with arrow bars. No significant difference is marked with bullet bars. No bar means, that there is no statistical evidence.

STIMULATION AND DISTRACTION

ACC-high has a significantly higher outcome in the stimulation than ACC-low and the instrument cluster. ACC-low is also significantly more stimulating than the instrument cluster (FIGURE 3 left). As a result, the design of the display concepts of ACC-high/-low shows the desired effects.

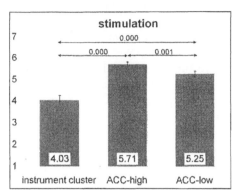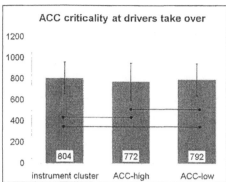

FIGURE 3: Stimulation of the display concepts from the AttrakDiff and the ACC-criticality at drivers take over

The ranking in subjective distraction is ACC-low < ACC-high < instrument cluster (FIGURE 4 left). This effect is supported by the subjective sense of safety and the subjective capacity of reaction (FIGURE 4 right). In both cases the ACC-low and the ACC-high showed significantly better results than the instrument cluster.

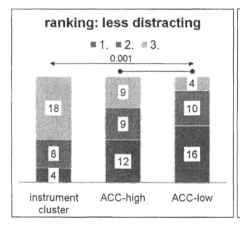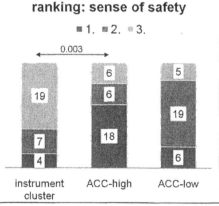

FIGURE 4: The participant's rankings of distraction and sense of safety

The objective driving performance (steering wheel standard deviation) shows no significant difference (FIGURE 5 left). Also the behavior at TOR has no significant difference in all concepts. The behavior at TOR is defined as the ACC-criticality when the participant takes over and the braking pressure at this time. This behavior shows if the participants take over earlier when having more information about the future TOR or if they brake with more force when the TOR happens surprisingly. The shown results indicate that the participants have their own level of risk in which they take over the driving (FIGURE 3 right). The contact analog display, stimulating or less distraction, has no influence to this effect.

WORKLOAD

The NASA TLX shows in every item the same effect: The two contact analog displays have no significant difference and both have significant less workload than the instrument cluster. The big difference in the Overall Workload Index (OWI) is also very interesting: instrument cluster 39, ACC-low 20, and ACC-high 21 (FIGURE 5 right). This experiment setup does not show any evidence that this effect is caused by the contact analog display or by the Head-up Display itself. The big difference in the OWI may result from the HUD where less gaze movement is necessary, but this has to be proven by a different experiment.

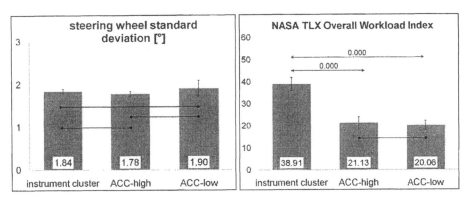

FIGURE 5: NASA TLX Overall Workload Index and steering wheel standard deviation

OVERALL IMPRESSION

The ranking for the overall impression of the display concepts is ACC-high > ACC-low > instrument cluster. The AttrakDiff shows the effect more precisely, in pragmatic quality and attractiveness, the HUD's displays were significantly better than the instrument cluster. The hedonistic quality shows significant differences between all concepts with ACC-high ranked best, followed by AAC-low and the instrument cluster (FIGURE 6).

The analysis of the participants' remarks about the preferred display concept reveals that user groups with different demands can be identified. While older participants and those without HUD-experience would prefer ACC-high because of improved security experience, long-term ACC users are more likely to prefer ACC-low due to less distraction and a better subjective driving performance. For these different user groups with their different preferences, the display design should allow using different configurations.

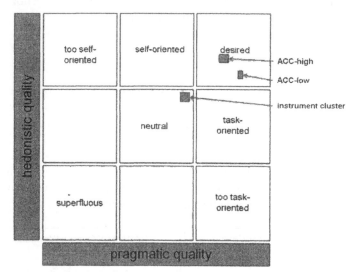

Figure 6: AttrakDiff matrix with medium value and confidence rectangle

CONCLUSION

This leads to the following conclusion: Since neither a lower stimulation nor an increased distraction seem to affect the driving performance negatively but reduce the workload significantly. These results pave the way to stimulate the driver with rich and animated display concepts. New designed augmented reality display concepts can be used to keep the driver in the loop and simultaneously lower the driver's workload. At the same time such display concepts can improve situation and system awareness.

The function of ADAS can be better explained to the driver, gets more attractive and the drivers feel safer and less stressed. As it is yet a long way to autonomic driving, the drivers are still responsible for driving and therefore need fast and supporting information for decision-making.

REFERENCES

Schneid, M. (2008), „Entwicklung und Erprobung eines kontaktanalogen Head-up-Displays im Fahrzeug", *Dissertation Lehrstuhl für Ergonomie TUM.*

Bubb, H., Bolte, U. (1986), „Head-up Display im Kraftfahrzeug", *Professur für Arbeitswissenschaften, Katholische Universität Eichstätt.*

Hassenzahl, M., Burmester, M., & Koller, F. (2003), AttrakDiff: Ein Fragebogen zur Messung wahrgenommener hedonischer und pragmatischer Qualität, in: J.

Ziegler & G. Szwillus (Hrsg.), Mensch & Computer 2003. Interaktion in Bewegung, 187-196, Stuttgart, Leipzig: B.G. Teubner.

Bergmeier, U. (2009) „Kontaktanalog markierendes Nachtsichtsystem Entwicklung und experimentelle Absicherung", *Dissertation Lehrstuhl für Ergonomie TUM.*

Nakamura, K., et al. (2004) „Windshield Display for Intelligent Transport System", *Proceedings of the 11thWorld Congress on Intelligent Transportation Systems, Nagoya, Japan*

Chapter 20

Usability of Approximated 3D Target Objects for Mission Preparation

Michael Kleiber, Carsten Winkelholz

Fraunhofer FKIE
Neuenahrer Straße 20
53343 Wachtberg
Germany

ABSTRACT

Using a 3D scene of the target area during mission planning has the advantage that the aircrew can prepare in a more exploratory manner in contrast to the traditional preparation using aerial images. An important aspect of the 3D scenes is the quality and the detail of the 3D models. Time, costs and risks to acquire 3D models of facilities and environmental features around the target are reciprocal to the detail and quality aimed for. Very basic 3D models can be created by using approximation methods. We evaluated the usability of 3D models with non-photorealistic, approximated appearance for training mission preparation.

Keywords: VR Usability, Spatial Perception, Desktop VR, 3D Objects, Flight Preparation

INTRODUCTION

One part of the preparation for training targeting missions of the German air force is the study of the available information about the broader target area and the target objects itself. This includes maps, pictures and intelligence data in general. In a two seat plane, it is the pilot's task to navigate to the target and it is the weapons officer's task to identify and mark the target. It is therefore imperative that both officers are able to infer from the data, they are provided with, the necessary information to accomplish their mission. Yet, it is not only the kind or amount of data that determines the success of the preparation process it is also the form the data is provided in. It is obvious that a bad presentation or display of the available information will be less effective and helpful in the preparation and the subsequent mission than a good one. Most of the information provided to the aircrew is geospatial in nature.

Exploration and map viewing are two distinct methods for the acquisition of geographical information and they also lead to different cognitive representations (Tversky, 2000, Mark, 1999). It is not clear whether the depiction of 3D scenes with 3D objects, which can provide a view similar to what the aircrew will encounter when approaching the target area, will lead to a better retention of the scene in general and therewith to a quicker and more reliable identification of the target objects (John, 2001). However, the aircrew and aerial image analysts already extract quantitative spatial information in three dimensions from aerial images to better understand a target scene. They use approximation methods such as measuring the lengths of shadows, comparing objects with other objects of known size and analyzing perspective distortions. Integrating these methods into a software application allows the quick creation of basic 3D scenes. It is debatable whether the aircrew itself should be given tools to analyze aerial images: on the one hand it takes additional effort and time but on the other hand it ensures that the aircrew knows the limitations of the approximate 3D models. Yet, a modeled 3D scene also depicts more details and could therefore lead to a concentration of the aircrew on unnecessary features. It might therefore be preferable to only show two dimensional aerial images of the target area (Shah, 2005). Independent of who actually creates the 3D scenes it is important to investigate whether the approximate 3D scenes can be used for mission preparation.

USE OF APPROXIMATED 3D SCENES

Currently, the data used in the preparation of a training targeting mission of the German air force is provided in a so called "electronic target folder" (ETF). The ETF basically consists of a set of interlinked electronic documents. Most of the documents are pictures or 2D maps of the target area. For the actual target and objects in the vicinity additional textual information is usually provided as well. In rare cases detailed 3D models have been constructed based on photographs and intelligence data. Some information in the ETF is created by aerial image analysts by integrating and aggregating information from different sources e.g. an analysis of the target area and vegetation maps will allow an estimation of the likelihood that target objects are occluded. Analysts also estimate object heights by measuring the lengths of their shadows, comparing them with objects of known size or by using stereoscopic image analysis tools. The process of extracting the dimensions of 3D objects based on existing information actually mimics a step in the internal visualization process of the aircrew during the preparation: given a 2D overview map or aerial image and oblique pictures of the target objects the aircrew tries to picture their actual approach by making use of the depth cues contained in the 2D image. The aerial image analysts can use the extracted dimensions to create 3D models of the objects at the target site. The advantage of using 3D models compared to the internal visualization is that the spatial relations in the created 3D scene will be correctly scaled and will not contain errors because of false mental rotations (McNamara, 2003). This also means that occlusions between objects will be displayed correctly.

Although the aforementioned techniques are only approximations they allow the quick creation of 3D scenes. In effect one always needs to decide between accuracy (details) and cost (time) in the creation of 3D models. By adopting the techniques which are used for interpreting aerial imagery the users are supported in their conventional method of operation. The results can not compete with detailed models created by 3D artists but can be used as approximations for such things as line of sight judgments, silhouette recognition or for the general mission preparation of pilots who need to quickly reorient themselves.

Figure 1. Creation of different approximate 3D objects based on the shadow length.

Although, it is tempting, from an aesthetic viewpoint, to enhance or improve the created 3D objects by using fake textures for the parts which are not covered by the aerial image it would lead to a false presentation of the 3D scene. The depiction of vegetation is problematic because of season changes and the subsequent drastic changes in appearance. We therefore decided to use only very simple shapes as shown in Figure 1 until a more viable approach is found. However, the usefulness of the display of the vegetation, although having a false color, for judging occlusions outweighs the potential interference. Furthermore, after the experimental validation, some of the pilots reported that the accentuation of the shapes of the vegetation areas conforms to their mental model since they memorize characteristic shapes of vegetation areas as landmarks to orient themselves. Nonetheless, these aspects might lead to a worse retention and recognition of target objects. Also, a modeled 3D scene depicts more details and could therefore lead to a concentration of the aircrew on unnecessary features. It is therefore imaginable that a 2D map, by filtering out unimportant details, emphasizes the landmarks and spatial relations which are used by the aircrew in the creation of a cognitive map of the target area (Klippel, 2005). A usability evaluation therefore is mandatory before recommending the use of approximate 3D scenes during mission preparation.

EMPIRICAL EVALUATION

In order to assess the developed system as a whole as well as the usefulness of the created 3D scenes a set of usability studies was carried out. In this paper we report on the usability of the approximated 3D target objects. We evaluated the system using objective and subjective measures.

HYPOTHESES

The general hypothesis for the empirical investigation was: Preparing for a flight mission using an approximate 3D scene will lead to a better target object identification. For the hypothesis we defined better to either mean more accurate (H_1) or quicker (H_2). A true field test in which one group prepares using the approximate 3D scenes and one using the traditional material and then actually carries out the mission was not an option because of the required number trials, the interference from non controllable factors and the associated cost. We therefore decided to use oblique photographs of the target areas to simulate the actual approach.

APPARATUS

When given an interactive 3D visualization of a target scene the aircrew can perceive spatial relations in three dimensions and therefore build a cognitive representation based on a preferred viewpoint (Roskos-Ewoldsen, 1998). Good spatial perception of the 3D scene is a precondition for building an accurate cognitive spatial representation. The 3D scene should therefore be presented using a 3D display. Yet, what constitutes a 3D display is ambiguous. Usually textured perspective renderings are considered to be three dimensional. However, through natural binocular sight, vergence and disparity provide important depth cues. Stereoscopic displays are required to provide these cues. Glassless stereoscopic displays, also called autostereoscopic displays, even allow an unobtrusive integration into the traditional workplace. By augmenting the preparation process with an interactive inspection of the target area on an autostereoscopic display we hoped to enhance the creation and retention of the cognitive representations.

The augmented workplace for the aircrew therefore consisted of a traditional 2D display and additionally of an autostereoscopic 3D display. We used an autostereoscopic display from SeeReal Technologies (Cn 3D Display Technology) which tracks the user and employs vertical lenticulars (Schwerdtner, 1998). The basis of the 3D visualization software was a VR prototyping system (Winkelholz, 2003) which allowed constantly adjusting the stereoscopic projection parameters like location of image plane, stereo base, and camera apex angle (Kleiber, 2008). Adapting these parameters continuously was necessary in order to ensure comfortable viewing of the stereoscopic images (Lambooij, 2007).

Figure 2. (left) The targeting widget and the text were stereoscopically in front of the rest of the scene. (right) The placement of widgets and overlaid information was limited to the upper half of the screen because of the stereoscopic visualization. The inset image shows how the scene is perceived stereoscopically with the bar representing the screen and projection plane.

Navigation and orientation was done using a SpaceMouse with six degrees of freedom. The navigation technique was designed so that the pilots could on the one hand control the parameters of interest and on the other hand always had an adequate stereoscopic presentation (for details see Kleiber, 2009). The standard 2D-PC mouse was only used for the selection and activation of objects in the scene. Additional information about a selected object was shown and a 3D targeting widget was displayed (see Figure 2). The widget provides an additional exocentric viewpoint which supports the pilots in judging their current orientation and heading towards the target (for a review of viewpoints and their implications see Wickens, 2005).

The material to be viewed on the stereoscopic system was generated using an application which is based on the same VR prototyping system. However, the authoring application is designed for use on a monoscopic display. The basis of the scenes to be visualized is formed by the terrain in a certain radius around the target area. Based on an aerial image of the target area objects can be extruded by first sketching their contour and then fitting a line along their shadow (see Figure 1). The height of the building is then inferred from the length of the shadow and the inclination of the sun. The sides of the extruded objects are textured with a regular grid texture to allow the viewer to judge or count its approximate height.

SUBJECTS

Overall, there were 28 pilots belonging to three different groups which took part in the experimental usability assessment. The experiment was carried out at the three locations where the pilots are stationed. The subjects of one group were retired pilots. Nonetheless, these pilots still have to successfully complete a number of flight hours per year. The pilots differed by age and number of flying hours. We did not test the visual acuity or the ability for stereoscopic vision of the pilots as they have to pass this examination regularly in order to be allowed to fly.

PROCEDURE

The pilots were seated in front of the autostereoscopic display at an approximate distance of 60 cm. The user centered projection parameters of the visualization system were set accordingly. This means that perspective distortions are introduced whenever the viewing position differs to the fixed position. However, we decided that the minor distortions introduced through seat posture changes did not warrant the use of a chin rest and subsequent discomfort.

We first showed the subjects a 3D test scene on the autostereoscopic display to acquaint them with the use of the application and the stereoscopic visualization. Since the parameters of the stereoscopic projection were adapted constantly, dependent on the visible depth range, the subjects were able to perceive the presented 3D scene without ghosting artifacts or visual discomfort. After explaining the procedure of the experiment, the subjects completed a test run with a scene which was not used in the subsequent trials. Since we were interested in the usability of the approximated 3D target objects we only varied their visibility as the only parameter. Therefore, the scenes were presented on the autostereoscopic display as 3D renderings for both conditions.

The procedure of the actual test consisted of two parts: first the subjects were shown a target scene in which two or three objects were labeled with different target indicators which had to be remembered; secondly the subjects were shown oblique photographs of the target area in which a single object was marked with a target indicator (see Figure 3). The subjects' task was to respond as quickly and accurately as possible whether the combination of target indicator and target was present in the scene during the preparation. About two thirds of all shown oblique photographs were false combinations, this means that either a wrong object was indicated or a wrong symbol was used. We randomized the order of the photographs across subjects to prohibit effects of sequence. This method of evaluation only allowed insight into one aspect of mission preparation namely of target object retention. Therefore a subjective evaluation was also carried out to find out what potential advantages or drawbacks a mission preparation using approximate 3D target objects entails.

Figure 3. (left) In the first phase of the experiment a scene was shown with two or three target indicators. (right) In the second phase multiple oblique photographs with one target indicator had to be judged.

The subjects were shown the traditional and the augmented scenes alternately. Since the scenes differed in the number of objects and in the similarity of objects only an assessment between subjects was possible. We recorded response times as well as accuracy of responses.

RESULTS

The results of the evaluation showed very large differences in error rate and reaction times between users and also between the different 3D scenes. Nonetheless, the results are distributed normally so that we were able to conduct analyses of variance (ANOVA). For the analyses of the correctness we differentiated between true and false oblique scenes. This means there were four possible combinations of answer and scene: true/false answer for true/false scene. The rate of correctness therefore includes the percentage of true images which were marked as true, as well as the percentage of false images which were marked as false. The reaction time was measured from the moment the oblique image was shown till the point the user clicked a mouse button to indicate his answer.

We used seven different scenes and overall recorded about 1500 answers. The scenes differed in complexity and also in the number of available oblique photographs. Of the 1500 images shown about 1000 were false images and 500 were true images. Half of the answers belonged to scenes which included the approximate 3D objects.

We first compared the three groups of pilots to assess whether the groups can be considered to be homogeneous. An analysis of variance with the factors target model type (?), query type (2), scene group (2) and pilot group (3) concerning the correctness of the answers given did only show weak significance for the correctness ($F_{2,88}$=3.26 with p<0.05). A pair wise comparison between the groups showed only weak significance between two groups. However, an analysis concerning the response time showed strong significance ($F_{2,88}$=9.25 with p<0.001). A pair wise comparison supported this and indicates that the group of retired pilots (M_r=4.7s, SE_r=0.2) was significantly slower as either of the other groups (M_1=3.5s, SE_1=0.2; M_2=3.7s, SE_2=0.2). For the following analyses we pooled the data of the active pilots into one group.

Since the scenes differed highly in the number and complexity of objects we considered this in the design of the evaluation. It is therefore not surprising that an analysis of variance of the response time considering scene (7), target model type (2), query type (2) and pilot group (3) as factors showed strong significance for the factor scene ($F_{6,1418}$=8.72 with p<0.001). The scenes also differed in the number of available oblique photographs so that an analysis of variance concerning the correctness was not sensible.

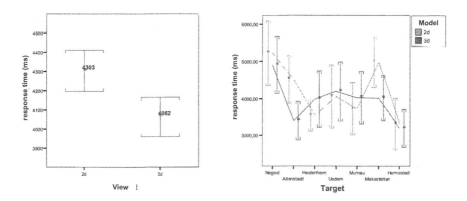

Figure 4. (left) Mean response time for all scenes as a function of view type. (right) Response time for each target as a function of view type.

An analysis of variance concerning the response time for the factors scene (7), query type (2) and model type (2) did not show significance ($F_{1,372}$=1.75 with p=0.08). However, analyzing the response times per scene and model type shows that for the demanding scenes, with a lot of similar objects, the response times for the trials with approximated 3D-models were up to one second faster than for the trials where only the pure terrain model textured with an aerial image was used for mission preparation (Figure 4 right).

CONCLUSION

Overall, the results show that the approximated 3D scenes do not confuse pilots in the target object retention task. The overall mean response times were actually even faster when the pilots inspected the target area with the added approximated 3D objects. Although this difference was not significant, the fact that for more complex target areas an effect was indeed visible supports the hypothesis that preparing for a flight mission using approximate 3D objects will lead to a better target object identification. It might therefore be reasonable to include the approximated objects into the scene especially since the subjective ratings of the stereoscopic display system and the visualization application overall were rather high. The pilots reported that especially the ability to clearly discern the silhouettes of the objects was very helpful. But it must be emphasized, that the usefulness of approximated 3D objects might strongly depend on the task they are used for. The 3D objects seem to be useful for the pilots to orient themselves, but for attack planning the quality criteria of the models probably needs to be higher. Yet, if no exact 3D data is available for attack planning, approximated 3D objects in the target area might still be useful provided that the pilot is able to asses the accuracy of the approximated 3D objects. This means that he needs to be able to consider the uncertainty within the data in his planning process. Presenting the uncertainty of the data in such a way that the pilot can integrate it during his internal visualization process is therefore one direction for future work.

REFERENCES

John, M.S., Smallman, H.S., Bank, T.E., Cowen, M.B. (2001). Tactical Routing Using Two-Dimensional and Three-Dimensional Views of Terrain. Human Factors and Ergonomics Society Annual Meeting Proceedings. Vol. 45, pp. 1409-1413

Lambooij, M. (2007). Stereoscopic displays and visual comfort: a review. SPIE Newsroom: http://spie.org/x14604.xml

Kleiber, M., Winkelholz, C. (2008). Distortion of depth perception in virtual environments using stereoscopic displays: quantitative assessment and corrective measures, in proceedings of the conference Stereoscopic Displays and Applications XIX. pp. 68030C

Kleiber, M., Winkelholz, C. (2009). Case study: using a stereoscopic display for mission planning, in proceedings of the conference Stereoscopic Displays and Applications XX, pp. 723704

Klippel, A., Winter, S. (2005). Structural salience of landmarks for route directions. Spatial Information Theory. Lecture Notes in Computer Science Volume 3693. pp. 347-362

Mark, D.M., Freksa, C., Hirtle, S.C., Lloyd, R., Tversky, B. (1999). Cognitive models of geographical space. International Journal of Geographical Information Science Volume 13 No. 8. pp. 747-774

McNamara, T.P. (2003). How are the locations of objects in the environment represented in memory? Lecture notes in computer science. pp. 174–191

Roskos-Ewoldsen, B., McNamara, T.P., Shelton, A.L., Carr, W. (1998). Mental representations of large and small spatial layouts are orientation dependent. Journal of Experimental Psychology – Learning, Memory and Cognition Volume 24 No. 1. pp. 215-226

Schwerdtner, A., Heidrich, H. (1998). Dresden 3D display (D4D), in proceedings of the conference Stereoscopic Displays and Virtual Reality Systems V. pp. 203-210

Shah, P., Freedman, E.G., Vekiri, I. (2005). The Comprehension of Quantitative Information in Graphical Displays, in The Cambridge handbook of visuospatial thinking. Cambridge University Press. pp. 426-476

Tversky, B. (2000). Remembering Spaces. Handbook of memory. Oxford University Press. New York. pp. 363-378

Wickens, C.D., Vincow, M., Yeh, M. (2005). Design Applications of Visual Spatial Thinking, in The Cambridge handbook of visuospatial thinking. Cambridge University Press. pp 383-425

Winkelholz, C., Alexander, T., Weiß, M. (2003). Open activeWrl: a middleware based software development toolkit for immersive VE systems, in proceedings of the workshop on Virtual environments. pp. 321-322

CHAPTER 21

Compensatory Tracking of a 3D Object Using 3D Technology

Jennie J. Gallimore[1], George Reis[2], Jeremy Warren[3], Mary Fendley[3], Julie Naga[3]

[1] HumanWise, Inc.

[2] 711th HPW/RHCV

[3] Booz Allen Hamilton

ABSTRACT

3D technologies might be beneficial for situation awareness of space operations given the 3D nature of the problem. There is little or no research on the effective use of 3D technologies for space operations, an area in which the visualization and interactions must consider the large volume of space across the different orbits (Low, Medium, and High Earth Orbits). The purpose of this study was to investigate human performance when using 3D input devices and 3D displays for a complex six degree of freedom (DOF) tracking task. Subjects were asked to track a 3D object that was orbiting in six DOF (translation and rotation) around the earth. The independent variables included display type and number of hands controlling the input devices. The displays were both 3D technologies. The input device was six DOF Connexion mouse. The results indicated that subjects could not perform simultaneous rotation and translation for the six DOF tracking task. A 3D display with passive glasses was preferred to the autostereoscopic Phillips 3D display but performance differences were negligible.

Keywords 3D displays, stereoscopic displays, 3D mouse, six degrees-of-freedom input devices, compensatory tracking.

INTRODUCTION

Optimal presentation of information to the space operators is crucial to the success of space-based operations because of the need to solve complex problems in fast-paced, information-dense, team environments. Three-dimensional (3D) visualization technology may provide dramatically improved situation awareness to our warfighter decision-makers. However, the value of this technology must be assessed and the development of the technology must be guided. To this end, we present a study investigating the effectiveness of six degree of freedom (DOF) input devices, 3D stereo displays, and number of hands controlling the input devices for a path tracing task for satellite operations. In order to make a useful comparison, we conducted an in-depth literature review of studies related to 3D technologies. To effectively design information on 3D displays, designers must have an understanding of the issues and solutions to possible problems. The investigation of presentation of 3D information on 3D displays requires consideration of how users will interact with the visualization. Users have specific goals with defined tasks in order to reach those goals. It is not unlikely that the way they interact affects how they reach their goals, and therefore can directly impact their perceptions of the 3D information. In other words, interacting with purpose may affect perceptions and mission accomplishment.

To date there is limited published research related to human performance in space operations. Research related to the use of 3D input devices and displays can provide structure for developing guidelines that carry over to the space domain. The use of 3D input devices with 3D environments can be difficult for the user and there has yet to be defined a "best" device for specific tasks.

3D INPUT DEVICES

Advances in 3D visualization and virtual environments have lead to the design of 3D input devices, which are now readily available on the market. However most studies involve the use of custom made research devices, and have not yet taken advantage of the commercial off the shelf 3D input devices that are now available.

Twenty different studies were identified that investigated the use of six DOF devices. From these studies, eight distinct tasks were used to measure performance in 3D environments. These performance tasks included point location and target selection, target docking, positioning, path tracing, position slice, psychophysical method of adjustment task for aligning the virtual pointer with designated targets, drag and drop, and grasping with hands (Kim, Ellis, Tyler, Hannaford, & Stark, 1987; McKenna, 1992; Zhai, 1993; Zhai & Milgram, 1994, 1998; Tachi, 1994; Ware, 1995; Zhai, Buxton, & Milgram, 1996; Boritz & Booth, 1997; Balakrishnan & Kurtenbach, 1999; Lindeman, Sibert, & Hahn, 1999; Ware & Rose, 1999; Hou, 2001; Hachet et al., 2003; Chittaro & Burigat, 2004; Fujimoto & Ishibashi, 2004; van Rhijn & Mulder, 2006). Currently there is not a good understanding of how best to designate the use of one vs. two hands on a given 3D input task. Balakrishnan &

Kurtenbach (1999) completed a study on one hand vs. two hand use of two integrated control tasks, raising the issue of how best to separate tasks for two-handed manipulation, and how well this works for 3D input devices versus the 2D mouse. Hinckley et al. (1998) summarize two-handed input devices that had not previously been tested through experimentation. Their work describes the theory behind bimanual manipulation and discusses two experiments to test the ideas. Other work by Lindeman, Sibert, & Hahn (1999) tested input devices using two hands; however, these devices were for 2D movements. In a series of four experiments, Ware & Rose (1999) examined how to determine the best way and time to rotate objects; however, they found no significant difference between one-hand and two-hands.

Several different dependent measures were used in six DOF performance research, including the total transport, which is defined as the line integral of the four vertices of the cursor tetrahedron (Zhai & Milgram, 1998), effective value, which measures increasing equivalent gain and reducing equivalent time delay (Tachi, 1994), mean score for capturing target (McKenna, 1992), memory of hand locations with closed eyes (Hinckley, Pausch, & Proffitt, 1997), coordination efficiency (Hachet et al., 2003), and subjective ratings of preferences or difficulty (Balakrishnan & Kurtenbach, 1999; Boritz & Booth, 1997; van Rhijn & Mulder, 2006). The most prevalent dependent measure were RMS error (Hou, 2001; Kim et al., 1987; van Rhijn & Mulder, 2006; Ware & Rose, 1999) and the length of time to track to an object, point, or rotate(Boritz & Booth, 1997; Chittaro & Burigat, 2004; Lindeman et al., 1999; Tachi, 1994; van Rhijn & Mulder, 2006; Ware & Rose, 1999; S. Zhai & Milgram, 1993; S. Zhai, Milgram, & Buxton, 1996). The use of RMS error is calculated for each axis separately, providing an indication of difficulty and dynamics of the input device by axis but does not tell much about coordinated movement.

3D DISPLAYS

There are a many studies investigating the effectiveness of stereoscopic displays. Performance with stereoscopic displays depends on the type of task. (Barfield & Rosenberg, 1995; Hendrix & Barfield, 1997; Hubona & Shirah, 2005; Lo & Chalmers, 2003; Sollenberger & Milgram, 1993; Todd & Norman, 2003; Van Orden & Broyles, 2000) The purpose of this study was not to compare stereoscopic with 2D displays, but to evaluate performance differences for two 3D display types; a projection stereoscopic display (Visbox) with passive glasses versus a Philips plasma autostereoscopic display (WOWvx).

OBJECTIVE

The objective of this study was to investigate how well novice users can track a 3D objective rotating and translating in 3D space using a six DOF input device and two types of 3D displays.

METHOD

The experiment design is a 2 x 2 within-subject full factorial design. The independent variables are 3D display type (VisBox, Philips WOWvx), and number of hands controlling the input devices (one or two). Each condition was repeated eight times after two practice trials. Each trial lasted 45 seconds. The dependent variables are RMS error for each axis.

SUBJECTS

Twelve people (11 men and 1 woman) from Wright Patterson Air Force Base (WPAFB) and the Air Force Institute of Technology (AFIT) volunteered to participate as subjects across two days. There was no compensation. The inclusion criterion was the ability to fuse stereo images using a TNO stereoscopic vision test. All subjects were novice users with respect to the 3D input devices.

APPARATUS

Two different 3D display systems were used. The VisBox™ Mini (by Visbox, Inc) includes two projectors mounted as a front projection system. The Visbox™ requires users to wear passive filter glasses by Infitec to see stereo images. The second display is the Philips 3D WOWvx, a 52-inch Plasma Display using slanted multi-view lenticular lens technology. This display is autostereoscopic; therefore, it does not require glasses.

The input devices were the SpaceNavigator and the SpaceTraveler by 3DConnexion (See Figure 1.1). Both devices are six DOF allowing simultaneous input of 3 translation movements (X, Y, Z) and three rotation movements (Roll, Heading, Pitch). The SpaceTraveler is slightly smaller than the SpaceNavigator but works the same way.

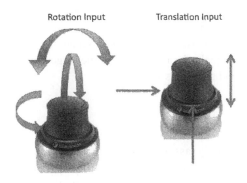

Figure 1.1 Connexion 3D mouse.

Several parameters can be set for the input devices including step size (how far the cursor travels given an input) and an input ignore range. A step size of 150 m was set for the translation movements based on trial and error to allow for users to be able to adequately move the cursor with the respect to the speed of the target and to minimize large jumps so that the tracking was continuous. (Real world distances were used). For rotation the step size was 2 degrees per second.

When users attempt to move in a translation direction only (e.g. X axis), a rotation input can also occur, that is, there is an unintentional cross coupling of movement when the translation and rotation ignore ranges are set to the same values. To limit this coupling an ignore range is included for rotation requiring greater input for the cursor to rotate. The ignore range for translation axes were set to 25, and the ignore range for rotation for each axis was set to 30.

A PC computer running the experimental software was attached to each 3D display. The custom task software was developed using OpenGL and C. The units used for distance are meters. The earth is 6378137 meters placed to the left of the screen. For the Visbox, the eye position was set to be 1300 m out of the screen and 100 m vertical up from the satellite symbol. For the WOWvx the eye position was 750 m out of the screen and 18.75 m vertical up from the satellite symbol. The differences between the Visbox and WOWvx were necessary to show the same information on each display given their different sizes.

TASK

Subjects performed a satellite compensatory tracking task that required subjects to keep a 3D tetrahedron cursor in alignment with a tetrahedron satellite target. The target moved in all six DOF and the subject was required to track the target and maintain the same orientation of the target. The target was placed on an orbit around the earth. Up to 10 separate orbits randomly generated for each trial were visible and the target stayed on one of the 10 orbits. The target moved across the screen, but the cursor tetrahedron always remained in the middle of the screen. Therefore, when the cursor was aligned with the target they were always in the middle of the screen. It was possible for the target to move off the screen if the subject did not keep up with it. Each face of the tetrahedron target was a different color so that orientation could be determined. The cursor had a similar color but was transparent so that the subject could tell when the cursor and target were aligned (Zhai,1994). A goal tolerance between the target and cursor was set to 150 meters for translation and 5 degrees for rotation.

PROCEDURE

Each subject was randomly assigned to begin with either the VisBox™ Mini (Visbox) or the Philips WOWvx. They participated in all conditions for the assigned display in one day. They used the second display on a separate day. Handedness was also randomly assigned, starting and completing all trials with one or two hands before switching. Upon completion of trials for each input device type subjects filled out a questionnaire to provide their impressions about the device.

To train subjects in the use of the input devices they were given a practice world to learn to move the tetrahedron cursor with the different input devices. Before each specific condition set subjects were presented with a scene that showed 3D objects placed in a 3D world. They were asked to move the cursor (tetrahedron) through the world to touch the various objects. They were also asked to rotate the tetrahedron. After performing the specific practice they were then allowed to move around in the world until they felt comfortable with the devices for 5 additional minutes. Note, this was not a compensatory tracking task.

Upon completion of the practice, subjects were told that the first two trials of the compensatory tracking task were practice trials, followed by eight experimental trials. They completed the 10 trials then moved to the next condition in which the procedure for practice in the 3D world followed by trials was repeated. Upon completion of all trials, subjects filled out a subjective ranking of the conditions.

RESULTS

Partial data for one subject was lost, so the total number of subjects included in the analysis was eleven. RMS error was calculated in order to examine performance for each axis separately. The Mean RMS Error across subjects was calculated and an ANOVA was conducted for each axis. Subject was treated as a random variable, and the independent variables were within-subject. Results showed that there were no significant differences in RMS Error for Display or number of Hands for translation axes (X,Y,Z).

For rotational movements there is a significant effect of Display for the Heading axes $F_{(1,10)}$) = 11.24, p = .0073. For Pitch and Roll, the effects are nearly significant (p = .0625 and .0760 respectively) and follow the same directional trend as Heading. That is, RMS error is greater when subjects used the WOWvx display (See Figure 1.2).

Results indicated a significant main effect for number of hands for each rotational axes (Heading: $F(1,10)$ =19.88, p=.0012; Pitch: $F(1,10)$ = 17.61, p = .0018; Roll: $F(1,10)$ = 19.42, p = .0014). In all cases the RMS Error was significantly higher when subjects used one input device controlling all six DOF at one time compared to controlling rotation in one hand and translation in the other (See Figure 1.3)

DISCUSSION

The compensatory tracking task was extremely difficult for subjects. Time on target data indicated that for the 45-second trials the average time they stayed on target when using one hand was only 1.2 seconds. Using two hands the average was

188

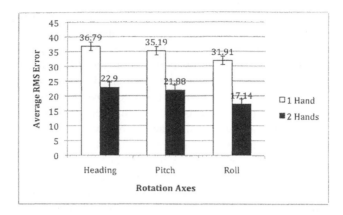

FIGURE 1.2 Average RMS Error by Display type for each rotation axis.

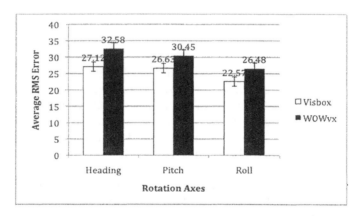

FIGURE 1.3 Average RMS Error by number of hands for each rotation axis.

2.0 seconds. It must be noted that time on target meant that they had to be within 150 meters in all three axes of translation as well as within 5 degrees for each axes of rotation. Subjects tended to focus on moving in translation first and if they had time they would rotate. In the case of one hand, rotation is cross-coupled with translation so there were more movements in rotation but they were primarily accidental even though the device had ignore ranges so that when translating, it would require more input for a rotation to take place. There are no documented studies that have evaluated step sizes, ignore ranges, and gain for these six DOF input devices. We chose one of the most difficult tracking tasks. Positioning tasks appear more easily accomplished. However, it is important to determine the correct settings for these devices if they are to be used properly and also the amount of training needed to become proficient. The results seem to indicate that placing rotation in one hand and translation in the other hand results in lower RMS for rotation. However, it is possible that subjects were not rotating and the error is related to drift from the target being less than the error introduced when trying to track with one hand. To investigate this possibility we examined whether subjects were providing input into each rotation axis. For each trial of 45 seconds there are 900 data points. We determined if there were any differences between each of the 900 values for rotation for every trial. We found that across all subjects, when two hands were used there was little or no input to any rotation axis, and these differences were statistically significant. The average number of zero inputs for the heading axis for 1 hand was 827, and for 2-hands 864. For pitch, 1-hand was 808 and 2-hands was 875. For roll, 1-hand was 825 and 2-hands was 874. Therefore, when separating out the roll and translation, the subjects still were concentrating on translation and making little or no inputs and the error is due to drift away from the target in each axis. The RMS error in rotational axes for one-hand is most likely due to inputs that were accidental while trying to translate given the high number of zero movements in each axis. In general, subjects could not move in all six DOF at the same time.

With respect to the two display types, only heading was statistically significant at 0.05, however, pitch and roll were nearly significant and in general subjects had higher RMS Error with the WOWvx than the Visbox. This is most likely due to the lower resolution of the WOWvx display. In questionnaires and ratings, subjects preferred the Visbox and indicated during the experiment that they did not like the WOWvx display. Although the WOWvx does not require glasses the perception of depth is not as great as the Visbox.

CONCLUSIONS

Overall, performance with the six DOF input device for a compensatory six DOF tracking task is very poor. Research is needed to determine the best settings for this device as a function of task type. The practice task used to familiarize subjects with the input device was a positioning task where they translated to an object or rotated to match a command. Subjects were able to execute this task more easily, but the task did not require movement in all six DOF at once. This study involved complete novice users to determine their ability to use a 3D device with minimal practice. Results indicate they would need much more practice to become sufficient; however, the amount of practice needed is unknown.

REFERENCES

Balakrishnan, R., & Kurtenbach, G. (1999). Exploring bimanual camera control and object manipulation in 3D graphics interfaces. *CHI '99: Proceedings of the SIGCHI Conference on Human Factors in Computing Systems,* Pittsburgh, Pennsylvania, United States. 56-62.

Barfield, W., & Rosenberg, C. (1995). Judgments of azimuth and elevation as a function of monoscopic and binocular depth cues using a perspective display.

Human Factors, 37(1), 173-181.

Boritz, J., & Booth, K. S. (1997). A study of interactive 3D point location in a computer simulated virtual environment. *VRST '97: Proceedings of the ACM Symposium on Virtual Reality Software and Technology,* Lausanne, Switzerland. 181-187.

Chittaro, L., & Burigat, S. (2004). 3D location-pointing as a navigation aid in virtual environments. *AVI '04: Proceedings of the Working Conference on Advanced Visual Interfaces,* Gallipoli, Italy. 267-274.

Fujimoto, M., & Ishibashi, Y. (2004). The effect of stereoscopic viewing of a virtual space on a networked game using haptic media. *ACE '04: Proceedings of the 2004 ACM SIGCHI International Conference on Advances in Computer Entertainment Technology,* Singapore. 317-320.

Hachet, M., Guitton, P., & Reuter, P. (2003). The CAT for efficient 2D and 3D interaction as an alternative to mouse adaptations. *VRST '03: Proceedings of the ACM Symposium on Virtual Reality Software and Technology,* Osaka, Japan. 225-112.

Hendrix, C., & Barfield, W. (1997). Spatial discrimination in three-dimensional displays as a function of computer graphics eyepoint elevation and stereoscopic viewing. *Human Factors, 39*(4), 602-617.

Hinckley, K., Pausch, R., & Proffitt, D. (1997). Attention and visual feedback: The bimanual frame of reference. *SI3D '97: Proceedings of the 1997 Symposium on Interactive 3D Graphics,* Providence, Rhode Island, United States. 121-ff.

Hinckley, K., Pausch, R., Proffitt, D., & Kassell, N. F. (1998). Two-handed virtual manipulation. *ACM Trans.Comput.-Hum.Interact., 5*(3), 260-302.

Hou, M. (2001). User experience with alignment of real and virtual objects in a stereoscopic augmented reality interface. *CASCON '01: Proceedings of the 2001 Conference of the Centre for Advanced Studies on Collaborative Research,* Toronto, Ontario, Canada. 6.

Hubona, G., & Shirah, G. W. (2005). Spatial cues in 3-D visualization. In Y. Cai (Ed.), *Ambient intelligence for scientific discovery* (pp. 104-128) Springer-Verlag Berlin Heidelberg.

Kim, W. S., Ellis, S. R., Tyler, M. E., Hannaford, B., & Stark, L. W. (1987). Quantitative evaluation of perspective and stereoscopic displays in three-axis manual tracking tasks. *IEEE Transactions on Systems Man and Cybernetics, 17*(1), 61.

Lindeman, R. W., Sibert, J. L., & Hahn, J. K. (1999). Towards usable VR: An empirical study of user interfaces for immersive virtual environments. *CHI '99: Proceedings of the SIGCHI Conference on Human Factors in Computing Systems,* Pittsburgh, Pennsylvania, United States. 64-71.

Lo, C. H., & Chalmers, A. (2003). Stereo vision for computer graphics: The effect that stereo vision has on human judgments of visual realism. Paper presented at the *Proceedings of the 19th Spring Conference on Computer Graphics,* Budmerice, Slovakia. 109-117.

McKenna, M. (1992). Interactive viewpoint control and three-dimensional operations. *SI3D '92: Proceedings of the 1992 Symposium on Interactive 3D Graphics,* Cambridge, Massachusetts, United States. 53-56.

Sollenberger, R. L., & Milgram, P. (1993). Effects of stereoscopic and rotational displays in a three-dimensional path-tracing task. *Human Factors, 35*(3), 483-499.

Tachi, S. (1994). Evaluation experiments of a teleexistence manipulation system. *Presence: Teleoperators, 3*(1), 35.

Todd, J. T., & Norman, J. F. (2003). The visual perception of 3-D shape from multiple cues: Are observers capable of perceiving metric structure? *Perception and Psychophysics, 65*(1), 31-47.

Van Orden, K. F., & Broyles, J. W. (2000). Visuospatial task performance as a function of two- and three-dimensional display presentation techniques. *Displays, 21*(1), 17-24.

van Rhijn, A., & Mulder, J. D. (2006). Spatial input device structure and bimanual object manipulation in virtual environments. *VRST '06: Proceedings of the ACM Symposium on Virtual Reality Software and Technology,* Limassol, Cyprus. 51-60.

Ware, C. (1995). Dynamic stereo displays. *CHI '95: Proceedings of the SIGCHI Conference on Human Factors in Computing Systems,* Denver, Colorado, United States. 310-316.

Ware, C., & Rose, J. (1999). Rotating virtual objects with real handles. *ACM Trans.Comput.-Hum.Interact., 6*(2), 162-180.

Zhai, S. (1993). Investigation of feel for 6DOF inputs: Isometric and elastic rate control for manipulation in 3D environments. *Proceedings of the Human Factors and Ergonomics Society 37th Annual Meeting,* 323-327.

Zhai, S., & Milgram, P. (1994). Asymmetrical spatial accuracy in 3D tracking. *Proceedings of the Human Factors and Ergonomics Society 38ʰ Annual Meeting,* 245-249.

Zhai, S., & Milgram, P. (1993). Human performance evaluation of manipulation schemes in virtual environments. *Proceedings of the IEEE Virtual Reality Annual International Symposium (VRAIS),* Seattle, WA, United States. 155-161.

Zhai, S., Buxton, W., & Milgram, P. (1996). The partial-occlusion effect: Utilizing semitransparency in 3D human-computer interaction. *ACM Trans.Comput.-Hum.Interact., 3*(3), 254-284.

Zhai, S., & Milgram, P. (1998). Quantifying coordination in multiple DOF movement and its application to evaluating 6 DOF input devices. *CHI '98: Proceedings of the SIGCHI Conference on Human Factors in Computing Systems,* Los Angeles, California, United States. 320-327.

Zhai, S., Milgram, P., & Buxton, W. (1996). The influence of muscle groups on performance of multiple degree-of-freedom input. *CHI '96: Proceedings of the SIGCHI Conference on Human Factors in Computing Systems,* Vancouver, British Columbia, Canada. 308-315.

Chapter 22

Design and Evaluation Approach for Increasing Compatibility and Performance of Digital Painter in Screen Drawing Tasks

Fong-Gong Wu, Chien-Hsu Chen, Li-Ru Lai

Department of Industrial Design
National Cheng Kung University
1, University Road, Tainan 70101, Taiwan

ABSTRACT

The study takes focus on improving weak points of present digital drawing system such as visual diversion and interruption during the function changing. For this purpose, the digital drawing system with tangible user interface (TUI) is developed. The methodology includes research, investigation, design and modification. As for the final conclusion, the study proves that operating drawing function with tangible user interface makes drawing actions better and more fluent than with graphical user interface. Moreover, it avoids creators' distraction and makes them concentrate on their works.

Keywords: Tangible user interface (TUI), digital pen, sketch, intuition, drawing

INTRODUCTION

The blooming development of computer technology has resulted in the common employment of digital creation in various design projects. Many designers have created high-quality digital works with pen-form digital input devices integrated with the present graphic application software. However, despite the indispensable strength digital illustration have over conventional paper-and-pen illustration, many designers still depend highly upon the conventional illustration methods during their concept development steps in the early stages of the design process. This is solid proof that the present illustration system hinders the process of design concept development. Design is a process of seeing-moving-seeing (Schon, 1992). Designers examine sketches to recognize, dissect, and solve problems (Simon, 1973). They focus their visions on the drawings to evaluate the feasibility of the practical functions and come up with the best solution. Hence, vision is an important part during the designing process. Excessive visual transference may distract designers' attention (Palmer, 1993). The present digital drawing systems focus on visual icon interfaces. During the process of function change, designers are required to break off from the creation process, work on visual search for the function icons, then resume the creation process. This controlling mechanism requires visual searching which distracts and interrupts the thinking process during the function changing process. On top of that, according to Fitts' law, moving distance increases control time, hence icon clicking interfaces decreases illustration efficiency (Fitts, 1954).

In order to create a fluent illustration process, this study employs a physical control interface as the control method. A physical control interface allows tactile and position recognition. With a fine coding compatibility (Norman, 1988), it should be easy for users to catch on and operate with fluency during focused creative processes. On top of that, a physical control interface decreases moving distances, hence operation time. Also, tactile control trains expected operations better than visual icon-clicking control modes, hence increases control speed and accuracy and decreases the data processing amount (Sanders, 1993) in the long run, which is beneficial to the illustration process.

This study introduces a physical control interface as the drawing system control mode in order to achieve fluency, speed, and concentration in the drawing process. The feasibility is also examined and tested using experiments.

WEAKNESSES IN THE PRESENT SYSTEM AND WAYS OF IMPROVEMENT

Present digital illustration employs visual icon interfaces with digital pens. The visual icon interface function changing task requires designers to first apply visual search to recognize the function icon required, move to click, and then back to the original creation process, as shown in figure 1. During the function change process, the visual searching and moving and clicking process often disrupts concentration and the creative process. Hence, physical control interfaces are brought forward in this research, as shown in figure 2, to replace the icon-clicking mechanism, in order to create a creative environment for fluent illustration activities.

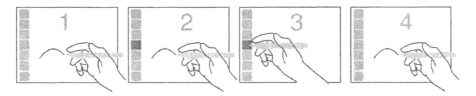

FIGURE 1. GUI drawing system function change procedure requires users to first apply visual search to locate function icons (1~2) then move to click (3~4).

FIGURE 2. Speculating that TUI drawing system may decrease the moving process as well as visual search time.

DESIGN DEVELOPMENT

ILLUSTRATION FUNCTION SETTING

Drawing systems focus on minimizing distractions during a creation process, allowing designers to concentrate during the concept development stage. Hence, the drawing system function settings should focus on allowing the expressing of creation concepts and integrating necessary illustration functions. Lim (2004) employed the observation and questionnaire survey methods to study literary sketch behaviors, procedures, and performances. He suggested that fundamental drawing system prototypes should include reference lines, grid lines, shades, shadows etc. In other words, the elements of line, hue, and shade should be sufficient to form a picture. The illustration process can be dissected into three major stages: 1. Sketching, 2. Coloring, and 3. Refining. Following the objectives tree method (as shown in figure 3), the finalized system functions include six parameters, they are: stroke thickness adjustment, brightness adjustment, hue options, brush tool options, and dropper tool.

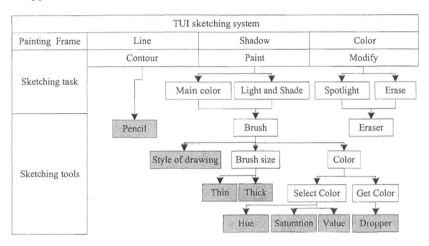

FIGURE 3. Integrating objectives tree method into the illustration function setting of the sketch writing system.

CONTROL COMPATIBILITY AND COMPONENT ARRANGEMENT

The controls of common uses are: buttons, toggle switches, rotation knobs, slide switches, wheel switches, keyboards etc. (Sanders, 1987). The toggle switches and keyboards are omitted from this study as they fall out of the control mode this study is aiming to study. Whereas buttons, rotation knobs, slide switches, and wheel switches are examined for compatibility.

In order to discuss the compatibility of control components and the corresponding illustration functions, designers were recruited as test participants. After describing system goals and functions, participants were given different control component flashcards and were asked to arrange controls according to compatibility to each illustration function. Table 1 shows the arranged data of priorities for illustration function versus controls as a reference for control options.

Table 1. Compatibility priorities of illustration function and controls.

Color group	Dropper	Button>Level slide switch>Vertic. 'on knob
	Value	Vertical slide switch>Level slide s' >Button
	Saturation	Level slide switch>Vertical slide s >Button
Line group	Hue	Rotation knob>Positioning rotat' .1>Button
	Size	Level slide switch>Button>Ve' tation knob
	Brush	Rotation knob>Positioning ro1 :lide switch

In order to increase user recognition speed of controls on a digital pen, the function setting considerations include the following points: 1. Form variation between control buttons to increase recognition. 2. Controls of similar characteristics (color, line) are placed in the same space to enhance coherency. Finally, considering collective arrangement, a prototype of the physical control digital pen is designed, as shown in figure 4.

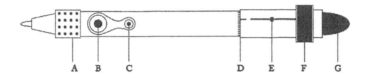

FIGURE 4. Control settings on a physical digital drawing pen.

A.   Rotating brush options: A rotation device is employed to change illustration brushes. Users choose their preferred brushes from the interface, increasing the speed to change brushes as well as avoiding pull-down menus.
B.   Brush thickness tool: Intuitively, the different size of circles represents the change of size. By touching the different buttons, users can adjust brush thicknesses directly.
C.   Brightness adjustment: The 180 degrees rotating button adjusts brightness with a half circle indicating brightness of 0%~100%. Similar to the concept of a sliding track, this allows adjustment of brightness with speed during illustration procedures.
D.   Saturation adjustment: The saturation sliding track adjusts saturation of colors.
E.   Dropper tool: The dropper tool is a commonly used tool in illustration which is difficult to reach in conventional illustration. This tool is designed to allow intuitive operation.
     a. Press on the button to activate the dropper tool function.
     b. Choose colors.
     c. Go back to brush mode.

FEASIBILITY ANALYSIS EXPERIMENT OF PHYSICAL CONTROLS

RECOLLECTION AND OPERATION EXPERIMENT OF PHYSICAL CONTROL INTERFACE ALLOCATIONS

The feasibility experiment consists of two stages, the recollection and the operation experiments. The first stage consists of the recollection experiment of the physical control pen allocation. The aim of the experiment was to examine whether the non-visual physical control device is easy to learn. All participants were first-time users of the device and were asked to state the function of each operating device and to perform correct operations after a brief lesson. Since intuition and recollection hold a positive relationship, the more intuitive an operation is, the easier it is to be memorized and learnt. Intuitive operation products also occupy a less amount of data, which is productive to the fluency of concept development illustration process. The experiment results show that after learning, users were able to memorize and correctly operate various functions with a perfect error rate even under a non-visual situation. This certifies that the control settings were easy to remember and recognize. The second stage consists of an operation experiment aiming at examining the fluency of function change, employing the physical control interface as the treatment group and the visual icon interface as the control group as shown in figure 5. in order to compare operation efficiency and to test whether the control function pair setting matches up to user expectations. During the experiment, participants were asked to complete tasks using the physical pen model and the icon interface respectively. The procedure consists of the participants receiving an task, adjusting to the task function, then draw a square in the designated location in order to model the process of switching between drawing and function changes.

The first part of the experiment consists of single operation tasks aiming at the comparison of performance of each function. The second part of the experiment consists of mixed-function tests aiming at testing for confusion, speed, and fluency during mixed-function changes. Function change during a creation process deeply affects the fluency of the thinking process. If the experiment is able to prove the higher efficiency of the non-visual physical input in comparison with the visual icon interface, then it would solidify the feasibility of the physical control interface drawing system.

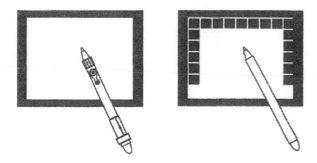

FIGURE 5. Employing the physical control interface as the treatment group and the visual icon interface as the control group.

RESULTS OF THE PHYSICAL CONTROL INTERFACE PERFORMANCE TESTS

Table 2 and Figure 6 show the experiment results. The blue line indicates the results of the icon interface while the red line indicates the results of the physical interface. The vertical axial shows the time consumed for each experiment task, while the horizontal axial A~F show the independent function experiment tasks and the horizontal axial G shows the intermix function integration task. From the experiment results we can see that the consumed time for the experiment tasks on the physical control interface were all less than that of the icon control interface, indicating a decrease of function change time during the illustration process by employing physical control systems. The results conform to the predication made by Fitts' law, that the decrease of moving distance decreases moving time, and hence increases control efficiency (Fitts, 1954). This proves the room for development for physical digital illustration pens.

Table 2. Comparison results with the GUI and TUI sketching system

		GUI	TUI	p
Size	Mean	17.80	9.20	.011*
	SD	9.41	1.75	
Brush	Mean	16.60	10.90	.001*
	SD	5.10	3.07	
Saturation	Mean	18.20	13.00	.000*
	SD	6.79	4.08	
Value	Mean	15.80	7.30	.007*
	SD	3.22	2.98	
Dropper	Mean	15.40	9.40	.000*
	SD	3.41	2.11	
Hue	Mean	14.40	9.00	.053
	SD	4.06	1.63	
Mix	Mean	59.30	32.50	.000*
	SD	18.91	6.31	

* $p < 0.05$

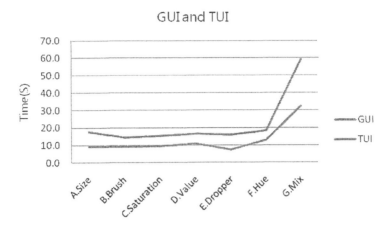

FIGURE 6. This figure shows shorter operation time and higher efficiency for the physical control interface than the icon control interface for all experiment tasks A~F and the mix experiment G.

PHYSICAL CONTROL ILLUSTRATION SYSTEM APPLICATION

The physical drawing pen model was built by placing control components according to the compatibility principle from the questionnaire survey, as shown in figure 7. It was matched with the corresponding feedback interface, integrating Arduino and flash to complete the physical control interface illustration system, as shown in figure 8.Designers were invited to test the drawing system and found that the drawing functions developed by this system were sufficient in the presentation of concept drawings. Participants were unfamiliar with the operations at first, however, operation fluency clearly increased after a period of time. This is solid prove that the physical control mechanism constructs expected sense perceptions hence increases operation speed.

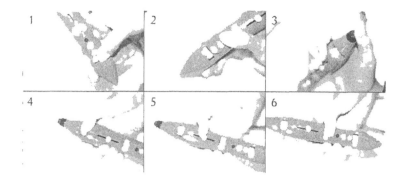

FIGURE 7. 1. Tool selection 2. Brush thickness adjustment 3. Dropper tool 4. Hue adjustment 5. Saturation adjustment 6. Brightness adjustment

FIGURE 8. The physical control digital pen operating with feedback interface. Practical tests proves that the illustration functions are sufficient to present drawing concepts.

CONCLUSIONS

The physical-control drawing system developed through this study has been through tests and a system prototype has been developed. The test results are as follow: 1. Subjects were able to grasp function control after a brief learning period. The speed and zero error rate during function operation indicate the high learning efficiency provided by this system. The fact that the subjects operated efficiently without visual, indicated the recognizable tactile position senses of this model, aiding users in "expected operation", becoming familiar with the changing mechanism. 2. The results of the function operation tests indicate the increase in operation speed of the physical control drawing interface system in comparison with the icon interface system. The causes include the decrease in moving time, visual search time, and the increase in other operation efficiencies.

Through experiment observation, the improvement direction of this system is also obtained. During the process of user operation observation, the tendency of single-handed operation has been found. Future research should focus on single-handed operation, reconsidering human engineering positioning, in order to increase operation fluency of the digital drawing pen. Also, the color-coding control, compared to other functions, is visual-judgment orientated. In order to employ this in a tactile-focused physical control system, visual feedback mechanism should be integrated.

In conclusion, the physical control interface drawing system is more efficient than the present system in many aspects. Within limited function ranges, this system employs physical tactile input instead of visual icon search to achieve the goal of picture focusing and operation fluency. Since the employment of physical control interface allows designers to focus on the picture and decreases distractions, the control mechanism of employing physical control interfaces in the place of icon interfaces brought forward by this study has proven to be feasible.

REFERENCES

Lim, S., Qin, S.F., Prieto, P., Wright, D., and Shackleton, J. (2004), "A study of sketching behaviour to support free form surface modelling from on-line sketching", *Design Studies*, 25(4),393-413.

Norman, D.A. (1988), "The Psychology of Everyday Things", Basic Books.

Palmer, J., Ames, C.T., and Lindsey, D.T. (1993), "Measuring the effect of attention on simple visual search", *J Exp Psychol Hum Percept Perform*, 19(1), 108-130.

Sanders, M.S., and McCormick, E.J. (1992), "Human Factors in Engineering and Design", McGraw-Hill Publishing Company.

Schon, D.A., and Wiggins, G. (1992), "Kinds of seeing and their functions in designing", *Design Studies,* 13(2), 135-156.

Simon, H A., and Gilmartin, K. (1973), "A simulation of memory for chess positions", *Cognitive Psychology*, 5, 29-46.

CHAPTER 23

Designing Small Touch Screen Devices: The Impact of Display Size on Pointing Performance

Michael Oehl, Felix W. Siebert

Institute of Experimental Industrial Psychology
Leuphana University of Lueneburg
Germany

ABSTRACT

The limited screen space in small touch screen devices imposes considerable usability challenges for human centered design. Thus, the tradeoff between object size and display size is crucial. In a previous study (Oehl, Sutter, and Ziefle, 2007), we found a strong effect of display size on pointing performance in a complex multidirectional serial pointing task. The very same task difficulty resulted in longer pointing times and a worse pointing accuracy on a smaller display compared to a bigger one. The central question of this current experimental study was, if this facilitating effect of display size is robust enough to appear and work even in a less complex and rather easy single pointing task. According to our previous experiment even for the less complex single pointing task, there was a significant interaction between display size and increased task difficulty for movement time as well as for pointing errors. Results allow applied ergonomic guidelines for an optimized design of touch-based screen devices in terms of human centered design.

Keywords: Touch Screen, Display Size, Task Difficulty, Task Complexity, Pointing Performance

INTRODUCTION

A broad user group increasingly uses small touch screen devices or touch-based mobile devices in business as well as private areas, e.g., GPS Navigators, PDAs, e-readers as well as the well-established iPhone as a prominent example for smartphones or the new promising iPad as a tablet PC, et cetera. The limited screen space in small touch screen devices imposes considerable usability challenges for human centered design. On the one hand, objects displayed on the small screens should be big enough to be hit successfully, but on the other hand they should be also small enough to house several objects on the screen at the same time. Thus, the tradeoff between object size and display size is crucial.

In a previous study (Oehl, Sutter, and Ziefle, 2007), we found a strong effect of display size on pointing performance in a complex multidirectional serial pointing task. The very same task difficulty resulted in longer movement times and a worse pointing accuracy on a smaller display compared to a bigger one. Similar preliminary findings were reported by Traenkle and Deutschmann (1991). The authors showed that pointing performance with a mouse was much more effective in a large display compared to a smaller display.

The current study focused on an evaluation of our previous new findings (Oehl, Sutter, and Ziefle, 2007) which are beneficial for the field of ergonomics and advanced human centered design. The central question of this current experimental study was, if the facilitating effect of display size is robust enough to appear and work even in a less complex and rather easy single pointing task (vs. more complex serial pointing task) again on a touch screen as a rather intuitive direct input device.

FITTS' LAW AND ERGONOMIC DESIGN OF INPUT DEVICES

A standard research paradigm for the evaluation of the ergonomic design of hard- and software is given by *Fitts' law* (Fitts, 1954). It predicts movement time (MT) as a function of task difficulty (ID = Index of Difficulty), determined by target distance (A = Amplitude) and target size (W = Width). The modification of the original law (Fitts, 1954) by MacKenzie (MacKenzie, 1992) is nowadays established in the EN ISO 9241-9 (2000) with MT = a + b $\log_2$ (A / W + 1). According to this function, bigger and nearer targets are easier to reach than smaller and farther away targets. A considerable amount of studies (e.g., Armbruester, Sutter, and Ziefle, 2004, 2007; Douglas, and Mithal, 1997; Sutter and Ziefle, 2004) supported the finding that motor behavior operating an input device follows the same fundamental psychomotor principle as found for manual aiming movements.

However, recent research reported some restrictions of Fitts' Law as well. Contrary to Fitts' law were findings of target size being of higher importance for the efficiency of aiming performance than target distance (Sheridan, 1979; Sutter and Ziefle, 2004). Sutter and Ziefle (2004) showed that target size and target distance did not equally contribute to task difficulty as proposed by Fitts. But target size had a more powerful effect on movement time insofar that movement time was disproportional longer for smaller targets. Preliminary findings of Traenkle and

Deutschmann (1991) pointed at another restriction of Fitts' law as a design tool. The authors showed that pointing performance with a mouse was much more effective in a large display compared to a smaller display. Beyond the significant effect of task difficulty the impact of display size might be interpreted as a cognitive effect: The processing of larger space in which the movement had to be executed effected a faster movement execution. Participants seemed to react more carefree and moved the mouse cursor much faster and snappier compared to the smaller display where movement space appeared more restricted. This was also observed for touch based interfaces operated with a stylus in our previous study (Oehl, Sutter, and Ziefle, 2007).

The discussion about task difficulty in modern applications might again be surveyed against this background with regard to display size in a more applied context. Displays in technical devices get smaller and smaller and their restricted movement space is contradictory to an efficient interaction. This study surveys again an optimized design for touch interfaces in small screen devices with a stylus as input device and a less complex applied pointing task in order to evaluate our previous new findings here in a more complex serial pointing task (Oehl, Sutter, and Ziefle, 2007). Our aim was to provide an insight into the interplay of display size and task difficulty to allow applied ergonomic guidelines for an optimized design of touch-based screen devices in terms of human centered design. Therefore, we addressed two central research hypotheses:

(1) The predicted effects of task difficulty on the basis of Fitts' law (MacKenzie, 1992) will be replicated ones more in this study. According to this, pointing time should be prolonged in high task difficulties being composed of small target sizes and target distances being farther away.

(2) The impact of display size and its interaction with task difficulty is again considered according to our previous findings with a complex serial pointing task (Oehl, Sutter, and Ziefle, 2007). If assuming that display size acts as perceptual frame of reference, determining the speed of movement due to its ballistic nature or to cognitive effects, it can be deduced that larger display sizes should result in faster movements, accompanied by a lower accuracy of the pointing movement. This should appeal to a less complex and easier single pointing task in this current study as well.

METHOD

According to our previous study (Oehl, Sutter, and Ziefle, 2007) the current experimental study was based on a two-factorial design with repeated measurements. The *independent variables* were the *display size* of the touch screen and the *difficulty of pointing task*.

Again we used three display sizes, covering a wide range of screen sizes present in real devices equipped with a touch interface (Figure 1: Display 1 = 6.00 x 8.00 cm, display 2 = 12.00 x 16.00 cm, and display 3 = 18.00 x 24.00 cm). We varied difficulty of the pointing task (ID) from 1.81 to 4.95 bits (MacKenzie, 1992). Depending on display size, a maximum of four target sizes (0.25, 0.50, 0.75, and 1.00 cm) and three target distances (2.50, 5.00, and 7.50 cm) were chosen. Due to the three different screen sizes and the spatial restrictions of the smallest display, the realized IDs differed across the display sizes. There were two IDs realized in all three display sizes, ID 2.58 bits and ID 3.46 bits. Thus, the research question at issue, the influence of display size on performance in a pointing task can be analyzed for these two IDs. Both IDs represent typical task difficulties for mobile devices. Exploiting the increasing display space with respect to the analysis of more IDs, five different IDs were realized for the medium display size (2.12, 2.58, 2.94, 3.46, and 4.39 bits) and in total nine IDs were possible to be realized for the large display size (1.81, 2.12, 2.58, 2.94, 3.09, 3.46, 4.00, 4.39, and 4.95 bits).

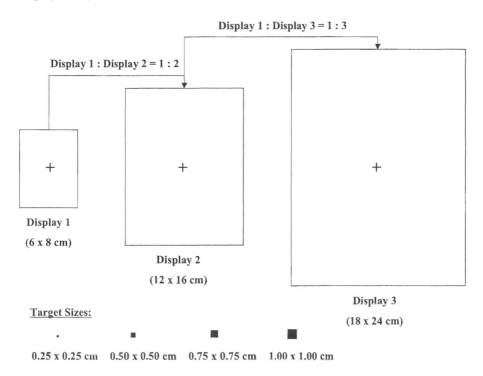

FIGURE 1. Schematic illustration of display size and target size.

Sixteen participants ($N = 16$, i.e., 8 male and 8 female) at the age of $M = 39.31$ years ($SD = 15.61$) accomplished each 320 less complex multidirectional single pointing tasks on a high precision touch screen with a stylus. In each task, one of nine possible square targets was presented. The users had to move the pen from a well-defined starting point inside the target area. Participants had to work on all display sizes and on all task difficulties (IDs). The order of display sizes was balanced over participants. The order of IDs within each display size was at random. To exclude confounding effects of movement direction, the target positions were placed in eight different directions (0°, 45°, 90°, 135°, 180°, 225°, 270°, and 315°) relative to the starting point of movement (EN ISO 9241-9, 2000). Time and spatial information of pointing movements (x- and y-coordinates) were recorded action-correlated with a logging and analyzing software tool (Oehl, Sutter, and Ziefle, 2007). A state of the art high precision 15'' (diagonal measurement) TFT touch screen (Iiyama AX 3819 UT) with a 1024 x 768 resolution and a "Touchlogic Twisted Nematic®" (RS 232C ELO Touchsystems) was used. The three different display sizes on the touch screen were created by the software program and represented simply a lighted area with the rest of the display kept dark and inactive. The stylus was a high precision professional touch input device for industrial applications (WES®).

Dependent variables were time and accuracy of pointing as objective measures as well as usability ratings in terms of task difficulty and their effort to complete the single multidirectional pointing task. The *time* measure of performance included the movement time to complete a whole task, which described the interval of time from correct pointing with the stylus on the onset point to final correct pointing on the target. For *accuracy*, the incorrect pointing was logged, defined as pointing error. This error occurred when participants did not hit the target itself, but pointed outside the target's boundaries.

Usability ratings were measured by *participants' judgments of task difficulty and effort* on a 4-point scale in a questionnaire (1 = low, 2 = reasonably low, 3 = reasonably high, and 4 = high). In order to facilitate ratings for participants, the actual task difficulties (IDs) composed of target sizes and distances were operationalized in either smaller or bigger targets for each display size. In doing so, the smaller targets comprised the higher IDs whereas the bigger targets comprised the lower IDs.

RESULTS

EFFECTS OF TASK DIFFICULTY

According to our first hypothesis and to Fitts' law a regression analysis comprising all realized task difficulties (IDs) across the three display sizes revealed again a high positive correlation ($r = .81$, $p < .001$, $R^2 = .67$) between task difficulty and movement time. Additionally, we found a high positive correlation ($r = .79$, $p < .001$, $R^2 = .60$) between task difficulty and pointing errors, too.

Analyzing only the two IDs (2.58 and 3.46 bits) present in all three display sizes, analyses of variance (ANOVA) for repeated measurements showed the validity of Fitts' law for movement time in pointing tasks on all three display sizes, i.e., for display 1 ($F_{(1,15)} = 18.99$, $p = .001$), display 2 ($F_{(1,15)} = 69.45$, $p < .001$), and display 3 ($F_{(1,15)} = 52.08$, $p < .001$). Movement time increased significantly with increased task difficulty (Figure 2). Again, a strong effect of task difficulty on pointing errors was found for display 1 ($F_{(1,15)} = 35.78$, $p < .001$), display 2 ($F_{(1,15)} = 59.82$, $p < .001$), as well as for display 3 ($F_{(1,15)} = 33.62$, $p < .001$), i.e., the higher ID of 3.46 bits induced significantly more pointing errors than the lower ID of 2.58 bits (Figure 3).

Usability ratings of task difficulty and effort confirmed the described findings in pointing time and accuracy across all display sizes. For the subjective difficulty of smaller and bigger targets per display size, a non-parametric Friedman-Test revealed significant ranks ($\chi^2_{(5)}$ = 53.64, p < .001). Generally, the smaller targets of each display size, which comprised actually the higher IDs, were rated as fairly "reasonably high = 3 of 4" in difficulty for the small display 1 (M = 2.56, SD = 0.89), for the medium display 2 (M = 2.75, SD = 0.93), and for the large display 3 (M = 3.00, SD = 0.63). The smaller targets on the large display were significantly rated as most difficult of all realized targets. Amongst them, the objective most difficult pointing task of the whole study with an ID of 4.95 bits was present, but on the other both display sizes not. This finding confirms the quality of participants' usability ratings. Additionally, this is confirmed by participants' difficulty ratings of the bigger targets (lower IDs) on the three display sizes as fairly "low = 1 of 4" for the small display 1 (M = 1.13, SD = 0.34), for the medium display 2 (M = 1.50, SD = 0.82), as well as for the large display 3 (M = 1.38, SD = 0.50).

Comparing the average difficulty ratings for the higher IDs (M = 2.77, SD = 0.65) and lower IDs (M = 1.33, SD = 0.40) a non-parametric Wilcoxon-Test showed significant differences in ratings (Z = -3.53, p < .001).

A similar pattern appeared for the estimated task effort of smaller and bigger targets per display size. The non-parametric Friedman-Test revealed likewise significant ranks ($\chi^2_{(5)}$ = 30.39, p < .001). The smaller targets of each display size were rated more or less as fairly "reasonably high = 3 of 4" in effort for the small display 1 (M = 2.31, SD = 1.01), for the medium display 2 (M = 2.69, SD = 0.95), and for the large display 3 (M = 2.69, SD = 1.01). The smaller targets on the medium and large displays, which comprised the highest IDs, were rated to be of highest pointing effort. Participants rated the pointing effort of the bigger targets (lower IDs) on the three display sizes in the range of fairly "low = 1 of 4" and "reasonably low – 2 of 4" for the small display 1 (M = 1.56, SD = 0.63), for the medium display 2 (M = 1.88, SD = 0.89), and for the large display 3 (M = 1.81, SD = 0.83).

Comparing the average effort ratings for the higher IDs (M = 2.56, SD – 0.87) and lower IDs (M = 1.75, SD = 0.73) a non-parametric Wilcoxon-Test showed significant differences in ratings (Z = -2.97, p < .005) just as well as for the usability ratings of task difficulty.

EFFECTS OF DISPLAY SIZE

According to our second hypothesis we analyzed the data for both IDs (2.58 and 3.46 bits) present in all three display sizes by analyses of variance (ANOVA) for repeated measurements. For the lower task difficulty (ID 2.58 bits) we found no significant effect of display size on pointing time ($F_{(2,14)}$ = 1.28, p = .293). Pointing times for the small display 1 (M = 893.08 ms, SD = 148.98 ms), for the medium display 2 (M = 821.60 ms, SD = 114.34 ms), and for the large display 3 (M = 884.27 ms, SD = 265.34 ms) were almost comparable. In direct contrast to the lower task difficulty (ID 2.58 bits), we observed for the higher task difficulty (ID 3.46 bits) significantly increased pointing times ($F_{(2,14)}$ = 8.85, p = .001) for the small display 1 (M = 2080.87 ms, SD = 1169.48 ms) compared to the medium display 2 (M = 1445.29 ms, SD = 373.51 ms) and to the large display 3 (M = 1286.65 ms, SD = 358.76 ms). The disadvantage of having a small display and a high task difficulty

(ID 3.46 bits) ranged at 61.73% compared with the same task difficulty on the large display. This interaction is illustrated in Figure 2.

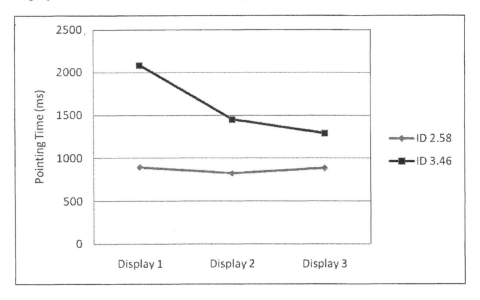

FIGURE 2. Effect of display size (small = display 1, medium = display 2, large = display 3) and ID (2.58 and 3.46 bits) on mean pointing times (ms).

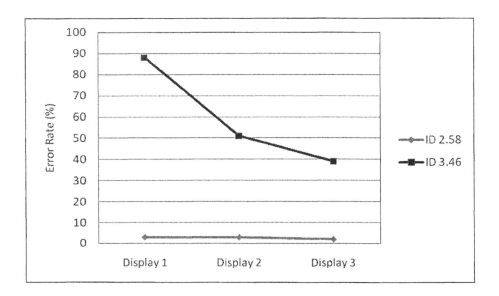

FIGURE 3. Effect of display size (small = display 1, medium = display 2, large = display 3) and ID (2.58 and 3.46 bits) on mean pointing error rate (%).

A similar pattern was found for pointing accuracy (Figure 3). The percentage of errors did not differ significantly ($F_{(2,14)}$ = 0.20, p = .821) due to display size for the lower task difficulty (ID 2.58 bits). We observed comparable low error rates for the small display 1 (M = 3.13%, SD = 5.10%), for the medium display 2 (M = 3.13%, SD = 5.10%), as well as for the large display 3 (M = 2.54%, SD = 4.60%).

On the other hand we found remarkable results for the higher task difficulty (ID 3.46 bits). In this case pointing error rates were significantly increased depending on the three different display sizes ($F_{(2,14)}$ = 12.17, p < .001). We observed decreasing error rates with increasing display size, i.e., for the small display 1 (M = 87.89%, SD = 55.92%), for the medium display 2 (M = 50.78%, SD = 23.51%), and for the large display 3 (M = 39.00%, SD = 23.05%). An error rate of 100% means that one error occurred in each single pointing task.

CONCLUSION

The design of efficient touch interfaces is becoming a challenge in applied human factors and ergonomics. Displays in technical devices are getting smaller and their restricted movement spaces are contradictory to an efficient interaction. In literature the impact of display size on motor performance in combination with an increase of task difficulty in smaller displays is reported (Oehl, Sutter, and Ziefle, 2007; Traenkle and Deutschmann, 1991). These results are theoretical unexpected, but can be interpreted in a cognitive or motor behavioral way. Nevertheless, the practical implications are manifold towards the design of small screen interfaces.

According to our first hypothesis, the findings of the present experimental study once more confirmed Fitts' law for pointing movements executed with a stylus. An increase of task difficulty (ID) in the applied multidirectional single pointing task resulted in a lower pointing performance, i.e., movement time and pointing errors rose distinctly. These results are consistent with many other studies regarding input device performance in dependence on Fitts' law (e.g., Armbruester, Sutter, and Ziefle, 2004, 2007; Douglas and Mithal, 1997; Oehl, Sutter, and Ziefle, 2007; Sutter and Ziefle, 2004). Moreover, whenever the task execution was complicated in the present experiment, i.e., for higher IDs, users clearly stated their higher effort and higher difficulty of solving the pointing task properly. These findings on objective as well as subjective measurements of usability are consistent with the results of our previous study with a more complex serial pointing task (Oehl, Sutter, and Ziefle, 2007).

According to our second hypothesis there was a significant interaction between display size and task difficulty (ID) for pointing time as well as for pointing errors. Findings of our previous study with a more complex serial pointing task were confirmed. But in contrast to performance in a more complex serial pointing task, here in a less complex single pointing task the facilitating effect of increasing display size appealed only to the higher task difficulty (ID), e.g., the disadvantage in pointing time of having a small display and a high task difficulty (ID) ranged at 61.73% compared to the very same task difficulty on the large display. The same pattern was found for pointing accuracy. We found no facilitating effect of a larger display size for the lower task difficulty (ID). This result is not consistent with our previous findings (Oehl, Sutter, and Ziefle, 2007). Remember that we did not find a facilitating effect of an increased display size for pointing accuracy in the more complex serial pointing task for lower task difficulty (ID), but we found this facilitating effect for pointing time. This might be due to the differences in movement complexity of the pointing tasks (single vs. serial).

The impact of display size and its interaction with task difficulty (ID) was considered with reference to two aspects.

First, from a theoretical point of view this result is rather surprising. So far, the interaction between display size and task difficulty (ID) was found under strict experimental conditions (Oehl, Sutter, and Ziefle, 2007; Traenkle and Deutschmann, 1991). When rapid aimed movements had to be executed to solve a simple point-click task with a mouse pointing performance was much more effective on a large display compared to a smaller display (Traenkle and Deutschmann, 1991). This was also confirmed for touch screens as rather simple and intuitive input devices in our previous study with a more complex applied multidirectional serial pointing task (Oehl, Sutter, and Ziefle, 2007) and again in our present study even with a less complex applied multidirectional single pointing task. The interaction was, until now, interpreted with regard to the perceived frame of reference, which allows a freehanded movement execution in larger spaces. However, the present results again hint at different strategies of motor behavior. Especially regarding the more difficult pointing tasks (higher IDs), in large displays the speed-accuracy tradeoff (Pachella, 1974) is shifted towards a fast and comparably accurate execution whereas in small screens a very inaccurate and time-consuming style is chosen. Thus, the cognitive processing of the frame of reference is very likely to affect the speed-accuracy tradeoff. However, this has to be addressed systematically in further research.

Second, from an ergonomic point of view the interaction between display size and task difficulty (ID) is very insightful. It shows that the very same task difficulty (ID) can result in a different performance depending on how large or how small the display is. Taking first attempts to define the optimal target size on touch screens additionally into consideration (e.g., Mizobuchi, Mori, Ren, and Michiaki, 2002; Sun, Plocher, and Qu, 2007), results of this current research allow ergonomic guidelines for an optimized usage and design of touch-based screen devices assuring efficient and effective handling. Therefore it is important to achieve an optimized balance of task difficulty (ID) and display size especially in small screen devices. We recommend an optimized balance of task difficulty (ID) and display size in small screen devices, i.e., especially small targets in small displays should be in easy reach for an effective interaction.

REFERENCES

Armbruester, C., Sutter, C., and Ziefle, M. (2004), Target Size and Distance: Important Factors for Designing User Interfaces for Older Notebook Users. In H.M. Khalid, M.G. Helander, and A.W. Yeo (Eds.), *Work With Computing Systems* (pp. 454-459). Damai Sciences, Kuala Lumpur, Malaysia.

Armbruester, C., Sutter, C., and Ziefle, M. (2007), Notebook input devices put to the age test: The usability of trackpoint and touchpad for middle-aged adults. *Ergonomics, 50*(3), 426-445.

Douglas, S.A., and Mithal, A.K. (1997), *The ergonomics of computer pointing devices.* Springer, London, UK.

EN ISO 9241-9 (2000), *Ergonomic requirements for office work with visual display terminals – Part 9: Requirements for non-keyboard input devices. International Organization for Standardization.* Beuth, Berlin, Germany.

Fitts, P.M. (1954), The information capacity of the human motor system in controlling the amplitude of movements. *Journal of Experimental Psychology, 47*(6), 381-391.

MacKenzie, I.S. (1992), Fitts law as a research and design tool in human-computer interaction. *Human-Computer Interaction, 7*, 91-139.

Mizobuchi, S., Mori, K., Ren, X., and Michiaki, Y. (2002), An empirical study of the minimum required size and the minimum required number of targets for pen input on small displays. In F. Paterno (Ed.), *Proceedings of the Mobile HCI 2002* (pp. 184-194). Springer, Heidelberg, Germany.

Oehl, M., Sutter, C., and Ziefle, M. (2007), Considerations on Efficient Touch Interfaces – How Display Size Influences the Performance in an Applied Pointing Task. In M.J. Smith, and G. Salvendy (Eds.), *Human Interface, Part I, HCII 2007, LNCS 4557* (pp. 136-143). Springer, Berlin, Germany.

Pachella, R. (1974), The interpretation of time in information-processing research. In B. Kantowitz (Ed.), *Human information processing: tutorials in performance and cognition* (pp. 41-80). John Wiley and Sons, Hillsdale, UK.

Sheridan, M.R. (1979), A repraisal of Fitts law. *Journal of Motor Behavior 11*(3), 179-188.

Sutter, C., and Ziefle, M. (2004), Psychomotor efficiency in users of notebook input devices: Confirmation and restrictions of Fitts law as an evaluative tool for user-friendly design. In *Proceedings of the Human Factors and Ergonomics Society 48th Annual Meeting* (pp. 523-528). Human Factors Society, Santa Monica, CA.

Traenkle, U., and Deutschmann, D. (1991), Factors influencing speed and precision of cursor pointing using a mouse. *Ergonomics, 34*(2), 161-174.

Sun, X., Plocher, T., and Qu, W. (2007), An Empirical Study on the Smallest Comfortable Button/Icon Size on Touch Screen. In N. Aykin (Ed.), *Usability and Internationalization, Part I, HCII 2007, LNCS 4559* (pp. 615–621). Springer, Berlin, Germany.

Chapter 24

Usability of Integrated Display Groupware in Collaborative Work

Tek Yong Lim[1], Halimahtun M. Khalid[2], Alvin W. Yeo[3]

[1]Faculty of Information Technology
Multimedia University
63100 Cyberjaya, Malaysia

[2]Damai Sciences Sdn Bhd
A-31-3 Suasana Sentral, Jalan Stesen Sentral 5
50470 Kuala Lumpur, Malaysia

[3]Faculty of Computer Science and Information Technology
Universiti Malaysia Sarawak
94300 Kota Samarahan, Sarawak, Malaysia

ABSTRACT

We investigated the usability of groupware in supporting collaborative work at the comprehension level of collaborative awareness. Two groupware, a non-integrated display groupware and an integrated display groupware, were evaluated in a controlled experiment. The results revealed that both groupware were effective, but the integrated display groupware improved efficiency in collaborative work, especially when it involved a two-person collaboration. This is because collaborators could focus their attention visually on the on-going work that enables task comprehension, while maintaining awareness of the cooperative process.

Keywords: Collaborative awareness, Usability, Integrated display groupware, Spatial-temporal component

INTRODUCTION

The collaborative awareness concept, as used in this paper, is akin to situation awareness (Endsley, 2004), comprising three levels, namely, perception, comprehension and projection. At the perception level, it has been established that both interpersonal space and shared workspace components are important basic data (Lim et al., 2006). To support collaborative work more efficiently, at the next level of comprehension, an integrated display would be beneficial. This is because providing data in multiple windows at the perception level is not sufficient given that collaborators may be constrained by their inability to focus their attention and to comprehend the status of on-going collaborative work. We therefore propose the use of an integrated groupware which warrants investigation into its usability on collaborative performance.

We hypothesize that the second level of collaborative awareness should enable collaborators to comprehend the on-going collaborative work better since they will be provided with a holistic view of the task. They will be able to comprehend the task easier on the basis of the overall meaning of the on-going collaborative situation. Moreover, the integrated display explains how the basic data fit together as a whole. A well designed user interface should offer a display where remote collaborators can focus their visual attention while performing the cooperative task, and that they can be involved in the process together (Lim et al., 2006). For example, to move an object from one location to another, a collaborator must know that the object may be heavy and that there are only two persons in the room to perform the task. Failure to perceive such requirements may result in poor mental models that cannot combine and interpret information (i.e. errors in human information processing) or several pieces of data are not properly integrated on the system display (i.e. technological design problem). Therefore, segregating data into multiple windows can cause individuals to become 'lost' or 'disoriented'. Common observations included: participants lost track where they were relative to other locations in the space, their mental workload increased, they searched fewer possibilities, missed 'interesting' events, took longer time to perform a task and performed with lower accuracy (Woods and Watts, 1997).

INTEGRATED DISPLAY GROUPWARE

An integrated display that combines interpersonal space and shared workspace would allow the groupware to support collaboration at the comprehension level (Ishii et al., 1993; Gutwin and Greenberg, 1999; Yang and Olson, 2002). However, the effectiveness of integrated display groupware has not been investigated much. On the contrary, efficiency of such displays has been increasingly reported. Gutwin and Greenberg (1999) claimed that integrated display groupware reduced completion time in two tasks and communication became more efficient in one task. Yang and Olson (2002) preferred the non-integrated display groupware using egocentric view, because it took less travel time and reduced total time on task. In terms of user satisfaction, Gutwin and Greenberg (1999) found that participants preferred the integrated display groupware over the non-integrated display groupware.

The integration of interpersonal space (subject-collaborator member) and shared workspace (object-outcome) into a seamless design has been studied by Ishii et al. (1993) and Gutwin and Greenberg (1999). The integration was implemented using spatial-temporal component. This component allows remote participants to move seamlessly in two different settings: (1) between participants' images and shared workspace images (Ishii et al., 1993), and (2) between individual workspace and group workspace (Gutwin and Greenberg, 1999). Normally, collaboration takes place at a certain time and space (Engeström et al., 2003). As time passes, the collaboration status could change or take place in a different location. The spatial-temporal component takes into consideration that the collaborative situation dynamically changes over the time and space. This component would provide information about who is doing what on which object at a particular time.

In our study, *groupware* refers to a computer software application that allows two remote participants to customize the interior of a virtual home. The radar that was used in Gutwin's (1997) experiment required participants to focus their attention on two different windows. The radar was considered an extra window. An egocentric view as used by Yang and Olson (2002) was also not suitable because the small screen limited viewing to first-person view only. Also, peripheral glancing as used by Hindmarsh et al. (2000) misled the participants, especially in relation to avatar pointing directions.

For developing the integrated display groupware, a modified version of zooming technique, as applied by Holmquist (1997), was used. The technique raised all walls in the room where the participant was working. This allowed the participants to be aware of the on-going group workspace, and at the same time they could view the wall details in their own individual workspace. This spatial-temporal component provides the context for individual workspace and group workspace at the same time (see Figure 1). Participants can work seamlessly from their individual workspace to group workspace (Gutwin and Greenberg, 1999) and they can use nonverbal cues such as an avatar pointing at a wall (Ishii et al., 1993). In other words, it can provide a clear focussed visual attention of the collaborative work being undertaken.

As participants perform the task, going from one room to another room, the avatars also move from the previous room to the current room. Both participants can work together in the same room or they can work individually in different rooms. This can be achieved by making the walls higher inside the room where the participants work, thus providing an easy access to the individual workspace. For the rest of the walls in the other rooms, the height is lowered to allow participants view the group workspace. This minimized visual search and information access effort (Wickens, et al., 2004). However, when all data are integrated into a single window, it will result in a cluttered display (Wickens, et al., 2004). Thus, some data will be provided to the participants when they require it at a particular time. To obtain the data, participants can mouse-over the objects in the virtual world. This technique reduced the costs of clutter.

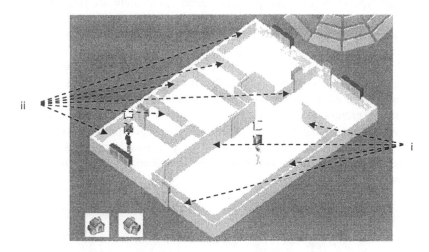

Figure 1 Partial Screenshot of Integrated Display Groupware: Individual Workspace (i), and Group Workspace (ii)

METHODOLOGY

A controlled experimentation was employed and two groupware were evaluated, namely non-integrated display groupware and integrated display groupware. The *integrated display groupware* is hypothesized to support collaborative work more efficiently and satisfactorily than a *non-integrated display groupware*. This means there is no difference in effectiveness since both groupware supported the same basic data.

Three hypotheses were tested:

Hypothesis 1: *There is no significant effect of groupware type on effectiveness.*
Since both groupware support the same basic data, it is expected that remote collaborators can perform effectively using either groupware.

Hypothesis 2: *There is a significant effect of groupware type on efficiency.*
Due to increased awareness in comprehending the on-going collaborative work, it is expected that remote collaborators can perform more efficiently using an integrated display groupware than a non-integrated display groupware.

Hypothesis 3: *There is a significant effect of groupware type on user satisfaction.*
An integrated display groupware that enables efficient collaboration is expected to increase user satisfaction.

GROUPWARE TYPE

The groupware type is the only independent variable, but there are two versions. System A (a non-integrated display groupware) had three small windows that supported interpersonal space, shared workspace, and modification tool (see Figure 2). Participants viewed the home (shared workspace) and customized it using a color wheel (modification tool). They could also hear, speak and see others (interpersonal space). The windows were deliberately made to be in the same size, given the importance of these components in collaborative work. System B (an integrated display groupware), is an extension of system A, which was enhanced with a spatial-temporal component (see Figure 2). Both groupware provided the same basic information but with different display representations. System B integrated the interpersonal space, shared workspace, and color wheel into a single large window.

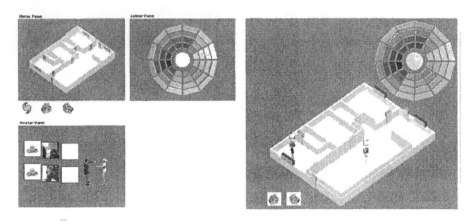

Figure 2 Screenshots for System A (left) and B (right)

USABILITY MEASURES

The usability measures constituted the dependent variables. There were 5 different measures: effectiveness, efficiency, user satisfaction, user preference, and user customization errors. All measures were based on joint responses of the collaborators. This means each participant pair gave their responses together and was evaluated as a single response. *Effectiveness* was measured in terms of the total number of rooms that were correctly customized with the given color schemes. A customized room was considered correct when the four-wall color combination matched the color schemes in the given task. *Efficiency* was measured in terms of the time taken to complete the given task. The collaboration ended when a participant logged out of the session. *User satisfaction* was measured on a rank order scale, where rank 1 indicated the high satisfaction and rank 2 as low satisfaction. *User preference* was also measured using rank order scale, where rank 1 means most preferred and rank 2 as least preferred. *User customization errors* were measured on the basis of, type of mistakes made during customization, namely, wall with wrong color and wall without any color.

EXPERIMENTAL DESIGN AND TASK

A univariate within-subjects design was used where all pairs worked together in all groupware conditions. To neutralize the potential effects of learning, each pair performed two different tasks using two different types of groupware. The order of groupware and tasks assignment was counterbalanced across participants and conditions.

Each task required each pair of participants to customize six rooms with six color schemes, and leave one specific room non-customized. This means for the latter task-item, they are not required to customize anything when given an instruction such as 'do not customize living room'. Some task-items required equal amount of effort from each participant in a pair. For example, task-item: complementary with key color f1. This task required each participant to customize two walls for each room. Some task-items do not require equal amount of effort. For example, task-item: triadic with key color c3. For this task, a participant customized three walls while the other customized (the remaining) one wall only of a particular room. (Triadic refers to the use of three colors which are equidistant from one another on a color wheel).

PARTICIPANTS

Student volunteers were recruited through various means such as bulletin boards, lecture announcements and referrals from colleagues. They were given drinks as a token of appreciation for their participation. Thirty-two university students (sixteen male and sixteen females), with a mean age of 22.2 years, participated in the experiment. They formed sixteen pairs of mixed genders, and participants in each pair knew each other as they were required to identify their own partner.

PROCEDURE

The experiment was conducted in a laboratory at Universiti Malaysia Sarawak. Participants were first briefed on the aims of the experiment. They were then introduced to the color schemes and subsequently tested on their knowledge of the color schemes. After that, the participants were assigned to an experimental condition. They signed an electronic consent form and completed an online questionnaire on their personal particulars. A tutorial on the groupware was then given, followed by a five-minute break. After the break, they performed the first and second conditions, followed by another five-minute break. The maximum time allowed for each session was ten minutes. During the break, the participants were asked about their experiences in using each of the groupware. Together, they completed an online user satisfaction questionnaire that evaluated their satisfaction and preference for the groupware.

RESULTS

The Statistical Packages for Social Sciences (SPSS) was used to analyze the data. The one way repeated measure ANOVA was used to test the hypotheses at 5% significance level. Univariate ANOVA revealed that there was no significant difference in effectiveness between system A and system B, $F(1,15) = 2.37$, n.s. The types of groupware, system A (Mean = 89%, SD = 12) and system B (Mean = 93%, SD = 13), did not influence effectiveness. Therefore, *hypothesis 1 was confirmed.*

Results of the one-way ANOVA showed a significant difference in efficiency between system A and system B, $F(1,15) = 8.47$, $p < 0.05$. A post ANOVA Least Significant Difference (LSD) test indicated that participants performed more efficiently using system B (Mean = 354 seconds, SD = 88) compared to system A (Mean = 432 seconds, SD = 91). *Hypothesis 2 was therefore accepted.*

All participants were highly satisfied with system B compared to system A, thereby *hypothesis 3 was supported.* All participants preferred system B compared to system A, thereby hypothesis 3 was further supported. Five types of customization errors were identified; three types with system A (one wall with wrong color, two walls with wrong color, and four walls without any color) and four types with system B (one wall with wrong color, two walls with wrong color, three walls with wrong color, and four walls with wrong color). Ten pairs committed customization errors when using system A, and five pairs with system B.

DISCUSSION

All hypotheses in this experiment were supported by the results. Clearly, participants could perform the home modification tasks using both types of groupware; they could customize the home up to 89% using system A and 93% using system B. Even though there was no difference in effectiveness, the integrated display groupware was more efficient (Mean = 354 seconds) than the non-integrated display groupware (Mean = 432 seconds). This indicated that participants were able to complete the task faster with an integrated display groupware because they could focus their attention visually on the on-going collaborative work in order to achieve the comprehension level of collaborative awareness. As such, they required less time in making sense of the group work. Therefore, undesirable actions such as customising the wrong room due to misunderstanding, and correcting the error, could be avoided. Gutwin and Greenberg (1999) also expressed the importance of focussed visual attention.

In this experiment, both participants used the exocentric view to move to a new room by clicking any walls within the room. This enabled them to have a good mental rotation in order to guide their partner (Wickens and Carswell, 2006). However, in the study by Yang and Olson (2002), the pairs used two different view modes; one person used the exocentric view, while the other used egocentric which resulted in poor performance when they navigated from one location to another.

IMPORTANCE OF SPATIAL-TEMPORAL COMPONENT

With system B, participants were given a single window that integrated all data of interpersonal space (virtual avatar), shared workspace (virtual home) and modification tool (color wheel). In other words, the integrated display illustrated how these basic data fit together as a whole and provided an overall meaning of on-going collaborative situation. The detailed view of the virtual home and the color wheel, as well as the virtual avatars' location, were found to be helpful in showing the working area of each participant. Through the integrated display, participants were able to comprehend the on-going collaborative work easily, and achieved the second level of collaborative awareness more efficiently.

The integrated display groupware did not require additional windows such as radar and peripheral glancing. It also supported two different modes in a seamlessness fashion: (1) participants could view seamlessly between the details of each wall in their working room (individual workspace) and the overview of their final home modification (group workspace), (2) they could also view seamlessly between the home (shared workspace) and the avatar (interpersonal space). This enabled them to find the right information at the right time as their collaboration progressed (Woods and Watts, 1997). As a result, they could customize the home effectively (Mean = 93%), and efficiently (Mean = 354 seconds). They were also more satisfied and preferred the integrated display groupware over the non-integrated display groupware. For example, Pair 2a (male) looked for the living room. Then Pair 2b (female) informed him that the living room was where her avatar stood. At the same time, Pair 2a also found the living room by moving the mouse over the wall. He then suggested that they go to the next room and clicked on the dining room wall. Pair 2b noted where her partner was and followed him into the dining room. They then customized the room immediately.

Without the spatial-temporal component, participants had to ask their partner in order to update the focussed visual attention of collaborative work. With system A, participants had to look into three different windows that contained the virtual home, the color wheel and the task. This increased their time on task. Due to the constrained window in the provided space, participants had to deal with a hardly visible virtual home and color wheel which was not satisfactory.

Additionally, the visual of the avatar was useless in system A as they could not view it seamlessly between the interpersonal space and the shared workspace. For example, Pair 4a (male) informed that the living room could not be customized based on the given task, and suggested that they go to the dining room together. Pair 4b (female) did not know where these rooms were and tried to look for the name of each room. Pair 4a customized the first wall in the dining room only to be recognized afterwards by Pair 4b. Pair 4b in fact had to customize the wall in the dining room, but clicked the wall in the third room, unaware that it did not belong to the dining room. After customising the wall, they both realized that the customized wall was wrong. Pair4b then corrected her mistake. This problem could be avoided if the avatar had been integrated into the virtual home in a single window.

However, simulating the real world in the groupware display does not ensure a good design. In the real world, collaborators move the color wheel as they move from one room to another. Whether the color wheel should follow the avatar in the virtual world depends on the location of the physical participants. As the participants are sitting in front of a computer in real-time distributed collaboration, the color wheel is a modification tool that should be located at a specific location in the window where the participants can reach it easily. In other words, the control (the color wheel) should be presented in a consistent location and should not follow the avatar.

CONCLUSION

This experiment has revealed that the integrated display groupware improved the efficiency of collaborative work, especially for two-person collaboration. If the groupware did not support the spatial-temporal component, participants have to spend more time comprehending the status of on-going collaborative work. With the spatial-temporal component, participants were able to understand the on-going collaborative work and were more satisfied in their collaboration outcome.

Although the participants could customize the home effectively using both groupware, they were still prone to commit customization errors. The errors could be avoided by providing additional information in the groupware design. The third level of collaborative awareness suggested that projection is the highest level that is valuable in decision-making. This additional information should help to project a shared goal for the participants, thereby helping them to coordinate their tasks more effectively. This includes identifying possible incomplete tasks and the person to complete those tasks. Data used in the projection level is not static (fixed) data, but data that is dynamically updated based on the on-going customizations made by the participants.

ACKNOWLEDGMENT

The research was partially supported by the Malaysian Ministry of Science Technology and Innovation, under the e-Science Fund, Grant No: 01-02-01-SF0112.

REFERENCES

Endsley, M.R. (2004). "Situation awareness: Progress and directions." In S. Banbury and S. Tremblay (Eds.), A Cognitive Approach to Situation Awareness: Theory and Application. Aldershot:Ashgate Publishing, 317-341.

Engeström, Y., Puonti, A. and Seppänen, L. (2003). "Spatial and temporal expansion of the object as a challenge for reorganizing work." In D. Nicolini, S. Gherardi and D. Yanow (Eds.), Knowing in Organizations: A Practice-based Approach. Armonk: M. E. Sharpe, 151-186.

Gutwin, C. (1997). Workspace Awareness in Real-Time Distributed Groupware. PhD Thesis. University of Calgary.

Gutwin, C. and Greenberg, S. (1999). "The effects of workspace awareness support on the usability of real-time distributed groupware." ACM Transactions on Computer-Human Interaction, 6(3), 243-281.

Hindmarsh, J., Fraser, M., Heath, C., Benford, S. and Greenhalgh, C. (2000). "Object-focused interaction in collaborative virtual environments." ACM Transactions on Computer-Human Interaction, 7(4), 477-509.

Holmquist, L.E. (1997). "Focus+context visualization with flip zooming and the zoom browser." In Extended Abstracts on Human Factors in Computing Systems, New York: ACM Press, 263-264.

Ishii, H., Kobayashi, M. and Grudin, J. (1993). "Integration of interpersonal space and shared workspace: ClearBoard design and experiments." ACM Transactions on Information Systems, 11(4), 349-375.

Lim, T.Y., Khalid, H.M. and Yeo, A.W. (2006) "Integrating Collaborative Awareness in Synchronous Distributed Groupware." In Proceedings of the 50th Annual Meeting of the Human Factors and Ergonomics Society, 594-598.

Wickens, C.D. and Carswell, C.M. (2006). "Information Processing." In G. Salvendy (Ed.), Handbook of Human Factors and Ergonomics, 3rd ed., John Wiley & Sons, 111-149.

Wickens, C.D., Alexander, A.L., Horrey, W.J., Nunes, A. and Hardy, T.J. (2004). "Traffic and flight guidance depiction on a synthetic vision system display: the effects of clutter on performance and visual attention allocation." In Proceedings of the 48th Annual Meeting of the Human Factors and Ergonomics Society, 218-222.

Woods, D.D. and Watts, J.C. (1997). "How not to have to navigate through too many displays." In M.G. Helander, T.K. Landauer, and P. Prabhu, (Eds.), Handbook of Human-Computer Interaction, Amsterdam, The Netherlands: Elsevier, 617-650.

Yang, H. and Olson, G.M. (2002). "Exploring collaborative navigation: The effect of perspectives on group performance." In Proceedings of the 4th International Conference on Collaborative Virtual Environments, New York: ACM Press, 135-142.

Chapter 25

Beauty and the Beast: Predicting Web Page Visual Appeal

Cathy Dudek, Gitte Lindgaard, Stephanie Pineau

Human Oriented Technology Lab
Carleton University
Ottawa, Ontario, K1S 5B6, Canada

ABSTRACT

This paper explores visual attributes that define web page visual appeal. A cluster analysis and subsequent regression analysis were undertaken to assess if ratings of appeal could be predicted from a particular combination of five visual attributes. Three samples of web pages used in different studies all seeking to understand how quickly visual appeal judgments are formed were used to support our argument that it is not only necessary to understand what makes web pages appealing but that it is equally important to understand what distinguishes those from unappealing web pages. We investigated if the attributes used to identify the highly appealing pages could also identify those rated lowest in appeal. That was supported by the results reported.

Keywords: Visual Appeal, Web Pages, Aesthetics, Design Attributes

INTRODUCTION

The purpose of this paper is to explore the attributes that define web page visual appeal as it relates to the concept of beauty. Beauty has been defined in many ways and is often considered part of the broader concept of 'aesthetics' (Lavie and Tractinsky, 2004;

Lindgaard and Whitfield, 2004). Over the years the term aesthetics has taken on a variety of meanings. One of these pertains to beauty as an objective property of the stimulus, and another defines beauty as a property of the subject, which is thought to be tied to emotion (see Lavie and Tractinsky, 2004 for an overview). In a series of studies, Lavie and Tractinsky (2004) designed and validated a scale to measure aesthetics in web pages and found that it comprised two components. *Classic* aesthetics maps onto beauty and what they call orderly and clean design. *Expressive* aesthetics encompasses judgments of creativity, which are an expression of the designer's breaking design conventions and of the opinion of the judge. Attempts to predict aesthetics from classical visual attributes alone would be a complicated undertaking, as the concept of aesthetics also includes a judgment about the expressive component. Therefore the definition of beauty used here is in keeping with Lavie and Tractinsky's (2004) *classical* aesthetics that focuses on those visual attributes that may contribute to clean and orderly design. The notion of beauty is considered synonymous with the concept of visual appeal, which was explored independently of the more complex term 'aesthetics'.

Visual appeal has been studied to understand the role it plays in initial impression formation (Adams and Huston, 1975; Dion, Berscheid and Walster, 1972; Eagerly, Ashmore, Makhijani and Longo, 1991). There is evidence suggesting that at least some initial impressions are formed very quickly (Bornstein, 1992; Zajonc, 1980), and it is likely that these impressions influence other types of judgments. This has led some HCI researchers to study the role visual appeal in guiding impressions of usability of ATM interfaces (Tractinsky, Katz and Ikar, 2000), for example.

Likewise, the goal for some researchers is to understand the role it plays in evaluating web sites (De Angeli, Sutcliffe and Hartmann, 2006; Hartmann, Sutcliffe, and De Angeli, 2008; Lindgaard and Dudek, 2002a, 2002b; Lindgaard, Dudek, Sen, Sumegi and Noonan, 2010). Other studies have shown that visual appeal is related to the formation of trust in web sites (Basso, Goldberg, Greenspan & Weimer, 2001; Kim and Moon, 1998), and some have found that the first impression may guide the evaluation of health sites that are subsequently selected or rejected (Sillence, Briggs, Harris and Fishwick, 2006). In a longitudinal study with participants seeking health advice, Sillence and colleagues found that web site selection occurs in stages, the first of which they call 'heuristic'. It is guided by the visual appeal of the web site, which includes design attributes such as amount of text, background color and layouts. Specifically, Sillence et al. found that cluttered, busy sites with unpleasant background colors were rejected within a few seconds, before the participants were able to assess the quality of the information on the site. Visual appeal has also been found to play a role in determining satisfaction with web sites (Lindgaard and Dudek, 2003). In those studies, the authors conducted interviews after participants browsed a web site and found that the satisfaction construct comprised five dimensions, one of which was visual appeal. Although the goal of Sillence et al.'s (2006) research was to understand what made people select a site and goal of Lindgaard et al.'s (2003) studies was to understand the role of aesthetics in defining another construct, the goal of Lavie and Tractinsky's (2004) research was to define the concept of aesthetics and the dimensions that comprise it. We expand that purpose here in exploring ways to produce definitive measures that can predict web page visual appeal.

Because visual appeal has been found to influence other judgments in a web context, and as it is part of the broader term aesthetics, it was worthwhile to explore objective attributes for measuring it. We argue that it is not only necessary to understand what makes web pages appealing but that it is equally important to understand what distinguishes appealing from unappealing web pages. Thus we also argue that any of the attributes used to identify the appealing pages should also identify those rated lowest in appeal. That was done here.

Although there are studies that explore the concept of visual appeal with interaction, and which account for user goals or context of use (See for example, Hassenzahl, 2004), this exploration is limited to ratings of web page visual appeal. In the present context, ratings were obtained after the stimuli had been exposed for only 50 ms, making interaction with the web site impossible. Employing this paradigm allowed us to isolate visual appeal from other concepts and to explore the objective attributes that comprise appeal ratings. There is no question that evaluations with interaction are valuable to understand the context of use (e.g., De Angeli, et al., 2006; Hartmann, et al., 2008) but here we were interested in assessing the criteria responsible for appeal judgments, free from other evaluations. This made it possible to develop a preliminary taxonomy of visual attribute combinations that could be useful in assisting designers to predict the relative appeal of their web pages.

VISUAL ATTRIBUTES

Lavie and Tractinsky's (2004) *classic* aesthetics appear to correspond to Park, Choi and Kim's (2005) set of 11 distinct visual attributes, which, in different combinations, were found to influence the perceived e-brand personality of web pages. Brand personality can be defined as, "the set of human characteristics associated with a brand" (Park, et al., 2005; p. 11). Although these authors make reference to Lavie and Tractinsky's (2004) work on aesthetics, they do not address visual appeal, but instead showed that the attributes of contrast, density, and simplicity defined the *analytic* personality and the dimensions of contrast, density, regularity and cohesion defined the *friendly* personality, for example. We wanted to know if different attributes and combinations of attributes could be related specifically to web pages rated high and low in visual appeal. When considering Park and colleague's (2005) results, it seems possible that if visual attributes can predict, for example, a *friendly* e-brand personality, these same visual attributes could also be used to predict visual appeal. It was possible to assess the web pages meaningfully according to only three (or four) of Park et al.'s 11 attributes and following definitions, namely density, balance, and contrast. Density refers to the ratio between the area of the background and the area covered by objects. Balance is the distribution of visual weight across the whole picture, and contrast is the degree of difference between elements. Symmetry was also assessed in our initial studies but the resulting rating was similar to that of balance, so it was not included here. Park and her colleagues used algorithms to both measure and vary these attributes. However, in our studies attribute ratings were made by the naked eye. Therefore, many of the attributes were difficult to assess. Given that in Park et al.'s studies 5/11 attributes did not contribute to defining e-brand personality, they were eliminated from this exploration. Since the ratio of graphics and the amount of text to background varied substantially in our samples, we also assessed the web pages according to these two attributes, which are consistent with some attributes Sillence and colleagues (2006) found guide the initial rejection of health web sites. In total, five visual attributes were employed to determine if pages previously rated high and low in appeal could be distinguished on the basis of these attribute combinations.

THE STUDY

The ratings presented here were compiled from three studies aimed at exploring the visual appeal of web pages. The experimental paradigm was the same in all studies and only the ratings of the pages were used in the following analyses. The first two samples of ratings came from previously published work but the final sample is new. Therefore, the former are only summarized briefly. The objective here was to compare the visual attribute combinations across several studies to determine if they consistently predict web page visual appeal. The method and results sections of this paper are therefore presented as one study and make reference to the previous studies.

METHOD

WEB PAGES

Three samples of web pages used in separate studies seeking to understand how quickly stable visual appeal judgments are formed were employed to test the five visual attributes. The first sample (hereafter, North American Pages) comprised 50 web pages, from different genres, used in our early studies exploring visual appeal and the immediacy of the first impression (Fernandes, Lindgaard, Dillon and Wood, 2003; Lindgaard, Dudek, Sen, Sumegi and Noonan, 2010; Lindgaard, Fernandes, Dudek and Brown, 2006). The second sample comprised 50 Chinese web pages (hereafter, Chinese Pages), also from different genres, used in a study that sought to uncover cultural differences in web page visual appeal (Lindgaard, Litwinska, and Dudek, 2008). The third sample comprised 30 North American web pages drawn from the genre of consultancy pages (hereafter, Consultant Pages).

VISUAL APPEAL RATINGS

The mean visual appeal ratings for the North American stimuli were gathered from a sample of 48 participants reported in Lindgaard, et al. (2010) and from 40 participants for the Chinese sample reported in Lindgaard, et al. (2008). Ratings in the Consultant sample, were obtained from 16 participants, in the latest study.

VISUAL ATTRIBUTE RATINGS

For the North American and Chinese samples, each of the 50 home pages was printed in color and affixed to a card displaying the mean appeal ratings on the reverse side. The five visual attributes were then assessed one at a time, by two researchers, for the North American sample, and 3 researchers for the Chinese sample, who rated each page on a 5-point scale, ranging from 1 = 'not at all' to 5 = 'vast' [amounts]. In both cases, disagreements were settled by negotiation. In the Consultant sample, however, participants rated the web pages on visual appeal and then on each of the five visual attributes. A data file was created for each sample. Each of the home pages was matched with its corresponding mean appeal rating and an additional five columns displaying each of the associated visual attribute ratings. In the North American and Chinese samples, only one attribute rating was assigned to each by the researchers, thus one whole number was assigned. In the Consultant sample, however, 16 participants offered ratings on each attribute for each web page, thus the mean value was calculated (rounded to the nearest whole number).

SUMMARY

A summary of the sample of the web pages, their genre, the number of appeal ratings, the number of attribute raters and where the studies were published is presented in Table 1.

Table 1: Web page sample, genre, number of appeal ratings, number of attribute raters and location of publication

Web Page Sample	Web Page Genre	Appeal Ratings (N)	Attribute Raters (N)	Publication
North American	Mixed	48	2	Lindgaard, et al. (2010)
Chinese	Mixed	40	3	Lindgaard, et al. (2008).
North American	Consultants	16	16	N/A

Two of the three samples comprised North American pages and one comprised Chinese pages. The genres were mixed in two samples and specific in the last. Mean appeal ratings were gathered from samples comprising 48, 40 and 16 participants, and two, three and 16 raters determined the attribute ratings, respectively. It is important to note that although the data from two of the three samples have been used in other publications, only the descriptions of the appealing web pages were included. Thus, new to this paper is the description of the results for the unappealing pages and the new study, not reported elsewhere.

CLUSTER ANALYSIS PROCEDURE

A series of cluster analyses were carried out for each sample. Each began with all five visual attributes included in the analysis. The purpose was to explore how well the mean ratings of appeal could be predicted from the various visual attributes and to better understand which attributes contribute most to appeal ratings. All three-, four- and five attribute combinations were explored. The best result was deemed to be one that resulted in the most optimal split between lowly and highly rated web pages. The analyses were performed using PAST, (Version 1.32), a program supporting multivariate non-parametric cluster analysis (Hammer and Harper, 2005). Paired linkages with Euclidean distances were measured in all comparisons.

RESULTS

The results are divided into three sections. Each describes the individual cluster- and regression analyses for each sample. Two dendograms are shown. An arrow indicates the point at which the web pages separate between high and low appeal ratings. A description of the attributes that define the most and least appealing sites is presented.

NORTH AMERICAN WEB PAGES

For North American Pages the visual attributes required to produce the clearest separation between the highest and lowest rated pages were the attributes of *density, contrast* and *graphics*. The results of the cluster analysis are shown in Figure 1.

The vertically represented numbers at the top of the dendogram represent the mean ratings on each of the 50 web pages, which are represented on the abscissa. The similarity of the clusters can be seen on the ordinate. As the Figure shows, the most highly rated pages achieving a mean rating of between 6.62 and 4.42 are clustered to the left of the dendogram and the lowest, achieving mean ratings between 2.21 and 5.06 clustered to the right (6.25 is one exception). When included in the analysis, the attributes of *balance* and *text* contributed little to the ratings of visual appeal, thus only the three were required to produce the desired separation of high and low ratings. Although a rather clean separation is clear from the resulting pairs, the meaning of it was interpreted cautiously because the dendogram shows somewhat complicated linkages.

To better understand the specific attribute ratings that characterized the pages with the highest and lowest appeal ratings, characteristics defining the five top and the five bottom pages were explored in more detail. Some of the highest rated pages are characterized by low-to-moderate density, low-to-moderate contrast and very high graphics. Other highly rated pages are characterized by moderate to high density, moderate to high contrast and high graphics. This indicates that two different combinations of density and contrast configurations coupled with high graphics defined the pages rated highest in appeal. On the one hand, those pages characterized by low to moderate density, very high contrast and very low graphics were amongst the lowest rated pages, indicating an inverse relation between density and contrast. On the other hand, very low density, very low contrast, and very low graphics also characterized some pages rated lowest in appeal. This was also the case when low density, low contrast and low graphics characterized the pages. This finding suggests that, in web design it is not any one of these attributes that characterize appealing or unappealing web pages but that at least two distinct patterns were found to be most influential in predicting web page visual appeal. To test this finding, a regression analysis, using the enter method, was conducted. The results showed that the linear combination of these three attributes accounted for 72% of the variance ($R^2 = .72$) in appeal ratings (F(3, 46) = 38.57, p > .001), with the graphics attribute accounting for the highest unique proportion of variance (partial r = .84, semi-partial r = .82). Given that two distinct combinations of density and contrast defined both the appealing and unappealing web pages, it is not surprising that these two attributes did not contribute more uniquely when the graphics attribute was already in the equation.

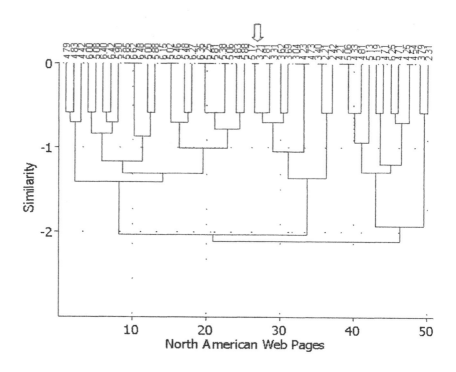

FIGURE 1. Mean appeal rating, similarity of attribute ratings for 50 North American pages

CHINESE WEB PAGES

In the Chinese sample, the results are similar to those of the North American sample. As before, *density*, *contrast* and *graphics* contributed to separating the high- from the low-rated pages. *Balance*, and relative amount of *text* to background did not contribute systematically to visual appeal ratings.

The dendogram for this cluster analysis looks very similar to that for North American pages, so it is not reproduced here. The split between high and low rated pages was not quite as clean as in the North American sample, but the pages were clustered in a very similar manner. Most of the lowest rated pages were clustered to the left, whilst the pages rated highest in appeal were clustered to the right. The split between appealing and unappealing pages appeared at approximately page 30 because 5 pages rated quite high in appeal were clustered on the low side, demonstrating that the separation for the Chinese pages was not as clean as those of the previous sample.

As before, there were two distinct combinations of visual attributes that defined the most- and the least appealing pages. Some pages rated highest are characterized by moderate density, moderate to high contrast and high graphics, as found in the North American sample. Other highly rated pages are characterized by moderate-to- high density, low-to-moderate contrast and moderate-to-high graphics. This finding indicates that a slightly different combination of density and contrast pairings separate visually appealing Chinese pages. Low-to-moderate- or moderate-to-high density and contrast pairings with very low graphics appear to characterize the lowest rated pages. It is important to note that the lowest rated pages in this Chinese sample were rated somewhat higher than those in the North American sample, which could explain why the separation demonstrated here was slightly less clear than in the previous sample. The results of a regression analysis, again using the enter method, showed that the linear combination of the three attributes accounted for 53% of the variance ($R^2 = .53$) in appeal ratings ($F(3, 46) = 17.27$, $p > .001$), with the graphics attribute accounting for the highest unique proportion of variance (partial $r = .71$, semi-partial $r = .70$).

CONSULTANT WEB PAGES

The cluster analysis for Consultant web pages showed an almost perfect split between the pages rated high and low in appeal, which is shown in Figure 2. This time the attributes of *density, contrast, graphics*, as well as *balance* contributed to this encouraging result. The *text* attribute contributed nothing.

The attribute ratings for the pages rated highest in appeal are quite consistent as are the attribute ratings characterizing the least appealing pages. Contrary to the two previous samples where two distinct combinations of attributes defined both the appealing and unappealing web pages, only one combination defines each of them here. Moderate-to-high balance, contrast, density and graphics consistently define the most appealing web pages. Low to moderate balance, contrast, and density, with very low graphics characterize the least appealing web pages. The results of a regression analysis showed that the linear

230

combination of the four attributes accounted for 83% of the variance ($R^2 = .83$) in appeal ratings ($F(4, 25) = 29.73$, p $> .001$). As before, using the enter method, the graphics attribute accounted for the highest unique proportion of variance (partial $r = .75$, semi-partial $r = .47$), but this, time the other attributes contributed more as evidenced by the somewhat lower partial and semi-partial correlations.

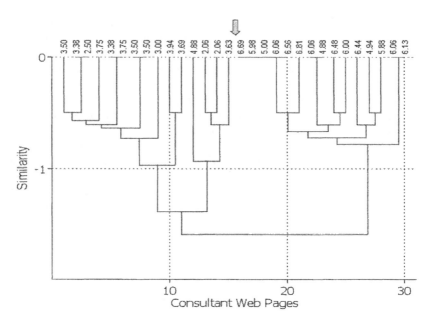

FIGURE 2. Mean appeal rating, similarity of attribute ratings for 50 Consultant web pages

RESULTS SUMMARY

A summary of the three analyses is shown in Table 2. The Table shows the two combinations of visual attributes that predicted the highest web page appeal ratings in the North American and Chinese mixed genre web pages and the only combination that predicted the same in the North American consultant genre. The two mixed genre samples produced similar results with some variation between both combinations. In the Consultant genre sample a consistent pattern was evident.

Table 2: Visual attributes for highest appeal ratings, by web page sample and genre,
combination number and associated attributes

Web Page (Genre)	Density	Contrast	Graphics	Balance
North American (Mixed)				
1.	LOW-MOD	LOW-MOD	HIGH	N/A
2.	MOD-HIGH	MOD-HIGH	HIGH	N/A
Chinese (Mixed)				
1.	MOD	MOD-HIGH	MOD-HIGH	N/A
2.	MOD-HIGH	LOW-MOD	MOD-HIGH	N/A
North American (Consultant)				
1.	MOD-HIGH	MOD-HIGH	MOD-HIGH	MOD-HIGH

The same summary was compiled for web pages rated lowest in appeal. The results are shown in Table 3. Like the web pages rated highest in appeal, both mixed genre sample web pages were characterized by two combinations of visual attributes that could predict low appeal ratings. Again, the Consultant pages rated lowest in appeal were consistently defined by only one combination of visual attributes. These are shown below.

Table 3: Visual attributes for lowest appeal ratings, by web page sample and genre,
combination number and associated attributes

Web Page (Genre)	Density	Contrast	Graphics	Balance
North American (Mixed)				
1.	LOW-MOD	HIGH	LOW	N/A
2.	LOW	LOW	LOW	N/A
Chinese (Mixed)				
1.	MOD-HIGH	MOD-HIGH	LOW-MOD	N/A
2.	LOW-MOD	LOW-MOD	LOW-MOD	N/A
North American (Consultant)				
1.	LOW-MOD	MOD	LOW	LOW-MOD

DISCUSSION

These results show support for Park, et al.'s (2005) attributes, in that density, contrast and balance, combined with graphics, were instrumental in predicting web page appeal in the genre specific sample and together accounted for 83% of the variance in visual appeal ratings. Density and contrast, combined with graphics were instrumental in predicting appeal in the two generic samples accounting for 73% and 53% of the variance, respectively. Specifically, when web pages are of mixed genre, low to moderate density and contrast is preferred to clutter; high impact color combinations and more graphic content is preferred to less. However, when web pages are representative of a specific genre, higher density and contrast are acceptable and graphic content is preferable at a lower level. Balance becomes a necessary element and according to these results, it was best at moderate levels. One possible reason for these results is that when a sample of web pages is taken from a specific genre, they look similar and thus more attributes are necessary to distinguish them from one each on visual appeal. When they are generic, the extra information provide by the additional visual attribute may not be required to separate the high and low appeal pages.

These results also show some support for Lavie and Tractinsky's (2004) classic aesthetics, which is similar to 'good design principles', outlined by others (e.g., Galitz, 2007). The five attributes used here are of the same nature as those defined as classic aesthetics, although it is unclear how the expressive aspects were accounted for by these results. Visual attributes that could define clean and orderly design were shown to have an impact on visual appeal ratings in these samples, regardless of web page genre. It is possible that the expressive aesthetics dimension may have been responsible for the somewhat less clear separation between high and low appeal pages in the two generic samples compared to those from the specific genre. It could also be that the somewhat less consistent pairings of contrast and density, with those of high graphics, evidenced in the Chinese sample particularly, was representative of the expressive dimension, not specifically accounted for in this series of explorations.

The clearer separation that was evidenced between appealing and unappealing Consultant pages could have been due to the larger number of visual attribute raters and also to the fact that all the web pages belonged to a single genre. Both factors could have been responsible for larger proportion of variance accounted for by this linear combination of four attributes than was evidenced in the previous two samples using only three of the attributes. The differences in the number of raters may also have affected the reliability of the results.

FUTURE RESEARCH

Further studies are needed to explore if these results hold for other genres, and how the attributes and attribute combinations might differ between genres. For example, balance may be more important for some genres than others. It would also be worthwhile to investigate if different combinations of attributes are necessary to define appealing and unappealing pages, when the context of use varies, as explored by some authors (De Angeli, et al., 2006; Hartmann, et al. 2008; Hassenzahl, 2004).

CONCLUSIONS

The above explorations are encouraging as they showed that visual appeal ratings could be predicted by four visual attributes, namely density, contrast, balance, graphs, in the consultant genre. The text attribute however played no role in defining the web pages rated high and low in appeal. However, these findings also show that it may be possible that different combinations of visual attributes can be used to predict web page appeal when genres differ. More research is needed to address this possibility.

REFERENCES

Adams, G.R. & Huston, T.L. (1975). "Social perception of middle-aged persons varying in physical attractiveness." *Developmental Psychology*, 11(5), 657-658.

Basso, A., Goldberg, D., Greenspan, S., & Weimer, D. (2001). First impressions: Emotional and cognitive factors underlying judgments of trust in e-commerce. *Proceedings Conference on Electronic Commerce (EC'01)*, Tampa, FL, pp. 147-143.

Bornstein, R. (1992). "Subliminal mere exposure effect." In F. Bornstein & T. Pittman, (Eds.), *Perception Without Awareness: Cognitive, Clinical and Social Perspectives* (pp. 191-210). NY: The Guilford Press.

De Angeli, A., Sutcliffe, A., & Hartmann, J. (2006). "Interaction, usability and aesthetics: What influences users' preferences?" *Proceedings of the 6th Conference on Designing Interactive Systems*, University Park, PA, pp. 271-280.

Dion, K., Berscheid, E., & Walster, E. (1972). What is beautiful is good. *Journal of Personality and Social Psychology*, 24(3), 285-290.

Eagerly, A.H., Ashmore, R.D., Makhijani, M.G. & Longo, L.C. (1991). "What is beautiful is good but..: a meta-analysis review of research on physical attractiveness stereotype." *Psychological Bulletin*, 110(1), 109-128.

Fernandes, G., Lindgaard, G., Dillon, R. & Wood, J. (2003). "Judging the appeal of web sites", *Proceedings 4th World Congress on the Management of Electronic Commerce*, McMaster University, Hamilton, ON.

Galitz, W.O. (2007). *The essential guide to user interface design. An introduction to GUI design Principles and Techniques*. 3rd Edition. NY: John Wiley Publishers.

Hartmann, J., Sutcliffe, A., & De Angeli, A. (2008). "Towards a theory of user judgment of aesthetics and user interface quality." *ACM Transactions on Computer-Human Interaction*, 15(4), Article 15, pp. 15:0-15:30.

Hassenzahl, M. (2004). "The interplay of beauty, goodness, and usability in interactive products." *Human-Computer Interaction*, 19(4), 319-349.

Kim, J. & Moon, J.Y. (1998). "Designing towards emotional usability in customer interfaces: trustworthiness of cyber-banking system interfaces." *Interacting with Computers*, 10, 1-29.

Lavie, T. & Tractinsky, N. (2004). "Assessing dimensions of perceived visual aesthetics of web sites." *International Journal of Human-Computer Studies*, 60(3), 269-298.

Lindgaard, G. & Dudek, C. (2002a). "User satisfaction aesthetics and usability: beyond reductionism." *Proceedings, International Federation of Information Processing*, (IFIP 2002), Montreal, PQ, pp. 231-246.

Lindgaard, G. & Dudek, C. (2002b). "High appeal versus high usability: implications for user satisfaction on the web." *Proceedings OZCHI*, Melborne, Australia.

Lindgaard, G. & Dudek, C. (2003). "What is this evasive beast we call user satisfaction?." *Interacting with Computers*, 15(3), 429-454.

Lindgaard, G., Dudek, C., Sen, D., Sumegi, L., & Noonan, P. (2010). "Visual appeal as a precursor to trustworthiness and perceived usability." *Manuscript under review*.

Lindgaard, G., Fernandes, G., Dudek, C., & Brown, J. (2006). "Attention web designers: You have 50 milliseconds to make a good first impression!." *Behaviour and Information Technology, Special Issue: User Experience – A Research Agenda*, 25(2), 115-126.

Lindgaard, G., Litwinska, J., & Dudek, C. (2008). "Judging web page visual appeal: Do East and West really differ?." *Proceedings of AIDIS, Multi Conference on Computers and Information Systems*, Amsterdam, Netherlands.

Lindgaard, G., Whitfield, T.W.A. (2004). "Integrating aesthetics within an evolutionary and psychological framework." *Theoretical Issues in Ergonomics Science*, 5, 73-90.

Park, S., Choi, D., & Kim, J. (2005). "Visualizing e-brand personality: Exploratory studies on visual attributes and e-brand personalities in Korea." *International Journal of Human-Computer Interaction*, 19(1), 7-34.

Sillence, E., Briggs, P., Harris, P., & Fishwick, L. (2006). "A Framework for Understanding Trust Factors in Web-Based Health Advice." *International Journal of Human-Computer Studies*, 64, 697-713.

Tractinsky, N., Katz, A. & Ikar, D. (2000). "What is Beautiful is Usable." *Interacting with Computers*, 13, 127-145.

Zajonc, R. (1980). "Feeling and thinking: Preferences need no inferences." *American Psychologist*, 35(2), 151-175.

Chapter 26

Testing Measures of Aviation Display Clutter for Predicting Pilot Perceptions and Flight Performance

Karl Kaufmann[1], David Kaber[1], Amy Alexander[2], Sang-Hwan Kim,[3] JT Naylor[1]

[1]North Carolina State University, Department of Industrial & Systems Engineering, Raleigh, NC 27695

[2]Aptima Inc., Boston, MA 01801

[3]University of Michigan-Dearborn, Department of Industrial & Manufacturing Systems Engineering, Dearborn, MI 48128

ABSTRACT

The objective of this research was to extend and validate a model of aviation display clutter including objective measures of visual display properties and pilot subjective assessments of display characteristics, as a basis for predicting pilot perceptions of clutter and flight performance. The focus of this study was on use of a Head-Down Display (HDD) in a vertical takeoff and landing (VTOL) aircraft simulator. Sixteen pilots flew a series of instrument approaches including an ILS (instrument landing system) segment to a hover phase involving vertical descent to a landing decision point. The HDD consisted of a monochrome primary flight display and navigation display along with combinations of advanced instrumentation, a highway-in-the-sky tunnel, a synthetic vision terrain depiction, and an infrared camera-based enhanced vision display. Pilots subjectively evaluated each display during the ILS segment and after completing the vertical descent on six characteristics (information redundancy, colorfulness, salience, dynamics, variability, and density) and for overall clutter. The luminance, contrast between information elements, occlusion of information elements, and global density of each display were also measured. Flight performance was measured as the root mean square error (RMSE) of deviations from the desired flight path. A factor analysis revealed ratings on the six display characteristics loaded on three latent variables comprising similarity, dynamics, and intensity. The three factors accounted for 72-76% of the variance in pilot ratings of overall display clutter. The measures of display properties accounted for 16-43% of the variance in clutter ratings. Regression models including both the latent variables and measures of display properties accounted for a small proportion of glide slope deviations under certain conditions as well as deviations in groundspeed and distance from touchdown point. The findings suggest the integrated model of display clutter, based on pilot ratings and visual properties, has utility across various display and aircraft types.

Keywords: clutter, cockpit displays, flight simulation, NextGen, vertical/takeoff and landing

INTRODUCTION

In order to increase the capacity and efficiency of the U.S. air traffic system, the NextGen (next generation) concept of operations suggests flight deck improvements, such as head up displays (HUDs) and enhanced and synthetic vision systems (EVS/SVS), may have a role in increasing airport throughput (Joint Planning and Development Office, 2009). One concern with integrating these new sources of information in the cockpit is the possibility of pilot information overload brought on by display clutter with detrimental effects on pilot performance and safety (Kaber, et al, 2007). However, defining and assessing display clutter is not a straightforward process.

Most definitions of clutter tend to emphasize the aspects of a display that interfere with gathering information, either due to the way information is presented or located on the display or the amount of information presented. The first includes a definition by Ewing, Woodruff, and Vickers (2006), which describes clutter in the context of target detection as imagery that masks a target or leads to confusion about its class or location. Miller, Grisedale, and Anderson (1999) also emphasize the way information is presented, defining clutter as the amount of overlap of objects in an image and the level of visual noise present. In the aviation domain, Prinzel and Risser (2004) defined HUD clutter as the superimposition of HUD instrumentation on the pilot's forward field of view, which is also an example of the first type of definition. In contrast, Ahlstrom (2005) provides an example of the second type of definition, saying air traffic control weather display clutter is the negative result of redundant information. The definition Verver and Wickens (1998) put forth in their summary of clutter research across domains also fits into this type; that is, clutter is the presence of information unrelated to the present task.

Methods for assessing clutter in a display, like the definitions, tend to focus on only one or two aspects of a display that might affect a viewer's experience of clutter (Kaber, et al, 2007). Some measures have employed subjective ratings of preference or workload (Haworth & Newman, 1993), target-to-background contrast (Aviram & Rotman, 2000), feature size and grouping (Muthard & Wickens, 2005), and display density (Ewing, Woodruff, & Vickers, 2006).

As a step in developing a more comprehensive method for assessing aviation display clutter, Kaber, et al (2008) developed a multidimensional model including objective measures of display properties, the information content present on the display, and subjective assessments of display by pilots. In a simulator study of HUD clutter using the Integration Flight Deck simulator (IFD) at NASA's Langley Research Center, configured as a Boeing 757, Kaber et al (2008) found objective measures of display luminance, contrast, and occlusion of features to account for up to 33% of the variance in pilot's assessments of overall clutter. When information content (basic instruments, a highway-in-the-sky tunnel, a wireframe SVS terrain depiction, and an EVS display) and pilot subjective ratings on display redundancy, colorfulness, salience, dynamics, variability and density were added to the model, 77% of the variance in overall clutter was accounted for (Kaber, et al, 2008).

The aims of the present research were two-fold. First, to validate and further refine the model of clutter developed by Kaber, et al (2008) and second to test its robustness for application to a head-down display (HDD) in a vertical takeoff and landing (VTOL) aircraft simulation. As a part of refining the model, the connection between display clutter and flight performance was also examined.

METHOD

Participants

Sixteen currently active fixed-wing pilots were recruited for the study. All had at least 15 years of line flying experience. Total flying time ranged from 4,500 hours to 22,424 hours (M=14,895 hours). Pilot ages ranged from 40-65 years (M=52.7 years old). Eight of the pilots had experience with a SVS, either in a simulator or actual aircraft, with a mean experience level of 368.5 hours. None had actual aircraft experience with an EVS, though five had a small amount of simulator experience with the system (M=12.4 hours).

Apparatus

The experiment was conducted in the Visual Imaging Simulator for Transport Aircraft Systems (VISTAS) at the NASA Langley Research Center. VISTAS is a fixed-base simulator that was configured as a generic VTOL aircraft. The simulator's out-of-cockpit visuals appeared on three display screens approximately 10 feet in front of the pilot. The HDD was presented on a single monitor, and included a combined monochrome primary flight display and navigation display (PFD and ND). In addition to the basic flight instrumentation, the HDD could also display a highway-in–the-sky tunnel, a wireframe SVS terrain model, and an EVS (infrared) image of the outside environment. These elements were combined in various combinations to vary the information content of the display from trial to trial. Figure 1 shows examples of a subset of the display configurations used. The simulator also recorded a wide range of performance and control input parameters.

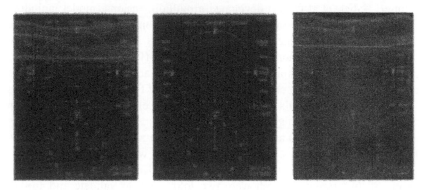

FIGURE 1 Example HDD, basic instrumentation and SVS (left); basic instrumentation only (center); basic instrumentation, tunnel, SVS and EVS (right).

The aircraft was controlled with a side-stick joystick operated by the pilot's left hand. The control system represented a relatively high-level of automation, and did not correspond to that of any current helicopter or fixed-wing aircraft. Left and right pressure on the stick controlled lateral flight path. Instead of changing pitch, fore and aft pressure affected only airspeed, increasing and decreasing, respectively. Vertical speed was manipulated with a hat-switch at the top of the side-stick. Forward movement of the switch increased rate of descent, while pulling the switch aft decreased the descent rate. Engine thrust and pitch were not directly controlled by the pilots, but were automatically adjusted in response to the airspeed and vertical speed commands from the side stick.

Procedure

The flight scenario was an instrument approach to a desert landing site under night instrument meteorological conditions (IMC) with an overcast cloud ceiling at 500 feet above the landing site. The approach began 7 NM from the landing site, with the aircraft at 2,460 MSL (1,168 AGL) traveling at an airspeed of 250 kts and on course and glide slope. The approach was flown in two phases. The first was a normal 3-degree glide slope instrument landing system (ILS) approach that ended at a hover point, 250 AGL directly above the landing site. The second phase was a vertical descent to the landing decision point 100 feet above the landing site. Figure

2 presents a schematic of the approach.

The entire approach was hand-flown by the pilots. The first phase differed from a normal fixed-wing ILS approach in that the aircraft was constantly decelerating to arrive at the hover point with zero airspeed. This required pilots to consistently reduce the rate of descent to maintain the aircraft on the glide slope. While in the vertical descent, the aircraft was beneath the cloud layer, but as it descended, it created a dust cloud that either partially or completely obscured the landing site.

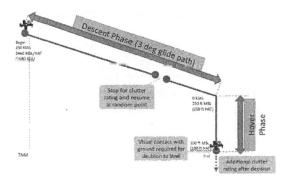

FIGURE 2 Schematic of the approach.

The pilots were also subjected to a failure in the onboard navigation system during the ILS phase on each trial. This failure caused the tunnel, SVS model of the landing site, and ND presentation of the landing site to drift either to the left or right of its actual position. The ILS localizer and glide slope indicator readings remained correct. Failures either occurred early or late in the approach at 2.39 NM or 0.9 NM from the landing site, respectively. In both cases, following the tunnel or ND course would result in a two-dot deflection of the ILS course indicator at maximum error, which occurred in about 1.5 min.

Pilot training consisted of a preflight briefing on the simulator and task, a 30 minute familiarization session in the simulator cab, and three daytime visual meteorological condition (VMC) approaches. The training approaches consisted of two part-task runs, where the pilot controlled only lateral path plus airspeed or vertical speed alone, followed by a third run in which the pilot had full control of the aircraft. Pilot performed 16 experimental trials, experiencing the eight combinations of display features twice. The experimental trials were defined by the display features (basic instrumentation, tunnel, SVS, EVS). In addition, landings were only possible in half the trials, while in the other half the landing site was completely obscured by the dust cloud. The order of experimental conditions was randomized across pilots and trials.

There was a pause during the ILS phase to collect pilot ratings of the display characteristics. Subsequently, the approach was resumed with the aircraft returned to the ILS course and glide slope. Ratings were also collected at the decision point, after the pilot announced whether or not a landing was possible.

Pilots rated the various HDD configurations on the dimensions of information redundancy, colorfulness, salience, dynamics, variability and density. Each rating was made on a 20-point scale. They also rated the overall clutter of the display on a similar 20-point scale. A rank-weighted sum of the ratings on the six dimensions, with weights based on rankings of the relative importance of each dimension, was computed as a 'clutter score' for each display in each phase.

The visual properties of the displays, including luminance, contrast between information elements, occlusion of information elements and global display density, were measured using a software image analysis tool. It was expected that these measures would be predictive of pilot overall clutter ratings and the calculated clutter scores.

Flight performance was assessed as the root mean square error (RMSE) of deviations from the localizer and glide slope during the ILS phase. These deviations were measured in dots of deflection on the localizer and glide slope indicators appearing on the PFD. In the vertical descent phase, performance was measured as the RMSE in distance from the touchdown point and RMSE from zero ground speed. It was expected that the visual display properties and pilot ratings of the displays on the six dimensions would be predictive of flight performance.

RESULTS

A total of 256 experimental trials were performed during the experiment. Several were excluded from analysis due to pilot performance outside of pre-established criteria (e.g., more than full scale deflection of the localizer or glide slope indicator). Other trials were terminated early due to loss of aircraft control resulting in crashes or flying backward. In total, 30 ILS phases and 15 vertical descent phases were excluded from analysis.

A factor analysis (FA) was conducted to determine whether there were latent variables driving pilot perceptions of the six display characteristics and whether the model of perceived clutter, based on the pilot ratings, could be made more parsimonious. The FA, using an orthogonal varimax rotation, revealed the ratings on the six display dimensions to load on three factors accounting for 79% of the variation in calculated clutter scores. Each factor was based on dimensions with loadings of 0.8 or greater. Table 1 shows the components and loading factors associated with each display dimension. Component #1, including ratings on the dimensions of redundancy, salience, and density, was labeled as "similarity." Component #2, composed of the dimensions of dynamics and variability, was labeled "dynamics." Component #3, consisting only of the dimension colorfulness, was labeled "intensity".

These three factors were then used in calculating a revised clutter score, basing the value of each factor on a weighted sum of the original dimensional ratings that loaded on each factor. Each dimension was weighted by dividing its loading value by the sum of loading values across dimensions for a particular factor. For example, the weighting factor for redundancy was 0.836/(.836+.831+.849). The resulting factor values were then weighted by the relative importance of the dimensions loading on each factor and summed to arrive at the revised clutter score.

Table 1. Rotated component matrix.

	Component		
	1 (36.705%)	2 (29.3%)	3 (13.388%)
Redundancy	0.836	0.041	0.045
Salience	0.831	-0.064	-0.032
Density	0.849	0.163	-0.074
Dynamics	0.121	0.874	0.146
Variability	-0.020	0.900	0.088
Colorfulness	-0.038	0.186	0.980

This new three-factor score was closely correlated with the original score calculated based on the rankings and ratings of the six display dimensions for the ILS phase (r=.98222, p<.0001), the vertical descent phase (r=.97833, p<.0001), and across phases (r=.98101, p<.0001). The three-factor score was also positively correlated with pilot ratings of overall clutter of displays in the ILS phase (r=.75526, p<.0001), in the vertical descent phase (r=.72281, p<.0001) and across phases (r=.76204, p<.0001)..

Regression modeling of clutter scores and ratings based on the measured display properties revealed significant predictors for both the ILS and vertical descent phases of the approach. Regressing clutter RATINGS from the ILS phase on visual properties resulted in a model accounting for 43.52% of the variance in the ratings, $F(2, 234)$=45.08, p<.0001, with the parameters shown in Table 2. Regressing clutter SCORES for the ILS phase on visual properties resulted in a model accounting for 29.59% of the response variance, $F(4,237)$=24.87, p<.0001, with the parameters shown in Table 3. The regression model for clutter RATINGS in the vertical descent phase, based on visual properties, accounted for 16.42% of the variance in ratings, $F(3, 236)$=15.46, p<.0001, with the parameters shown in Table 4. The model of vertical descent phase clutter scores only accounted for 7.33% of the response variance, $F(3, 237)$=6.25, p=.0004, with the parameters shown in Table 5.

Table 2. Descent phase clutter rating regression parameters.

Term	Estimate	Std Error	t Ratio	Prob>|t|
Intercept	-159.68	48.13	-3.32	0.0011
Luminance	14.46	1.56	9.29	<.0001
Contrast	2901.23	733.28	3.96	0.0001
Occlusion	4.031	0.88	4.56	<.0001
Density	27.37	7.34	3.73	0.0002

Table 3. Descent phase clutter score regression parameters.

Term	Estimate	Std Error	t Ratio	Prob>|t|
Intercept	-86.74	34.88	-2.49	0.0136
Luminance	7.299	1.120	6.52	<.0001
Contrast	1938.01	531.42	3.65	0.0003
Occlusion	2.776	0.6431	4.32	<.0001
Density	18.67	5.318	3.51	0.0005

Flight path performance was also modeled in terms of the overall clutter ratings, three-factor clutter score, and display visual properties. Because the measures of flight path deviation were different for the ILS and vertical descent phases, separate models for each measure were constructed within phases. Further, since the information acquisition and interpretation demands differed in the ILS phase before and during the navigation failure, separate models were constructed for each period.

Table 4. Hover phase clutter rating regression parameters.

Term		Estimate	Std Error	t Ratio	Prob>ltl
Intercept	Biased	11.444029	4.983806	2.30	0.0225
Luminance (Measured)	Biased	8.8659616	2.044094	4.34	<.0001
Contrast (Measured)	Biased	118.07951	23.98766	4.92	<.0001
Occlusion (Measured)	Biased	1.2606117	1.826808	0.69	0.4908
Density (Measured)	Zeroed	0	0	.	.

In the ILS phase, a regression model predicting glide slope (G/S) RMSE based on the three-factor clutter scores and display properties was significant for the late navigation system failure conditions, $F(6, 234)=2.46$, $p=.0252$; however, it only accounted for 6% of the variance in the RMSE response. Regression parameters for this model are shown in Table 6.

Table 5. Hover phase clutter score regression parameters.

Term		Estimate	Std Error	t Ratio	Prob>ltl
Intercept	Biased	34.119649	3.470966	9.83	<.0001
Luminance (Measured)	Biased	3.0630575	1.426673	2.15	0.0328
Contrast (Measured)	Biased	41.238669	16.74194	2.46	0.0145
Occlusion (Measured)	Biased	1.2486978	1.272552	0.98	0.3275
Density (Measured)	Zeroed	0	0	.	.

Table 6. Regression model of G/S RMSE in descent (late NAV failure).

Term	Estimate	Std Error	t Ratio	Prob>ltl
Intercept	1.8130943	0.600623	3.02	0.0028
Similarity (3 Factor Score)	0.0542225	0.019237	2.82	0.0052
Dynamics (3 Factor Score)	0.0300809	0.010336	2.91	0.0040
Clutter Score (3 Factor)	-0.016474	0.005778	-2.85	0.0047
Contrast (Measured)	-25.82127	9.30327	-2.78	0.0060
Density (Measured)	-0.257157	0.092696	-2.77	0.0060
Luminance (Measured)	-0.040942	0.021435	-1.91	0.0573

Table 7. Hover phase groundspeed RMSE regression parameters.

Term	Estimate	Std Error	T Ratio	Prob>ltl
Intercept	1.8927662	0.140976	13.43	<.0001
Clutter Ratings	0.0118393	0.002767	4.28	<.0001
Similarity (3 Factor Score)	-0.098824	0.016755	-5.90	<.0001
Dynamics (3 Factor Score)	-0.025314	0.009141	-2.77	0.0061
Density (Measured)	0.0007716	0.000843	0.92	0.3610

For the vertical descent phase, it was possible to create regression models of groundspeed and range RMSEs from the pilots' ratings of overall clutter, a subset of the factors in the three-factor clutter score, and display properties. The model of ground speed RMSE accounted for 15% of the variance in RMSE, $F(4, 233)=10.10$, $p<.0001$, with the parameters shown in Table 7. Range from touchdown RMSE was also predicted by the parameters, $F(4, 233)=7.74$, $p<.0001$, accounting for 12% of the variance in RMSE. The parameters are shown in Table 8.

Table 8. Hover phase range RMSE regression parameters.

Term	Estimate	Std Error	T Ratio	Prob>ltl
Intercept	55.915851	6.113913	9.15	<.0001
Clutter Ratings	0.4473533	0.120002	3.73	0.0002
Similarity (3 Factor Score)	-3.544948	0.726635	-4.88	<.0001
Dynamics (3 Factor Score)	-1.185273	0.396442	-2.99	0.0031
Density (Measured)	0.0406353	0.036562	1.11	0.2675

DISCUSSION

The results of this study suggest that the three factor variables (similarity, dynamics and intensity) represent a parsimonious set of display qualities that predict pilot experiences of clutter. The variance in overall clutter ratings accounted for by the three-factor clutter score was comparable to the variance accounted for by the original six-dimension score. In addition the utility of the three-factor score for predicting perceived clutter was comparable to the utility of the six-dimension model developed for HUD evaluation in a normal fixed wing aircraft by Kaber et al. (2008). In this prior study, the six-dimension model accounted for 65-77% of the variance in pilot overall clutter ratings (Kaber et al., 2008). The three-factor model developed here accounted for between 72-77% of the variance in pilot overall clutter ratings.

The connection between clutter and flight path deviations, while present, does not appear to be as strong as that between the three–factor score, display properties, and clutter ratings. This too, is similar to the prior findings of Kaber et al. (2008). The low predictive power of the clutter models for explaining performance in both studies may have been due to the global nature of the clutter measures. In performing the simulated approaches, pilots most likely did not make equal use of all the information available on the display, instead concentrating their attention on its most relevant portions. Thus, performance would have been dependent on small subset of the display, while the overall clutter ratings, individual display dimension ratings, and visual property measurements were based on the appearance of the entire display (HUD and HDD), leading to the low proportion of performance variance accounted for by the regression models.

ACKNOWLEDGEMENTS

This research was supported by NASA grants NNL06AA21A and NNX09AN72A. The opinions expressed are those of the authors and do not necessarily reflect the views of NASA. Lance Prinzel led the NASA Langley Research Center team including Trey Arthur, Steve Williams, and Mike Norman. Elliot Entin of Aptima conducted the FA work. We also thank Emily Stelzer for her extensive work planning and conducting the experiment. We thank Randy Bailey and Steve Young for committing NASA Langley Research Center resources to this project.

REFERENCES

Ahlstrom, U. (2005). Work domain analysis for air traffic controller weather displays. *Journal of Safety Research, 36,* 159-169.

Aviram, G., & Rotman, S. (2000). Evaluating human detection performance of targets and false alarms, using a statistical texture image metric. *Optical Engineering, 39,* 2285–2295.

Ewing G. J., Woodruff, C. J., & Vickers, D. (2006). Effects of 'local' clutter on human target detection. *Spatial Vision, 19,* 37-60. Haworth, L. A. & Newman, R. L. (1993). Test Techniques for Evaluating Flight Displays (Tech. Memo. 103947). Washington, D.C.: NASA.

Joint Planning and Development Office (2009, December). Concept of operations for the next generation air transportation system (version 3.0). Retrieved February 26, 2010, from http://jpe.jpdo.gov/ee/request/folder?id=28445.

Kaber, D., Alexander, A. L., Stelzer, E., Kim, S-H, Kaufmann, K., Cowley, J., Hsiang, S., & Bailey, N. (2007). *Testing and validation of a psychophysically defined metric of display clutter.* (NASA Langley Research Center Grant Number NNL06AA21A). Hampton, VA: NASA Langley Research Center.

Kaber, D. B., Kim, S-H., Kaufmann, K., Veil, T., Alexander, A., Selzer, E., Hsiang, S., & Bailey, N. (2008). *Modeling the effects of HUD visual properties, pilot experience, and flight scenario on a multi-dimensional measure of clutter* (NASA Langley Research Center Grant Number NNL06AA21A). Hampton, VA: NASA Langley Research Center.

Miller, G., Grisedale, S. & Anderson, K.T. (1999). 3Desque: Interface elements for a 3D graphical user interface. *Journal of Visualization and Computer Animation*, 10(2): 109-119, Apr-Jun.

Muthard, E. K. & Wickens, C. D. (2005). *Display size contamination of attentional and spatial tasks: An evaluation of display minification and axis compression* (Tech. Rep. AHFD-05-12/NASA-05-3). Savoy, IL: University of Illinois, Aviation Human Factors Division.

Prinzel, L. J. III., & Risser, M. (2004). *Head-up displays and attention capture.* (NASA Technical Report NASA/TM-2004-213000 Corrected Copy). Hampton, VA: Langley Research Center.

Verver, P. M., & Wickens, C. D. (1998). Head-up displays: effects of clutter, display intensity, and display location on pilot performance. *The International Journal of Aviation Psychology, 8,* 377-403.

CHAPTER 27

Operator Display Issues: Examples from Hydrocracking

Steven Underwood, Jennifer Shinkle, Jennie J. Gallimore

Wright State University

ABSTRACT

Hydrocracking is a refining process that converts gasoil into higher grade fuels. This complex process requires human supervision twenty-four hours, seven days a week and is high-risk; therefore, safety is a primary concern. Operators supervise this process in a control room using information presented through visual displays, visual alarms, and auditory alarms. They require access to information quickly, accurately, and in formats that support rapid decision-making. Using interviews, observations, and surveys we evaluated the visual displays used during hydrocracking to begin improving the displays. Petrochemical plants primarily use mimic graphic displays, which are diagrams of the physical plant with data values related to the process integrated into the schematic. There is no evidence to support that this format is best for providing situation awareness or decision-making support. Hundreds of displays and hundreds of alarms are available to the operator during the process supervision. This paper summarizes our findings related to display issues that are relevant not only to hydrocracking but also to other refinery and chemical process systems.

Keywords: displays, operator control, mimic displays, refineries, decision-making

INTRODUCTION

Hydrocracking, the refining process that converts gasoil into higher grade fuels, is a high-risk operation because it involves high pressures, high temperatures, and volatile compounds. Petrochemical facility safety is the top concern for operators, engineers, management, and the surrounding communities. Events that seem innocent and unrelated can accumulate to create malfunctions with disastrous results (Perrow, 1999). In addition to human injury, the monetary costs of accidents and interruptions to production are staggering. Nimmo (1995) stated that abnormal situations cost the petrochemical industry an estimated ten-billion dollars per year. In principle, the simple distillation techniques of oil refining from years ago are still used in the present complex oil petrochemical plants of 2009. However, increased demand for higher efficiency and productivity resulted in tremendous increases in the sophistication of process control systems (Cochran, Miller, Bullemer, 1996). Unchecked technological upgrading contributed to the technology-centered instead of user-centered interfaces, which contributes to cognitive overload and information excess. Typically, experienced control room operators, rather than human factors professionals, have designed control interfaces by replicating schematics; therefore, human factors and human-computer interface standards were not employed (Gallimore and Shinkle, 2008; Jamieson, 2007).

STATEMENT OF THE PROBLEM

The specific problem being addressed is the analysis of the information presentation to process operators. The large number of displays and presentation formats must be addressed to provide better support for situation awareness and decision-making.

RESEARCH METHODS

As an exploratory study to investigate current display technology in the refinery industry, we performed literature reviews, concept mapping exercises to understand the domain, subject matter interviews, observations and a simple survey. During this time we visited five different refinery plants throughout the U.S. and Canada.

SURVEY

The purpose of the survey was to obtain information about operator experience and opinions of the visual displays. A web-based survey was created and made available using Survey Monkey (SurveyMonkey.com). The 41 questions were broken into the following areas of interest: Demographics, Environment, Display

Effectiveness, Task-related, and Physical. Participation in the survey was voluntary and anonymity was guaranteed. Volunteers were initially recruited through an email sent by the Center for Operator Performance (COP) to refineries. An email containing a link to the survey was sent to those wishing to participate. The survey was open for approximately one month before results were collected and analyzed.

FINDINGS

DISPLAY TYPES AND ISSUES

Though exact graphic methods differed by location, there were hundreds and sometimes thousands of display screens available to a single operator. Overall, there were five types of control-board interface displays in use.

Mimic Displays

The most frequently used display types were mimic displays; these are schematic diagrams that show the plant or equipment layout, similar to a map or piping and instrumentation diagram. The mimic displays allow the operator to go from a high-level view of the plant (showing all stages of hydrocracking) down to individual hydrocracking equipment. Specific data related to temperatures, pressures, etc are integrated into the mimic display. The mimic display provides spatial layout information and matches the operator's mental model of how the hydrocracking process works. Before working on the hydrocracker console, board operators gain experience working as outside field operators; therefore, their mental models are developed based on plant and equipment layout. Some operators indicated that the mimic display helped them to relate the information to the equipment in the field. For example, if they got an alarm on a specific tray, they could call up the schematic that provided relevant spatial information showing the location and surroundings of the temperature indicators. Their cognitive map based on their field experience supported their ability to mentally navigate through the system in an organized manner to consider potential sources of error.

During a complex system upset requiring very rapid response, effective navigation through mimic displays is difficult. In addition, processing the large amount of information presented on mimic displays can be overwhelming. Hydrocracking board operators are presented with specific data values for Present Value (PV), Set Points (SP) Output Values (OP), valve states (% open), temperatures, and pressures. They are required to cognitively process the differences between values to determine system state. The dynamic process with different types of quantity information requires operators to develop an understanding of the range and variation of these quantities. When asked how they "knew" what the value should be their answers tended to be "you just know" or "it's common sense." One operator's facilitating strategy was to print all the displays at the beginning of his shift in order to compare the current state to the previous values from the start of his shift. The printout, as much as 60 pages, required the operator to shuffle through display screens as well as paper. Humans can cope with processing information by applying mechanisms to facilitate memory retrieval and comparative evaluation, but having hundreds of indicators across hundreds of screens and thousands of alarms is beyond the capability of humans even when using memory enhancement strategies.

Novice board operators (those learning or with 1-2 years of experience) indicated a preference for the schematic mimic screens over other display types. Some of the more experienced operators indicated that they preferred group or trend displays that showed them information about the overall process; however they were required to train new operators to use the mimic displays. In general, less experienced operators have not yet developed a complete mental model of how the whole system operates, while more experienced operators understand system nuances that allow them to quickly pick up on system changes by looking across groups of data and trends. These differences have impact on display design and training. Jamieson, Miller, Ho, & Vicente (2007) concluded that the mimic displays were the primary means by which operators controlled the hydrocracker. Currently, most process control companies focus on development of mimic displays.

Benefits of mimic displays are documented in the literature. In 1993, the Idaho Engineering National Laboratory listed in their guidelines that mimic displays are a positive feature of digital process control (DPC) systems (Wilhelmsen, Gertman, Ostrom, Nelson, Galyean, & Byers, 1993, 281). Builemer & Nimmo (1998) found that schematic mimic displays inherently contain prompts to cue the operator to action and stated that the loss of mimic boards negatively impacted understanding of disturbances and their effects on the system. They proposed that mimic boards provided integrated configuration needed to cue the operator through the big picture gained from schematically-presented data elements.

Integrated Group Displays

Integrated group displays were used to allow the operator to view information across equipment on one screen. The designs of these displays vary across facilities in terms of layout and information presented. Some experienced operators preferred these display to schematics, especially if they were trained on these displays. However, they did not perform control inputs from these displays.

Trending Graphs

Line graphs that showed system trends over time were reported to be very important, and they were prominently displayed and available at all times to the operator. Trending information was used to provide the operator with overall high-level system situation awareness.

Tabular Text Lists

Tabular lists were used to present alarm information. While auditory and other flashing icons were also used, the text list was used to scroll through all the alarms. The importance of the tabular listings varied across locations. At some facilities, a text listing of alarms was at all times displayed on a monitor directly in front of the operator. On some systems, clicking on the alarm served as quick navigation to the relevant data items.

Bar-Overview Displays

Bar-overview displays allow operators to determine if the process state has changed from the original set point using a graphic bar. The operator usually sets the initial system states at the beginning of the shift. When a status bar changes in size, they look through other displays to determine why the state has changed. In several facilities, the bar-overview display was always visible on one monitor. However, the bars were not normalized; the critical range varied across all the bars, so the operator had to know what the size of each individual bar should be.

Survey Results

A total of 28 participants, 26 males and 1 female, 1 unknown completed the survey (answering at least 50% of the questions). Participants ranged in age from 27 to 60 years with a mean age of 46. Experience as a board operator ranged from 2 years to 33 years, with the average being 16 years. Of the 28 participants, more than half indicated having some form of corrected vision (contact lenses, bifocals, or other glasses). One participant indicated that they have a form of color deficiency. Not all participants answered every question.

Figure 1 summarizes issues related to fatigue and stress. Approximately half the operators polled indicated they felt pain during their shift. With respect to eye strain and fatigue, more than half indicated moderate to severe levels, and for stress, nearly half indicated moderate to severe levels. Figure 2 summarizes responses related to eight questions on display effectiveness: trend information, information organization, providing situation awareness (SA), easy-to-make decisions, easily understood plant information, prediction support, decision-making support, and easy-to-find data. Interestingly, about half the responses rate the displays as effective or very effective, and half of the responses fall from slightly ineffective to ineffective. Note that the sample size is small (n=21) and operators use different systems. However, it appears that improvements can be made.

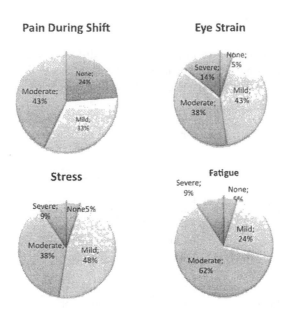

Figure 1. Percentage of people reporting pain, eye strain, stress, and fatigue during a shift.

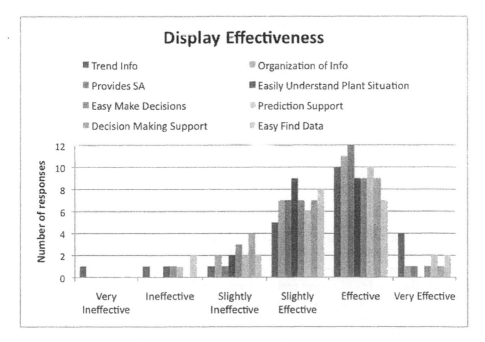

Figure 2. Subjective impressions of effectiveness of visual displays.

Operator Strategies

Hydrocracking control board operators deploy different strategies depending on their level of training and personal preferences. Table 1 illustrates a three-step strategy for supervising the plant offered by a few experienced operators during interviews.

Table 1. Three-step strategy for supervising hydrocracker during a 12-hour shift.

Make Relief or *"start the coffee intravenous line"*	Obtain high-level SA through overview & trend screens
	Information transfer between shift operators
	Inspect key indicators: Limits, constraints, product characteristics vs. specifications, lab results
	Examine critical items: H2 levels, feed section
Virtual Walk Through or *"BR break and coffee refill"*	High pressure separator
	Outputs, limitations, valve positions
	Debutanizer
	Off gasses
	Liquid flow
	Furnace O2
Supervise or *"try to chill out - enjoy the shift"*	Perform benchmarking
	Monitor system at high level
	Overview screen
	Trend screen
	Respond to alarms, drill down when needed
	Maximize charge to the limits
Critical Situation	Respond and get help!

NEXT STEPS

Based on our exploratory study, there is a concern related to information presentation. The issue is whether current display techniques provide appropriate decision-making support and actionable information. Many questions may be asked. What are the benefits of current representations? What are the limitations? What is the best way to combine the information so that it is easily available, given the decisions being made? How do operator strategies affect how information should be displayed? What information do expert systems controllers look for and focus on throughout the day under normal and abnormal circumstances?

The next step of this research focuses on determining the specific data elements and information required to support decision-making. The data and information will be mapped to the decisions. The displays will then be evaluated to see where the information is currently located across the display screens and monitors. The purpose of the evaluation is to determine the extent to which operators must traverse multiple screens and rely on their memory to make decisions. If the mapping of information to decisions is poor, concepts will be offered to support redesign.

An ongoing research area is the use of configural displays, which use geometric shapes as global indicators of system state. Within normal operating ranges, the display appears as a symmetrical, geometric polygon. With abnormal ranges, the shape appears as an irregular polygon, helping the operator to instantly recognize the problem. The symmetrical, normal state shape is considered an emergent feature (Jenkins 2007). Configural displays show promise of equipping the operator with high-level SA but still allow the user to obtain specific data values. An example of a configural display with emergent features is shown in Figure x.3.

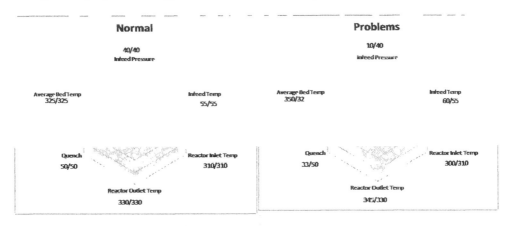

Figure 3. The left-side of the figure shows an emergent hexagonal shape because the system is in normal state. On the right-side figure, abnormal system status make she hexagon become an irregular polygon.

CONCLUSIONS

The refinery industry has recognized the need for improvement of process displays. This research was an exploratory study to evaluate the current state of refinery displays using hydrocracking as an example. There is significant need for the development of interactive displays that support decision-making, SA, and agile and responsive operations; provide visibility of system status and problems before they impact operations; support proactive versus reactive responses and multi-task collaborative environments; and provide the right information at the right time.

ACKNOWLEDGEMENTS

This research was sponsored by the Center for Operator Performance. (www.operatorperformance.org).

REFERENCES

Builemer, P., and Nimmo, I. (1998). Tackle abnormal situation management® with better training. *Chemical Engineering Process, 94*, 43-54.

Cochran, E. L., Miller, C., & Bullemer, P. (1996). Abnormal situation management in petrochemical plants: Can a pilot's associate crack crude? In *Proceedings of the IEEE National Aerospace and Electronics Conference, 2*, 806-813.

Gallimore, J.J., and Shinkle, J. (2008). *Color usage for graphic displays in process control*. Technical Report, Wright State University, Dayton, OH 45345.

Jamieson, G.A. (2007). Ecological Interface Design for Petrochemical Process Control: An Empirical Assessment._*IEEE Transactions On Systems, Man, And Cybernetics—Part A: Systems and Humans, 37*(6), 906-920.

Jamieson, G.A. Miller, C.A., Ho, W.H. and Vicente, K.J. (2007). Integrating task- and work domain-based work analyses in ecological interface design: A process control case study. *IEEE Transactions On Systems, Man, And Cybernetics—Part A: Systems And Humans, 37*(6) 887-905.

Jenkins, J.C., and Gallimore, J.J. (2008). Configural features of helmet-mounted displays to enhance pilot situation awareness. *Aviation, Space, and Environmental Medicine, 79*, 397-407.

Perrow, C. (1999). *Normal accidents: Living with high-risk technologies*. New Jersey: Princeton University Press

Nimmo, I., (1995). Adequately address abnormal situation management. *Chemical Engineering Progress, 91*, 36–45.

Wilhelmsen, C.A., Gertman, D.I., Ostrom, L.T., Nelson, W.R., Galyean, W.J. and Byers J.C. (1993). *Reviewing the impact of advanced control room technology*. Technical Report No. DE-AC07-76ID01570, Idaho National Engineering Laboratory, Idaho Falls, ID 83415.

Chapter 28

Automatic Generation of Graphical User Interface using Genetic Algorithm

Takanori Mori, Takako Nonaka, Tomohiro Hase

Ryukoku University

ABSTRACT

This paper proposes a method to arrange icons with different size and shape on an audio visual control screen efficiently and to create graphical user interfaces automatically using a genetic algorithm. In this paper, genes of the genetic algorithm mean icons, and individuals mean their alignments. The proposed system was fitted to an embedded system, employing a microcomputer with an operation clock of 100 MHz and a touch panel. As a result, a graphical user interface was created within 30 seconds.

Keywords: Genetic Algorithm, AV Remote Controller, Graphical User Interface, Embedded System

INTRODUCTION

Recent audio visual (AV) devices are becoming more multi-functional, and AV systems consisting of such AV devices are becoming larger in scale. Consequently, an AV remote controller requires a number of buttons and icons. On the other hand, AV remote controllers are required to be small enough to fit in the operator's palm, and therefore, icons are desired to be suitably arranged on them.

This paper proposes an automatic generation method to arrange the icons to be displayed as components of the graphical user interfaces (GUI) on the LCD screen of an AV remote controller by means of a genetic algorithm (GA) (Grefenstette, 1987).

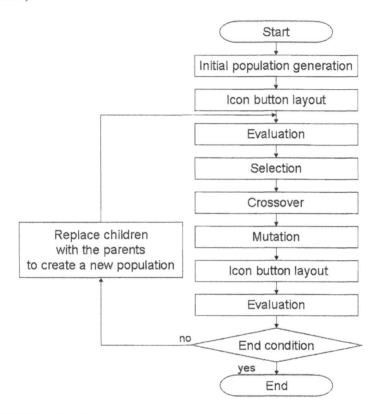

FIGURE 1 Flowchart of the GA of our proposed system.

PROPOSED SYSTEM

In the proposed idea, the GA is performed in two stages. The first stage is to minimize the number of display screens as much as possible, and the second stage is to minimize the gaps as much as possible. In general, the flow of a GA is composed of five processes, creation of parent population, assessment, selection, crossing, and mutation. Figure 1 shows a flowchart of the GA of our proposed system. In this system, the gene is defined as an AV equipment icon. The individuals are assumed to be combinations of arrangements in which all the gene icons are adequately located. A set of individuals, which is a combination of arranged icons, is assumed to be a group of the parental generation or the child generation. The GA repeats crossing, selection and/or mutation. For selection, we adopt a tournament system in synchronization with the specified assessment function. The number of individuals to be used in the tournament is set to 7. In crossing, the two individuals that survive the tournament are used and the crossing ratio is set to 100 %. For mutation, the highest value of a reasonable range times 0.01 is adopted to promote quicker calculation. The calculation was stopped when the number of display screens and the number of gaps, specified beforehand, were reached.

OPERATION VERIFICATION

For verification, an experimental evaluation device was created for trial by using a embedded microcomputer and a embedded OS (Renesas Solutions Corp, 2005, paras. 18-19), supposing that the device would be used in consumer appliances.

When the above process was finished, the results of the GUI, in which icons were laid out properly, were displayed on an LCD screen. These processes were executed on the experimental device within 30 seconds regardless of the number of icons and other conditions. Figure 2 shows the appearance of the selection screen, and Figure 3 shows the GUI displayed as a result of GA.

As a result, it was confirmed that it is possible and effective to use GA to achieve an adequate arrangement of GUI displayed icons on the LCD screen of an AV remote controller.

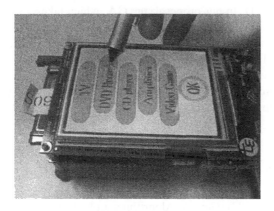

FIGURE 2 Selection of device.

FIGURE 3 The GUI created by GA.

CONCLUSIONS

This paper has proposed a method of using GA to display all icons of an AV remote controller on an LCD. In our proposal, the GA, which takes an icon as a gene, and an icon array as an individual, is embedded in a two stage structure. A prototype, working like a remote control, was used in order to verify the proposed technique. When the GA ended, the GUI results, in which the icons were arranged adequately, were displayed on the LCD screen within 30 seconds.

The above results suggest that it is possible and effective to use GA to achieve an adequate arrangement of GUI displayed icons on the LCD screen of an AV remote controller.

REFERENCES

Grefenstette, J. J. (1987). Genetic Algorithms and Their Applications: Proceedings of the Second International Conference on Genetic Algorithms. *Lawrence Erlbaum Assoc Inc.*

Renesas Solutions Corp; Microcomputer Tool Marketing Department. (2005, Jul). R0P7727TH002TRK General Information Manual Rev. 1.00. Retrieved from http://documentation.renesas.com/eng/products/tool/rej10j1008_7727_g.pdf

Chapter 29

A First Step Towards an Integrated Warning Approach

Kathrin Maier, Heike Sacher*, Jürgen Hellbrück

Catholic University Eichstätt-Ingolstadt, Germany
*Audi AG Ingolstadt, Germany

ABSTRACT

Aiming to address future HMI requirements, a new warning concept for Advanced Driver Assistance Systems focusing on high-priority braking events was evaluated. Therefore, a driving study was conducted testing the potential of two innovative visual warning approaches, a peripheral LED-strip warning and a windshield display warning, in comparison to an auditory alarm. All warnings were presented both unimodal as well as multimodal along with a haptic warning element. The findings on peripheral visual warnings are encouraging in terms of brake-onset reaction-time, especially in the context of multimodal warning strategies. Further implications will be discussed.

Keywords: Advanced Driver Assistance Systems, human-machine-interface, warning strategy, multimodal warnings, peripheral warnings

INTRODUCTION

Advanced Driver Assistance Systems (ADAS) are playing a crucial role in directing drivers' attention towards accident-critical traffic events by providing additional information, warnings and action recommendations. Yet, driver support and accident prevention can only be as good and effective as warning concepts are intuitive and informative to the driver.

To address future requirements on the human-machine-interface (HMI), a multistage research project was undertaken aiming to develop an innovative display concept for ADAS. Today's traditional warning philosophy is system-specific by providing characteristic warnings for each single ADAS. Yet, with the ongoing increase of ADAS this feedback strategy will result in a burst of different warnings and thus might lead to heavy information overload for the driver (Cummings, Kilgore, Wang, Tijerina, & Kochhar, 2007). Therefore, a new integrated warning approach relying on a context-based warning strategy is introduced. Instead of displaying numerous signals conveyed by different ADAS, warnings in a context-based setting are comprised across single systems. Concentrating on accident prevention, they offer action recommendations based on information about proximity and location of hazardous traffic situations. Thereby, the number of warning signals can be significantly reduced to two key messages: recommendations to brake and recommendations to steer. As part of this research project, the present study examines different warnings to optimally inform drivers about *high-priority braking events*.

Several studies have already assessed warning effectiveness as a function of warning modalities, including different forms of visual, auditory and haptic warnings (see for review Spence & Ho, 2008; Ho & Spence, 2008). Claiming that at least 90 per cent of the driving task is based on visual information, research on warning displays concentrated primarily on auditory and haptic warnings (e.g. Ho & Spence, 2005, 2008; Scott & Gray, 2008; Ho, Reed, & Spence, 2007; Lee, Hoffman, & Hayes, 2004; Brown, 2005). Non-visual warnings indeed revealed important reaction-time benefits over visual warning signals in cars, yet these findings largely relied on studies testing foveal visual warnings (Nikolic & Sarter, 2001; Hameed, Ferris, Jayaraman, & Sarter, 2009; see for example Scott & Gray, 2008; Lee, McGehee, Brown, & Reyes, 2002; Liu, 2001).

Contrary to foveal warnings, peripheral warnings are unspecific in nature, often realized by means of luminance change, colour contrasts or burst of motion. Thus, peripheral signals do not require foveal focusing in order to be perceived. They rather support the driver to keep his primary attention focus on the driving task (Hameed et al., 2009; Regan, 2000). Located in the driver's right and/ or left peripheral visual field, they additionally offer spatial cues about the critical driving event (Wright & Ward, 2008; Wickens & McCarley, 2008; Calvert, Spence, & Stein, 2004). Despite these promising features, until today hardly any study has evaluated the potential of peripheral visual warnings in complex driving situations (Doshi, Cheng, & Trivedi, 2009; Nikolic & Sarter, 2001 in the context of aviation).

Considering that the auditory modality is also very much engaged in the driving process (Wiese & Lee, 2004; Scott & Gray, 2008; Lee et al., 2004; Ferris, Penfold, Hameed, & Sarter, 2006), innovative peripheral visual warning strategies should not be ignored by research.

Due to promising findings of synergistic multisensory integration effects in cognitive neuroscience (Calvert, Spence, & Stein, 2004; Stein, Stanford, Ramachandran, Perrault Jr., & Rowland, 2009; Ferris & Sarter, 2008), recent research has primarily concentrated on multimodal warnings. Until today, a great deal of studies has revealed significant benefits of multimodal warnings in comparison to unimodal warnings, especially in complex and demanding driving situations (e.g. Ho et al., 2007; van Erp & van Veen, 2004; Brown, 2005; Ho & Spence, 2008; Ferris et al., 2006; Santangelo, Ho, & Spence, 2008). In consequence, research on peripheral warnings represents a promising approach, especially in a multimodal presentation mode.

In search of a context-based intuitive warning strategy to recommend imminent braking events, the present study compared two peripheral visual warnings to an auditory signal. All warnings were displayed unimodal as well as multimodal. Concerning multimodal warnings, former studies had concentrated predominantly on audiovisual signals (Chan & Chan, 2006; Selcon, Taylor, & McKenna, 1995; Lee et al., 2002). Yet, referring to recent results indicating significant benefits of haptic elements within multimodal warnings (Spence & Ho, 2008; Brown, 2005; Fitch, Kiefer, Hankey, & Kleiner, 2007; Ho et al., 2007), in this study visuo-haptic respectively audio-haptic signals were provided.

METHOD

PARTICIPANTS

Forty participants with an average driving experience of 19 years took part in this study. The sample consisted of 17 females and 23 males between the ages of 18 and 57 (mean age of 37 years; $SD = 12.7$).

APPARATUS

As warning concepts for imminent braking events were to be tested, critical almost-collision scenarios were required. To guarantee maximum safety and yet benefit from the advantages of a real driving study, the "Vehicle in the Loop", a test and simulation environment, was used (Bock, Maurer, & Färber, 2007). While driving a real car on a test route, the participants were wearing a non see-through head-mounted display (HMD), which builds up a fictitious driving environment. In this Virtual Reality (VR), the depiction of a city with hazardous events such as pedestrians suddenly crossing the street was simulated. The virtual driving scene was precisely adjusted to the dimensions of the test route, so that the appearance points of crossing pedestrians in virtual reality was accurately matched to predefined points on the driving ground. In this way, the triggers for the virtual pedestrians crossing the street and subsequent warning signals could be manually released by the experimenter whenever the vehicle was crossing certain coordinates on the test route (e.g. street points in the virtual city).

INDEPENDENT VARIABLES – WARNINGS

Six different warning strategies were evaluated in the test scenario, including unimodal as well as multimodal warnings.

The unimodal signals comprised of two peripheral visual warnings and an auditory alarm. The first peripheral warning was realized by a *LED-strip* located in the windshield root. Red LEDs lighted up to signal the detection of crossing pedestrians. To additionally offer spatial cues, the signal was located on the driver´s respectively passenger´s side only in respect to the hazardous situation. Due to the use of a non see-through HMD the LED-strip was included into the virtual driving environment. The second visual warning, a *windshield display*, also needed to be virtually depicted. Presenting information close to the driver´s field of view, windshield displays minimize eye movement and accommodation time and additionally provide peripheral spatial cues by directly highlighting hazardous objects on the windshield. In the present study, hazardous objects (e.g. pedestrians) were framed with a red triangle. The *auditory signal* was a 55-db, 1.800-Hz tone issued from loudspeakers located on both sides of the vehicle´s dashboard.

For the multimodal warnings, an audio-haptic and two visuo-haptic signals were presented by combining each of the three unimodal warnings with a haptic element in form of a *brake pulse*, lasting for 300ms with a peak deceleration of 0,3g.

EXPERIMENTAL DESIGN

The experiment was conducted in a 2 x 3 mixed factorial design. The two-level factor "*warning mode*" (unimodal vs. multimodal warnings) was a between-subject factor. While half of the participants received only unimodal warnings (e.g. auditory and visual signals), the other half experienced exclusively multimodal warnings (e.g. audio-haptic and visuo-haptic signals). In the within-subject factor "*warning modality*" each participant experienced three warnings: an auditory alarm, a LED-strip warning and a windshield-display warning.

To assess an overall efficacy of warning signals, each test drive additionally included two baseline conditions without warning support. In these trials, participants had to master critical pedestrian-crossings without driver assistance. All in all, each participant was confronted with five critical driving events during the test drive. Both the assignment of the participants to the experimental groups as well as the individual order of warnings and baseline trials during the experiment was completely randomized.

PROCEDURE

Before starting the test drive, the participants were instructed about the "Vehicle in the Loop" and informed about settings and aims of the study. Yet, they were unaware about the kind of warnings they were going to be exposed to. After a practice drive to get used to the unfamiliar driving situation, the 40-minute experimental session started. Participants were advised to drive safely through the virtual city, keeping an average speed of 30km/h and following the route-instructions of the experimenter.

With a time to collision (TTC) of 1.5s, from time to time pedestrians popped-up at one side of the street and started running across. Appearance and timing were manually triggered by the experimenter. In the warning trials, the auditory respectively visual warnings were presented simultaneously with the pedestrian's appearance (TTC = 1.5s), while the haptic warning was triggered half a second later (TTC = 1.0s). After each braking event participants evaluated the experienced warnings.

DEPENDENT VARIABLES

For the objective data assessment, reaction time (RT) was the main dependent variable in this analysis. Assessed in two steps, *accelerator-release-time* was measured as the time between the appearance of a crossing pedestrian and the release of the accelerator-pedal. *Brake-onset-time* corresponded to the time period between the appearance of the pedestrian and the application of the brakes. Additionally, *total braking-time* as a measure of the time interval between brake-activation and brake-release as well as different driving parameters were surveyed.

Furthermore *subjective ratings* were assessed to evaluate the experienced warnings from the driver's point of view using a 7-point rating scale.

RESULTS

Performance results presented in this paper include driver brake-onset-times, total braking-times as well as data of maximum brake pressures. In addition, subjective evaluations of warnings are provided.

EFFECTS OF WARNING MODE AND WARNING MODALITY

Considering *brake-onset-time*, ANOVAs revealed significant main effects both for warning mode ($F[1,38] = 20.33$, $p < .01$) and warning modality ($F[2,76] = 4.86$, $p = .01$). Brake-onset-time tended to be shorter for multimodal warnings ($M = 716$ms) compared to unimodal warnings (M = 864ms). Post-hoc comparisons of warning modalities showed that overall participants responded more rapidly following either the peripheral LED-strip warning (M = 747ms, p' < .01) or the auditory alarm (M = 777ms, p < .05) compared to the windshield display warning (M = 846ms). In addition, there was a significant interaction effect between warning mode and warning modality ($F[2,76]$ = 4.3, p < .05). The ordinal interaction pattern revealed that brake-onset-time for both the peripheral LED-strip warning (unimodal: M = 832ms; multimodal: M = 662ms, p < .01) and the windshield display warning (unimodal: M = 961ms; multimodal: M = 731ms, p < .01) was shorter in multimodal compared to unimodal presentation mode, whereas for auditory warnings reaction time was affected only marginally by warning modes (unimodal: M = 798ms; multimodal: M = 755ms, p > .05).

While no differences in *total-braking-time* could be found for warning modes ($F[1,38]$ = 0.60, p > .05), a significant main effect was obtained for warning modalities ($F[2,76]$ = 3,28, p = .04). Peripheral LED-strip warnings (M = 5,86s) caused significantly shorter total-braking-times compared to both the auditory tone (M = 7,47s, p = .05) and the windshield display warning (M = 7,46s, p < .05). *Maximum brake pressure* in contrast did not reveal any significant main effects (warning mode: $F[1,38]$ = 0.07, p > .05; warning modality: $F[2,76]$ = 0.90, p > .05).

COMPARISON TO BASELINE

To assess baseline comparisons, overall data on visual and auditory warnings were considered. Pairwise t-tests revealed significant shorter *brake-onset-times* when either presenting a peripheral LED-strip warning (M = 747ms; t[39] = 5.09, p < .01) or an auditory ton (M = 777ms; t[39] = 4.05, p < .01) compared to baseline data without warnings displayed (M = 893ms). The windshield-display warning in contrast (M = 846ms) did not prove overall reaction-time benefits. Also, for *total-braking-time* and *maximum brake pressure* no differences between assisted and non-assisted driving could be found.

SUBJECTIVE RATINGS

Subjective ratings were measured on a 7-point rating scale (0: "not noticed", 1-2: "(very) negative", 3: "rather negative", 4: "rather positive", 5-6: "(very) positive"). In the participants' *overall evaluation of warnings*, the auditory alarm and in particular the windshield display warning received top grades, both in unimodal (alarm tone: M = 4.32, SD = 1.4; windshield display: M = 4.95, SD = .9) and multimodal warning modes (alarm tone: M = 4.55, SD = 1.1; windshield display: M = 4.80, SD = 1.4). The peripheral LED-strip warning in contrast was rated rather negatively (unimodal: M = 3.38, SD = 1.5; multimodal: M = 3.40, SD = 1.5), while almost 58% of the participants in the unimodal and 25% of the participants in the multimodal warning condition had not even noticed the LED-warning.

DISCUSSION

Drivers in this study initiated significantly faster braking responses when *multimodal rather than unimodal warning signals* were displayed. The multimodal warning was composed of an auditory respectively visual signal followed by a haptic brake jerk. These results build upon a great deal of previous findings on facilitatory effects of multisensory warning signals in cars (e.g. Ho & Spence, 2008). If at all, former studies considered the potential of haptic elements in multimodal warnings primarily in terms of vibrotactile warnings like seat vibration (e.g. Fitch et al., 2007), seat belt vibration (e.g. Ho et al., 2007; Scott & Gray, 2008), vibrating gas pedals (e.g. De Rosario et al., 2009) or vibrations in the steering wheel (e.g. Suzuki & Jansson, 2003). Hardly any study has evaluated the intuitive action-recommending potential of a haptic brake jerk (e.g. Tijerina, Johnston, Parmer, Pham, Winterbottom, & Barickman, 2000; Shutko, 2001), especially in multimodal settings (Brown, 2005). Hence the present findings not only suggest a further implementation of haptic components into multimodal warnings, but especially encourage a sharper focus on the brake jerk as an important element of future multisensory warning design.

Drivers´ braking reactions in terms of brake-onset-times revealed to be influenced by warning modalities, too. Both the peripheral LED-strip warning and the auditory alarm contributed to enhanced braking reaction times compared to the baseline condition and the windshield-display warning. These findings concerning the *windshield-display warning* are somewhat surprising, as both the windshield display and the LED-strip warning should minimize driver distraction due to the presentation of peripheral spatial cues. However, the flashing triangles of the windshield-display warning still had a rather icon-like character. They therefore seemed to have encouraged rather foveal fixations than a peripheral signal detection (Doshi et al., 2009). In addition, the simulation of the windshield-display warning was very striking and might have led to further driver distractions. These suggestions seem plausible when considering the participants´ enthusiastic preference ratings on the windshield-display warning. This also indicates a high salience of the warning that might have turned out to be rather distractive than actually assistant.

As expected, in general the peripheral visual LED-strip warning proved to be just as effective as the auditory tone. In the unimodal presentation mode, no differences in brake-onset-time could be revealed. In the multimodal condition, the peripheral LED-strip warning even contributed to significantly shorter reaction times than the auditory signal. Additionally, results on total-braking-times and maximum brake pressures indicate that the maximum brake pressure was built up significantly faster when a peripheral LED-strip warning was displayed compared to the presentation of auditory- respectively windshield-display warnings. Thus, these driving data reveal an excellent and even enhanced reaction-performance based on peripheral LED-strip warnings compared to other warning signals. These promising results on peripheral warnings correspond with previous findings concerning the attention-capturing potentials of peripheral visual cues with little or even without any conscious mental efforts (Hillstrom & Yantis, 1994; Sarter, 2000). Yet, in contrast to these driving data, subjective ratings reveal rather negative evaluations. Almost 60% of the participants in the unimodal and 25% of the participants in the multimodal presentation mode claimed not to have noticed the LED-warning at all. Based on these findings an interesting conclusion can be drawn: due to its non-intrusive character and its location in the driver's peripheral visual field, the LED-strip warning not only supports information perception and processing while keeping the primary attention focus concentrated on the driving task. It furthermore seems to support an implicit and subliminal information perception.

Future research is needed to validate these findings and assumptions, especially with respect to the peripheral LED-strip warning and the windshield display warning both required to be simulated. Underlying mechanisms need to be specified. Yet, these results are of significant importance: they might introduce new ways of designing non-distractive but highly efficient in-vehicle warnings in applying the paradigm of subliminal priming into vehicle-design. Implications of these findings for the overall research framework of developing a new and integrated warning philosophy are meaningful and encourage an important pre-selection of promising warning strategies to recommend high-priority braking actions.

REFERENCES

Bock, T., Maurer, M., and Färber, G. (2007), "Validation of the Vehicle in the Loop (VIL). A milestone for the simulation of driver assistance systems." *Proceedings of the IEEE Intelligent Vehicles Symposium*, 612–617.

Brown, S.B. (2005), *Effects of haptic and auditory warnings on driver intersection behaviour and perception*. Unpublished Master's thesis. Department of Industrial and Systems Engineering, Virginia Tech.

Calvert, G.A., Spence, C., and Stein, B.E. (eds.). (2004), *The handbook of multisensory processes*. MIT Press, Cambridge, MA.

Chan, A.H.S., and Chan, K.W.L. (2006), "Synchronous and asynchronous presentation of auditory and visual signals: Implications for control console design." *Applied Ergonomics*, 37(2), 131–140.

Cummings, M.L., Kilgore, R.M., Wang, E., Tijerina, L., and Kochhar, D.S. (2007), "Effects of Single versus Multiple Warnings on Driver Performance." *Human Factors*, 49(6), 1097–1106.

De Rosario, H., Louredo, M., Díaz, I., Soler, A., Gil, J.J., Solaz, J.S., and Jornet, J. (2009), "Efficacy and feeling of a vibrotactile Frontal Collision Warning implemented in a haptic pedal." *Transportation Research F*, 13(2), 80–91.

Doshi, A., Cheng, S.Y., and Trivedi, M.M. (2009), "A novel active heads-up display for driver assistance." *IEEE Transactions on systems, man, and cybernetics B*, 39(1), 85–93.

Ferris, T.K., and Sarter, N.B. (2008), "Cross-modal links among vision, audition, and touch in complex environments." *Human Factors*, 50(1), 17–26.

Ferris, T., Penfold, R., Hameed, S., and Sarter, N. (2006), "The implications of crossmodal links in attention for the design of multimodal interfaces: A driving simulator study." *Proceedings of the Human Factors and Ergonomics Society 50th Annual Meeting*, 406–409.

Fitch, G.M., Kiefer, R.J., Hankey, J.M., and Kleiner, B.M. (2007), "Toward developing an approach for alerting drivers to the direction of a crash threat." *Human Factors*, 49(4), 710–720.

Hameed, S., Ferris, T., Jayaraman, S., and Sarter, N. (2009), "Using informative peripheral visual and tactile cues to support task and interruption management." *Human Factors*, 51(2), 126–135.

Hillstrom, A.P., and Yantis, S. (1994), "Visual motion and attentional capture." *Perception and Psychophysics*, 55(4), 399–411.

Ho, C., and Spence, C. (2005), "Assessing the effectiveness of various auditory cues in capturing a driver's visual attention." *Journal of Experimental Psychology: Applied*, 11(3), 157–174.

Ho, C., and Spence, C. (2008), *The multisensory driver: implications for ergonomic car interface design*. Ashgate Publishing, Aldershot.

Ho, C., Reed, N., and Spence, C. (2007), "Multisensory in-car warning signals for collision avoidance." *Human Factors*, 49(6), 1107–1114.

Lee, J.D., Hoffman, J.D., and Hayes, E. (2004), "Collision warning design to mitigate driver distraction." *Proceedings of the SIGCHI conference on Human Factors in Computing Systems*, 6(1), 65–72.

Lee, J.D., McGehee, D.V., Brown, T.L., and Reyes, M.L. (2002), "Collision warning timing, driver distraction, and driver response to imminent rear-end collisions in a high-fidelity driving simulator." *Human Factors*, 44(2), 314–334.

Liu, Y.-C. (2001), "Comparative study of the effects of auditory, visual and multimodality displays on drivers' performance in advanced traveller information systems." *Ergonomics*, 44(4), 425–442.

Nikolic, M.I., and Sarter, N.B. (2001), "Peripheral visual feedback: A powerful means of supporting effective attention allocation in event-driven, data-rich environments." *Human Factors*, 43(1), 30–38.

Regan, D. (2000), *Human Perception of Objects*. Sinauer Associates, Sunderland, MA.

Santangelo, V., Ho, C., and Spence, C. (2008), "Capturing spatial attention with multisensory cues." *Psychonomic Bulletin and Review*, 15(2), 398–403.

Sarter, N.B. (2000), "The need for multisensory interfaces in support of effective attention allocation in highly dynamic event-driven domains: The case of cockpit automation." *The International Journal of Aviation Psychology*, 10(3), 231–245.

Scott, J.J., and Gray, R. (2008), "A comparison of tactile, visual, and auditory warnings for rear-end collision prevention in simulated driving." *Human Factors*, 50(2), 264–275.

Selcon, S.J., Taylor, R.M., and McKenna, F.P. (1995), "Integrating multiple information sources: Using redundancy in the design of warnings."

Ergonomics, 38(11), 2362–2370.

Shutko, J. (2001), *An investigation of collision avoidance warnings on brake response times of commercial motor vehicle drivers.* Unpublished Thesis, Virginia Polytechnic Institute and State University, Blacksburg, Virginia.

Spence, C., and Ho, C. (2008), "Multisensory warning signals for event perception and safe driving." *Theoretical Issues in Ergonomics Science*, 9(6), 523–554.

Stein, B.E., Stanford, T.R., Ramachandran, R., Perrault, T.J. Jr., and Rowland, B.A. (2009), "Challenges in quantifying multisensory integration: Alternative criteria, models, and inverse effectiveness." *Experimental Brain Research*, 198(2-3), 113–126.

Suzuki, K., and Jansson, H. (2003), "An analysis of driver's steering behavior during auditory or haptic warnings for the designing of lane departure warning system." *JSAE Review*, 24(1), 65–70.

Tijerina, L., Johnston, S., Parmer, E., Pham, H.A., Winterbottom, M.D., and Barickman, F.S. (2000), *Preliminary studies in haptic displays for rear-end collision avoidance system and adaptive cruise control system applications.* Report DOT HS 808 (TBD), National Highway Traffic Safety Administration, Washington, DC.

Van Erp, J.B.F., and van Veen, H.A.H.C. (2004), "Vibrotactile in-vehicle navigation system." *Transportation Research F*, 7(4-5), 247–256.

Wickens, C.D., and McCarley, J.S. (2008), *Applied Attention Theory.* CRC Press Taylor & Francis Group.

Wiese, E.E., and Lee, J.D. (2004), "Auditory alerts for in-vehicle information systems: The effects of temporal conflict and sound parameters on driver attitudes and performance." *Ergonomics*, 47(9), 965–986.

Wright, R.D., and Ward, L.M. (2008), *Orienting of Attention.* Oxford University Press, New York.

Chapter 30

Citarasa Engineering for Modeling Affective Product Design

Halimahtun M. Khalid

Damai Sciences Sdn Bhd,
A-31-3 Suasana Sentral, Jalan Stesen Sentral 5
50470 Kuala Lumpur, Malaysia

ABSTRACT

Citarasa Engineering is a method for achieving affective design. It is inspired by the *citarasa* concept which means emotional intent. The method was developed to assist vehicle designers in analyzing and utilizing customers' affect in the design of vehicles. A vehicle buyer is assumed to seek and identify design features that are emotionally satisfying. The method is a 5-step process, starting with Citarasa modeling as a theoretical basis for affective design. This was followed by gathering customer needs using a detailed questionnaire. The data were then analyzed using content analysis method to generate data tables that contributed towards WEKA data mining. In this process affective and functional requirements were mapped to design parameters that were formalized in Design Equations for Citarasa Analysis (DECA). Design of a truck living cab is used to illustrate the method.

Keywords: Citarasa Engineering, Affective Design, Human Factors

INTRODUCTION

Affective design has emerged as a challenge to product developers as more customers now look beyond functionality in the products they purchase (Norman, 2004; Khalid, 2005; Helander & Khalid, 2006). Their desire is towards a product that can satisfy both functional and emotional needs. An object with a strong identity and styling – good quality, design and functionality – can induce pleasure. However, many affective needs are unspoken and need to be identified by probing user emotions. Designers must understand how customers interact with products, both from a functional and emotional perspective. Functional requirements are easy to understand, but affective requirements are subtle and puzzling, and often difficult to identify. They vary over time, and customers often have difficulties in explaining what the requirements are. They may say "I like it," which is a statement at a high level of abstraction and not useful for design. We need to break down such statements to lower levels of abstraction, where design elements can be identified in terms of preferred color, shape, material and so forth (Khalid et al., 2007a). Below we describe a method that can help designers conceptualize affective design. The method emerged from the CATER research project (Computerized Automotive Technology Reconfiguration) (Khalid et al., 2007b).

FRAMEWORK FOR CITARASA ENGINEERING

There are five steps in *Citarasa* Engineering: (1) modeling, (2) data gathering, (3) text and semantic analysis, (4) data mining, and (5) deriving design solutions, as shown in Figure 1.

Step/Activity	Tool	Output
1. Model Citarasa	Theories of affect	Citarasa Model
2. Gather data	Field survey- probe elicitation interview	Questionnaire
3. Analyze Citarasa	Content analysis	Data Tables
4. Mine Data	Data mining	WEKA Rules
5. Derive Design Solutions	Citarasa Analysis Module	DECA Mapping Matrix

Figure 1. CATER approach to Citarasa Engineering (Khalid et al 2008)

MODELING CITARASA

To model citarasa, a framework was created as shown in Figure 2. There are three subsystems: Designer Information and Constraints, Formulation of Design

Equations for Citarasa Analysis (DECA), and Summative Evaluation of Design. The first two subsystems are intended to be used by the vehicle planner in conceptualizing and evaluating product design, and the third subsystem explains the steps that customers go through in their decision making.

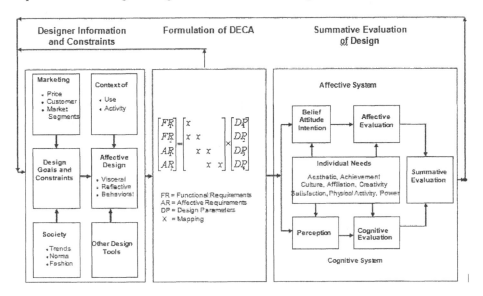

Figure 2. Citarasa engineering model in vehicle design (Khalid et al 2007a)

Prior to product conceptualization, designers need to understand user requirements, which are classified here as functional and affective. Customer needs are captured in a concept called *Citarasa* (see Khalid et al., 2007a). In using this concept we assume that customers have an emotional intent when they purchase a vehicle (e.g. truck), and truck design must therefore also address customers' affective needs. The emotional intent is driven synergistically by separate brain mechanisms – the affective part in the limbic system, and the cognitive part in the frontal lobe. The affective part is necessary not only for evaluation of affective design but also for decision making between alternatives. Citarasa is different from the rather passive emotions that one may experience in purchasing regular or less costly products, such as clothes. In Citarasa, there is a strong *intentional* component; customers are actively seeking design features that are important for their emotional satisfaction. To investigate intent it is possible to ask customers what they are looking for, and thereby obtain an explicit identification of the user's affective requirements when buying a truck.

Designer Information and Constraints. In designing a new vehicle, a product planner/designer uses information from many sources, including: Marketing, Context of Use/Activity of a vehicle, and Society Norms and Fashions. These three factors broadly determine the Design Goals (what one should design) and the Constraints (what one should not design). The Marketing department will offer information about customer needs and future markets, which will determine design goals. In addition, Marketing will often determine economic parameters – including the price – thereby, imposing important constraints.

Affect is elicited not only by the product, but also by the context of use and the activity itself. A safe truck can have a stable and reliable look, and this pleasure will be enhanced when driving. Hence, product planners must analyze the context of activity. This will determine many affective and functional requirements of the vehicle. Some design factors are influenced by Social Norms and Fashions. For example, functional requirements for high fuel efficiency will be regulated by governments as well as by customers. After they are adopted by Marketing, they will appear to the designer as Design Constraints. Trends and fashions in design are important and play a decisive role in Affective Design.

Product planners and designers need to understand how affective design can be derived. Norman (2004) noted that there are three types of affective design features, namely: visceral, behavior and reflective. *Visceral design* refers to the visual aspects of the design, such as shape, color, materials, ornamentation, and texture. *Behavior design* has to do with the pleasure in using the object, such as steering a smooth and well-balanced vehicle, driving at a very high speed, and intuitively finding controls. *Reflective.design* has to do with things that have been learnt over the years. A vehicle buyer may take much interest in an aesthetic design if he/she was surrounded by beautiful objects from early childhood.

Reflective design is important to Citarasa Engineering, which assumes that the emotional intent (affective requirements) drive customer choice, together with functional requirements. Customers reflect on the suitability of various design options and select a vehicle that suits both types of requirements. It is important that designers are well informed about these different and complementary design options and understand how to implement them. This will require training, so that designers can make conscious decisions and fully understand when and how to utilize visceral, behavior and reflective design.

Formulation of Design Equations. The principles of DECA are illustrated in the centre box in Figure 2. This design stage should be undertaken iteratively with the other two design stages, see Figure 2. It will make the designer be aware of the importance of affective design and instruct him/her to make trade-off decisions that can promote affective design. It is therefore considered a helpful design tool.

Summative Evaluation of Design. The third subsystem in Figure 2 illustrates customer decision making and the important principle of the cross-coupling between affect and cognition. Emotions are now recognized and product designers are keen to develop design models that can incorporate emotions as well as cognitive considerations in product design.

Affective Systems. The prospect of owning an expressive or expensive truck can generate a variety of emotions. Deep-seated desires for individuality, pleasure, and aesthetics can bring about strong emotion in product evaluation. In the beginning, there are hypothetical attitudes about product ownership, such as personal image, utility and vehicle performance. Later, as the result of social consideration, family and friends, the customer will moderate his initial assessment. This will be the basis for the formulation of an intention. In the end, the customer may not follow through with the purchase. Although there may be positive inclinations to buy the truck, practical considerations may postpone or cancel the purchase.

Cognitive Evaluation. The cognitive system in Figure 2 follows the human information processing model. The perception of the artifact triggers cognition and memory recall which leads to evaluation and decision making. The final decision comes about through a negotiation between the affective and the cognitive system.

GATHERING CUSTOMER CITARASA

To capture the citarasa of vehicle customers, a field survey was conducted using questionnaire. The survey applied probe interview technique which required trained interviewers to ask follow-up questions in the form of why-why-why (Helander & Khalid, 2009). About 177 owners/drivers from Europe (n=137) and Asia (n=42) were interviewed. Among them, 99 were male and 80 female. The subjects were sampled using purposive sampling to represent three age groups: 18-24, 25-54, and 55-65 years. The surveys were conducted in 12 countries; China, Finland, France, Germany, Greece, India, Italy, Sweden, Switzerland, UK, Malaysia, and Singapore.

The Truck Needs Tracker (TNT) questionnaire was constructed for trucks to capture design features and product attributes associated with different customer functional and affective needs of the cab interior, especially the living area. The functional needs relate to activities in the cab when not driving, while the affective needs refer to impressions of the cab interior design. The TNT questionnaire comprised open-ended questions that required probe interview. The interviewer probed with follow-up questions to get a deeper understanding of the respondents' needs and impression of the cab interior. The probe generated the following information:

- **Affective and functional needs**: What the drivers/owners experienced or would like to do experience when interacting with the truck.
- **Product features**: Equipment, component or area in the truck that is related to an impression or an activity.
- **Design parameters**: Attributes and physical elements that contribute to an impression or to an activity.

Examples of probe interview items:

Question: *"When entering the cab, what kind of feelings would you like to experience?"*

Answer: *" I want to have an interior that feels safe and cosy"*

Follow-up question: *"Do you have some suggestions to what it is that could create a feeling like that in a cab?"*

Answer: *"I want it to feel like home"*

Follow-up question: *"What kind of materials or colors do you think would create that feeling?"*

Answer: *"In my living room I have a black leather sofa"*

The above example provided the following input to the Citarasa database:

- Affective requirements (AR): A cosy impression
- Product features (PF): A living compartment or seat similar to sofa
- Design parameters (DP): Black color and leather material

CITARASA ANALYSIS

The qualitative data from the TNT survey were transformed into quantitative data. The analysis was performed using SPSS Textsmart, SPSS Advanced Models 13.0, and Microsoft Excel. The data analysis was carried out in four steps:

1. Data imported using SPSS and recoded into Textsmart format;
2. Responses refined, such as spell checked responses, and data structured;
3. Results categorized according to three predefined categories: Needs, Product features and Design parameters. This categorisation was made in two steps: (a) The elicited words were grouped into activities (e.g. brush teeth, wash hands, categorised as *hygiene*; expensive, pricey, over-priced, classified as *costly*); (b) Irrelevant words excluded from non-related questions (e.g., words related to impressions were removed from activity related categories).
4. Derived results exported to MS Excel and data compiled into matrices for data mining purposes.

A sample of an Excel matrix is shown in Figure 3.

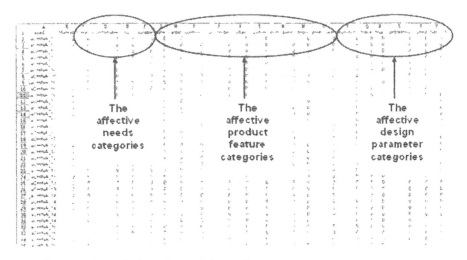

Figure 3. Excel result matrix from the truck data collection survey

In Figure 3, the three categories are highlighted in red ellipses. For example, if a respondent mentioned a specific need, product feature or design parameter, the response was assigned "1". If there was no response, "0" was given. This matrix became the building block for the Citarasa database.

MINING RULES

From the data tables, WEKA (2008) mining was performed on the data sets. The purpose was to discover the relationship between affective needs and design parameters. The data from the TNT survey were analyzed and pre-processed using Excel and Matlab to organize into a standard format for WEKA mining. The variables used to present the data are shown below:

Variable	Usage
Region	To present the region
Occupation	To present the occupation
Descriptor	To describe affective needs
Component	To present product features
Attribute	To present design parameters

Table 1 provides a sample of derived WEKA rules for steering wheel design.

Table 1: Sample of WEKA rules for steering wheel (Khalid et al 2008)

1. /* 0,467556 0,050000 */ REGION = Asia ==> ATTRIBUTE = Flexible
2. /* 0,454017 0,000000 */ REGION = Asia and OCCUPATION = Truck driver ==> ATTRIBUTE = Flexible
3. /* 0,348068 0,450000 */ REGION = Eur and OCCUPATION = Truck driver ==> ATTRIBUTE = Rounded
4. /* 0,348068 0,450000 */ REGION = Eur and OCCUPATION = Truck driver ==> ATTRIBUTE = Soft
5. /* 0,348068 0,450000 */ REGION = Eur and OCCUPATION = Truck driver ==> ATTRIBUTE = Design
6. /* 0,342194 0,000000 */ OCCUPATION = Truck driver and DESCRIPTOR = Ergonomics ==> ATTRIBUTE = Flexible
7. /* 0,342194 0,000000 */ OCCUPATION = Truck driver||Truck owner ==> ATTRIBUTE = Flexible
8. /* 0,286374 0,550000 */ REGION = Eur ==> ATTRIBUTE = Rounded
9. /* 0,286374 0,550000 */ REGION = Eur ==> ATTRIBUTE = Soft
10. /* 0,286374 0,550000 */ REGION = Eur ==> ATTRIBUTE = Design

DESIGN EQUATION FOR CITARASA ANALYSIS

The processed data from WEKA mining were used to illustrate the DECA method. The data were analyzed and transferred to a Design Matrix. Detailed descriptors were extracted from the TNT data. They were compiled from eight data matrices covering: General interior impression, Color impression, Material impression, Shape impression, Comfort impression, Fit & Finish impression, Durability impression, and Clean impression. The number of Citarasa descriptors was reduced to 16, see Customer Needs column in Table 2. They describe affect information as perceived by different truck users. The data described the following:

- **Affective and functional needs**: What the drivers/owners experienced or would like to do experience when interacting with the truck.
- **Product features**: Equipment, component or area in the truck that is related to an impression or an activity.
- **Design parameters**: Attributes and physical elements that contribute to an impression or to an activity.

Table 2. Summary of customer citarasa and design

	Customer Needs (Affective & Functional)	Product Feature	Design Parameters
General impression	Spacious Functional Comfortable Clean	Whole-cab Seat Bunk	Square Material Stain-resistant Leather
Color impression	Cosy Soft	Work-area Bunk Seat	Black Grey
Material impression	Comfortable Durable Easy-to-clean	Bunk Dashboard Seat	Leather Fabric Plastic
Shape impression	Soft Good looking Spacious	Shape Dashboard	Rounded Curved Squared
Comfort impression	Spacious Ergonomic	Bunk Seat Controls	Ventilated Adjustable Within-reach
Fit & Finish impression	Quality Solid Comfortable Well-made	Seat Dashboard Controls	Smooth Material Color Seamed
Durability impression	Hard Long-lasting	Dashboard Seat Whole-cab	Metal Plastic Leather
Clean impression	Easy-to-clean Stain-resistant Clean	Seat Dashboard Bunk	Color-dark Leather Smooth

The data summarized in Table 2 was analyzed and mapped on design parameters in the design matrix. The values in a design matrix will be either 'x' or '0', where 'x' represents the existence of a correlation between the corresponding vector components while '0' signifies no correlation. Figure 4 (see Appendix) shows the DECA mapping.

The DECA mapping of the truck living cab was derived by domain experts, by analyzing the data from the TNT survey and considering trade-offs in design. As shown in Figure 4, the customers mentioned one AR "cosy," which can be associated with the whole cab, and the seat, which was covered with leather, thereby comfortable; the main color is black, and it was equipped with high quality TV and audio devices. These customer requirements and design parameters were stored in the citarasa database. In this case study, the DP matrix is not very detailed. In order to extend the design parameters into more detailed level, a more specific and in-depth interview/survey should be carried out.

CONCLUSION

The citarasa engineering is a systematic approach towards identifying customer needs, and mapping them to design parameters for generating vehicle design solutions. The method is driven by the Citarasa model that is based on affect-behavior-cognition paradigm. The method is a five-step process to arrive at design solutions. It relies on the customer's expertise and expressions of affective and functional needs, which are elicited by means of probe interviews, and mined using WEKA. The derived citarasa descriptors from customers were then mapped to design parameters to generate design solutions. The DECA method helps to structure the design process, and prompts designers to consider both affective and functional design. It will also make designers aware of the importance of affective design and instruct them to make trade-off decisions that can promote affective design. Citarasa Engineering is still new and has not been applied to many other application domains. The method is promising and may be extended to other consumer products with similar affective characteristics.

ACKNOWLEDGEMENT

The work reported here was undertaken with partial support from the grant provided by the European Commission under Framework 6 Call 5 (CATER IST 5 035030 STREP), in collaboration with 14 partners of the CATER consortium.

REFERENCES

Helander, M., and Khalid, H. (2006). "Affective and pleasurable design," in: Handbook on Human Factors and Ergonomics, Salvendy, Gavriel (Ed.), New York: Wiley, pp. 543-572.

Helander, M., and Khalid, H. (2009). "Citarasa engineering for identifying and evaluating affective product design," Proceedings of IEA Congress 2009, Beijing, China (CD-ROM).

Khalid, H.M. (2005). "User emotion in product design. Bridging users and designers," Proceeding of SEAES-IPS 2005, Vol. I, Bali, Indonesia, pp. 11 – 21.

Khalid, H.M., Lim, T.Y., Opperud, A., Helander, M.G., Hong, P., Xi, Y., Suardo, G.S., and Gemou, M. (2007a). Deliverable 2.1 on customer citarasa needs data. Technical Report, Available at CATER Website: http://www.cater-ist.org

Khalid, H.M., Dangelmaier, M., and Lim, T.Y. (2007b), "The CATER approach to vehicle mass customization," Proceedings of IEEM 2007, Singapore (CD-ROM).

Khalid, H.M., Helander, M.G., Jiao, R., Xu, Q., Hong, P., Mavridou, E., and Opperud, A. (2008). Deliverable 2.2 on citarasa engineering methodology. Internal Consortium Report, Unpublished.

Norman, D.A. (2004). Emotional design: why we love (or hate) everyday things. New York: Basic Books.

WEKA (2008), Available at WEKA Website: http://www.cs.waikato.ac.nz/ml/WEKA/

$$
\begin{bmatrix}
\text{spacious} \\
\text{functional} \\
\text{comfortable} \\
\text{clean} \\
\text{cosy} \\
\text{soft} \\
\text{durable} \\
\text{easy - to - clean} \\
\text{good looking} \\
\text{ergonomic} \\
\text{quality} \\
\text{longlasting} \\
\text{eat} \\
\text{watch} \\
\text{sleep} \\
\text{listen}
\end{bmatrix}
=
\begin{bmatrix}
x & x & 0 & 0 & 0 & 0 & 0 & 0 & 0 & 0 & 0 & 0 & 0 & x & 0 & 0 & 0 & 0 & 0 & 0 & 0 & 0 \\
0 & 0 & x & x & 0 & 0 & 0 & x & 0 & 0 & 0 & 0 & 0 & 0 & 0 & 0 & x & x & x & 0 & 0 & 0 \\
x & x & x & 0 & x & x & x & x & 0 & 0 & 0 & 0 & 0 & x & x & 0 & 0 & 0 & 0 & 0 & 0 & 0 \\
x & x & x & 0 & 0 & 0 & x & 0 & x & 0 & 0 & 0 & 0 & x & 0 & 0 & 0 & 0 & x & 0 & 0 & x \\
x & 0 & x & 0 & 0 & x & 0 & 0 & x & 0 & x & 0 & 0 & 0 & 0 & 0 & x & x & 0 & 0 & 0 & 0 \\
0 & 0 & 0 & x & 0 & x & x & 0 & 0 & x & x & x & 0 & x & 0 & 0 & 0 & 0 & 0 & x & 0 & 0 \\
0 & 0 & x & x & 0 & 0 & x & 0 & 0 & 0 & 0 & 0 & 0 & 0 & 0 & 0 & 0 & 0 & 0 & x & x & 0 \\
0 & x & x & x & 0 & 0 & x & 0 & x & x & 0 & 0 & 0 & x & 0 & 0 & 0 & 0 & x & x & x & x \\
x & 0 & x & 0 & 0 & 0 & 0 & 0 & 0 & 0 & 0 & x & x & 0 & 0 & x & 0 & 0 & 0 & 0 & 0 & x \\
0 & x & 0 & 0 & x & x & 0 & x & 0 & 0 & 0 & 0 & 0 & 0 & x & 0 & 0 & 0 & 0 & 0 & 0 & 0 \\
x & 0 & x & x & 0 & 0 & x & 0 & 0 & 0 & 0 & 0 & 0 & x & 0 & x & 0 & 0 & 0 & x & 0 & x \\
x & 0 & x & 0 & 0 & 0 & x & 0 & 0 & 0 & 0 & 0 & 0 & 0 & 0 & 0 & 0 & 0 & 0 & x & x & 0 \\
0 & 0 & 0 & 0 & x & x & 0 & 0 & 0 & 0 & x & 0 & 0 & 0 & 0 & 0 & 0 & 0 & x & 0 & 0 & 0 \\
0 & 0 & 0 & 0 & x & x & 0 & x & 0 & 0 & 0 & 0 & 0 & 0 & 0 & 0 & 0 & x & 0 & 0 & 0 & 0 \\
0 & 0 & 0 & x & x & x & x & x & 0 & 0 & 0 & 0 & 0 & 0 & 0 & 0 & 0 & 0 & 0 & 0 & 0 & 0 \\
0 & 0 & 0 & 0 & x & x & 0 & x & 0 & 0 & 0 & 0 & 0 & 0 & 0 & 0 & x & 0 & 0 & 0 & 0 & 0
\end{bmatrix}
\times
\begin{bmatrix}
\text{whole - cab} \\
\text{whole - cab : ventilated} \\
\text{seat : leather} \\
\text{material : fabric} \\
\text{seat : adjustable} \\
\text{seat : comfortable} \\
\text{bunk : material} \\
\text{Bunk : adjustable} \\
\text{Black colour} \\
\text{grey colour} \\
\text{work - area : flat surface} \\
\text{dashboard : curved/rounded} \\
\text{dashboard : squared} \\
\text{dashboard : smooth} \\
\text{controls : within - reach} \\
\text{seamed finish} \\
\text{Audio equipment} \\
\text{TV equipment} \\
\text{stain – resistant} \\
\text{material : metal} \\
\text{material : plastic} \\
\text{color - dark}
\end{bmatrix}
$$

Figure 4. DECA mapping of truck living cab (Khalid et al 2008)

Chapter 31

Modeling Mobile Devices for the Elderly

Martina Ziefle

Communication Science, Human Technology Centre
RWTH University Aachen
Germany

ABSTRACT

Increasingly more and older adults are confronted with mobile devices. Although applications should be accessible to everyone, a gap between those, who are "computer-literate" and those who are not (predominantly older users) is emerging. An age-sensitive display design is needed, which allows users of all ages and ability levels to interact with mobile technology. The paper addresses factors to be considered for age-sensitive mobile device design and identifies usability problems that are age-exclusive (causing interaction difficulties only for older users) or age-specific (problems that cause serious interaction difficulties preferably for older people).

Keywords: Aging, mobile devices, information design, navigation performance

INTRODUCTION

The rapid proliferation of ubiquitous computing and mobile information access is accompanied by the development of devices, which promise to facilitate the daily living activities of people. The effective integration of mobile information technology (ICT) and its broad acceptance impose considerable challenges. Increasingly, older adults are confronted with technical devices and urged to learn and use them. As opposed to the past, when mostly technology prone professionals were the typical end-users of technical products, today, broad user groups have access to ICT (e.g., Arning & Ziefle 2007 a +b). In order to include all users, age-sensitive display designs are needed, which allow users of all age groups and ability levels to interact with technology. Design approaches should consider older users' abilities and barriers, which affect the interaction with mobile devices. This is of specific importance as mobile technologies are expected to specifically support older adults, applicable for a wide range of functions (medical aid, navigation and memory aids (e.g., Leonhardt 2006).

ERGONOMIC CHALLENGES OF MOBILE DEVICES FOR OLDER USERS

The mobile character of small screen devices represents especially high usability demands. Mobile devices are equipped with a specific display technology which- from the visual ergonomic perspective- has negative effects on information processing, especially for older adults (e.g., Oetjen & Ziefle 2007, 2009). Also, mobile devices are equipped with a miniaturized window, which is extremely problematic for providing optimized information access (e.g., Bay & Ziefle, 2004; Bernard et al. 2001; Omori et al. 2002; Zhao et al. 2001). Beyond visibility, cognitive aspects of information display-ing need to be considered. The restricted display space allows only a few items to be seen at a time. Users navigate through a menu whose complexity, extension and structure is not transparent (e.g., Han & Kwahk 1994; Sein et al. 1993; Ziefle, 2006,2008). Users report to not know what to do next, when to do it and how to complete a targeted action successfully. Disorientation in the menu is a frequent problem (e.g., Edwards & Hardmann 1989; Kim & Hirtle 1995; McDonald & Stevenson 1998), especially for aged users or those with little computer-related knowledge (e.g. Arning & Ziefle 2009, Westerman 1997, Willis & Schaie 1986).

Aging affects several characteristics of information processing that are crucial for the interaction with technology. The slowing down of functions regards sensory (e.g. Kline & Scialfa 1997; Schieber 2005), psychomotor (e.g. Vercryssen 2007), and cognitive performance (e.g. Park & Schwartz, 1999). Among cognitive factors, age-related declines of memory and spatial abilities are well-known   (e.g. Craik & Salthouse 1992; Ziefle et al. 2007). Both abilities considerably impact navigation in hierarchically structured menus (Bay & Ziefle 2003; Ziefle, 2005, 2008). Older adults are penalized when confronted with tasks that require navigation through menu hierarchies of different depths and breadths (e.g., Sein et al. 1993; Westermann 1997; Ziefle & Bay 2006). Aggravating, older users were educated in times when technical devices were far less complex than current devices (e.g., Ziefle & Bay 2004, 2005). A mental model of how technology works, built in a former time, should interfere with, or at least should not be sufficient for, proper interaction with devices currently available. Though, older adults report to be interested in technology, but they do not feel that devices meet their demands for a usable design (e.g., Arning & Ziefle 2009; Brodie et al. 2003; Melenhorst et al. 2006; Noyes et al. 2003).

QUESTIONS ADDRESSED

This paper addresses older users' requirements when using mobile small screen devices (taken the mobile phone as example). It reports a survey of studies dealing with different aspects of aging and mobile devices. The first study addresses the "navigation costs" of cognitive complexity of phones' menus in terms of effectiveness and efficiency. The second study considers language issues and shows the impact of suboptimal functions' naming on detouring within the menu. The third study is concerned with navigation aids that may reduce older users' menu disorientation. Concluding, age-specific as well as unspecific usability factors are identified which are to be considered for an age-sensitive information design of mobile devices.

MENU COMPLEXITY

Using a technical device requires basic procedural knowledge and refers to knowledge procedures of how and when to do something in the dialogue between human and device. Especially in complex technical menus this is a prominent issue. Menu complexity can be formally described by means of the Cognitive Complexity Theory CCT (Kieras & Polsen 1985). It deals with the cognitive complexity of the interaction between user and device by describing the user's goals on the one hand and the reaction of the computer system on the other by means of production rules. Production rules can be expressed as the sequence of rules in the form of IF condition (display status) and THEN-action (keystroke or input by the user). For the definition of complexity of mobile phones, the CCT approach is helpful, as the definition of production rules (the specification of what the system says and how users react) comprises the factors that may contribute concurrently to a phone's complexity (menu structure, keys, functions' names, menu location).

Method: The first independent variable refers to user age, comparing the navigation performance of younger and older users. The second variable is the cognitive complexity, defined by the number of production rules to be applied when solving four tasks with two mobile phones. Navigation effectiveness (tasks solved) and efficiency were measured. For efficiency (1) the time on tasks (2) the number of detour steps (i.e. the difference between the number of keystrokes actually effectuated and the number of keystrokes that were necessary to solve the task) were measured as well as (3) the number of returns to higher levels in menu hierarchy, indicating that users in the belief of having taken the wrong path, go back to a known menu position, consequently reorienting themselves. Mobile phones were simulated as software solutions and displayed on a touch screen. Four phone applications (calling s.o., send text message, hide own number, edit entry in the phone book) had to be solved. 32 users, 16 younger (20–30 years) and older (50–65 years) took part in the study

Results: The MANOVA yielded significant main effects of cognitive complexity ($F_{(1,28)} = 4.6$; $p<0,05$) and age ($F_{(1,28)} = 6.4$; $p < 0.05$). Further, the interaction of both variables was significant ($F_{(1,28)} = 4.5$; $p < 0.05$). Results are visualized in Figure 1.

Participants using the less complex phone solved the tasks 14% more effectively than those using the complex phone. Considering efficiency measures, the advantage of the less complex phone was even more convincing: comprising all four tasks, participants spent 40% less time on task, making 50% less detour steps and disorienting less often, which is taken from the lower number (44%) of returns to higher levels in the menu hierarchy. Though, effects of cognitive complexity were clearly stronger for the younger adults who needed three times more detour steps and returns to solve the tasks on the more complex phone. In contrast, the older adults' loss in efficiency by the complex phone was "only" factor 1.5. This result is noteworthy as it may prove the widespread prejudice wrong that younger adults master technological demands anyway, independently of how complexly they may be structured.

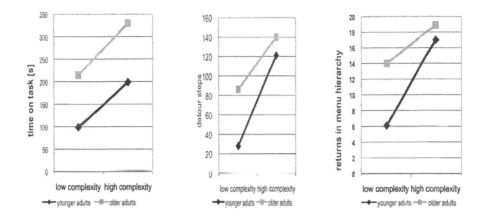

FIGURE 1: Interaction of phone complexity and age for efficiency measures (left: mean time on task(s); center: detour steps; right: number of returns) (Ziefle & Bay 2005)

NAMING OF MENU FUNCTIONS

While the first study examined cognitive complexity as a whole, the second study focused specifically on shortcomings in language design of mobile phones. Poor categorization and ambiguous naming increase the risk of getting lost within the menu. Findings from memory research (e.g. Chang 1986; Noordman-Vonk 1979) assume that well-defined semantic categories should represent the basic or prototypic idea of the concept they stand for. Accordingly, "typical" items lead to benefits in information processing as they allow more reliable predictions about the concept's meaning than less typical ones (e.g., Rosch 1975; McCloskey & Glucksberg 1978). As choices of menu options mostly involve inferences about logical or categorical relationships among items, urging the users to draw a deduction from a category name to the corresponding sub-functions, it is a crucial demand of usable designs to implement unequivocal naming of functions and categories and a clear-cut assignment of subordinate terms to categories. Optimally, any menu category should unmistakably include all its sub-functions and, at the same time, unequivocally exclude other functions placed elsewhere in the menu. It is examined, if the menus are designed in a way that users are directed through the menu by an adequate naming and allocation of functions to categories (Schröder & Ziefle 2005).

Method: Simulations of the menus of two real mobile phones were used, which had to be navigated through in order to search for specific functions. As tasks participants had to search for particular functions within common phone applications (looking up a number in the phone book, reading and sending a text message, hiding own number, call divert, activating the mail box, setting the alarm clock). As the functions searched for were not to be mentioned in the instruction, ruling out any biases from priming effects, circumscriptions of the scenarios were used. In the simulations of the phones' menu each menu item was represented by one file, which was inter-connected with other files according to the hierarchical structure of each of the two models.

Participants navigated through the menu structure by selecting one file after another entering different submenus within the hierarchy. To ensure that all items displayed on one level were considered participants had to read aloud the whole list first before making a decision. 20 older adults (41- 65 years) took part as well as 20 teenagers (control group). As dependent variable, the number and type of navigation errors were investigated. Since incorrect decisions had different consequences, three errors types were defined: (1) *False turn-offs.* The most severe error occurs if users select one of the alternative options instead of the correct function. (2) *Keeping the false track.* This error occurs when users do not realize that they have chosen a wrong path, thus going deeper within the hierarchy by selecting another incorrect function. (3) *Erroneous returns.* These errors occur when the targeted function is not recognized as such, though users are on the correct path.

Results: Analysis of users' actions indicated that usage of functions' naming is a critical factor affecting navigation performance. It was shown that language based design features in current mobile phones are in fact sub-optimally implemented, failing to meet the requirements due to cognitive functioning of users (Figure 2).

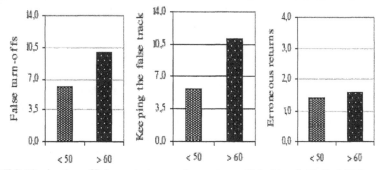

FIGURE 2 Navigation efficiency in terms of error types (Schröder & Ziefle 2005)

False turn-offs resulted mainly from a mismatch between category name and subordinate function, a high distracting value of rather generic category terms and a low distinctiveness of different options within one menu level. Keeping false track errors resulted primarily from semantically highly similar functions located on different menu levels, and, second, from terms with a very generic scope located on lower menu levels. Errors by erroneous returns emerged from semantically incoherent function sequences within mobile phones' menu. Insightfully, the very same problems were found in a technology-experienced teenager group (not visualized here, Schröder & Ziefle 2005). It shows that suboptimal language design affects navigation performance independent of age.

Summarizing, the central problems within language-based designs of mobile phones result from poor categorization of functions, low distinctiveness of different options on the same menu level, assignment of semantically similar functions to different submenus, logically incoherent item sequences and the usage of very generic terms.

THE UTILITY OF NAVIGATION AIDS

The third study examines the utility of two navigation aids, which were implemented into the phones' display. The hierarchical menu structure, still the most common form of interaction in communication technology devices, is advantageous as the functions are organized in smaller groups, thus rationing the multitude of options and keeping the information manageable. On the other hand, hierarchical menus also hold great potential for confusions and disorientation when menu structures are complex. The screen only provides the functions currently available with only little information given about previous or subsequent menu levels. As depth of the phones' menu continuously increases with increasing number of functionalities, still more function labels intervene along the path and seduce the users to take the wrong turnoff in the menu. Users report to loose their bearings in the menu, and, subsequently delve into distraction. The knowledge of how a menu is spatially structured guides users in their search through different levels in different menu depths. Following (Thorndyke & Goldin 1983) orientation includes three major spatial knowledge types: route knowledge (sequence of actions required to get from one point to another), landmark knowledge (salient features on the route), and survey knowledge (overall menu structure). The navigation performance was best when users had built up survey knowledge (hierarchical menu structure), route knowledge (route to take through the menu), and also landmark knowledge (at which crossings to turn 'left' or 'right') (Ziefle & Bay 2004)

Method: The first independent variable was the type of navigation aid. Two aids were compared, implemented into a computer simulation of a real mobile phone. In the first interface, the name of the current category was shown, as well as a list of its contents. This aid delivers mainly landmark knowledge (Figure 3, right). The other interface was identical to the first except that it showed the parents and parent-parents of the current of the category and that sub-categories were indented to emphasize the hierarchical structure (Figure 3, left). This aid contained spatial cues, providing survey knowledge.

FIGURE 3 Navigation aid with survey knowledge (left) and landmark knowledge (right) (Ziefle 2008)

The second independent variable was users' age, comparing navigation performance of younger adults to that of older adults. For dependent variables, we measured the effectiveness and efficiency of navigation. For efficiency, two disorientation measures were taken: The first was the number of hierarchical steps back to higher levels in the menu hierarchy, indicating that users, believing they had taken the wrong path, went back to a known position within the menu, consequently re-orientating themselves. The second measure was the number of returns to the top, which was assumed to reflect utter disorientation and complete helplessness of users. To get out of the menu maze, users had to go back to the start level, beginning completely from scratch. 32 users took part in the study: 16 younger adults (23-28 years; M = 25.1; SD = 1.4) and 16 older adults (46-60 years M = 52.9, SD = 3.7). Users had to solve 9 common phone tasks. Half of the sample (8 young, 8 older) was supported by the landmark aid, the other one by the aid delivering survey knowledge.

Results: The aid providing survey knowledge performed much better than the aid delivering landmark knowledge (F (1,21) = 6, p =.00). Also younger adults performed better than older adults (F (1,21) = 8.4, p = .00). It is of high ergonomic interest that the older adults performed significantly better when users were supported by an aid providing for survey knowledge of the menu, and reached the performance of the younger adults (F (1,21) = 6.2, p = .00 (Figure 4). The relative benefit from the survey knowledge was much higher for older adults. Their profit from the aid with the survey compared to the aid with the landmark knowledge was 53%, in contrast to only 26% in the younger group with respect to the number of returns in menu hierarchy (Figure 4, left). With regard to the number of returns to the top (Figure 4, right), the older group showed 77% better performance with the aid providing survey knowledge (in contrast to only 36% for the younger group) than the aid with landmark knowledge.

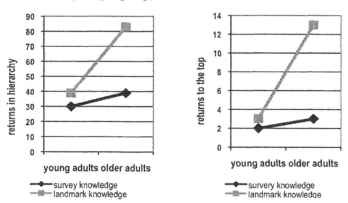

FIGURE 4 Interacting effect of navigation aid (Ziefle & Bay 2006)

The outcomes indicate that an appropriate navigation aid can remedy disorientation for older users and make them catch up their lower performance. Thus older adults can reach the performance of younger adults when given appropriate support: a navigation aid providing survey knowledge.

DISCUSSION

In this paper three studies were reported addressing the question if current mobile technology meets demands of universal access including older adults as a major user group. This is of specific impact as mobile technology specifically assumed to allow older adults to keep up mobility and an independently life.

Mobile devices are ubiquitous, context-adaptive and invasive, providing on-the-go lookup and entry of information, and quick communication. Mobile devices and increasingly upcoming services deliver a huge potential of assistance, in all kind of areas of life, be it as navigational assistants (implemented in automotives or serving as electronic companions for disabled persons, e.g. blind people), as medical assistants (monitoring vital data, and calling emergency if necessary), as technical devices in the context of sports and rehabilitation sector, as information and communication assistants (providing quick entry of information and communication with social partners) or as interacting device for electronic services (mobile purchase goods and electronic deliveries). But the effective integration of mobile information technology and its broad acceptance imposes considerable challenges. Contrary to current popular stereotypes, older users do take great interest in computer and mobile technology (e.g., Arning & Ziefle 2006; Brodie et al. 2003; Melenhorst et al. 2006), but they are often confronted with the problem of a diminished usability, as was shown in recent studies (Arning & Ziefle 2009; 2007a+b; Ziefle & Bay, 2004, 2006, 2008; Ziefle et al. 2008). Therefore many older users reject the utilization of ICT (Arning & Ziefle 2006, 2007a+b).

According to Melenhorst et al. (2006) the reluctance of older adults to weigh technological devices as "useful" is referred to the lack of the perceived advantages. The expected gain of a device may be perceived as not worth the trouble in terms of learning cost, frustration and anger about a suboptimal usability. Thus, for older adults, a high usability of a device is decisive.

In order to understand pitfalls and shortcomings older users experience and to clarify whether usability problems are age-exclusive (causing interaction difficulties only for older users) or age-specific (problems that cause serious interaction difficulties preferably for older people) it is necessary to study older adults' interaction with technical devices. This was undertaken here. First, menu complexity was experimentally addressed and navigation costs in terms of effectiveness and efficiency determined. Second, language issues of mobile phones' menus were investigated aiming at the effects of imprecise functions' naming and illogical categorization of functions. Findings show that a high menu complexity and a suboptimal language design have detrimental effects for navigation performance but they are not age-exclusive as also younger adults are also hindered by a low usability (though not as much than older adults are). Age-related declines in spatial abilities result in a higher proneness of older users to delve into distraction and to experience disorientation in the menu. Supporting the older group by adequate navigation aids compensates for age-related disadvantages and allows universal access.

Concluding, mobile devices should be developed in a way that users want to use it, and, still more important, that they are able to use it (Arning & Ziefle 2007; Ziefle & Ba, 2008). As long as mobile interface and information designs are not easy to use, technical innovations will not have sustained success.

REFERENCES

Arning, K. & Ziefle, M. (2009). Cognitive and personal predictors for PDA navigation performance. *Behaviour & Information Technology.* 28(3), 251-268.

Arning, K. & Ziefle, M. (2007a). Understanding differences in PDA acceptance and performance. *Computers in Human Behaviour*, 23(6), 2904-2927.

Arning, K. & Ziefle, M. (2007b). Barriers of information access in small screen device applications. In: C. Stephanidis & M. Pieper (Eds.), *Universal Access in Ambient Intelligence Environments* (pp. 117-136). Berlin, Germany: Springer.

Arning, K. & Ziefle, M. (2006). What older adults expect from mobile services. In: R.N. Pikaar, E.A. Konigsveld, & P.J. Settels, P.J. (Eds.), *IEA 2006: Meeting diversity in ergonomics.* Amsterdam: Elsevier.

Bay, S. & Ziefle, M. (2004). Effects of menu foresight on information access in small-screen devices. In: *Human Factors and Ergonomics Society 48th Meeting* (pp. 1841-1845). Santa Monica, CA: Human Factors.

Bay, S. & Ziefle, M. (2003). Design for All: User Characteristics to be considered for the Design of Devices with Hierarchical Menu Structures. In: H. Luczak & K. Zink (Eds.). *Human Factors in Organizational Design and Management* (pp. 503-508). Santa Monica: IEA.

Bernard, M.L., Chaparro, B.S., Russell, M.C. (2001). Examining automatic text presentation for small screens. In: *Human Factors and Ergonomics Society 45th Meeting* (pp. 637-639). Santa Monica, CA: Human Factors.

Brodie, J., Chattratichart, J., Perry, M. & Scane, R. (2003). How age can inform future designs of the mobile phone experience. In: C. Stephanidis (Ed.). *Universal Access in Human Computer Interaction* (pp. 822 – 826). Mahwah, NJ: LEA.

Chang, T.M. (1986). Semantic memory: Facts and models. *Psychological Bulletin*, 99(2), 199-220.

Craik, F.I.M. & Salthouse, T.A. (1992). Handbook of Ageing and Cognition. Hillsdale, N.J.: LEA.

Edwards, D. & Hardmann, L. (1989). Lost in Hyperspace-Cognitive Mapping and Navigation in a Hypertext Environment. In: R. McAleese. (Ed). *Hyperspace: Theory into practice* (pp.105-125). Oxford: intellect unlimited.

Han, S.H. & Kwahk, J. (1994). Design of a menu for small displays presenting a single item at a time. In: *Human Factors and Ergonomics Society 38th Meeting* (pp. 360-364). Santa Monica, CA: Human Factors.

Kieras, D. & Polson, P.G. (1985). An approach to the formal analysis of user complexity. *International Journal of Man–Machine Studies*, 22, 365 – 394.

Kim, H. & Hirtle, S. (1995). Spatial metaphors and disorientation in hypertext browsing. *Behaviour & Information Technology.* 14, 239-250.

Kline, D.W. & Scialfa, C.T. (1997). Sensory and perceptual functioning. In: A.D Fisk & W.A. Rogers. (Eds.). *Handbook of Human Factors and the Older Adult* (pp. 27-54). San Diego: Academic.

Leonhardt, S. (2006). *Personal health care devices.* In: S. Mukherjee et al. (Eds.) AmIware: Hardware Technology Drivers of Ambient Intelligence (pp. 349 – 370). Dordrecht: Springer.

McCloskey, M.E. & Glucksberg, S. (1978). Natural categories. Well-defined or fuzzy sets? *Memory and Cognition*, 6, 462-472.

McDonald, J. & Stevenson, R. (1998). Navigation in Hyperspace. *Interacting with Computers*, 10, 129-142.

Melenhorst, A.-S., Rogers, W.A. & Bouwhuis, D.G. (2006). Older adults motivated choice for technological innovation: evidence for benefit-driven selectivity. *Psychology of Aging*, 21(5), 190-195.

Noordman-Vonk, W. (1979). Retrieval from Semantic Memory. In: W.J.M. Levelt (ed.). *Language and Communication 5*. Berlin: Springer.

Noyes, J.M. & Sheard, M.C.A. (2003). Designing for older adults – are they a special group? In: C. Stephanidis C. (Ed.). *Universal Access in Human Computer Interaction* (pp. 877-881). Mahwah, NJ: LEA.

Oetjen, S. & Ziefle, M. (2007). Effects of LCD's Anisotropy on the Visual Performance of Users of Different Ages. *Human Factors*, 49(4), 619-627

Oetjen, S. & Ziefle, M. (2009). A visual ergonomic evaluation of different screen technologies. *Applied Ergonomics*, 40, 69-81.

Omori, M., Watanabe, T., Takai, J., Takada, H. & Miyao, M. (2002). Visibility and characteristics of the mobile phones for elderly people. *Behaviour & Information Technology*, 21 (5), 313-316.

Park, D. & Schwarz, N. (1999). *Cognitive Ageing*. Philadelphia, Buchanan.

Rosch, E. (1975). Cognitive representations of semantic categories. *Journal of Experimental Psychology, General*, 104 (3), 192-233.

Schieber, F. (2005). Vision and Aging. In: J.E. Birren & W. Schaie (Eds.). *Handbook of the Psychology of Aging*. San Diego: Academic.

Schröder, S. & Ziefle, M. (2005). Semantic transparency of cellular phones menu. Comparing users from different age groups. In: H.-C. Fisseni, B. Schmitz, B. Schröder & P. Wagner. (Eds). *Computer Studies in Language and Speech* (pp. 302-315). Frankfurt: Peter Lang.

Sein, M.K., Olfman, L., Bostrom, R.P. & Davis, S. (1993). Visualization ability as a prediction of user learning success. *International Journal of Man–Machine Studies*, 39, 599–620.

Thorndike, P. W., & Goldin, S. E. (1983). Spatial learning and reasoning skill. In: H. Pick & L. Acredolo (Eds.). *Spatial orientation: Theory, research and application* (pp. 195–217). New York: Plenum.

Vercruyssen, M. (1997). Movement Control and Speed of Behavior. In: A.D Fisk & W.A. Rogers (Eds.). *Handbook Human Factors and the Older Adult* (pp.55-86). San Diego: Academic.

Westermann, S.J. (1997). Individual differences in the use of command line and menu computer interfaces. *International Journal of Human-Computer-Interaction*, 9 (2), 183-198.

Willis, S.L. & Schaie, K.W. (1986). Training the elderly on the ability factors of spatial orientation and inductive reasoning. *Psychology of Aging,* 1(3), 239-247.

Zhao, C., Zhang, T., and Zhang, K. (2001). User interface design for the small screen display. In: *Human Factors and Ergonomics Society 45th Annual Meeting* (pp. 1548-1550). Santa Monica, CA: Human Factors.

Ziefle, M. (in press). Information presentation in small screen devices: The trade-off between visual density and menu foresight. *Applied Ergonomics*.

Ziefle, M. (2009). Spatial cues in small devices - benefit or handicap? In: T. Groß, J., Gulliksen, P. Kotzé, L. Oesterreicher, P. Palanque, R. Prates & M. Winckler (Eds). *Human Computer Interaction-Interact 2009* (pp. 620-633). Berlin: Springer.

Ziefle, M. (2008). Instruction format and navigation aids in mobile devices. In A. Holzinger (Ed.). *Usability and Human Computer Interaction for Education and Work* (pp. 339–358), Berlin: Springer.

Ziefle, M. & Bay, S. (2008). Transgenerational Designs in Mobile Technology. In J. Lumsden (Ed.). *Handbook of Research on User Interface Design and Evaluation for Mobile Technology* (pp. 122-140). Hershey, PA.: IGI Global.

Ziefle, M., Schroeder, U.; Strenk, J., Michel, T. (2007) How young and older users master the use of hyperlinks in small screen devices. *In: SIGCHI Conference on Human Factors in Computing Systems* (pp. 307-316). Association for Computing Machinery.

Ziefle, M, & Bay, S. (2006) How to overcome disorientation in mobile phone menus. *Human Computer Interaction*, 21(4), 393-432.

Ziefle, M. & Bay, S. (2005). How older adults meet complexity. *Behaviour & Information Technology*, 24(5), 375-389.

Ziefle, M. & Bay, S. (2004). Mental Models of a Cellular Phone Menu. Comparing Older and Younger Novice. In: S. Brewster & M. Dunlop (Eds.). *Mobile Human Computer Interaction* (pp. 25-37). Berlin: Springer.

Chapter 32

Exploring What Young Consumers Want in Cell Phones

Fethi Calisir, Cigdem Altin Gumussoy

Industrial Engineering Department, Faculty of Management
Istanbul Technical University
34367, Macka-Istanbul,
Turkey

ABSTRACT

As the competition among cell-phone manufacturers becomes fierce, they give more importance for the consumers' preferences of cell phones. In this study, using data collected from a sample of 378 young consumers, we explore the most important features and factors of cell phones. The study reveals that "battery duration," "voice quality," "duration of charging," and "design image" are the most important features for young consumers. In addition, seventeen factors were derived, which accounts for 70.23% of the total variance. We conclude with recommendations for future research.

Keywords: Cell Phone, Customers, User Preferences

INTRODUCTION

Cell phones are widely used all over the world and they have become an integral part of people's everyday life. It is expected that mobile accounts around the globe would grow from 1.7 billion in 2004 to 2 billion in 2008 (Boretos, 2007). In Turkey the number of mobile phones' users has grown by over 208 percent over the past six years according to the market research firm Euromonitor International. In addition, "Turkey currently has the sixth-largest young mobile subscriber base in the world, with more than 11 million subscribers under the age of twenty-five, providing a very lucrative market for mobile phone companies" (Aliprandini, 2009).

People want to stay in touch and have easy access to information anywhere and anytime (Ling et al. ,2006). Nurvidathi (2002) stated that the most frequently cited reasons for having a mobile phone is its mobility, usefulness in emergency, and the feeling of self-assurance. This enhances users' feelings of family security and personal security. Mobile phones also have facilitated the formation of social networks for their users (Srivastava, 2005) and have increased social inclusion and connection (Mathews, 2004; Wei and Lo, 2006). Young users believe that using a technologically advanced mobile phone improves their status among peers (Ozcan and Kocak, 2003). Consumers use cell phones not only for communication but also for facilitating universal information access (Ling et al., 2006), quick communication through instant messaging, and voice calling (Weiss, 2002), or technological features of cell phones such as camera and mobile Internet (Ling et al., 2007). Besides these factors, Walsch et al. (2008) revealed that some young people are extremely attached to their mobile phone with symptoms of behavioral addiction revealed in participants' descriptions of their mobile phone use. Therefore, impact of mobile phone on users' daily life cannot be ignored (Ling et al., 2006).

Different expectations and needs of consumers orient cell-phone manufacturers to produce a wide variety of cell phones. Cell phones differ from each other in terms of design features such as shape, size, color, or material. However, as the cell phones gain more features, they may become more complicated for consumers. So manufacturers have to determine the needs of consumers that are more important than the others and aim at designing phones that are easy to use (Bay and Ziefle, 2008). The better a company can satisfy one's needs, the better it manages itself in the competition (Solomon, 2001). So what features are more important is a critical question to be answered for cell-phone manufacturers.

However to our knowledge, there are only a limited number of studies that examine the important features. Han et al. (2004) identified design features of mobile phones critical to user satisfaction in terms of luxuriousness, attractiveness, harmoniousness, and overall satisfaction. By comparing the values of the critical design features, design properties common to "desirable" and "undesirable" phones were extracted. Ling et al. (2006) conducted a survey among 2571 college students. They asked to evaluate their preference of five design features (color screen, camera, voice-activated dialing, Internet browsing, and wireless connectivity) and the overall satisfaction level regarding their phone. The results show that color screen, voice-activated dialing, and Internet-browsing features can strongly affect users' satisfaction level. Ling et al. (2007) investigated the relationship among the design features of the cell phone and identified the most important design features and design factors. The most important design features are physical appearance, size, and menu organization.

This study attempts to identify critical features for consumers in terms of ergonomic, technological, or other features such as price, brand, and environmental effect. The next section discusses the proposed methodology which is based on a survey of 378 users. This is followed by the results of the survey. The article concludes with a discussion of the findings that discuss the most important features of cell phones.

METHODOLOGY

A survey methodology was used to gather data. A questionnaire was constructed based on an extensive review of the literature in the areas of cell phones' ergonomic factors and technology-related factors. Many survey questions were adopted from previous literature and from suggestions from academics (Seva et al., 2006; Schröder and Ziefle, 2006; Lin and Kang, 2006; Yucel and Aktas, 2006; Yun et al., 2003; Han et al., 2004). After the initial development of the survey questionnaire, it was completed by 15 users to ensure the understandability. On the basis of the information provided by the users, the instrument was "fine-tuned" and finalized. The target population for the study was young customers in Turkey. A total of 378 questionnaires were collected.

The final questionnaire consists of two main parts. The first section involved demographic questions designed to solicit information about the respondents. Table 1 presents a summary of the demographic characteristics of the sample. The second section asked respondents to indicate the effect level of 70 factors on the selection of a cell phone. All items in the second section were measured with five-point Likert scales, where 1 represents "not important at all" and 5 represents "very important."

Table 1: Demographic profiles of participants

Variables	Description	Frequency	Percentage
Gender	Male	221	58.5
	Female	157	41.5
Age (years)	14-18	150	39.7
	19-22	93	24.6
	23-27	113	29.9
	28-32	20	5.3
	33 and higher	2	0.5
Education level	High school	162	42.9
	Undergraduate	151	39.9
	Master	61	16.1
	Doctorate	4	1.1
Cell phone usage	Yes	360	95.2
	No	18	4.8

RESULTS

STATISTICS FOR PREFERENCES AND CELL PHONES

The internal consistency for the 70-item measure was determined using the Cronbach's alpha calculation (Cronbach, 1984). The value of Cronbach's alpha calculation for 70 items is 0.94, indicating a high level of reliability. The mean and standard deviation of the impact for each of the 70 factors arranged from the highest to the lowest value are shown in Table 2. As seen in Table 2, battery duration has the highest mean effect level of 4.37 and the second lowest standard deviation of 0.88. It means that most users unanimously give high importance to this feature while selecting a cellular phone. In contrast, family advice has the lowest impact level of 2.97 and one of the highest standard deviation of 1.28. The impact level of family advice is just below neutral, and it shows that family advice does not have an impact on young customers' selection of a cell phone. The large standard deviation shows that users have different opinions about whether family advice has an impact.

A t-test is performed to see which factors are perceived to be "very important." The results show that the features "battery duration," "voice quality," "duration of charging," and "design image" are perceived to be very important factors. Karjaluoto et al. (2003a) explored consumer motives in a mobile phone industry in Finland. The results show that the most explicit reasons for buying a new mobile phone were that the mobile phone did not work, the calls were interrupted due to weak audibility, battery run too fast, the screen was broken, or the keypad was so much used up that the numbers were invisible. Also, product image plays an important role in the consumers' preference and choice of the product (Chuang et al., 2001).

In addition, "phones' alarm clock," "price," "ease of use," "reliability," "Mp3," "attractiveness," and "size" are perceived to be important when selecting a cell phone. This finding is consistent with that of (Karjaluoto et al., 2003a) and (Karjaluoto et al., 2003b), indicating that phone price has been identified as a critical factor in the choice of the mobile phone model, especially among younger people. In addition to these findings, (Karjaluoto et al., 2005) identified factors that influence intention to acquire new mobile phones and factors that influence the change of the mobile phone. The results show that while technical problems are the basic reason to change mobile phone among students, price, brand, interface, and properties are the most influential factors affecting the model selection.

Table 2: Preferences of cell phone features

	Mean	Std.	
Battery duration	4.37	0.88	A
Voice quality	4.26	0.89	A
Duration of charging	4.18	0.92	A
Design image	4.16	0.99	A
Phones' alarm clock (have or setting time)	4.06	1.04	B
Price	4.05	1.11	B
Ease of use	4.02	1.06	B
Reliability	4.01	1.02	B
Mp3	3.98	1.13	B
Attractiveness	3.98	1.05	D
Size	3.97	0.96	B
Organization	3.90	0.89	C
Shape	3.90	0.87	C
Internal memory	3.89	1.05	C
Functionality	3.88	1.12	C
Color	3.87	1.09	C
Regularity	3.86	0.93	C
Time required to read a SMS	3.83	1.12	C
Time required to look up a number in the phone book	3.83	1.08	C
Video	3.83	1.17	C
Clarity	3.82	0.93	C
External memory	3.81	1.08	C
Simplicity	3.81	0.99	C
Bluetooth	3.80	1.22	C
Brand	3.80	1.16	C
Learnability	3.79	1.01	C
Time required to send a SMS	3.78	1.10	C
Model	3.76	1.15	C
Display area	3.76	0.98	C
Luminous mechanism of display	3.76	1.08	C
Weight of cell phone	3.75	1.08	C
Ease of use of operation of buttons and interface design	3.74	0.96	C
Balance	3.72	1.05	C
Buttons and digital interface design can easily be used	3.69	1.07	C
Camera phones	3.68	1.23	C
Width of body	3.67	1.09	C
Thickness of body	3.67	1.10	C
Buttons and digital interface design allow for errors in user	3.66	1.01	C
Size of buttons, screen and symbols and use space are suitable	3.63	1.01	C
Wireless connectivity	3.62	1.24	C
Buttons and digital interface design are distinctive	3.62	1.03	C
Shape of body	3.60	1.06	C
Vertical length of the body	3.59	1.10	C

Radio	3.59	1.16	C
Calculator	3.59	1.20	C
Degree of button softness	3.56	1.09	C
Digital interface design/ buttons are suited to the use of each group	3.54	1.04	C
Height of display	3.54	1.02	C
Infrared	3.54	1.28	C
Promotional campaign	3.52	1.21	C
Instructions for buttons are suited to the use of each group	3.51	1.07	C
Internet accessibility	3.49	1.23	C
Ratio of display width to height	3.47	1.14	C
Hiding the own number	3.46	1.18	C
Luminous color of display	3.45	1.12	C
Advanced SMS options	3.44	1.13	C
Phone material	3.43	1.12	C
GSM operator	3.37	1.39	C
Voice-activated dialing	3.34	1.32	C
Size of navigation button	3.31	1.04	C
Number of fonts in display	3.26	1.14	C
Backlight mechanism of button	3.26	1.09	C
Number of colors in the body	3.22	1.20	C
Setting up a call divert	3.19	1.17	C
Number of buttons	3.18	1.07	C
Activating the mailbox	3.13	1.21	C
Number of differently shaped buttons	3.13	1.08	C
Friends	3.03	1.31	C
Environmental effect	3.00	1.24	C
Family	2.97	1.28	C
Average	3.66	1.10	

FACTOR ANALYSIS

A principal components analysis (PCA), with a Varimax rotation, was used to reduce the data. For the purpose of describing the underlying factor structure, the 'eigen value-one criterion' was used to determine the number of components. The PCA produced seventeen factors that accounted for 70.23% of the total variance. For interpretation purposes, a general cutoff point of 0.50 was chosen, and a factor loading greater than 0.5 for a phone feature is considered to be significant (Ling et al., 2007).

The name of each component is given on the basis of the factor loadings. We have chosen to name component 1 (F1) 'technological features', component 2 (F2) 'description of image', component 3 (F3) 'buttons and digital interface related features', component 4 (F4) 'functional and design features', component 5 (F5) 'aesthetic aspect', component 6 (F6) 'influential persons and external effects', component 7 (F7) 'durable aspect', component 8 (F8) 'phone dimensional features', component 9 (F9) 'calling-related features', component 10 (F10) 'display dimensional features', component 11 (F11) 'memory related features, ease of use and learn', component 12 (F12) 'model characteristics', component 13 (F13) 'length of body and shape of buttons', component 14 (F14) 'calculator', component 15 (F15) 'radio', component 16 (F16) 'backlight mechanism of button', and component 17 (F17) 'phones' alarm clock'.

For those factors that included more than one feature, Cronbach's alpha values were calculated. All variables achieved a high level of reliability. The seventeen factors with their corresponding features, percentage variance explained, cumulative percentage variance explained, and their internal consistency values are listed in Table 3 (see Appendix).

CONCLUSION AND FUTURE STUDIES

Cell phone market has become fierce as cell phones have become an integral part of everyday life. Manufacturers offer a wide variety of products. However, offering a cell phone in the direction of consumers' preferences may be a critical point for success in the rapidly changing cellular-phone market. The approach used in this study may help cell-phone manufacturers to design cell phones with critical features for young consumers. A survey of 378 young consumers examined the importance level of 70 cell-phone features. The most noticeable aspect of this research is that "battery duration", "voice quality", "duration of charging", and "design image" features are the most important features from the young consumers' views. These findings are consistent with those of Isiklar and Buyukozkan (2007). The aim of their study is to propose a multi-criteria decision-making approach to evaluate the mobile phone options with respect to the users' preferences order. The results show that technical features (talk time, stand by time, international roaming, safety standards) are the most important factors in product-related criteria and that functionality (ease of use) is the most preferable factor. In addition, consumers' choice of a product depends largely on their perception of product image (Hsiao and Liu, 2002).

"Ease of use" is perceived to be important for consumers. As cell phones expand their functionality, use of cell phones becomes more complicated. In addition manufacturers may not have time to conduct usability tests, because cell phone market becomes more competitive, and the fast release of product may be a key factor. However, to develop an easy-to-use cellular-phone user interface, an effective usability evaluation is essential in the product development process (Sears and Jacko, 2002).

Although the findings of the present study contribute to a better understanding of the features that are more important for young consumers, there are several limitations to this study. First, cell phones are not designed only for young consumers. A similar study may focus on consumers in different age groups. Second, the results of this study are far from providing implications for other countries. A similar study may be applied to different parts of the world in order to investigate whether critical features differ among cultures. Finally, a longitudinal research design is essential to confirm the results of the study.

REFERENCES

Aliprandini, M. (2009), "Preparing for the next generation: ICT set to become a major driver of the Turkish economy." *Foreign Affairs*, 01/01/2009, 8, 1, 7.

Bay, S., and Ziefle, M. (2008), "Landmarks or surveys? The impact of different instructions on children's performance in hierarchical menu structures." *Computers in Human Behavior*, 24(3), 1246–1274.

Boretos, G.P. (2007), "The future of the mobile phone business." *Technological Forecasting and Social Change*, 74(3), 331–340.

Chuang, M.C., Chang, C.C., and Hsu, SH. (2001), "Perceptual elements underlying user preferences toward product form of mobile phones." *International Journal of Industrial Ergonomics*, 27(4), 247–258.

Cronbach, LJ. (1984), *Essentials of psychological testing*. New York: Harper & Row.

Han, S.H., Kim, K.J., Yun, M.H., Hong, S.W., and Kim, J. (2004), "Identifying mobile phone design features critical to user satisfaction." *Human Factors and Ergonomics in Manufacturing*, 14(1), 15–29.

Hsiao, S.W., and Liu, M.C. (2002), "A morphing method for shape generation and image prediction in product design." *Design Studies*, 23(5), 497–513.

İsiklar, G., and Buyukozkan, G. (2007), "Using a multi-criteria decision making approach to evaluate mobile phone alternatives." *Computer Standards & Interfaces*, 29(2), 265–274.

Karjaluoto, H., Karvonen, J., Pakola, J., Pietilä, M., Salo, J., and Svento, R. (2003a), *Exploring consumer motives in mobile phone industry: an investigation of Finnish mobile phone users*. Proceedings of the 1st International Conference on Business Economics, Management, and Marketing, June 26-29, Athens, Greece, 335–342,

Karjaluoto, H., Karvonen, J., Kesti, M., Koivumäki, T., Manninen, M., Pakola, J., Ristola, A., and Salo, J. (2005), "Factors affecting consumer choice of mobile phones: Two studies from Finland." *Journal of Euromarketing*, 14(3), 59–82.

Karjaluoto, H., Pakola, J. Pietilä, M., and Svento, R. (2003b), *An exploratory study on antecedents and consequences of mobile phone usage in Finland*. Proceedings of the AMA Summer Marketing Educators' Conference, Chicago, USA, 14, 170–178.

Lin, R., and Kang, Y. (2006), *The usability evaluation of mobile-phone*. IEA 2006 The 16th World Congress on Ergonomics, July 10-14, Maastricht, the Netherlands.

Ling, C., Hwang,W., and Salvendy, G. (2006), "Diversified users' satisfaction with advanced mobile phone feature." *Universal access in the Information Society*, 5(2), 239–249.

Ling, C., Hwang,W., and Salvendy, G. (2007), A survey of what customers want in a cell phone design." *Behavior & Information Technology*, 26(2), 149–163.

Mathews, R. (2004), "The psychosocial aspects of mobile phone use amongst adolescents." *InPsych*, 26(6), 16–19.

Nurvidathi, E. (2002), *A study of mobile phone usage in the united states and Japan: An international undergraduate research experience*. Thesis 2002, Electrical Engineering and Computer Science Department, Oregon State University, Corvallis, OR, USA.

Ozcan, Y.Z., and Kocak, A. (2003), "Research note: A need or a status symbol? Use of cellular telephones in Turkey." *European Journal of Communication*, 18(2), 241–254.

Schröder, S., and Ziefle, M. (2006), *Evaluating the usability of cellular phones' menu structure in a cross-cultural study*. IEA 2006 The 16th World Congress on Ergonomics, July 10-14, Maastricht, the Netherlands.

Sears, A., and Jacko, J. (2002), *The human–computer interaction handbook: Fundamentals, evolving technologies and emerging applications 2002*. Lawrence Erlbaum Associates, Inc., Mahwah, NJ, 1091–1092.

Seva, R.R., Duh, H.B.L., and Helander, M. (2006), *Predicting affect from phone features using discriminant analysis*. IEA 2006 The 16th World Congress on Ergonomics, July 10-14, 2006, Maastricht, the Netherlands.

Solomon, M.R. (2001), *Consumer behavior. Buying, having, being*. 5th ed./New Jersey. Prentice-Hall; 2001.

Srivastava, L. (2005), "Mobile phones and evolution of social behavior." *Behaviour & Information Technology*, 24(2), 111–129.

Walsch, S.P., White, K.M., and Young, R.M. (2008), "Over-connected? A qualitative exploration of the relationship between Australian youth and their mobile phones." *Journal of Adolescence*, 31(1), 77–92.

Wei, R., and Lo, V.H. (2006), "Staying connected while on the move: Cell phone use and social connectedness." *New Media and Society*, 8(1), 53–72.

Weiss, S. (2002), *Handheld usability*. West Sussex, England: John Wiley & Sons; 4–6.

Yucel, G., and Aktas, E. (2006), *Multi-attribute comparison of ergonomics mobile phone design based on information axiom*. Applied Artificial Intelligence Proceedings of the 7th International FLINS Conference, Genova, Italy, 29-31 August, 351–358.

Yun, M.H., Han, S.H., Hong, S.W., and Kim, J. (2003), "Incorporating user satisfaction into the look-and-feel of mobile phone design." *Ergonomics*, 46(13/14), 1423–1440.

Table 3: Factors and internal consistencies

Factors	Cell phone features	Variance explained (%)	Cum. variance explained (%)	Cronbach's alpha
Factor 1 Technological features	Wireless connectivity Infrared Camera phones Bluetooth Internet accessibility Video Voice-activated dialing Advanced SMS options Mp3	9.37	9.37	0.92
Factor 2 Description of image	Clarity Simplicity Organization Regularity Shape	6.03	15.4	0.911
Factor 3 Buttons and digital-interface-related features	Instructions for buttons are suited to the use of each group Buttons and digital interface design are distinctive Buttons and digital interface design can be easily used Digital interface design/ buttons are suited to the use of each group Buttons and digital interface design allow for errors in user operations	5.96	21.36	0.88
Factor 4 Functional and design features	Time required to look up a number in the phone book Ease of use Time required to read a SMS Functionality Time required to send a SMS	5.20	26.56	0.85
Factor 5 Aesthetic aspect	Design image Color Attractiveness Number of colors in the body Luminous mechanism of display	4.81	31.37	0.82
Factor 6 Influential persons and external effects	Family Friends Environmental effect Promotion	4.71	36.08	0.87
Factor 7	Reliability	4.31	40.39	0.80

Durable aspect				
	Battery duration Balance Duration of charging	4.21	44.6	0.79
Factor 8 Phone dimensional features	Size of phone Weight of phone Degree of button softness Phone material Width of body	4.16	48.76	0.78
Factor 9 Calling-related features	Setting up a call divert Hiding the own number Activating the mailbox	3.97	52.73	0.79
Factor 10 Display dimensional features	Area of display Ratio of display width to height Height of display	3.50	56.23	0.81
Factor 11 Memory related features, ease of use and learn	Internal memory External memory Learnability Ease of use of operation of buttons and interface design	3.50	59.73	0.81
Factor 12 Model characteristics	Brand Model Price	2.46	62.19	0.61
Factor 13 Length of body and shape	Vertical length of the body Number of differently shaped buttons	2.25	64.44	-
Factor 14 Calculator	Calculator	1.95	66.39	-
Factor 15 Radio	Radio	1.95	68.34	-
Factor 16 Backlight mechanism of button	Backlight mechanism of button	1.89	70.23	-
Factor 17 Phones' alarm clock	Phones' alarm clock (have or setting time)			

CHAPTER 33

Tactile Buttons Used in Non-Visual Cell Phone

Nai-Feng Chen, Fong-Gong Wu, Chun-Heng Ho

Department of Industrial Design
National Cheng Kung University
No.1 Daxue Rd., Tainan City, 701 Taiwan (R.O.C)

ABSTRACT

Although cell phone has become very popular and much more innovative these days, most input interfaces still rely on the sense of sight. In order to improve the usability and convenience of using a cell phone for the visually impaired and average people under special circumstances, this research take telephone call as example and develop two sets of tactile buttons that can be used under non-visual situation by touch alone. The first set contains two kinds of function buttons and the second set has two kinds of number buttons. In the research we use non-visual experiment, think-aloud protocol and interview to test the design of tactile buttons and analyze the cognitive behavior of users. The result shows that participants can do better with tactile buttons comparing with traditional ones. They tend to recognize shapes according to clear edge outlines and be able to distinguish several basic features, which could be lead to the guidelines for tactile buttons design.

Keywords: Tactile button Design, Tactual identification, Prototype, Cognitive behavior

INTRODUCTION

In the research, we want to discover two main centre questions: can people identify cell phone buttons by pure tactile touch? And how do people recognize the buttons without seeing them? We use quantitative method to exam the first question and use qualitative methods to discuss the second one.

Cell phone nowadays has several functions including telephone call, message, e-mail, clock, calendar, camera, mp3 player and so on. The most basic, frequently used and representative function still remains on telephone call, which involves with two main processes: dialing and hanging up. There are two kinds of buttons that would be used in the process: the function and the number buttons. In this research we use index code to modify and simplify the function buttons and use analogy and location points to design the number buttons. This kind of buttons could provide tactile cues that could be used as substitutes for sense of sight.

Human explore the world through five senses, which are sight, hearing, taste, smell and touch. To recognize the shapes of objects, we mostly rely on sense of sight and touch. There is analogy between the movements of fingers touch, which is called "tactile scanning", and movements of eyes, which is "ocular scanning" (Gibson, 1962). By touch alone, we could still receive information of an object such as shape, texture, and the information that vision could not provide us such as temperature and pressure. An editorial discussed about this: "the hand can often see more than the eye can visualize (P.R.M., 1999)." Tactile touch could really help a lot when recognizing objects for visually impaired or average people as well in some special circumstances. For instance, in the dark theater we can hardly see the interface and buttons of a cell phone clearly. And the sounds come from the movie are usually too loud that it is impossible to use voice guided operation either. With the aid of tactile sense, such as haptic cues, contrasting shapes and surface textures, we can correctly match shapes, recognize between objects, reduce confusion, prevent error, avoid danger, minimize mistakes and even speed up the identification time (Chamberlain, Gardner, & Lawton, 2007; P.R.M., 1999; Walters, Chamberlain, & Press, 2003). Taking an example from our daily life, when taking a shower we can use sense of touch to identify the content of each bottle and reduce the error. So we won't use shampoo to wash our body and use shower gel to clean our hair when our eyes are covered with bubbles and water.

Additionally, Norman (1988) suggested that a good design can be operated correctly even in the dark without seeing it. In the example of beer-bump handles of a nuclear power plant, tactile cues help to prevent misidentification (Norman, 1988). Tactile sense has been commonly used when designing this kind of control devices. Jenkins (1947) identified two sets of knobs, which contain eight different coded shapes in each set, and found that knobs in each set are rarely confused with each other. During the process, participants could only rely on sense of touch to recognize (Jenkins, 1947). Similar to Jenkins's research, fifteen knob designs, which are seldom confused with one another by touch, have been developed by the U.S. Air Force (Hunt, 1953). Later on, U.S. Air Force developed another set of knob controls, which have symbolic meanings associated with their intended functions so that the learning process could be easier and quicker (Air-Force-System-Command, 1980, June). Another research about tactual identification is that sixteen differently shaped handles were tested and compared with two methods: "find" and "learn", which are alike with "multiple matching" and "learning efficiency." The result came with ten handles, which were selected to prevent confusions, be more homogeneous and be learned quickly (Green & Anderson, 1955).

In our research, we also used tactile codes to design cell phone buttons, however, they are smaller in size comparing to previous examples. Although the tactile codes have features such as shape, color, texture, location and size, we would pay more attention on the shape and texture in this research. To test the possibility that different shape-coded controls might be confused with one another, researchers suggested letting participants indicate, only by hands, between all possible controls to tell if there is any confusion (Sanders & McCormick, 1993). In this research, non-visual experiment would measure several items, including multiple matching, reaction time, accuracy, learning efficiency and typical task.

DESIGN AND PROTOTYPE

FUNCTION BUTTONS DESIGN

We use index code to design function buttons and simplify into two types (see Figure 1). The dialing button is shaped like a telephone transmitter. Two circles at ends could provide location hints to help recognizing the shape. And the hanging up button, which owns one circle in the middle, is complementary to the dialing button.

(a)   (b)

FIGURE 1. The two sets of function buttons are designed with (a) index code modify and (b) simplify. The left ones are dialing buttons and the right ones are hanging up buttons.

NUMBER BUTTONS DESIGN

We analyze basic structure of Arabic numbers and use analogy method to transfer them into tactile number buttons (see Figure 2). There are five features, which are circle, arc, horizontal line, vertical line and oblique line, and each number contains less than two features. Take number two as an example that it is composed of one arc at the top and one horizontal line in the bottom.

FIGURE 2. The analogy number buttons. From left to right show number one to zero.

We also create a new form of number button by using location points to suggest the arrangement of numbers (see Figure 3). The numbers arrange into a concentric circle in a clockwise direction. Number zero is the location circle and number five is the location line. These two numbers divide others into two groups: 1~4 and 6~9. Each group arranges from one point to four points in order. In the middle there is a drop-like button (point to zero) to hint the direction.

FIGURE 3. The location points number buttons.

PROTOTYPES

We made four 1:1 tactile cell phone prototypes (see Figure 4, a~d) and three tactile icon boards (e~g), which would be used in the non-visual experiment. The main bodies of the cell phones are paper prototypes and the buttons are made by acrylic with laser engraving.

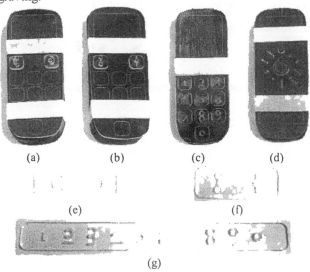

FIGURE 4. There are prototypes of (a)(e) 1st type index code function buttons , (b)(f) 2nd type index code function buttons, (c)(g) analogy number buttons, and (d) location points number buttons.

EXPERIMENTS

In the experiment, prototypes are covered with black cloth so that participant cannot see the buttons (see Figure 5). There are 15 participants, who are all around age 20 to 25. Each of them uses cell phone more than five times a day during the past year. The experiment time is around thirty minutes.

FIGURE 5. Pictures show the experiment situation. Prototypes are covered with black cloth.

The experiment has six parts. The first part is the basic measurement to collect basic information and hand sizes of participants. We paint black pigment on participant's hand, thumb and forefinger and then print on a white paper to measure the size A to F. (see Figure 6)

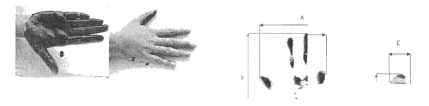

FIGURE 6. Pictures show the situation of the first part of experiment: basic measurement.

In the second part (multiple matching), we let participant touch the function tactile icon boards (see Figure 4, e & f) and ask participant which one is for dialing or hanging up. The reaction time and accuracy are recorded. Then we tell participant the answer, and after learning we repeat the test again to measure learning efficiency.

In the third part, we test function buttons with two cell phone prototypes (see Figure 4, a & b). We give participant prototype and let participant do the task: dialing and hanging up. We record the time and accuracy as well.

Alike with the second part, in the fourth part we give participant number tactile icon board (see Figure 4, g) and randomly ask participant each icon's meaning from number 0~9.

And the fifth part is alike with the third part that we test number buttons with cell phone prototypes (see Figure 4, c & d) and change the task into: dialing phone numbers.

The sixth part tests traditional cell phones, which are considered as a contrast to the tactile ones. We use participant's own cell phone to test the typical task: dialing phone numbers.

During the experiment, we measured several items as follows:

a. Multiple matching: to let participant choose the most appropriate meaning among a lot.
b. Reaction time: to record the time it takes when participant identifying each button.
c. Accuracy: to count errors during the process. To see if participant would confuse one button with another.
d, Learning efficiency: after learning, to record reaction time and accuracy again and compare with previous test.
e. Typical task: to give participant a certain goal to achieve and record the time and accuracy.

RESULTS & DISCUSSION

We use ANOVA and confound matrix to analyze the result of non-visual experiment.

BASIC MEASUREMENT

From the first part, we know the average hand sizes are: A: 141.46mm, B: 166.35mm, C: 14.2mm, D: 24.75mm, E: 17.46mm, F: 29.77mm.

FUNCTION BUTTONS

We compared tactile function buttons with traditional ones. Participants spent less time on identifying traditional buttons but the accuracy is better with tactile ones. In the confound matrix, dialing and hanging up has three times error in traditional function buttons, once in 1st type index code function buttons, and none in 2nd type. From ANOVA, we know that there is no statistically significant difference in two kinds of tactile buttons (p value>0.05). But the accuracy of tactile buttons and traditional buttons are highly significant different (p value=0.000385, <0.001), which means these tactile buttons can efficiently improve the identification. As we know, dialing and hanging up button are at two sides of a cell phone so that it's easy to confuse people without tactile cues. However, adding tactile cues could reduce error but may take more time to recognize.

NUMBER BUTTONS

We compared analogy and location points tactile number buttons with traditional ones. Analogy number buttons perform the best in reaction time, location points the second, and traditional ones the last. The accuracy of two kinds of tactile buttons are both better than traditional ones. In the confound matrix, there are twelve errors in traditional number buttons, four errors in analogy tactile buttons, and one in location points. From ANOVA (see Table 1), there was significant different in accuracy of two kinds tactile number buttons comparing with traditional ones ($0.01 > p$ value > 0.001). There was little significant different ($0.05 > p$ value > 0.01) in reaction time of analogy number buttons and traditional ones.

Table 1 Result from ANOVA

Number buttons	Item	P value	Item	P value
Analogy & traditional	Time	0.029400	Accuracy	0.001108
Location points & traditional	Time	0.866051	Accuracy	0.001108

Most traditional number buttons are lack of tactile hints that only number 5 has one location point. The analogy number buttons arrange in the traditional way and provide Arabic numbers' strokes as tactile hints. Location points number buttons arrange into a concentric circle and use points, line and circle as tactile cues. After learning, participants could perform better in identification and spend less time on tactile buttons than traditional ones when doing the typical task: dialing phone numbers.

THINK-ALOUD PROTOCOL

In this part, four participants were involved. In the non-visual situation, the participants are not allowed to see the buttons when touching. During the think-aloud protocol process, participants were asked to speak out what they are thinking and how do they feel. There is an example segmentation of one participant talking about function buttons:

About 1st type index code dialing button: "it's round. Two big points...the area is bigger. It's an arc with two points at the end. It feels like the shape of an ear or a telephone."

About 1st type index code hanging up button: "there is an arc, too. In the middle, there's a circle. Umm, because the arc let me think that in the middle must be a circle."

About 2nd type index code dialing button: "it feels like a line. And there are two circles at two ends."

About 2nd type index code hanging up button: "they feel alike. But two ends are sharper. There is a circle in the middle and it is been penetrated with one line. You know, the difference is that one is sharp and the other is not."

INTERVIEW

After think-aloud protocol, there is an interview to ask participant open-end questions as follows:
1. How do you recognize the buttons?
2. Do you use any strategy to help you?
3. How do you feel when you touch the tactile buttons?
4. How do you think about the sharp edges?
5. And how do you think we can improve tactile button to be better?

RESULTS

The results of think-aloud protocol and interview are organized from four aspects: hand-movements, identification, design suggestions and materials.

ABOUT HAND-MOVEMENTS

1. Participants would touch with whole finger to feel the rough shape, with finger tip to detect the details, and with fingernail to distinguish an interval.

ABOUT IDENTIFICATION

2. Participants are able to recognize line, arc and circle. They can recognize whether the line is straight or curved.
3. Participants can recognize whether a line is horizontal, vertical, or oblique.
4. It's easy for participants to recognize the direction of a line but hard to know the length, distance and gradient without comparing.
5. Participants can recognize whether there is a gap in a circle and it's ok to feel the circumference and compare the radian by touching along the edge.
6. The sharp edges and corners could help participants recognizing shapes.
7. The clear edge line provides a track for participants to follow along it and recognize the shape. And corners can help to understand the details.
8. Participants can easily find the order of points and guess the meaning in the location points number buttons.
9. The interval between two units of one icon cannot be too close that hard to distinguish or too far that might look like two separated icons.
10. There is a strong learning effect that participants can do much better after learning.

ABOUT DESIGN SUGGESTIONS

11. If the intervals of number points are too close to each other, it's hard to count how many points are there. (location points number buttons)
12. Participants would image the strokes of Arabic numbers in mind to recognize which number it is. (analogy number buttons)
13. If there is a circle, like 6 and 8, it might be better to be solid without a hole so that it would be easier and quicker to distinguish a circle from an arc.

ABOUT MATERIALS

14. The material is one of the factors that might affect the recognition.
15. Acrylic is hard so the edge is sharp and clear. This may help recognizing a lot but it may not be comfortable to participants.
16. If the material changes into rubber, which is much softer, the edge would be filleted so the area for touching has to be bigger and the difference has to be more obvious.

DISCUSSIONS

There are six features that have been used in the tactile buttons design in this research (see Figure 7). Each of them is easy to recognize and seldom confuses participants by tactile touch alone.

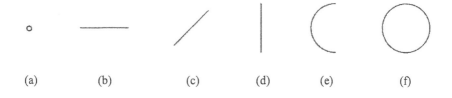

| (a) | (b) | (c) | (d) | (e) | (f) |

FIGURE 7. The six features, from left to right are: (a) point, (b) horizontal line, (c) oblique line, (d) vertical line, (e) arc, and (f) circle.

According to result 9, the interval between two units in one icon cannot be too close that participants may get confused or increase the reaction time. But it still have to maintain a near distance to show an union but not too far that it feels like two separated icons. The proper distance of interval in an icon still remains unknown and could be an issue for further studies. How close does an interval be people would feel two units are in the same union? How far do people think they are independent icons? Would the distance change according to the complexity of design?

And according to result 14~16, designers may face a dilemma of focusing on "easy to recognize" or "comfortable to touch." With hard material like acrylic, tactile buttons can provide clear edges, which do help a lot for recognizing. However it can't satisfy the emotion aspect that Participants don't like the stabbing feelings. Our suggestion is that we can use hard material but fillet the edge a little bit or choose a softer material but make the difference of each button more obviously.

As a result, we provided three basic design guidelines for tactile buttons:

1. Use simple shapes, which can be six features in Figure 7, to design icon.
2. Use location points or lines to hint the arrangement if there has one.
3. The edge of the icon in a tactile button has to be clear and easy for participants to follow the outline and recognize the shape.

REFERENCES

Air-Force-System-Command. (1980, June). *Design handbook 1-3, human factors engineering* (3d ed.): U.S. Air Force.

Chamberlain, P., Gardner, P., & Lawton, R. (2007) Shape of Things to Come. In R. Michel (Series Ed.), *Board of International Research in Design* (pp. 99-116): Birkhäuser Basel.

Gibson, J. J. (1962). Observations on active touch. *Psychological Review, 69*(6), 477-491.

Green, B. F., & Anderson, L. K. (1955). The tactual identification of shapes for coding switch handles. *Journal of Applied Psychology, 39*(4), 219-226.

Hunt, D. P. (1953). *The coding of aircraft controls* (Tech. Rept. 53-221): U.S. Air Force, Wright Air Development Center.

Jenkins, W. O. (1947). The tactual discrimination of shapes for coding aircraft-type controls. In I. P. Fitts (Ed.), *Psychological research on equipment design*: research report 19, US Army Air Force Aviation Psychology Program.

Norman, D. A. (Ed.). (1988). *The Psychology of Everyday Things*: New York: Basic Books.

P.R.M. (1999). The Sense of Touch. *The Journal of Hand Surgery, 24*(2), 213-214.

Sanders, M., & McCormick, E. (Eds.). (1993). *Human Factors in Engineering and Design*: McGraw-Hill INC., N.Y.

Walters, P., Chamberlain, P., & Press, M. (2003). In Touch: an investigation of the benefits of tactile cues in safety-critical product applications. *Proceedings of the Fifth European Academy of Design Conference, Barcelona University*.

CHAPTER 34

Ergonomic Evaluation of Brazilian Interfaces of Mobile Banking Sites Developed for iPhone

Vanessa Kupczik, Stephania Padovani

Universidade Federal do Paraná
Curitiba – PR, Brazil

ABSTRACT

The objective of this research, based upon HCI (human-computer interaction), was to evaluate the Brazilian interfaces of m-banking sites developed for iPhone based on ergonomic criteria and cognitive stages. The following techniques were used in this method: literature research, heuristic evaluation, cognitive walkthrough and check-list. Once the problems were identified, some suggestions to improve the websites interface design were proposed by ergonomic reports. From the obtained data, the conclusion was that the Brazilian interfaces of m-banking sites developed for iPhone present ergonomic and usability problems and the use of an evaluation method without involving users and great costs can contribute to enhance the ergonomics and usability of the site systems.

Keywords: Human Computer Mobile Interaction, Interface, M-banking, iPhone

INTRODUCTION

With the advancement of information technology, the amount of mobile digital services that are incorporated into the lives of ordinary people is growing at an accelerated pace. Among these services is the mobile banking or m-banking in which there is self-service through a mobile device.

The banks, both in Brazil and abroad, are making major investments in this new channel, betting on growth in this market segment. The idea behind the self-service is to cater to a wide audience (through standardized services) with greater speed and convenience, optimize the point of sale (branch) with extension of opening hours and still reduce costs rates (Neves, 2006; Salerno Jr., 2008).

However, studies indicate that the adoption of m-banking is still small and there are consumer groups who resist the adoption of this technology. Among the factors listed in the literature is the perceived cost, the complexity versus the experience and skill required for their use, the difficulty of using mobile interfaces (small screen, small keyboard and multiple passwords).

In turn, the efficiency, convenience and security are key factors in the use of m-banking. In this sense, new devices with larger screens, new keyboard options and design custom interfaces for mobile devices promise to address these issues and provide a satisfying experience to the user.

An example of these new devices is the Apple iPhone: a hybrid device that combines several functions and that is attracting consumers around the world. Aware of this fact, in 2008, some major retail banks in Brazil launched their mobile banking sites to the iPhone (figure 1).

This study was conducted to find out if Brazilian banks have taken into account in interface design of their systems for mobile banking: the users, the benefits of mobile banking and the limitations of humans as information processing. Thus, the main objective of this study was to evaluate the design of interfaces Brazilian sites of m-banking for iPhone on the basis of ergonomic and cognitive stages.

FIGURE 1. Brazilian m-banking sites to iPhone (Kupczik, 2009)

INTERFACE DESIGN FOR MOBILE DEVICES

The use of mobile devices is changing the way people interact with information and services, previously only available on fixed computers (Cybis et al, 2007). To meet the needs of the mobile user, new equipments, applications and services are emerging as well as a new area of study for the Human Computer Interaction (HCI): mobile interaction (Cybis et al, 2007).

According to Love (2005), the human-machine interaction in mobile context can be defined as the study of the relationship (interaction) between people and mobile computing systems. To Gorlenko and Merrick (2003), the concept of mobile interaction regards to the mobility not only of the device, but also the user: the equipment must be portable and should allow the mobility of the user during the interaction.

For Love (2005), the individual characteristics of users and their attitudes impact the use of the applications, devices and mobile services. The author cites the following characteristics of the mobile user: spatial ability, personality, memory, verbal ability, prior experience and elderly users. As for Ballard (2007) these features are: mobility, the facility of interruption and distraction, the availability, the sociability, the context and the identification.

Cybis et al (2007) show that it is important to understand the context of the mobile user and its dynamics: basically handheld computers are meant for quick applications, performed during a shorter time, are extremely focused, occur in a unpredictable environment in which the user's attention may be divided on other tasks and also be interrupted (it should be possible for the user to resume a task in where it was interrupted).

Kiljander (2004) identifies four types of user context: the physical context, the social context, the mental context and context of mobile infrastructure. For example, users use the iPhone while they are moving, probably in an environment full of distractions (Apple, 2008). This does not mean that an application for iPhone can not perform important tasks that require user concentration, however, this means that the application should take into account that users will not give their full attention to content, at least not for long (Apple, 2008).

The success of a user interface mobile device is affected by all these dimensions of context of use (Kiljander, 2004). To enhance the ergonomics and usability of the interfaces between user and system, there are criteria, principles and heuristics proposed by several authors and institutions in recent decades (Cybis et al, 2007). According to the project, different levels of guidelines should be used: general guidelines apply to any user interface, category-specific guidelines for the type of system development and product-specific guidelines for an individual product (Nielsen, 1993).

The following principles were identified in the literature:

- **General principles:** provide a simple and natural dialogue, minimize memory usage (reduce cognitive load), be consistent, provide feedback, manage errors, provide help, consider the user experience, provide user control, be compatible (Nielsen, 1993; Shneiderman, 2004; Bastien and Scapin, 1993)

- **Category specific principles:** create dialogues with completion, reduce the memory load, be consistent, provide feedback, manage errors, consider the experience, allow control by the user, provide compatibility, consider the limitations of the device, designing for attention limited and divided, considering the design speed and recovery (rapid application and support interrupts), using hierarchy by top to down design, creating a nice design, offer a return button, provide search options (default options settings), designing for mobility (principle of transport), designing for multitasking (allowing the use of other functions) (Gong e Taracewich, 2004; Chan et al, 2002; Weiss, 2002; Ballard, 2007; Love, 2005 e Cybis et al, 2007).

- **Product Specific Principles:** design interfaces in a simple way and easy to use, design focused applications, provide communications and feedback to the user, provide a consistent interface, designing applications with receptivity (respond quickly to the user), designing intcroperable applications (in such way that supports interrupts and use the native features) and provide adaptability (the user can choose the format screen: landscape or portrait, and the type of connection: Wi-Fi or EDGE) (Apple, 2008).

METHOD

This research can be considered qualitative and exploratory. It uses the Study Laboratory, because this method facilitates the collection of data and allows easy replication (Love, 2005).

Among the techniques associated with this method, were selected the heuristic evaluation and cognitive inspection (cognitive walkthrough), and also the checklist to complete the data collection.

It was created an instrument to collect data based on the model proposed by Love (2005) for the organization of the heuristics. Thus, the ergonomic principles cited in the literature review were classified into three categories: general heuristics, specific category heuristics and product specific heuristics (Nielsen, 1993).

The cognitive inspection (cognitive walkthrough) was conducted on the interfaces of the sites of the Banco do Brasil, Bradesco, Itaú and Unibanco based on the model of Warton (1994) and Rocha and Baranauskas (2003). This model divides the cognitive inspection in two phases: the preparatory phase and the phase of analysis.

At the preparatory stage, after the identification of system users, the definition of the task to be done, and the definition of the interface to be used, an analysis was made of the task, its decomposition in a sequential manner and the creation of a flowchart functional action-decision according to the model proposed by Moraes and Mont'Alvão (2008), this flowchart describes the correct sequence of actions to correct task performance.

For the task analysis of this study, it was selected a task transaction: transfer money between bank accounts of different banks. All the banks offered this option in their sites. This was a critical task, the incorrect use of it could bring undesirable results to the user.

At the phase of analysis it was applied a checklist at the decomposition (to evaluate each step). This checklist was represented by the letters A, B, C and D in a red or green circle, according to the answer to question (eg. Figure 2). In the sequence, were created believable stories of success and failure according to the results.

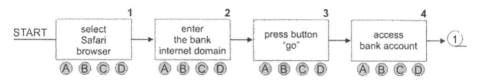

FIGURE 2. Example of sequential task decomposition + checklist application (Kupczik, 2009)

Next, it was applied the check-list prepared in accordance with the guidelines for the design of human-computer interfaces proposed by Apple (2008) the "iPhone Human Interface Guide for Web Applications".

Finally, to analyze the data collected, the problems that were founded were classified at ergonomic reports according to the design process model of Garrett (2003). Garrett (2003), proposed a conceptual model composed of five levels - strategy, scope, structure, skeleton and surface - to discuss the problems of user experience when using a site and the tools available to solve them.

In this research, the ergonomic reports were made as a table, containing information about the problem classification, description, constraints or difficulties, suggestions for improvements and restrictions on suggestions (Padovani, 2008). Through this classification, it was possible to analyze the problems encountered, identify in which plans the problems concentrate and also suggest improvements to the interfaces design. Figure 3 shows a general representation of the research method.

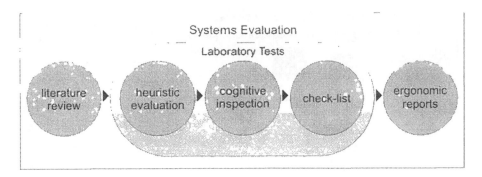

FIGURE 3. General outline of the research method (Kupczik, 2009)

A convenience and non-probability sample was selected for this study. In 2009 four banks with headquarters in Brazil had for mobile internet applications developed for iPhone: Banco do Brasil, Bradesco, Itaú and Unibanco.

The interfaces of the Brazilian sites of m-banking for iPhone targeted to users of electronic banking were the objects of this research (the restricted area to account users with access via bank password). To collect the data it was used an iPhone 3G and its connection to the wireless network. It was used the Safari browser (standard iPhone) for the display of web pages. For security reasons (transactions involving real money), only one evaluator was used in this study.

RESULTS

As pointed Chan et al (2002), due to the subjective nature of heuristic evaluation and cognitive inspection, the findings of this study are derived from subjective observations of an evaluator and may not reflect all the ergonomic and usability problems that the interfaces for iPhone web sites presents.

HEURISTIC EVALUATION

The main problems noted in the heuristic evaluation were inconsistent with the **general principles** of: conduct (a recurring problem: lack of legibility), the workload, the explicit control (when an action was selected, the user couldn't cancel or stop it), adaptability (beginners and experienced users must use the same interface), management errors (problems with the quality of the messages), consistency (Itaú had inconsistencies throughout the site: a screen with different procedures, labels, commands, text fields - fig. 4), the meaning of codes and compatibility (large amount of banking terms).

FIGURE 4. Inconsistencies in the interface of the Itaú site (Kupczik, 2009)

Considering the **specific category principles**, the following heuristics were partially infringed: the adequacy of the mobile context (using other objects besides the phone, many pages of information), the interface is miniaturized (inadequate sizes of controls and letters), ease of navigation, support for selecting options (lists and links inadequate), the scrolling (lots of scrolling in Bradesco), support for interrupts (Unibanco could not bear the disruption which requires the user to enter all data to return to site navigation) and customization of the interface (the only forms of personalization are placing the icon on the main screen of the iPhone and add the site to favorites).

When considering the **product specific principles** were partially breached guidelines on: the content simple and easy to use (lots of request of data input from the user), communication and feedback with the user (lack of clarity in communication), consistent interfaces (lack of integration with the layout of native applications on iPhone), the responsiveness of the application (Unibanco took several attempts to finally interact with the client area), interoperability of the application (partial integration with the keyboard, the form assistant and the pop-up menu of iPhone) and the adaptability of the content (horizontal problems viewing web sites).

The only principle that has been followed on all interfaces was the **focus** of the application on the main functionality. It hasn't been identified any serious problem, only some medium seriousness problems and lots of problems with low gravity. In some banks were not even reported problems with some heuristics.

COGNITIVE INSPECTION

The results of the questions in the analysis phase indicate that, somehow, the interfaces of the banks sites failed with respect to any stage of the cognitive process of the user, it failed at: the creation of a goal, the creation of a interaction, the specification of an interaction, the execution of an action, the perception of the system, the interpretation of the state of the system or the assessment of output produced by the system (Norman, 1998).

To successfully complete the task of transferring money, the user needed to: decode the information correctly, learn by exploring the screen, memorize various data and understand the navigational structures and dialogues (which were not always clear in the interface), in order to resolve problems and to make the right decisions.

The cognitive inspection also indicated that if a user is accessing the site, for the first time, to make a transfer between banks, he probably wouldn't accomplish the task due to the large number of prerequisites (account number, branch, passwords, value limits, time transaction, cost, use of security card or token, subscribe the other account via other channels) for it to be successful.

The number of steps to complete the task also indicated the degree of difficulty encountered: 20 steps in Itaú, Unibanco in 22 steps, 24 steps in the Banco do Brasil and 31 steps at Bradesco.

CHECKLIST

The results indicated that the guidelines and metrics for the iPhone interface to internet sites are partially addressed. The main problem identified was about the size of the iPhone screen (320x356 pixels) that is not respected by any interface.

In addition, the pattern of Apple's iPhone is not followed. In this sense there are indications that the interfaces attempt to maintain consistency using external elements of the site's default database to the web site interface for the iPhone.

Navigation was another topic that had problems: it not complied with the experience gained by the user of the iPhone. And yet, there were problems of interaction of the bank site with the native applications on iPhone. And finally, texts that did not follow the metrics recommended by Apple (2008) and so, caused problems of legibility.

ERGONOMIC REPORTS

The results of ergonomic reports indicated that the Brazilian interfaces of mobile banking sites for iPhone had problems at all levels of the conceptual model of the design process of Garrett (2003) (especially the plans of the skeleton and structure). These problems were common, occurred in more than one bank even in the same plane (see example of ergonomic report in Table 1).

Table 1: Example of Ergonomic Report (Kupczik, 2009)

Problem Classification	Problem Description	Constraint or Difficulty	Suggestion	Restriction
Strategy Plan: customization problem	Its wasn't possible to customize the sites interfaces	An experienced user has to go through the same interface for a beginner	Create options for customizing the interface	Technology
Strategy Plan: usability probelm	Lots of form fields must be filled in to accomplish the task	The user must memorize lots of data: branch number, agency, check digit, passwords	Create a easy login system. For financial transaction ask for aditional information	Security

From the problems identified were proposed suggestions for improving the interfaces between the sites through ergonomic reports. Among the suggestions are:

- **Strategy plan:** create options for customization, the site must be available 24 hours a day, 365 days a year and avoid the use of a second object to perform operations (eg. token).
- **Scope plan:** reduce the content, using appropriate vocabulary (options for help) and building applications with session time explicit for the user.
- **Structure plan:** one should follow the established conventions, improve error messages, perform the recovery of data already entered, indicate to the user the need of other objects such as security cards and tokens.
- **Skeleton plan:** organize or group the information according to guidelines, minimize the use of automatic zoom, cut labels, use the same nomenclature for labels alike, reduce data entry, matching default settings and checkboxes, use appropriately links.
- **surface plan:** minimize the use of automatic zoom (so that the user is not lost in the application), group the information as recommended in the guidelines, use buttons size and letters size appropriate to maintain consistency between the screens of the same site.

For most of the suggestions there were no restrictions. The restrictions that were found were linked to issues of technology and security.

CONCLUSION

The objective of this study was to evaluate the design of Brazilian interfaces of m-banking sites for iPhone on the basis of ergonomic criteria and cognitive stages. Three techniques of data collection were used to diagnose problems and, subsequently, carried out analysis and classification of results under the contextual model of the design process proposed by Garrett 2003).

Through the proposed method it was possible to achieve the goal of research and to suggest improvements through ergonomic reports. From the obtained data, the conclusion was that the interfaces of Brazilian m-banking sites developed for iPhone present ergonomic and usability problems.

As a general conclusion of this research, it can be inferred by the results that the use of a method of evaluation without involving users and not spending a lot of resources, can help improve the ergonomics and usability of web site applications.

REFERENCES

Apple (2008). *iPhone Human Interface Guidelines for Web Applications*. Apple Inc.

Ballard, B. (2008). *Designing the Mobile User Experience.* West Sussex: John Wiley & Sons.

Bastien, C.; Scapin, D. (1993). *Ergonomic Criteria for Evaluation of Human-Computer Interfaces.* INRIA Institut National de Recherche en Informatique et en Automatique.

Chan, S. et al. (2002). "Usability for mobile commerce across multiple form factors". *Journal of Eletronic Commerce Research*, Volume. No. 2.

Cybis, W. et al. (2007). *Ergonomia e Usabilidade: conhecimentos, métodos e aplicações.* São Paulo: Novatec Editora.

Garrett, J. (2003). *The Elements of User Experience: User-Centered Design for the Web.* Berkeley: New Riders.

Gong, J.; Tarasewich, P. (2004). "Guidelines for handheld mobile device interface design" proceedings of the *Decision Sciences Institute 2004*. Annual Meeting (DSI04 Paper). Boston, MA.

Gorlenko, L; Merrick, R. (2003). "Usability Challenges in the wireless world". *IBM System Journal.* Volume 42 No. 4.

Kiljander, H .(2004). *Evolution and Usability of Mobile Phone Interaction Styles.* Helsinky University of Technology. (thesis)

Kupczik, Vanessa. *Pesquisa exploratória sobre avaliação ergonômica de interfaces de sites de mobile banking brasileiras para iPhone.* Curitiba, 2009.

Love, S. (2005). *Understanding Mobile Human-computer Interaction.* Oxford: Elsevier.

Moraes, A. & Mont'alvão, C. (2000). *Ergonomia: conceitos e aplicações.* Rio de Janeiro: 2AB.

Neves, J. et al. (2006). "Estratégias de auto-atendimento no serviço bancário: o caso da agência Alfa". *Revista Eletrônica do Mestrado em Administração da UNIMEP.*

Nielsen, J. (1993). *Usability Engineering.* California: Morgan Kaufmann.

Norman, D. (1998). *The Design of Everyday Things.* London: The MIT Press.

Padovani, S. (2008). *Síntese dos problemas identificados* (Slide).

Rocha, H.; Baranauskas, C. (2003). *Design e avaliação de interfaces humano-computador.* Campinas, SP: NIED/UNICAMP.

Salerno Jr. E. (2008). *As salas de auto-atendimento bancário, os caixas eletrônicos e suas interfaces gráficas: usabilidade, funcionalidade e acessibilidade.* Universidade de São Paulo. São Carlos.

Shneiderman, B. (1998). *Designing the user interface: strategies for effective human-computer interaction.* 3rd edition. Addison-Wesley.

Wharton, C., et al (1994) "The Cognitive Walkthrough: A Practitioner's Guide".In: J. Nielsen (ed.) *Usability Inspection Methods.* John Wiley, New York.

Weiss, S. (2002). *Handheld Usability.* England, John Wiley & Sons, Ltd.

CHAPTER 35

Design Comparison of Commercial TV Remote Controls

Hayden Beauchamp, Young Sam Ryu

Ingram School of Engineering
Texas State University-San Marcos
San Marcos, TX 78666, USA

ABSTRACT

In this paper, we classify the groups of controls and identify physical attributes of each group of controls on TV remotes. Then, we analyze and compare the current designs of TV remote controls of digital cable and satellite TV service providers in United States. In doing so, we identify usability issues and provide design recommendations. Further, we identify the current trends in TV remote design, and determine which variations more effectively promote usability.

Keywords: TV interface, remote control, interactive program guide

INTRODUCTION

The first electronic televisions made their way into households in the late 1930's. Remote controls have been the primary input device for TVs since Zenith introduced the first primitive TV remote control in 1950, which was attached to the set by a cord (Bellis, 2010). The drive for bringing remote control technology to television was to allow the viewer to turn the volume off during commercials (Letourneau, 2009). The first practical wireless remote was introduced in 1956, with only four controls: POWER, CHANNEL▲/▼, and MUTE (Letourneau, 2009). Modern digital TV services provide universal remotes with almost 50 buttons, with which multiple devices can be controlled.

In 1985, the first Electronic Program Guide (EPG) channel began broadcasting, showing viewers on-screen program listings ("Electronic program guide," 2010). In 1996, Prevue Networks offered the first Interactive Program Guide (IPG) service in the United States ("Electronic program guide," 2010). The IPG required more effort for the user to operate, using more complex functions than before (Kang, 2002). Before IPGs, traditional TV controls were used to toggle features on/off or vary one-dimensional settings, like POWER and VOLUME +/–. Using an IPG required the viewer to navigate through a two-dimensional page of program listings. This 2D interactivity necessitated a new type of interaction. Following this paradigm shift, most remotes were equipped with a set of navigation controls.

Consisting of four directional arrows and an OK/SELECT button, navigation controls allow the user to interface with a two-dimensional menu. Similar to the IPG, the user interacts with Pay-Per-View (PPV), Digital Video Recorder (DVR), and Video on Demand (VOD) services, as well as numerous settings, preferences, and supplementary features available via a menu system. All of these menus are integrated together to form the graphical user interface (GUI). In addition to navigation controls, other controls have been added to allow the viewer to fully utilize these services. For example, the user can direct the playback of digital video with the DVR controls. Special purpose buttons have been added to grant the user access to particular features, like interactive TV.

To promote maximum usability, remote control devices should have an intuitive design which allows ease of use for all kinds of television viewers. Users often have difficulties adapting to a new GUI or TV remote which is unfamiliar or different. Although the remote controls look similar, there are numerous differences in design that can prevent users from being able to complete well-known tasks using an unfamiliar service provider. Minimal research is available in the literature which addresses the physical design of TV remote controls supporting these popular and relatively new interactive services.

In this paper, we analyze and compare the current designs of TV remote controls available in United States which support interactive applications. We conduct an in-depth evaluation of the designs of three different remote controls in order to identify usability issues. First, we categorize control groups based on their use. Then, we define physical characteristics and relationships of each set of controls. We analyze the design factors of each control group in order to identify design trends and how these mechanisms aid interaction with the GUI. We compare control groups between remotes to identify contrasting aspects of design. Where differences exist, we attempt to discover which design is more effective and provide appropriate recommendations.

CONTROL GROUPS OF REMOTES

To provide the design specifications of currently available TV remotes that support IPG, PPV, DVR, and VOD applications, we investigated the remotes of DirecTV,

Dish Network, and AT&T U-verse. Although the designs of the three remotes are not identical, they share similar controls (buttons) and organization. Table 1 shows the general categories of the controls on the remotes.

Table 1. Control groups on TV remotes

Control Groups	Usage	Number of Buttons
Equipment Controls	Power, Mode, TV/Video Input etc.	5 - 7
Basic TV Controls	Volume +/–, Channel ▲/▼, etc.	4
Alpha/Numeric Controls	Numbers and text keys	12
DVR Controls	Play, Pause, Rewind, etc.	8
Navigation Controls	Arrows and Select	5
Special Controls	Dedicated buttons for special functions, color soft keys	4 - 8

DESIGN FACTORS OF REMOTE CONTROLS

The design of remote controls includes the physical design of each button, the layout of the buttons, and interaction design for each button. The physical design attributes of each button include shape, size, color, and label. A well-designed layout of buttons will increase the effectiveness and efficiency of conducting routine tasks. The layout design consists of two layers; layout within each control group and layout between control groups. Interaction design involves the allocation of control functions to each button depending on application.

PHYSICAL ATTRIBUTES OF BUTTONS

Shape

The shape of a button is the product of two defining characteristics: surface and cross sectional outline. These two components of tactile feedback establish how easily individual buttons can be distinguished based on the user's sense of touch. Additionally, the shape of a button can communicate the relative position of other buttons. The surface of the DISH navigation arrows, for example, is tilted toward the SELECT button. FIGURE 1 shows a similar example with DirecTV.

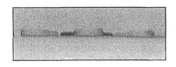

FIGURE 1. DVR controls: slanted button surface indicates location of PLAY button.

The user can also receive valuable feedback based on a button's cross section. For example, buttons placed in a circle may be shaped such that a taper in the cross section of each button will point toward the center of the button group (FIGURE 2).

FIGURE 2. Dish Network teardrop button shape indicates location of PAUSE button.

The role of a tactile feedback system is to reduce the need to interrupt the main task of viewing TV in order to seek visual cues from the remote. Tactile feedback is valuable once the user has been trained on the layout and use of the controls. The tactile feedback of a button with distinct surface characteristics among surrounding buttons provides relative information. For example, the Dish VOLUME + and − buttons are concave and convex shaped respectively, to indicate opposing use. Another example of tactile coding is the BACK and FWD buttons (FIGURE 2). In this example, these buttons are separated by many other buttons with concave surface to indicate similarity in function. This relative information is combined with the user's knowledge of the control layout to supply him/her with the identity of a control without looking at it.

Size

An accepted convention for determining the size of each button is the use of scale hierarchy. The more frequently used or important buttons are usually larger. The larger the button, the more visible and available for interaction. The difference in size among surrounding buttons can provide the identity of a control from visual cues as well as tactile cues.

Color

Color coding of each button can provide visual cues for the user to infer its use. Assigning a red color to the power button is a typical example. Also, color coding between control groups is used to indicate disparities in function between the groups. For example, navigation controls are black and numeric keys are white on DirecTV remotes. The recent introduction of color coded soft keys (red, green, yellow, and blue) supports numerous functions and interactions. Basically, each color key represents the function indicated on the bottom of the TV screen. To avoid interference with the four colors of soft keys, other colors or shades of grey (white, grey, and black) may be used for other buttons or groups of controls.

Label

Each button, except for the color soft keys, is labeled with a name, symbol, or both. The name of a button is a short description of the primary function of the button. Sometimes short labels inadequately describe the functions they represent. Symbols can be more effective at communicating intended usage. Additionally, a language barrier can be overcome with the use of universal symbols.

Layout

Relative placement of buttons can communicate the use of buttons. For example, left and right placements for rewind and fast forward, respectively, are compatible with our mental model. Similarly, top and bottom placements for up and down functions, such as channel and volume, match the users' expectations. Also, buttons of similar control are grouped together to reduce the distance of travel when they are used sequentially.

INTERACTION DESIGN

Soft keys are buttons that are assigned different functions depending on what part of the user interface is being used. Certain menus or applications have auxiliary functions that cannot be controlled by dedicated buttons. To bypass the need for menu-type control, these functions are assigned to soft coded buttons to make the user interface easier to control. By using a soft coding system these four buttons can serve many different purposes, eliminating the need for more buttons.

DESIGN COMPARISON OF THE CONTROLS

EQUIPMENT CONTROLS

The equipment controls are always placed at the top of these universal remotes. Equipment control buttons include POWER, input source selection, and device selection buttons (TV, set-top box, DVD player, etc.). The device selection buttons, which choose the device the user is operating, can be integrated into one multiple choice switch (FIGURE 3). Powering on the system (set-top box) and TV is usually the first task to complete in order to watch TV. Some remotes have shortcut power buttons that bypass the use of device selection buttons by powering on both the TV and set-top box simultaneously (FIGURE 3).

FIGURE 3. DirecTV device selection switch and integrated device power buttons.

Recommendation: A Multiple choice switch reduces the number of device selection buttons. Shortcut power buttons for the two most frequently used devices (i.e., TV and set-top box) are effective because they reduce the number of steps to complete the task of turning the devices on and off.

BASIC TV CONTROLS

Several sets of controls have been universally adopted from traditional TV remotes: VOLUME +/–, MUTE, CHANNEL ▲/▼, PREVIOUS, and the numeric keypad. Both DIRECTV and AT&T U-verse have integrated the channel ▲/▼ and page ▲/▼ functions into one set of buttons (FIGURE 4). Because the tasks of navigating channels and navigating pages never overlap, the two sets of controls can share one set of buttons.

FIGURE 4. DirecTV dual purpose buttons with proximity of PREV button.

The MUTE button and PREVIOUS/RECALL buttons are consistently placed adjacent to the VOLUME +/– and CHANNEL ▲/▼ buttons respectively (FIGURE 4). While designers can agree on the general placement of the PREVIOUS button, they do not use the same term to label it. Terms and abbreviations used to label this button include: RECALL, RCL, PREV, PRE-CH, LAST, RETURN, R-TUNE etc. Because any of these labels can be ambiguous, the placement of the button helps to clarify its use.

   Recommendation: Integrating multiple sets of functions into one set of controls is a practical way to achieve simplicity.

DVR CONTROLS

The DVR controls (BACK, FORWARD, STOP, SKIP BACK, SKIP FORWARD, and RECORD buttons) are typically placed around the center button in a layout representing the opposition of the functions. Either the PLAY or PAUSE button is at the center of the DVR controls. Complementary functions like BACK and FWD are placed on opposite sides (left, right) of the central button. AT&T U-verse also physically connects the paired complementary controls (FIGURE 5).

The record function is the keystone feature for a DVR system. The record symbol (red circle) is traditionally the only colored label, which sets it apart from the other DVR controls. The AT&T and Dish RECORD buttons are both distinctively sized compared to the DVR buttons that surround them. This difference in size allows the user to easily detect the RECORD button using either visual or tactile senses. The placement of the RECORD button varies because it does not have an obvious complimentary function. Because of this, Dish places it on the center line, instead of on either side (FIGURE 2). Similarly, AT&T places the RECORD button apart from other seven DVR controls.

Recommendation: Some DVD player remotes lack a PLAY button because the play function has been integrated into the PAUSE or SELECT button. Because the PAUSE button performs the function of toggling on/off the playback of media, a separate PLAY-only button is unnecessary. We recommend that the PLAY and PAUSE buttons be integrated into one central (▶ / ‖) button to promote simplicity.

NAVIGATION CONTROLS

Navigation controls consist of four directional buttons and a SELECT/OK button at the center, similar to cell phones, for users to navigate through a menu-type user interface. Due to the widespread use of IPG with digital television, the navigation controls have become ubiquitous to interactive TV services.

The place where the thumb naturally rests when a user holds the remote is called the "sweet spot". Because the navigation controls are used more frequently, they are usually placed at the sweet spot of the remote. However, the sweet spot may vary for each remote according to grip design and center of gravity. While DirecTV and AT&T place the DVR controls above the navigation controls, Dish does not.

Recommendation: Since the use of numeric keypad is more closely associated with the use of navigation controls than DVR controls, the layout of Dish Network causes more travel distance and difficulty of reaching between navigation controls and numeric keypad. In general, the arrangement of control groups should be balanced around the sweet spot to reduce travel distance and maximize comfort.

SPECIAL CONTROLS

We designated the buttons with dedicated functions, which do not belong to any of the control groups mentioned above, as special controls. Because the use of the special controls follows and/or precedes the use of the navigation controls, the two groups are usually placed within immediate proximity to each other. Because the functions of the special controls are complicated and rarely exist in other devices, labeling them properly is critical to communicate the use of each control to the user. For example, most users will have difficulty in guessing the function of the ACTIVE button on DirecTV remote. The straightforward common special controls are MENU, GUIDE, INFO, BACK, CANCEL, and EXIT. Table 2 lists the common special controls that are inconsistently labeled.

Table 2. Common special control buttons labeled differently

Function	DISH Network	DirecTV	AT&T U-verse
Go to previous channel	RECALL	PREVIOUS	LAST
Load interactive TV	DISH	ACTIVE	GO INTERACTIVE
View DVR playlist	DVR	LIST	RECORDED TV

Currently, DISH button to load interactive TV on Dish Network is placed far away from the navigation controls. Because interactive TV requires the use of navigation controls, they should be placed together.

DirecTV and DISH Network use four color soft keys (red, green, yellow, and blue) to support numerous functions with the limited number of buttons. In each page of a specific application, there are function indications of each button at the bottom of the screen. Color soft keys are consistently placed below the navigation controls because they are most commonly used in a menu environment. While Dish and DirecTV both have four color soft keys, AT&T U-verse only has three (yellow, green, and red). Additionally, the AT&T U-verse soft keys are shape-coded (triangle, square, and circle) and labeled (A, B, and C).

FIGURE 5. The REPLAY and skip FWD controls are bridged.

Instead of grouping them together, special controls are usually placed adjacent to the buttons they are most closely associated with, usually the navigation controls. Furthermore, AT&T groups all three of the menu buttons together (FIGURE 5). The large amount of empty space of the remote body surrounding each of these buttons could possibly serve to emphasize this control group.

To support the third item of Nielsen's ten usability heuristics titled "User Control and Freedom" (Nielsen, 1994) every remote has an exit or cancel button. To facilitate user control and freedom the user must always have a means to exit the current menu (EXIT) and undo an action (BACK). Both DIRECTV and AT&T have EXIT and BACK buttons. The EXIT button is used to egress any onscreen menu to return to viewing TV, while the BACK button is used to undo and action. With Dish these functions are called VIEW LIVE TV and CANCEL.

Recommendation: While the Dish CANCEL function is similar to the

DirecTV's BACK function, it does not always return the user to the previous screen. Dish should change the CANCEL control to fully support the undo function. The AT&T remote has the fewest number of buttons because the number of equipment controls and special controls were reduced. Designers should strive to eliminate unnecessary buttons as much as possible in order to achieve simplicity.

ALPHA/NUMERIC CONTROLS

Virtually all remotes share the same layout of numeric buttons. There are always 12 buttons in a 3x4 array, containing the numbers 0 to 9. The two remaining buttons on either side of zero, which are typically star (*) and pound (#) on telephone keypad, have different functions allocated to them. All three remotes assign different functions to these two auxiliary buttons (FIGURE 6).

FIGURE 6. Different functions and labels assigned to the two auxiliary buttons

Both DirecTV and AT&T use the bottom right number button as ENTER, to speed up the channel entry process. Because a digital cable or satellite channel can be up to five digits long, the receiver cannot immediately tune after only three digits have been entered. Thus, when a channel number is entered, a short delay (1 second timeout) allows for the entry of additional digits before the receiver tunes to that channel. Pressing ENTER after inputting a channel number directs the receiver to tune the specified channel without delay. Although the SELECT button within the navigation controls could be designated to carry out this function, like it is with Dish, it is too far away to be reached when the numeric controls are being used.

DirecTV's bottom left numeric control button, the DASH (–), is used to enter a dash during channel number entry to access HD channels. AT&T designates the bottom left button as the DELETE function. The DELETE function deletes one character to the left of the cursor in an alpha/numeric input field. Dish's bottom left numeric control button is used to move the cursor one space left without deleting any characters. Although the left arrow of navigation controls is also designated as the delete function, it is often too far away to be reached during text entry.

Recommendation: It is important to have a delete function within the user's reach while using the alpha/numeric controls because typing errors are common during text entry due to the difficulty of the multi-press input with timeout method (Butts & Cockburn, 2002). The DELETE function can be assigned to a number key as a character, like SPACE (␣) is with the zero (0) button. During text entry, the 0 button is used to enter a SPACE character before the number 0. Like the 0 button, the 1 button can be assigned an additional function because it is not used to enter letters. With this integration every remote can have the delete function located in the alpha/numeric controls, even if the current design doesn't accommodate it. A DELETE function should be placed within the alpha/numeric controls, regardless of which button it is assigned to.

CONCLUSION

In this paper, we defined and categorized the design factors of remote controls. We compared the design factors of the remotes of three different TV service providers to identify usability issues. We also provided recommendations to designers to make alterations for the identified design issues. In conclusion, we expect that superior usability with future remotes will be achieved through effective control layout, comprehensible labeling, intuitive control integration, and overall simplicity.

It is important to acknowledge that this usability inspection was conducted by two usability experts. Future work should involve user studies collecting both objective and subjective data. This data may include: assessment of the effectiveness of various layouts between control groups based on travel distance evaluation, identification of unnecessary controls based on task analysis, and exploration of further integration of controls. Finally, more opinion should be represented through user-centered focus groups and questionnaires.

REFERENCE

Bellis, M. (2010). History of the Television Remote Control. Retrieved February 3, 2010, from http://inventors.about.com/od/rstartinventions/a/remote_control.htm.

Butts, L., & Cockburn, A. (2002). An evaluation of mobile phone text input methods. *Australian Computer Science Communications*, *24*(4), 55–59.

Electronic program guide. (2010). *in Wikipedia, the free encyclopedia*. Retrieved February 11, 2010, from http://en.wikipedia.org/wiki/Electronic_program_guide.

Kang, M. (2002). Interactivity in Television: Use and Impact of an Interactive Program Guide. *Journal of Broadcasting & Electronic Media*, *46*(3), 330-345. doi: 10.1207/s15506878jobem4603_2.

Letourneau, K. (2009, August 20). From Lazy Bones to RedEye: A Brief History of the TV Remote : MoreControl: Universal remote control, iPhone remote control, and home automation. Retrieved February 3, 2010, from http://morecontrol.com/2009/08/lazy-bones-to-redeye-a-brief-history-of-the-tv-remote/.

Nielsen, J. (1994). *Usability inspection methods*. New York: Wiley.

Chapter 36

Can a "Tactile Kiosk" Attract Potential Users in Public?

Y. Yasuma, M. Nakanishi, Y. Okada

Fac. of Science & Technology
Dept. of Administration Engineering
Keio University
3-14-1, Hiyoshi, Kohoku
Yokohama 223-8522, Japan

ABSTRACT

The use of kiosk terminals, which provide diverse services allow and users to obtain information easily, depends on the spontaneity of users. As per some psychology studies, the spontaneous use of such terminals depends on intrinsic motivation. In the present study, we propose a method of measuring the intrinsic motivation of users by subjectivity evaluation. We doubt whether the present kiosk terminals attract potential users effectively because use only physical appearance (shape and color), although numerous communication methods that employ sight and sound are available today. Therefore, we pay attention to the tactile sense and examine the appeal of touch screens that have tactile feedback by an evaluation method proposed in the present study. This result suggests that tactile feedback may raise a user's intrinsic motivation. We expect that measuring intrinsic motivation will also be useful in marketing the kiosks, e.g., in grasping users' potential requirements and trends, if we can, in addition to a subjectivity evaluation, measure intrinsic motivation by studying user behaviors.

Keywords: Kiosk terminal, Tactile feedback, Intrinsic motivation, Potential users

INTRODUCTION

Today, we see numerous computer terminals with touch screens. In particular, interactive kiosk terminals, which provide diverse services and allow users to obtain information easily, are increasing in popularity. The main advantages of kiosk terminals are as follows: for vendors, (a) reduction in space requirements (only kiosk terminals can supply a variety of information to a large number of users), and (b) reduction of personnel expenses (setting up kiosk terminals can avoid the use of window clerks.), and for suppliers, (a) users can receive services whenever required regardless of whether it is day or night, and (b) because users can use kiosk terminals without human assistance, strangers do not gain access to their personal information. To achieve these advantages, an individual user must use kiosk terminals spontaneously. The most critical problem of kiosk terminals is how to attract both potential and existing users. The popular usability evaluation methods for computer terminals assume previous use by users, and their principal objective is to evaluate usability and simplicity for existing users. Therefore, such evaluation techniques may not always be appropriate when evaluating the appeal for potential users.

In this study, we aim to evaluate kiosk terminals in terms of whether potential users want to use them, and whether existing users want to use them again. First, based on existing psychology studies, we develop the evaluation method of the motive. Because the present kiosk systems aim to attract users only based on physical appearance (shape and color), although numerous communication methods that employ sight and sound are available today, we doubt whether they are sufficiently effective in this regard. Therefore, using the method devised in the present study, we next evaluate touch screens that use tactile feedback in addition to sight and sound. We determine the spontaneity of using kiosk terminals that employ the tactile sense; we believe that these panels have potential to advance human–machine interaction.

ESSENCE OF FEELINGS WHEN USERS ACT SPONTANEOUSLY

OBSERVATION FROM PSYCHOLOGICAL KNOWLEDGE

Motivation involves the internal and external processes that drive a human being to an action, and motive involves the conscious or unconscious factors that drive a human being to an action. Drive reduction theory is a traditional theory about motivation which explains that human beings are essentially passive and does not take action without disagreeable strain. As per this theory, the motives of human beings are defined by the following three conditions: (a) impetuses to give pain, (b) requirements of the body, and (c) psychological traumas. Because many psychologists doubted the validity of this theory, numerous experiments have attempted to produce evidence against it by proving that a human being does not always follow it (Heron, 1957; Harlow, Harlow, & Meyer, 1950; Harlow, 1950; Hebb, 1949; Butler, 1953). These experiments have made the classifying of motivation as extrinsic and intrinsic an accepted practice.

Extrinsic motivation refers to human beings acting according to factors outside of themselves, e.g., pain and reward and punishment. Intrinsic motivation refers to human beings acting according to internal factors such as curiosity and mental inquiry. In other words, the actions resulting from intrinsic motivation are spontaneous actions of human beings. Experiments by Deci and Berlyne are representative experiments on intrinsic motivation (Deci, 1972; Lepper, Greene, & Nisbett, 1973; Berlyne, 1957). Deci proved that giving monetary rewards reduced the intrinsic motivation of human beings. Further, he explained that intrinsic motivation consists of self-determination and competence—self-determination involves depending on self-choice, and competence involves feelings of self-worth. He points out that when both these feelings arise, they promote intrinsic motivation (Deci, 1975).

Berlyne (1957) conducted an experiment to investigate the relationship between sight stimulus and intrinsic motivation and found that intrinsic motivation is affected by various characteristics of figures, such as complexity, vagueness, wonder, novelty, and indistinct discord. He explained that a gap, "unexpectedness," between an existing individual experience and the stimulus affected intrinsic motivation. Many psychologists observed the inverted-U plot of "unexpectedness" versus intrinsic motivation and indicated that there exists a single value of unexpectedness for which intrinsic motive is maximum (Dember, & Earl, 1965; Berlyne, 1965; Hunt, 1965). This result signifies that moderate novelty raises intrinsic motive more than immoderate novelty, which may produce terror, or "thin" novelty, which may produce tedium (Giyoo, & Kayoko, 1972).

CAUSE AND EFFECT OF INTRINSIC MOTIVATION

According to the definition in the previous paragraph, the feeling that users want to use a kiosk terminal is a type of intrinsic motivation. We classify intrinsic motivations into four categories: (i) self-determination, composed of interest and enthusiasm for other users; (ii) competence, which involves pride and envy of users; (iii) unexpectedness, which is concerned with the shock and freshness of terminals; and (iv) non-displeasure, which represents moderate novelty. Satisfaction and feeling good are inferred as factors of non-tedium, and anxiety and enjoyment as factors of non-terror.

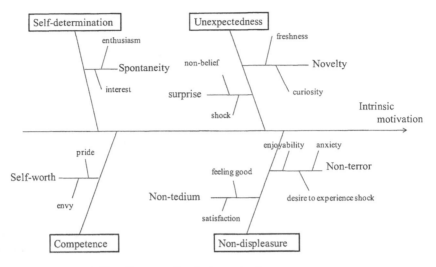

FIGURE 1 Cause and effect diagram of intrinsic motivation.

COMPOSITION OF EVALUATION QUESTIONS

Based on Figure 1, we composed questions that investigate whether users who have not used terminals yet (potential users) want to use them, or if users who have already used them (existing users) want to use them again. We aim to measure the intrinsic motivation of potential and existing users with these questions. Table 1 shows 10 questions asked to potential users, and Table 2 shows 14 questions asked to existing users.

Table 1 Evaluation questions for potential users

Category	Questions
Self-Determination	Did you feel a desire to touch terminal immediately?
	Did you feel interested?
	Did you have good expectations?
Competence	Were you envious of users who had operated the terminal?
Unexpectedness	Did you feel anything unexpected?
	Did you feel it was a novel experience?
Non-displeasure	Would you like to experience shock?
	Did you feel any anxiety?
	Did you expect that the terminal would be comfortable?
	Did you expect that the terminal would be entertaining?

Table 2 Evaluation questions for existing users

Categories	Questions
Self-Determination	Did you feel a desire to touch the terminal over and over again?
	Did the terminal absorb your attention?
	Did you expect to have a good experience when using the terminal?
	Were you excited?
	Did you feel interested?
Competence	Did you feel the desire to boast about this experience?
	Did you feel affection for the terminal?
Unexpectedness	Were you surprised?
	Did you feel freshness?
	Did you get a strong impression?
Non-displeasure	Did you feel comfort?
	Did you feel satisfaction?
	Did you feel entertained?
	Was it hard for you to be satisfied?

EXPERIMENT ON TERMINAL WITH TACTILE FEEDBACK

For the study, we used a touch screen that vibrates when users touch it (8.4-inch Tactile Immersion Unit, made by Immersion). The subjects were 40 students, who ranged from 18 to 26 years old. Subjects visited the terminals in pairs and each one experienced it .The subject who entered the terminal first is called "preceding subject" and the one who entered it later is called "subsequent subject." First, the preceding subject took a seat in front of the terminal (Figure 2), and the subsequent subject sat next to him or her. We displayed a landscape picture on the left side of the screen, and there was a button on the right side of the screen (Figure 3), which when pressed would randomly display one of 17 landscape pictures. (The same picture did not appear in succession.) Before the experiment started, we instructed the subjects to continue pushing the button until they felt satisfied. The subjects could finish using the terminal only if they told us that they felt satisfied. Afterward, the subjects answered 14 questions for existing users using a seven-point scale (0–6).

338

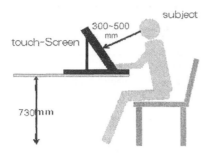

FIGURE 2 Experimental setup.

FIGURE 3 Touch screen.

The subsequent subject answered 10 questions for potential users using a seven-point scale (0–6) after observing the actions of the preceding subject. Further, the subsequent subjects also experienced the terminal and answered 14 questions for existing users after answering the 10 questions for potential users. Each subject performed the abovementioned process in two conditions (terminals). In each condition, we prepared different pictures, and one of the nine vibration types appeared at random when the subject touched the screen (Table 3). We controlled the two conditions (image and tactile feedback) closely to prevent an order effect.

Table 3 Nine vibration types.

Base effect	Characteristics	Relative Magnitude	Duration	Repeat Buffer
Pulse1	Pulse-type click with standard repeat	10	10 ms	90 ms
Pulse1	Pulse-type click with standard repeat	6	10 ms	90 ms
Crisp 1	Crisp-type click with standard repeat	8	11 ms	90 ms
Crisp 2	Alternative crisp-type click with no repeat	10	105 ms	0 ms
Smooth 1	Smooth-type click with standard repeat	10	50 ms	90 ms
Smooth 2	Alternative smooth-type click with standard repeat	10	10 ms	90 ms
Double 3	Alternative double-click with standard repeat	10	20 ms	90 ms
Pulse Vibration	Pulse vibration, medium-low frequency, standard repeat	Medium	100 ms	90 ms
Mag Ramp Down	Decreasing magnitude over short duration, no repeat buffer	Various	320 ms	0 ms

RESULTS

RESULTS OF SUBJECTIVITY EVALUATION OF POTENTIAL USERS

Figure 4 shows the results of the evaluation questions for potential users. The value of each axis is an average score of the questions that correspond to each category. The values of all four categories were higher when tactile feedback was used. This seems to be because the neighboring users who had not yet operated the terminal felt an appeal toward the tactile feedback. This suggests that tactile feedback can raise intrinsic motivation in potential users.

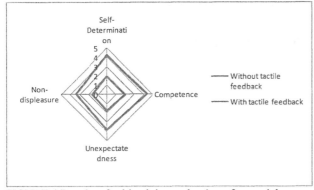

FIGURE 4 Results of subjectivity evaluation of potential users.

RESULTS OF SUBJECTIVITY EVALUATION OF EXISTING USERS

Figure 5 shows the results of evaluation questions for existing users. The value of each axis is an average score of the questions corresponding to each category. The values of all four categories were higher tactile feedback was used. This seems to be because the existing users felt an appeal for tactile feedback. This suggests that tactile feedback can increase intrinsic motivation in existing users and lead them to use the terminal again.

The operation frequency was significantly high when using tactile feedback. In addition to the subjectivity evaluation, actual actions showed that tactile feedback increases a user's intrinsic motivation.

340

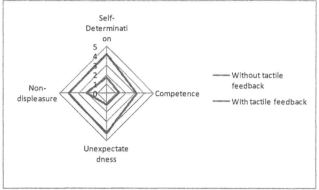

FIGURE 5 Result of subjectivity evaluation of existing users.

Results of number of times the button was pushed

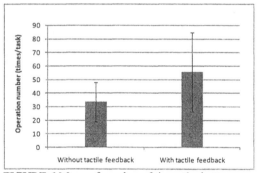

FIGURE 6 Mean of number of times the button was pushed when operating the terminal.

DISCUSSION

The above results suggest that tactile feedback may induce spontaneous actions in users. This seems to be because it has peculiar characteristics that visual and auditory feedback lack. Unlike visual and auditory stimulation, only a user who personally uses the terminal can experience tactile feedback. In other words, unlike visual and auditory stimulus, which users can receive even if they are in a passive state, the spontaneous action of users is indispensable for experiencing tactile stimulus. We guess that this is a specific factor that contributes to inducing the actions of potential users, because they cannot experience the stimulus without using the terminal spontaneously.

It is also difficulty to make the tactile stimulus persist in a user's memory. Moreover, it is a unique experience for an individual user, which may give rise to a sense of superiority and strong emotion more than that possible with visual and auditory stimuli. These are specific factors that induce the actions of the existing users. Besides, because it is evident that the vibration of the touch screens involved in the tactile feedback does not harm or place an excessive burden on the human body, tactile feedback gives users only moderate surprises.

CONCLUSION

In this study, using a terminal that uses tactile feedback, we examined whether tactile feedback effectively attracted users in kiosk terminals, which depend on to the spontaneity of users. In particular, we assumed that the desire to use the terminals was a type of intrinsic motivation and proposed an evaluation method for it. From the results, the following became clear:

1. Tactile feedback raises intrinsic motivation in potential users. This seems to be because, passive users cannot experience the sense of tactile stimulus.

2. Tactile feedback raises the intrinsic motivation of existing users. This seems to be because they feel that they want to experience it again, since with tactile feedback, stimulus information is hard to preserve in one's memory.

Thus, these results suggested that when tactile feedback is given to users, it raised their intrinsic motivation. As a future problem, we plan to develop an evaluation method for intrinsic motivation by directly observing user behaviors in addition to subjectivity evaluation. If we could measure intrinsic motivation by observing user behaviors, we might grasp the potential requirements and trends in marketing.

REFERENCES

Heron, W. (1957). The pathology of boredom. *Scientific American*, January.

Harlow, H.F., Harlow, M.K. & Meyer, D.R. (1950).Learning motivated by a manipulation drive. *Journal of Experimental Psychology, 40*, 228-234.

Harlow, H.F. (1950). Learning and satiation of response in intrinsically motivated complex puzzle performance by monkeys. *J. comp. physiol. Psychol, 43*, 289-294.

Hebb, D.O. (1949). *The organization of behavior.* Wiley.

Butler, R.A. (1953) Discrimination learning by rhesus monkeys to visual-exploration motivation. *Journal of Comparative and Physiological Psychology, 46*, 95-98.

Deci, E.L. (1972). Intrinsic motivation, extrinsic reinforcement and inequity. *Journal of Personality and Social Psychology, 22*, 113-120.

Lepper, M.R., Greene, D. & Nisbett, R.E. (1973) Undermining children's intrinsic interest with extrinsic rewards : A test of the overjustification hypothesis. *Journal of Personality and Social Psychology, 28*, 129-137.

Berlyne, D.E. (1957). Conflict and choice time. *British Journal of Psychology, 48*, 106-118.

Deci, E.L. (1975). *Intrinsic motivation.* Plenum Press.

Dember, W.N., & Earl, R.W. (1965). Analysis of exploratory, manipulatory, and curiosity behaviors. *Psychological Review, 64,* 91-96.

Berlyne, D.E. (1965). *Structure and direction in thinking.* John Wiley & Sons.

Hunt, J.McV. (1965). Intrisic motivation and its role in psychological development. *Nebraska symposium on motivation, 13,* 189-282.

Giyoo, H., & Kayoko, I. (1972). *Intellectual curiosity.* Tyukou shinsho.

Chapter 37

The Improvement of Beijing Digital Information Booth Interface Design

P.L. Patrick Rau, Yue Sun

Institute of Human Factors & Ergonomics
Department of Industrial Engineering
Tsinghua University, Beijing, China, 100084

ABSTRACT

This article mainly addresses two issues related to the interface design of Beijing Digital Information Booth (BDIB): poor navigation system design and layout and look&feel design. With the methods of checklist and Navigation Stress Test, this article evaluates the interface design of BDIB and summarizes 13 problems from the evaluation result. In order to improve the usability of BDIB, this article also recommends concrete solutions to each detail problem.

Keywords: Beijing Digital Information Booth; Usability Evaluation; User Interface Design, Navigation; Layout and look&feel

INTRODUCTION

RESEARCH BACKGROUND

Beijing Digital Information Booth (BDIB) is a touch screen web information system established by Beijing government for the convenience of Beijing citizens. Its basic concept is multimedia self-service terminals and multifunction service platform. The applications of BDIB can be divided into convenient service, E-government and E-commerce. The interaction map of BDIB is shown in the figure 1.1. The homepage interface of BDIB and each function are shown in figure 1.2.

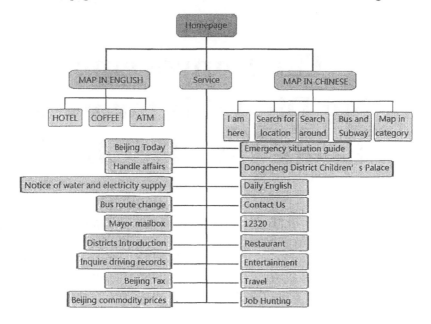

Figure 1.1. Interaction Map of BDIB

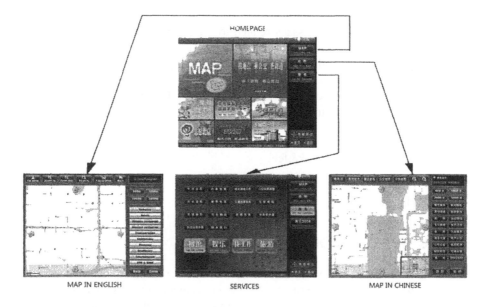

Figure 1.2. the interface of BDIB (Beijing Digital Information Booth website)

For a preliminary knowledge of ergonomics aspect related to BDIB, we made a field survey on it, from which we found two types of deficiencies of BDIB, including:

- Poor navigation support through the system
- Inappropriate layout and look&feel design.

To improve the Interface design of Beijing Digital Information Booth (BDIB), we need to find an analytical way to identify the most critical issues of BDIB and give practical improvement measures.

WHY DO WE DO IT

It costs the government millions of RMB to build up the BDIB every year. The object of the government is to make life in Beijing more convenient. But the result is not satisfying—the utilization of BDIB is quite low due to its poor usability.

Normally, users that come to the information booth are motivated by their needs for information. However, in most cases, the users are in a hurry or have little time to use the system. Without well-designed layout and navigation support system, the users often get lost through the system and fail to find the information quickly. Then they will easily feel frustrated and give up. Therefore, "it is important that a kiosk system presents a clear and simple structure to the user" (Maguire M.C., 1999), which encourages them to retrieve Information and makes the retrieving process easier and efficient.

We hope that after the evaluation of the user interface design of BDIB, the main problems could be found and corrected to improve the usability of BDIB and make it a good tool for the daily life in Beijing.

LITERATURE REVIEW

In order to design appropriate guidelines for our research, we refer to similar works including website interface design guidelines (Apple Computer Inc., 1996) (Lynch, P. J, Horton, S., 2009) (IBM Web Design Guide, 1998), touch-screen design guidelines (Waloszek, G., 2000), interactive multimedia design guidelines (Schofield, J., Flute, J., 1997) (Borda, A., 2003), and public information booth design guidelines (Maguire, M.C., 1999). We summarize some of the guidelines that concerning navigation and layout and look&feel of interface design from these references in the following tables.

Table 2.1. Guidelines for Website Design Guideline (Apple Computer Inc., 1996) (Lynch, P. J, Horton, S., 2009) (IBM Web Design Guide, 1998)

Navigation and Way Finding	Layout and Look&feel
Provide search engine and search scope options.	Put navigational links in the left column.
Don't provide too many choices on major menu pages.	Provide redundant navigation links for long pages in the page footer.
Present basic navigation links in consistent locations on every page.	Display contact information in a consistent location on every page.
Enable users to return to home page and other major navigation points in the site at anytime.	Basic layout grids, graphic themes, editorial conventions, and organization hierarchies should be consistent.
Keep users oriented by using consistent landmarks in site navigation.	Put the search box on the upper right areas.
Show the results of search query on a page that looks like the rest of the web site.	Provide breadcrumb navigation in the content area.

Table 2.2. Guidelines for Touch-screen Design Guideline (Waloszek, G., 2000)

Navigation	Layout and Look&feel
Don't make scroll buttons too small.	Use 3*3 layout for number entry.
Use Tabstrip or Stack for presenting hierarchies, instead of conventional trees.	Divides the screen into a fixed number of cells with constant size with girds.

Table 2.3. Public Information Booth Design Guideline (Maguire, M.C., 1999)

Navigation and Way finding
Set a "go back" option, so that users can return to previous page.
Show the path followed by users or their position in a path when presenting a sequence of screens, such as "screen 2 of 5".
Provide clear, short and distinctive title in each page.
Present no more than 12 options on one menu.
Sort menu lists with a certain order, such as alphabetical order.
Auto-reset the system if no input is made within a few minutes.

Layout and look&feel
Provide multilingual interfaces.
Provide help function throughout the system.
Highlight the input position clearly.
Use no more than 5 colors in a page.

We abstract good and bad design examples and key points of navigation and layout and look&feel design from the guidelines and apply them in the following checklists and test.

METHOD

Based on guidelines summarized above, we design two checklists and a navigation stress test to evaluate the ergonomics design of BDIB.

CHECKLIST

The two checklists we designed are Layout and Look-and-feel Checklist and Universal Detailed Checklist. They focus on layout and Look&feel and navigation support issues of BDIB respectively. Both of the two checklists contain objective questions summarized from related parts of similar studies which are detailed and systemic, and are evaluated by 6~7 experts.

LAYOUT AND LOOK-AND-FEEL CHECKLIST

This checklist mainly focuses on the layout and look&feel design of BDIB interface, which is composed of four columns. The first column consists of various categories with standard items to be evaluated, concerning the layout and look&feel aspect of BDIB, including feathers, menu, layout and design, content, forms and interaction, graphics, color, help system and icons, pointer, text. In the second column, the experts can choose the grade of fulfillment from three degrees of severity: "Bad", "Fair" and "Good". In the third column, the explanations of every single choice should be noted. Experts should give suggestions on the topic which are evaluated as "Bad" or "Fair" in the fourth column.

UNIVERSAL DETAILED CHECKLIST

The aim of Universal Detailed Checklist is to evaluate the navigation support of BDIB Interface. It has evaluated two systems in the Beijing Digital Information Booth: Nested Navigation System (Figure 3.1) and Assistant Navigation System.

The Nested Navigation System is based on the appearance of a webpage, which is divided in global, regional and context parts. The global panel contains information such as homepage and context that depends on the users' needs, which is usually located on the top of each page. The regional panel is normally located on the left or right side, whose function is to provide navigation support by showing the subcomponents of the webpage. As the main part of the page as well as the smallest and most detailed unit in this structure, the context panel contains quick links and a special area for important context.

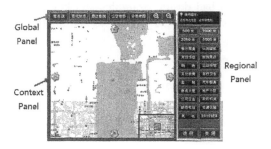

Figure 3.1. Nested Navigation System

The Assistant Navigation System comprises all assistant functions, including help function, search engine, a sitemap, an index and a guide who leads the users through the system and explain the usage of BDIB.

NAVIGATION STRESS TEST

The questionnaire we used, Navigation Stress Test was developed by Keith Instone (Keith Instone, 1997), which was used to test the mind workload while viewing a website. The test measures the navigation support of BDIB from the structural feeling and a subjective view.

EVALUATION RESULTS AND RECOMMENDATIONS

From the results of the checklists and the Navigation Stress Test, we summarize 13 major problems and give improvement recommendations respectively.

PROBLEM 1: LACK OF HELP FUNCTION

Analysis: Tasks for Novice users are similar and concentrated in some aspects, so instruction can be designed to include the most popular topics. To sum-up, we can get the following solutions:

- Add help function with a "help" link or a simple "?" button in the fix position on each page.
- Two ways to input the topic: select one popular topic & input the topic
- A short introduction of the system is also in need to help the user to understand the structure.

PROBLEM 2: LACK OF SEARCH ENGINE

Analysis: The search engine could only be nested in a help system or web index which is enough to find important and frequent using functions and also is not time or money consuming. Also, it is suggested to offer search scope options with list to let users know the scope of what they are searching. The results of search query should appear on a new page, so that users won't be confused with the sense of place. Besides, according to Yale web style guidelines, the search engine should be put on the upper right areas of the page.

PROBLEM 3: LACK OF LANGUAGE OPTION

Analysis:

- Set language option with clear icon which is easy to be noticed.
- At least an English option should be implemented.
- Display languages with different icons to distinguish from each other.
- The position on the language option should be eye-catching, normally in the top right area of the page.

PROBLEM 4: INAPPROPRIATE AND INCONSISTENT LAYOUT AND INSUFFICIENT FUNCTIONS OF THE NESTED NAVIGATION SYSTEM

Analysis: We suggest that in the Nested Navigation System the following functions should be added: popular searches and topics, contact and copyright (Figure 4.1). More importantly, the locations of global panel, regional panel and context panel should keep the same throughout the system.

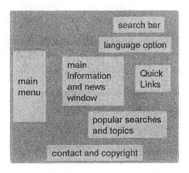

Figure 4.1 Recommending Layout

PROBLEM 5: TOO MANY COLORS USED ON THE INTERFACE

Analysis: Too many colors on the interface make it difficult to separate between different parts of the page and makes the interface looks gaudy. We suggest that no more than five colors should be used to keep the interface clear.

PROBLEM 6: LACK OF BREADCRUMB NAVIGATION

Analysis: According to the canonical page design example provided by the Web Style Guide, 2009, the breadcrumb trail often appears in small type on the top of the content area. It not only clearly shows users their position and the stage of the hierarchical which the page lies in, but also enables the users to go back to any stage of the hierarchy of the system directly.

PROBLEM 7: THE GUIDE TO SHOW HOW TO USE HANDWRITING FUNCTION OF THE TOUCHSCREEN IS NOT EASY TO BE NOTICED.

Analysis: When the user is asked to input some information, there is a guide to remind people to use touchscreen writing on the upper right corner of the page. However, the guide to introduce how to use the touchscreen input is on the bottom of the keypad, which lies in the left side of the page (Figure 4.2). Besides, the words of the guide are quite small. Therefore, users can hardly notice it, which causes greatly confusion. We suggest that the guide to introduce how to use touchscreen input should be duplicated in the guide on the upper right corner of the page.

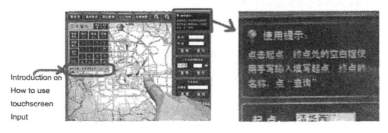

PROBLEM 8: THE FUNCTIONAL BUTTONS ARE EITHER MISSING OR CONFUSING.

Analysis: There are only two functional buttons including homepage and "go back" button on the page, while the previous button is missing. On the page of the subcomponent "Map", there is even no homepage button. The existed buttons also have some problems. The back button sometimes serves as the previous button, which greatly confuses the users. The homepage button does not link to the same homepage. Sometimes it goes back to the primary homepage, while sometimes it goes back to the subcomponent homepage. Based on this, we give the following recommendations:

- Previous button should be added, with conventional icon;
- All of the "homepage" buttons link to the primary homepage;
- Enable the users to go back to the homepage of subcomponent by providing a breadcrumb trial.

PROBLEM 9: DIFFERENT SIZES OF BUTTONS IN THE SAME LAYER OF THE HIERARCHY UNDER THE SUBCOMPONENT "SERVICE" CAUSE CONFUSION.

Analysis: In the regional panel of the subcomponent "Service", the buttons are divided in to two kinds of sizes, 14 small buttons with dark blue and 4 big buttons with different light colors individually. Although the aim of difference is to highlight the four most popular services: Restaurant, Entertainment, Job hunting and Traveling, it gives users the illusion that the four items are on an upper stage beyond the 14 items.

We suggest keeping all the buttons in the same size to eliminate confusion and using color difference to highlight the popular services, or setting up a new component in the global panel and assigning the four items in the new group.

PROBLEM 10: THE WORD WRITTEN ON THE SCREEN CANNOT BE DISTINGUISHED FROM THE BACKGROUND CLEARLY.

Analysis: The word written on screen uses various colors, with the map as its background. Too many colors are used in the screen, which cause great trouble for the users to recognize the word, especially those with visual deficiencies. Besides, there are lots of lines on the map, while the Chinese characters are also made of lines. This makes the word difficult to be recognized even more.

We suggest that when touchscreen writing is needed, a translucent dark area should be shown in the screen as the background of the word. This also tells users on where to input the words. Meanwhile, the color of the word should be light and consistent.

PROBLEM 11: CONFLICTS OCCUR BETWEEN NAVIGATING AND PROTECTING PRIVACY WHEN MARKING THE USED HYPERLINKS.

Analysis: In order to help users navigate through this complex structure and show the path they followed, the used hyperlinks usually are marked by color changes. In a personal computer it would be fine, but conflicts occur while marking hyperlinks in an information kiosk, due to the public usage of the kiosk. When people wait in line in front of the BDIB, if the first user finished and the next one starts to use the system, this user will probably get confused if some paths are already marked. It will even become a legal problem if the second user follows the paths of the first user, which definitely violates the privacy of the first user. As it mentioned earlier, the BDIB offers many functions, with the range from government facilities, with detailed information on where you can pay taxes and information on taxes, to medical health advisory. In many cases the user does not want other people to gain insight in their privacy. Therefore, the challenge is to find a right balance between these conflicts, which could be considered as a question of how to reset the system and delete the marks as well. In the following, we provide four common methods to reset the marked hyperlink.

- Reset every time when return the homepage.
- Reset when the interface is left still without any operation for some time.
- Set a button to reset the hyperlinks manually.
- Set a camera on top of the BDIB. Return to the homepage automatically when the user left.

PROBLEM 12: INAPPROPRIATE LAYOUT OF THE INPUT KEYPAD

Analysis: The input keypad lies on the left of the screen, while the input column is on the other side. Users have to look back and forth across the screen to input a word. Besides, since most people are right handed, when users are pointing to the keypad, their arm and hand may block them from seeing the whole screen. And the position is not comfortable at all. We suggest putting the input keypad near the column, and move it to the right side of the screen instead.

PROBLEM 13 TAGLINES ARE INCONSISTENT ON THE PAGES

Analysis: In the main pages of the subcomponents of "Services", which are in the same stage of the hierarchy in the system, the tagline of each page lies in different place and uses different fonts, color, and style.

A clear tagline is very important in the guidance through the system as it could show the intention of the system directly to the users. Besides, the tagline can also show the stage which the page lies in of the hierarchy. Therefore, it should be easy to notice by the users. Thus a clear tagline should be added in the fixed place in the main page.

CONCLUSIONS

The aim of the research is to improve the interface design of Beijing Digital Information Booth. In the research, we designed two checklists and a Navigation Stress Test, based on guidelines we summarized from references, to evaluate the interface design of BDIB. After the main problems are found, we give analysis toward the problems individually and raise recommendation on each problem.

We hope the recommendations could improve the usability of BDIB and attract more people to use it and enjoy the benefit it brings. We believe with good usability, Beijing Digital Information Booth has great potential and will play an important role in people's life in the future.

ACKNOWLEDGEMENTS

We gratefully acknowledge the support from No. 70771059 of National Natural Science Foundation of China and No. 0821 of China — EU S&T Cooperation Special Funding, as well as the German students in the course of Ergonomics and Working Orgniztion for the data collection and orgniztion work they have done in 2007.

REFERENCES

The Beijing Digital Information Booth website,
   http://www.touchbj.com/web/website/web_news/dt/xxtdt.html
Apple Computer Inc. (December, 1996). The Apple Web Design Guide,
   http://applenet.apple.com/hi/web/web.html
Lynch, P. J, Horton, S. (2009, January 15), Web Style Guide, 3rd edition: Basic Design Principles for Creating Web Sites (Web Style Guide: Basic Design Principles for Creating Web Sites), Yale University Press
IBM Web Design Guide. (1998). The IBM web site:
   http://www.ibm.com/ibm/easy/design/lower/f030100.html
Waloszek, G. (December, 2000), Product Design Center, Interaction Design Guide for Touchscreen Applications, Version 0.5, from Resources on the SAP Design Guild Website: www.sapdesignguild.org
Schofield, J., Flute, J. (1997), Use and Usability, A guide to Designing Interactive Multimedia for the Public, Multimedia Victoria
Borda, A. (January, 2003), Guidelines for multimedia development (With Emphasis on Browser-Based Media), Science Museum
Maguire, M.C. (March, 1999). A Review of User-Interface Design Guidelines for Public Information Kiosk Systems, International Journal of Human-Computer Studies, Volume 50, Issue 3, March 1999, pp. 263 - 286
Keith Instone, 1997, from: http://keith.instone.org/navstress

Chapter 38

Emotions Invoked by Overall Shape of Robot

Jihong Hwang[1], Wonil Hwang[2], Taezoon Park[3]

[1]Seoul National University of Technology
Seoul, Republic of Korea

[2]Soongsil University
Seoul, Republic of Korea

[3]Nanyang Technological University
North Spine, Singapore

ABSTRACT

In the present study, emotion varied with the overall shape of robot is experimentally investigated using the surface-modelled images of robots and the real models fabricated using these images. In total, 27 robot shapes were used for the experiment, which are combinations of three different shapes of head, trunk and limb (legs and arms), respectively. During the experiment, the images and real models of these robot shapes were presented to 20 subjects, and their answers to the questionnaire relating to the emotions invoked by the images and models were collected. Then, statistical analysis was performed to estimate the matching levels between the overall shapes of robot and the emotions. The result showed that the overall shape most suitable for the service robots is the one which has a cylindrical shape for all of head, trunk and limb.

Keywords : Service robot, Robot shape, Emotion

INTRODUCTION

With the recent exponential increase of interest in robots, researches on robots have been very actively conducted in various areas. Traditionally, researches on robots had been focused on the technological issues to realize the essential functions of robots such as kinematic mechanism, actuating, sensing, control, etc. In these days, however, these have been expanded to include the affective interaction between human and robots (Breazeal, 2002; Fong et al., 2002; Shin 2008), which is associated with the high expectation for the rapid growth in the service robot market (United Nations Economic Commission for Europe, 2004). The service robots for which the affective interaction between human and robots should be considered important include reception robots in convention centers or airports, nursing robots in hospitals or facilities for old people, teaching robots in schools, and entertaining robots in theme parks. For these service robots, the appropriate affective interaction between human and robots would allow for increased satisfaction to the users (Bartneck, 2004; Kirby, 2010; Woods, 2006).

The affective interaction between human and robots could be influenced by various aspects of the service robots, which are the overall shape, countenance, gesture, voice, display panel, etc. (Goetz, 2003; Nehaniv et al., 2005; Weinschenk, 2000). Among these, relatively little is known of the relation between the overall shape of robot and the ensuing emotions, while the overall shape of robot could play a big role in creating desired emotions for the users. For an example, the more hospitable the overall shape of nursing robots is, the more the patients might rely on them. However, it is not clear yet what shape of robot could provide such emotion with the patients.

In this context, the present study intends to determine the overall shapes of robot that match best with given emotions and to possibly provide guidelines for the affective design of the overall shape of service robot. In doing so, about fifty images of humanoid robots or their neighborhoods are collected from various sources such as cartoon, movie, internet, etc. Then, the shapes of the head, trunk, arm and leg are classified into 3-5 categories, respectively. Among them, three standard shapes with higher frequency are chosen, surface-modeled using a 3-D CAD program, and fabricated using a rapid prototyping machine for the head, trunk, arm and leg, respectively. The parts which take the same place in the overall shape of robot are designed to be interchangeable when they are assembled.

The experiments are conducted by displaying the real models and the 3-D images of different shapes of robot in front of the subjects. The subjects are asked to mark the level of matching between the given shape of robot and the emotions specified in the questionnaire. By statistically analyzing the questionnaire, the shapes of robot which match best the specified emotions are determined. Besides, by comparing the results from the real models and the 3-D images, suitability of the use of 3-D images for the study of the affective interaction between human and robot through the overall shape of robot is verified.

CLASSIFICATION OF OVERALL SHAPES OF ROBOT AND FABRICATION

For the classification of overall shapes of robot, about one hundred of robot images were collected without any constrains from a variety of sources - cartoon, movie, internet, etc. It was found that majority of these images are on humanoid robots or their neighborhoods whose bodies consist of head, trunk and limb although some of them miss arms or legs. This would be attributed to the robot users' preference for the robots sharing morphology similar to human (Breazeal 2003; Duffy 2003; Reeves and Nass, (1996); Severinson-Eklundh, 2003). So, the classification was focused onto the humanoids robots and their neighborhoods. In doing so, fifty images of humanoid robots or their neighborhoods were randomly selected again, and the shapes of the parts (head, trunk, arms and legs) were classified into 3-5 categories. Then, the frequency that each of these shapes was observed in these images was counted. The results are shown in Table 2.1.

Table 2.1 Classification of the shapes of the parts comprising the overall shape of robot

Head		Trunk		Arms		Legs	
Shape	Frequency	Shape	Frequency	Shape	Frequency	Shape	Frequency
Near-spherical	40	Near-spherical	6	Cylindrical	25	Cylindrical	8
Rectangular Parallepiped	7	Cylindrical	22	Rectangular Parallepiped	9	Rectangular Parallepiped	15
Free-formed	3	Rectangular Parallepiped	11	Free formed	10	Free-formed	8
		Free-formed	11	None	6	Wheel	6
						None	13

When considering all the shapes listed in Table 1, it is possible to have 240 different overall shapes of robot (3 for head × 4 for trunk × 4 for arms × 5 for legs). However, due to the constraints in time and cost expended for fabricating the real models and conducting the experiments, this was reduced to 27 using the following rules.

- For the shapes of each part of a robot, only three shapes are considered.
- Two of the three shapes should be those with highest frequencies for the shape of each part in Table 1.
- The rest of the tree shapes should be the one that is most favored by the authors and the subjects excluding the shapes determined by the above rule.
- The same shapes are considered for arms and legs (limb).

After the shapes for each part of a robot had been determined by applying the above rules, they were surface-modeled and optimized using a 3D-CAD program so that all the parts are in good harmony when they are assembled to construct the overall shape of a robot. Then, the surface-modeled shapes were transformed to a file format that can be accepted by rapid prototyping machines for the fabrications.

The height of the real model was about 70cm when all the parts were assembled. The Surface-modeled shapes of the parts comprising the overall shape of robot and an example of their assembly are shown in Table 2.2.

Table 2.2 Surface-modeled shapes of the parts comprising the overall shape of robot and an example of their assembly

		Shape		Assembly
	Rectangular parallepiped ID #: 1	Cylindrical ID #: 2	Near-spherical/ free-formed ID #: 3	Robot ID # 132

EXPERIMENT

METHOD

For the experiment, 27 images and 27 real models were prepared. The total number of subjects was 20. All of the subjects were university students, and their ages are in the range of 20-26. The numbers of male and female subjects were 10, respectively. The subjects were divided into 4 groups, each group having 5 subjects. The subjects who belong to the same group participated in the experiment at the same time. So, the experiment was repeated four times with changing the subject group.

During the experiment, all of the 3-D images and all of the real models were displayed one by one in front of the subjects. For the first two groups, the images were displayed first while, for the other two groups, the real models were displayed first. The order in which the images or the real models were displayed was randomly given prior to the experiment to eliminate its influence on the results.

Throughout the experiment, a care was taken to make sure that the conditions such as viewing angle, illumination, background color, and screen size were maintained same. For the 3-D images, these conditions were adjusted using the 3-D CAD program while, for the real models, these conditions were adjusted physically. Table 3.1 summarizes the conditions used for the experiment.

Table 3.1 Summary of the experimental conditions

	3-D Images	Real Model
Viewing Angle	15°	15°
Image/Model Color	White	White
Illumination	White light	White light
Background color	Black	Black
Height	70 cm	70 cm

QUESTIONNAIRE

The questionnaire consisted of 15 questions as listed in Table 3.2. The questionnaire was designed to cover a variety of emotions, based on the earlier studies relating to the measurements of emotions (Hwang, 2009; Khan, 1998; Scopelliti, 2005)

Table 3.2 Questionnaire used for the experiment

Robot ID #_____   Image_____/Model_____

The presented robot looks _____.	Strongly Disagree (1)	Disagree (2)	Slightly Disagree (3)	Neutral (4)	Slightly Agree (5)	Agree (6)	Strongly Agree (7)
(1) interesting							
(2) amusing							
(3) useful							
(4) relaxing							
(5) scary							
(6) dangerous							
(7) out of control							
(8) embarrassing							
(9) overwhelming							
(10) safe							
(11) pretty							
(12) accessible							
(13) exciting							
(14) complex							
(15) amiable							

During the experiment, the subjects were asked to evaluate the level of matching between the overall shapes of robot presented by the images or the real models and the emotions specified in the questionnaire. The highest matching level was 7, while the lowest was 1. The subjects were given enough time to see an image or real model and complete the questionnaire (about 90 seconds on average). The next image or real model was presented after all the subjects finished answering the questionnaire for the previous image or model. So, by the time the experiment was completed, a subject answered the questionnaire for all of 27 images and 27 real models. That is, 54 sheets of the questionnaire as shown in Table 3.2 were collected for a subject when the experiment was over.

ANALYSIS & RESULTS

The experimental data were digitally saved using a spreadsheet program for the statistical analysis. Then, the average values of the matching levels between the robot shape and the emotions specified in the questionnaire were calculated. This resulted in 810 average values (27 robot shapes × 15 emotions × 2 types (image and real models)) in total. Based on these values, the combinations of robot shape and emotion which match relatively well were selected. In this case, the matching level of 4 was employed as a threshold. That is, the combinations whose average matching level is higher than 4 were considered to match relatively well. Then, ANOVA and Duncan analysis ($p = 0.05$) was performed to compare the average values of matching level and thus to determine the group of robot shapes which has the highest matching level for a given emotion. The results are shown in Table 4.1.

Table 4.1 Group of robot shapes with the highest matching level

Emotions		Robot ID # * (ascending order in the matching level)	Range
Interesting	Images	(113),(132),(323),(332),(223),(123),(333),(133),(233)	4.2~4.8
	Models	(211),(223),(123),(323),(121),(233),(332),(113),(122),(232),(212),(213),(112),(133),(132),(333),_(222)_	4.1~4.9
	Common	(113),(132),(323),(332),(223),(123),(333),(133),(233)	
Amusing	Images	(113),(323),(223),(132),(123),(332),(333),(133),(233)	4.05~5.0
	Models	(123),(323),(212),(223),(121),(113),(231),(332),(233),(122),(133),(112),(132),(232),(333),_(222)_	4.05~5.05
	Common	(113),(323),(132),(123),(332),(333),(233)	
Useful	Images	(123),(133),(331),(132),(211),(131),(122),(221),(323),(112),(113),(212),(232),(231),(332),(222),(111),(233),(333),(223)	4.2~5.0
	Models	(213),(223),(311),(312),(321),(231),(131),(211),(233),(112),(132),(212),(221),(111),(121),23,(232),(323),(122),(332),(133),(333),_(222)_	4.35~5.35
	Common	(231),(131),(211),(233),(112),(132),(212),(221),(222),(111),(232),(323),(122),(332),(133),(333),(222)	
Relaxing	Images	(111),(122),(332),(223),(132),(323),(233),(112),(232),(333),(222)	4.05~4.75
	Models	(111),(321),(133),(323),(112),(132),(211),(231),(322),(121),(221),(212),(332),(333),(122),(232),_(222)_	4.25~5.35
	Common	(111),(323),(112),(332),(333),(122),(232),(222)	
Safe	Images	(131),(123),(133),(211),(331),(111),(212),(221),(112),(233),(121),(332),(223),(232),(333),(231),(122),(222)	4.05~4.9
	Models	(121),(132),(321),(323),(331),(133),(212),(312),(231),(111),(221),(232),(311),(322),(112),(332),(333),(222),(211),_(122)_	4.2~5.3
	Common	(121),(331),(133),(212),(231),(111),(221),(232),(112),(332),(333),(222),(211),(122)	
Pretty	Images	(223),(132),(323),(222),(332),(233),(133),(333)	4.1~4.7
	Models	(311),(323),(133),(211),(132),(231),(333),(212),(332),(322),(221),(122),(112),_(222)_	4.2~5.3
	Common	(132),(323),(222),(332),(133),(333)	
Accessible	Images	(332),(121),(112),(223),(122),(323),(131),(133),(232),(222),(233),(333)	4.05~5.05
	Models	(132),(321),(323),(133),(221),(322),(231),(211),(212),(333),(332),(112),(232),(122),_(222)_	4,25~5.45
	Common	(332),(112),(122),(323),(133),(232),(222),(333)	
Amiable	Images	(112),(113),(123),(332),(133),(223),(323),(233),(222),(333)	4.2~4.95
	Models	(231),(133),(221),(322),(332),(211),(212),(112),(333),(122),(232),_(222)_	4.35~5.45
	Common	(112),(332),(133), (222), (333)	

*The first digit: head, the second digit: trunk, the third digit: limb (1: rectangular paralleopiped, 2: cylindrical, 3: near-spherical/free-formed)

As shown in Table 4.1, the emotions that match well the robot shapes presented turned out to be *"interesting"*, *"amusing"*, *"useful"*, *"relaxing"* *"safe"*, *"pretty"*, *"accessible"* and *"amiable"*. In general, these are considered as positive emotions. This is contrasted to the observation that the negative emotions such as *"scary"*, *"dangerous"*, *"out of control"*, *"embarrassing"*, *"overwhelming"*, and *"complex"* match none of the robot shapes presented during the experiment. The better matching with the positive emotions than the negative emotions may be due to the ignorance of other features of robot such as countenance, voice, and motions. That is, the overall shape of robot in itself doesn't seem to be able to invoke the negative emotions.

In Table 4.1, it is also interesting to note that more robot shapes match a given emotion when they are presented by the real models than by the images. That is, the real models could invoke more emotions than the images could. In Table 4.1, the overall shapes of robot whose matching levels for the real models were highest are highlighted by the underlined, italic, bold fonts. Interestingly, the robot shape which consists of a cylindrical head, a cylindrical trunk, cylindrical arms and cylindrical legs (robot ID number: 222) shows best matching with most of the positive emotions. This indicates that, when designing service robots for which the affective interactions with their users are very important, this shape of robot would be most appropriate.

CONCLUSION & FUTURE WORK

In the present study, the overall shapes of robot that match best with given emotions were investigated. The result indicated that the robot shape which consists of a cylindrical head, a cylindrical trunk, cylindrical arms and cylindrical legs shows best matching with most of the positive emotions, and thus, for affective design of the service robots, this shape would be most appropriate. Also, it was found that the overall shape of robot in itself doesn't seem to be able to invoke the negative emotions under conditions employed for the present study. Therefore, in order to creative negative emotions, other features such as countenance, voice, and motions need to be added. The present study also indicates that the real models could invoke more emotions than the images could. So, when investigating the relations between the shapes of robot and emotions, the images would not be as efficient as the real models.

In the present study, when determining the overall shapes of robot that match best with given emotions, the influence of the subjects' preference or experience was not investigated. So, these aspects will be included in our future work. Also, a study of the robot's personality varying with the overall shape of robot is currently undergoing.

REFERENCES

Brazeal, C. (2002), *Designing Sociable Robots*. MIT Press, Massachusetts.

Brazeal ,C. (2003), "Emotion and sociable humanoid robots." *International Journal of Human-Computer Studies*, 59, 119-155.

Barneck, C. and Forlizzi, J. (2004), "A design-centered framework for social human-robot interaction." Proceedings of IEEE ROMAN2004 Workshop, London, UK.

Duffy, B. R. (2003), "Anthropomorphism and the social robot." *Robotics and Autonomous Systems*, 42, 177-190

Fong, T., Nourbakhsh, I., and Dautenhahn, K. (2002), *A survey of socially interactive robots. concepts, design, and applications* Technical Report No. CMU-RI-TR-02-09, Robotics Institute, Carnegie Melon University, Pittsburgh.

Goetz, J., Kiesler, S., and Powers, A. (2003), "Matching robot appearance and behavior to tasks to improve human-robot cooperation." *Proceedings of the 12th IEEE International Workshop on Robots and Human*.

Hwang, J. and Hwang, W. (2009), "Vibration perception and excitatory direction for haptic devices." *Journal of Intelligent Manufacturing* , On-line Available.

Kirby, R., Forlizzi, J. and Simmonsa, R. (2010) "Affective social robots." *Robotics and Autonomous Systems*, 58, 322-332.

Khan, Z. (1998), *Attitudes towards intelligent service robots*, Technical Report No. TRITA-NA-P9821, Interaction and Presentation Laboratory, Royal Institute of Technology, Stockholm.

Nehaniv C. L., Dautenhahn, K., Kubacki, J., Haegele, M., Parlitz, C., and Alamic, R. (2005), "A methodological approach relating the classification of gesture to identification of human intent in the context of human-robot interaction." *Proceedings of IEEE ROMAN 2005*, 371-377.

Reeves, B. and Nass, C. (1996) *The Media Equation: How People Treat Computers, Television and New Media Like Real People and Places*, Cambridge University Press, Cambridge.

Scopelliti, M., Giuliani, M. V. and Fornara, F. (2005), "Robots in a domestic setting: a psychological approach." *Univ Access Info Soc*, 4, 146-155.

Severinson-Eklundh, K., Green, A. and Hüttenrauch, H. (2003), "Social and collaborative aspects of interaction with a service robot." *Robotics and Autonomous Systems*, 42, 223-234.

Shin, E. (2008), *A Study on the dsign of human-robot interaction based on the classification of social robot appearances – with emphasis on the body features of a robot and its interaction types*. MS Thesis, Korea Advanced Institute Science and Technology, Daejeon.

United Nations Economic Commission for Europe (2004), *2004 World Robotics Survey*. ECE/STST/04/P01, Geneva.

Weinschenk, S. and Barker, D. T. (2000), *Designing Effective Speech Interfaces*. MIT Press, Massachusetts.

Woods, S. (2006), "Exploring the design space of robots: Children's perspectives." *Interacting with Computers*, 18, 1390–1418.

Chapter 39

User-Centered Smarter Robotic Medical Systems

T. Ahram[1, 2], *W. Karwowski*[1, 2], *B. Amaba*[3], C. Andrzejczak[2]

[1] Institute for Advanced Systems Engineering
University of Central Florida
Orlando, FL 32816, USA

[2] Department of Industrial Engineering and Management Systems
University of Central Florida
Orlando, FL 32816, USA

[3] IBM Industry Solutions
Miami, FL 33178, USA

ABSTRACT

This paper introduces a user-centered design framework derived from systems engineering practices and used for modeling the next-generation of robotic medical devices. Human systems considerations from systems engineering knowledge areas aim to ensure that human capabilities and limitations have a prominent place in the integrated design and development of safe medical devices throughout the total system lifecycle. Future challenges that collaborative user-centered systems engineering techniques are likely to face in this domain are also discussed.

Keywords: Smarter products, Systems Engineering, Robotic Medical Devices, Human Systems Integration

INTRODUCTION

Robotic medical systems are finding their way into many new application areas. Due to this proliferation of use, medical robot design has come under increased scrutiny, with safety a critical issue. These robots can have a direct interaction with patients and medical specialists. A medical robot such the *da Vinci Si Surgical System®*, developed by *Intuitive Surgical*, is an example of such a close interaction. This robot receives manual inputs from a surgeon, then scales, filters, and translates those movements to precise robotic appendages that perform minute movements directly on the patient's body. During the design process of medical robots, systems engineers must pay much attention to aspects of safety, which are currently an essential requirement for commercialization of robots in medical applications (Tondu and Lopez, 2000).

Systems Engineering (SE) concepts and principles are an integral part of the contemporary engineered world (Hitchins, 2007). Such concepts are also used to create smarter products, produce food, protect human health, enable travel over great distances, and allow for instant, ubiquitous communication. These principles are also used to build houses, design workplaces, and develop an infrastructure that society can safely employ. The SE principles are used to make products and services cheaper, more functional, and get them to the market faster.

Systems engineers apply and integrate concepts and rules derived from multiple knowledge domains in an effort to appropriately employ them in application domains. For example, the energy used to heat, cool, and light residential or industrial dwellings is typically generated hundreds of miles away from where it is used and needs to be transferred over long distances. This requires understanding of domains including but not limited to: electricity, current, voltage, material science, appropriate distribution methods, network/graph theory, industrial engineering, and building development/population forecasting. Systems engineers must use knowledge from these domains to successfully complete the task in an efficient manner.

SMARTNESS IN MEDICAL SYSTEMS

This section presents a synthesis and summary of the most innovative work that influenced research in this field. Allmendinger and Lombreglia (2005) highlighted smartness in a product from a business perspective. They regard *"smartness"* as the product's capability to predict business errors and faults, thus *"removing unpleasant surprises from the users' lives."* Ambient Intelligence (AMI) group describes a vision where distributed services, mobile computing devices, and embedded devices functioning in almost any type of environment (e.g., homes, offices, cars), all integrate seamlessly with one another using information sharing to enhance user experiences (Weiser, 1991; Ahola, 2001; Arts and de Ruyter, 2009). Rapid technological advancements and agile manufacturing coined the term "smart environments."

Definitions of smart environments may be taken into account as a first reference point, since smart products have to be considered in the context of their environment. For example, Das and Cook (2006) define a smart environment as the one that is able to acquire and apply knowledge about an environment and adapt to its inhabitants in order to improve their experience in that environment. It is noticed that the knowledge aspect has been recognized as a key issue in this definition. Mühlhäuser (2008) refers to smart product characteristics that are attributed to future smart environments: i.e., *"integrated interwoven sensors and computational systems seamlessly embedded in everyday systems and tools of our lives, connected through a continuous network"* (Mühlhäuser, 2008). In this respect, smarter products can be viewed as those products that facilitate daily tasks and augment everyday experiences. In 2007, AMI identified two motivating goals for building smart products (Sabou et al. 2009):

1) Increased need for simplicity in using everyday products, as their functionalities become more complex. Simplicity is desirable during the entire life cycle of the product to support manufacturing, use, or repair.

2) Increased number, sophistication, and diversity of product components (for example, in the aerospace industry), as well as the tendency of the suppliers and manufacturers to become increasingly independent of each other, which requires a considerable level of openness on the product side.

Mühlhäuser (2008) observed that smart product characteristics can now be developed due to recent advances in information technology afford real world awareness in these systems through the use of sensors, smart labels, and wearable, embedded computers. According to Mühlhäuser (2008), product simplicity can be achieved with improved product to user interaction (p2u). Furthermore, openness of a product requires an optimal product-to-product interaction (p2p). Knowledge intensive techniques enable better p2p interaction through self-organization within a product or a group of products. Smart products also require some level of internal organization by making use of planning and diagnosis algorithms, following is the definition of smart products given by (Mühlhäuser, 2008):

"A Smart Product is an entity (tangible object, software, or service) designed and made for self-organized embedding into different (smart) environments in the course of its lifecycle, providing improved simplicity and openness through improved p2u and p2p interaction by means of context-awareness, semantic self-description, proactive behavior, multimodal natural interfaces, AI planning, and machine learning." (Mühlhäuser, 2008)

Recent research on semantic web service description, discovery, and composition may enable self-organization within a group of products and, therefore, reduce the need for top-down constructed smart environments (Ahram et al. 2009, Karwowski and Ahram 2009). Major characteristics of smart products are illustrated by comparing their essential features. For example, Maass and Varshney (2008) define six major characteristics for smart products illustrated in Table 1.

Table 1. Smart Products Characteristics (Maass and Varshney, 2008)

Characteristic	Description
Personalization	Customization of products according to user preferences.
Business-awareness	Consideration of business and legal constraints.
Situated ness	Recognition of situational and community contexts.
Adaptive ness	Change product behavior according to user responses to tasks and needs.
Networkability	Ability to communicate and bundle with other products.
Pro-activity	Anticipation and prediction of user's plans and intentions.

In addition to the above characteristics, smart products should be devoted to offering multimodal interaction possibilities with their users, as multiple interaction possibilities increase the simplicity characteristics of the products; some users may prefer one method of interaction than another.

For medical robots, safety is a major concern. Specialists sharing the working area with the robot necessitated the need for integration of the field of human factors in the development of medical robots, especially those robots designed for delicate and accurate tasks. Hence, the human component has to be integrated in the early steps of the development process.

One tool employed by systems engineers is Systems Modeling Language (SysML), a subset of the unified modeling language (UML). Originally created for use in software engineering, these modeling languages provide visual representations of concepts and the relationships between them, their users, and other systems. For the purposes of this paper, it is used to communicate the critical need to consider human capabilities and aspects within the design process. This paper will focus on the SysML/UML contribution to the human systems integration and analysis of a medical robot and presents functions allocation and task analysis to leveraging human error in systems engineering.

The medical robot discussed in this paper is a robotic system operated by an expert surgeon who remotely performs a tumor scan examination; a general system overview is shown in Figure 1. A virtual probe is mounted on the master interface device. The real probe is placed on the slave robot end-effecter. The focus is on the computer control system of the slave site, where safety and human systems integration is critical. Medical robot functional allocation determines the distribution of work between human actors as part of the SysML and machines. It is particularly important to define non-ambiguous and consistent tasks for humans who are using the robot. Task analysis is conducted to identify the details of specified tasks, including the required knowledge, skills, attitudes, and personal characteristics and capabilities required for successful operation and task performance.

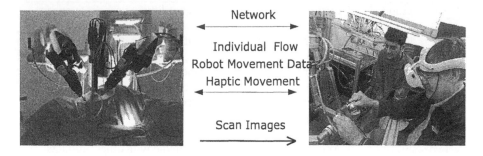

Figure 1. Tumor scanning robot system overview

A significant barrier in current complex robotic medical applications is that many formal languages and formal analysis techniques are unfamiliar, difficult to understand, or are not included in the requirements analysis process. SysML affords the integration of medical knowledge and practices into systems design requirements, and is now a standard in system and software engineering. These practices are easily transferrable to robotic systems design.

SYSTEMS ENGINEERING APPROACH TO DESIGN AND MODELING OF SMARTER ROBOTIC MEDICAL SYTEMS

The contemporary systems engineering process is an iterative, hierarchical, top down decomposition of system requirements (Hitchins, 2007). The hierarchical decomposition includes Functional Analysis, Allocation, and Synthesis. The iterative process begins with system-level decomposition and then proceeds through the functional subsystem level, all the way to the assembly and program level. The activities of functional analysis, requirements allocation, and synthesis will be completed before proceeding to the next lower level.

Modeling Systems Engineering Process Activity can be augmented and improved through the use SysML. SysML provides visual semantic representations for modeling system requirements, behavior, structure, and parametric data, which are then used to integrate with other engineering analysis models (Friedenthal et al., 2008).

Systems engineering teams along with product designers are responsible for verifying that systems under development meet all requirements defined in the system specification documents. The following procedures outline the relevant systems engineering process steps (DAU Guidebook, 2004):

- Requirements analysis - review and analyze the impact of operational characteristics, environmental factors, functional requirements and develops measures suitable for ranking alternative designs in a consistent, objective manner. Each requirement should be re-examined for consistency, desirability, applicability, and potential for improved return on investment (Ahram and Karwowski 2009).

- Functional analysis - systems engineers and product designers use the input of performance requirements to identify and analyze system functions in order to create alternatives to meet system requirements.

- Performance and functionality - systems engineering allocates design requirements and performance to each system function. Performance and functionality allocation process identifies any special personnel skills or design requirements.

- Design Synthesis - designers and other appropriate engineering specialties develop a system architecture design to specify the performance and design requirements, which are allocated in the detailed design.

- Documentation - the primary source for developing, updating, and completing the system and subsystem specifications. Smart product requirements and drawings should be established and maintained.

- Specifications – to transfer information from the smart product systems requirements analysis, system architecture design, and system design tasks.

- Specialty engineering functions - participate in the systems engineering process in all phases. They are responsible for system maintainability, testability, producibility, human factors, safety, design-to-cost, and performance analysis to assure design requirements are met.

- Requirements verification - systems engineering and test engineering verify the completed system design to assure that all the requirements contained in the requirements specifications have been met.

Model-based user-centered systems engineering (SE) approaches for design and modeling of smart systems and products differentiate between human performance and effectiveness criteria. These criteria define a total system mission performance level and tolerances that are directly attributable to specific actions allocated to human performance metrics. These are indicators measure which performance effectiveness criteria are met (Ahram et al. 2009, Karwowski and Ahram 2009). The smarter products user-centered SE knowledge integration framework can be used to develop a system where the human and machine synergistically and interactively cooperate to conduct the mission, and the "low hanging fruit" of performance improvement lies in the human–system interaction block.

THE USER-CENTERED SMART SYSTEM DEVELOPMENT CYCLE

The ISO 13407 human-centered design framework is considered the cornerstone for incorporating different design techniques, all of which can be merged to support a user-centered design process. According to the ISO 13407 standard (ISO, 1999), appropriate user-centered smart systems (USS) processes are composed of five iterative steps, which will aid in the fulfillment of all requirements into the medical robotic system design process as follows:

1) Planning robotic system design processes
2) Smart system context of use
3) Requirements specification
4) Integration of design solutions
5) Smart medical robot evaluation and assessment

The five iterative user-centered systems engineering design steps are based on the ISO 13407 framework and are depicted in Figure 2. The first step in planning smart system design processes is to communicate smart system needs with stakeholders and users to gain agreement on how user-centered design techniques can contribute to the smart system objectives (Karwowski, Salvendy, Ahram, 2009). In addition, the planning process prioritizes smart product requirements and highlights potential benefits gained from including user-centered smart system activities within the system development process.

Experts have studied the scan examination to determine all the interactions between the doctor manipulating the scanner probe and the patient (context of use), particularly pressures and movements on the patient's body, which are critical safety aspects (see Figure 3). The main safety criterion was to limit the work envelope and limit unnecessary, unplanned, and uncontrolled collisions with human tissue (Davies, 1993). The difficulty of modeling the working area (i.e. the patient's body and tumor) leads to the choice of a compliant operating robot, with an actuation by precision servos, motors, or actuators (Tondu and Lopez, 2000) (Booch, Rumbaugh, and Jacobson, 1999).

The SysML use-case diagram in Figure 3 presents the main steps undertaken during a scan examination. This use-case diagram depicts the relationships between process steps and actors performing them during a robot scan examination process, particularly between engineers and doctors. This diagram facilitates communication about basic requirements. These requirements then follow system development and are later integrated in requirement modeling in subsequent diagrams.

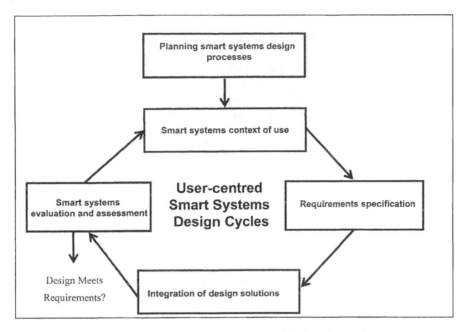

Figure. 2. User-centered smart system (USS) design cycle.
Modified from original ISO 13407 framework (ISO, 1999).

The use case patient management analysis provides scenarios of communication between the patient and the medical experts, who are essential to successful tumor, scan examination. This naturally led to the choice of a two-way video-audio synchronous communication system as indicated in the use-case diagram shown in Figures 4 and 5. The Equipment service provider is the person (actor) in charge of the Robot Management and involved in task completion.

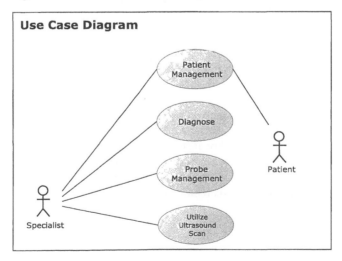

Figure 3: Use case diagram: tumor scan examination

370

SysML modeling is very useful for modeling the tumor robot examination system for many reasons (Guiochet, Tondu, and Baron, 2003). SysML diagrams show all interactions between actors and the system, as well as between actors themselves. An actor characterizes an outside user or related set of users who interact with the system (BaochBooch, Rumbaugh, and Jacobson, 1999). It is possible for an actor to be a human user or an external system; such modeling allows the interactions to be handled for safety and regulatory studies. In this case, we have a robotic system with two components, a robotic system controller and the robot scanner or probe. The robotic system controller interacts and takes actions from the actor specialist (see Figures 4 and 5) who is in the charge of the examination.

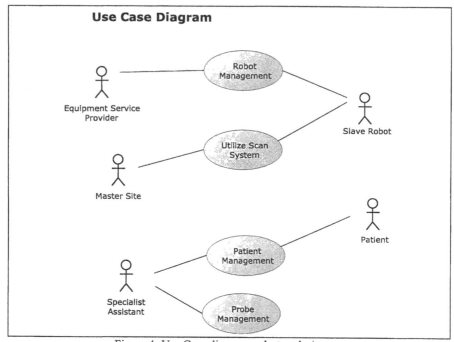

Figure 4: Use Case diagram: robot probe/ scanner system

Most of the interactions between the human and technology occur during probe management tasks. Sequence diagrams help in describing what the tasks actors (humans) have to accomplish. For instance, the operator has to prepare the patient, which requires such steps as positioning the patient, application of scan gel on patient's body, communicate information to the patient, monitor the patient, etc. Therefore, sequence diagram greatly aids in task representation.

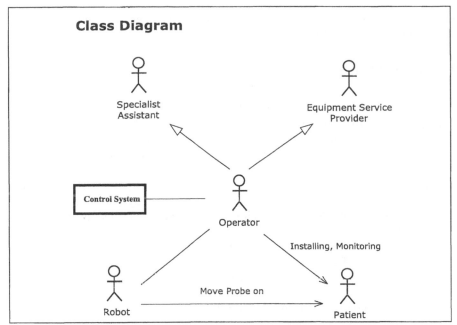

Figure 5: Class diagram: overall system view for tumor scanning robot

Human factors safety aspects and human systems integration requirements are a major concern during safety analyses of medical robotic systems. In this case study, SysML was used in order to provide consistent, accurate representations of information between engineers, doctors, human factors specialists and other professionals throughout the development process with a focus on requirements analysis. Throughout the human factors analysis major advantages and benefits of using SysML are apparent. The power of SysML lies in integrating safety concerns during system design and describing interfaces between systems and actors that are used to identify and analyze areas of concern regarding safety.

CONCLUSIONS

As an introductory contribution to the application of user-centered Systems Engineering (SE) and human factors processes for the design and development of smarter robotic medical systems, this paper provides a motivation and rationale for integrated user-centered human factors knowledge management approach to systems engineering using SysML. While a large number of disciplines and research fields must be integrated towards development and widespread use of smarter medical systems, considerable advancements achieved in these fields in recent years indicate that the adaptation of these results can lead to highly sophisticated yet widely useable smart systems. It is supported by the research that an integrated SE and human factors approach to design and modeling of smarter products should prove useful in supporting and facilitating the development and applications of these systems in the near future.

REFERENCES

Ahola, J., (2001) Ambient Intelligence. ERICM News, (47).

Ahram, T. Z., Karwowski, W., Amaba, B., Obeid, P. (2009). "Human Systems Integration: Development Based on SysML and the Rational Systems Platform", Proceedings of the 2009 Industrial Engineering Research Conference, Miami, FL. USA. (pp. 2333-2338)

Ahram, T. Z., Karwowski, W., (2009) "Measuring Human Systems Integration Return on Investment" The International Council on Systems Engineering – INCOSE Spring 09 Conference: Virginia Modeling, Analysis and Simulation Center (VMASC), Suffolk, VA. USA.

Allmendinger, G., and Lombreglia, R. (2005). Four Strategies for the Age of Smart Services. Harvard Business Review, 83(10):131-145.

Arts E. and de Ruyter B. (2009). New research perspectives on ambient intelligence. Journal of Ambient Intelligence and Smart Environments, 1:5-14.

Booch G., Rumbaugh J., and Jacobson I. (1999). Unified Modeling Language Users Guide. Addison Wesley Longman.

Das, S., Cook, D. (2006). Designing Smart Environments: A Paradigm Based on Learning and Prediction. In: Shorey, R., Ananda, A., Chan, M.C., Ooi, W.T. (eds.) Mobile, Wireless, and Sensor Networks: Technology, Applications, and Future Directions, pp. 337–358. Wiley, Chichester

Davies B. (1993). Safety of medical robots. ICAR'93, pages 311-313.

Defense Acquisition University (DAU) Guidebook (2004). Chapter 4: Systems Engineering.

Friedenthal, S. Moore, A. Steiner, R. (2008). A Practical Guide to SysML: The Systems Modeling Language, Morgan Kaufmann; Elsevier Science.

Guiochet J., Tondu B. and Baron C. (2003). Integration of UML in Human Factors Analysis for Safety of a medical robot for tele-echography, Proceedings of the Intl. IEEE Conference on intelligent Robots and Systems Las Vegas. Nevada.

Hitchins, D. K. (2007). Systems Engineering: A 21st Century Systems Methodology. Chichester, UK: John Wiley & Sons.

ISO 13407: Human-centered Design Processes for Interactive Systems. International Standards Organization, Geneva, 1999. Also available from the British Standards Institute, London.

Karwowski, W., Ahram, T. Z. (2009). Interactive Management of Human Factors Knowledge For Human Systems Integration Using Systems Modeling Language. Special Issue for Information Systems Management. Journal of Information Systems Management. Taylor and Francis.

Karwowski, W., Salvendy, G., Ahram, T. Z., (2009). Customer-centered Design of Service Organizations. In: G. Salvendy, W. Karwowski (eds). Introduction to Service Engineering, Chapter 9 (pp.179-206). John Wiley & Sons, NJ (ISBN-10: 0470382414).

Maass W. and Varshney U. (2008). Preface to the Focus Theme Section: 'Smart Products'. Electronic Markets, 18(3):211-215.

Mühlhäuser, M. (2008). Smart Products: An Introduction. In Constructing Ambient Intelligence - AMI 2007 Workshop, pages 154 - 164.

Sabou, M., Kantorovitch, J., Nikolov, A., Tokmakoff A., Zhou, X., and Motta, E., (2009). Position Paper on Realizing Smart Products: Challenges for Semantic Web Technologies, Report by Knowledge Media Institute : http://people.kmi.open.ac.uk/marta/papers/ssn2009.pdf

Tondu B. and Lopez P. (2000). Modeling and control of McKihhen artificial muscle robot actuators. IEEE Control Systems, 20(2): 15-38.

Weiser M. (1991). The computer of the 21st century. Scientific American, 265(3):66-75.

Chapter 40

Development and Pilot Test of a Usability Evaluation Protocol for an Ergonomic Device in a Hospital

*So TW Patrick, *Cheng WC Stella, *So HP, *Chan HS Alan

*Kowloon West Cluster
Hospital Authority

*Department of Manufacturing Engineering & Engineering Management
City University of Hong Kong

Hong Kong SAR, PR China

ABSTRACT

This report is about a pilot study of a usability evaluation protocol in a Hong Kong hospital setting. A very common patient care activity, known as the "spike-in procedure", which involved the insertion of a tubing spike into intravenous fluid bottles, was chosen for the study. Tailor-made ergonomic device to ease the spike-in procedure was designed with active participation of frontline nurses. A protocol for product evaluation was developed and pilot tested in various work units. The findings indicate that the musculoskeletal problems of the nurses and specific job demands in performing the procedure in different work settings are essential factors to consider in future usability evaluations. By refinement of the evaluation protocol, the provision of more user-centered devices may be facilitated through systematic methods of collecting essential information for product design, and this also serves as a major step to promote wellness at work.

Keywords: ergonomics, human factors, usability, comfort, safety

INTRODUCTION

BACKGROUND

An in-house Ergonomics Consultation Service has been introduced into the Kowloon West District of Hong Kong to promote the concept of wellness at work to more than 12000 staff working in a cluster of 7 hospitals and 19 out-patient clinics. There is a clear need for the development of special tools and devices to decrease the risk of workers suffering from musculoskeletal discomfort and/or disorders and to increase in safety while carrying out their duties. Ergonomic risk factors have often been associated with hand tool design and therefore this must be viewed as an essential focus of interest in any attempt to produce a general reduction in musculoskeletal discomfort and injuries (Li, 1998). Nursing staff often have to perform a wide variety of demanding physical tasks and were selected here as a major group for test purposes. The general aim was to develop low cost, user friendly tools and devices, to not only satisfy the frontline users, but also satisfy the budget concerns of senior management which is an important consideration in most circumstances.

As with any new system, there was a real need to justify the value of this in-house service, and therefore user opinions on ergonomic devices developed or to be developed were sought as necessary evidence of the value of the service. The consequence of this approach was that here a standardized protocol for product usability evaluation was developed and pilot tested in a real life situation. The work reported here resulted from a request from a hospital work unit that had a large proportion of the nursing staff suffering from musculoskeletal discomfort and/or disorders (MSD) in performing what is known as the "spike-in procedure". The spike-in procedure involved the insertion of the "administration tubing spike" into the intravenous solution port of a medication fluid bottle, and was considered by frontline nurses to be one of the most common patient care activities. The unit requested an ergonomic device to improve comfort and reduce MSD risks. This request was considered to be a good example to illustrate the nature of most service requests and was therefore selected as the test case for the newly developed protocol for usability evaluation.

The aim of this study was to design an ergonomic device to assist nursing staff to insert administration tubing spikes into intravenous solution ports and to evaluate the usability of such a device for nursing staff working in a variety of settings.

PROTOTYPE DESIGN AND USABILITY EVALUATION

The active involvement of frontline workers, with support from an ergonomics practitioner, in the development and implementation of the ergonomics improvement program constituted the core of this in-house service. The prototype design was developed with reference to guidelines by Sauer & Sonderegger (2009) which advocated consideration of functionality, interactions and breadth of features. A participatory approach in designing tools and devices aimed at reducing physical strain should employ expert ergonomics knowledge at all stages of the process (Devereux et al, 1998). Here, an ergonomics practitioner provided support in identification of worker needs, analysis of physical demands in work environments and, in the design and user-testing of tool and devices throughout the entire process of prototype development.

The International Standards Organization (ISO DIS 9421-11) defined usability as "... the effectiveness, efficiency and satisfaction with which specified users can achieve specified goals in particular environments" (Jordan, 2002). One framework for usability evaluation of a product is the four-component model which consisted of user, product, activity, and environment. This is a model which has long been accepted as incorporating the principal components of good human-machine system design (Shackel, 1984). More recent literature has added three more major factors that influence the usability of a product, and these are; guessability, learnability and experienced user performance (Jordan, 1994). Buurman (1997) further commented that guessability facilitated ease of learning. Evaluation techniques of such models have included observation, inquiry and empirical testing (Kwahk & Han, 2002). In this study, these essential concepts of usability have been integrated into the user opinion survey and the resulting parameters were rated during the process of prototype testing.

METHODOLOGY

The process of protocol development and pilot testing was conducted in 2 phases, which were:

I.   Prototype development
II. Usability evaluation by prospective users

I.   PROTOTYPE DEVELOPMENT

A focus group was formed which consisted of four nursing staff with more than five years of working experience from the work unit that requested help, one member of technical staff with extensive experience in fabrication of tools and devices and one Occupational Therapist with expertise in Ergonomics. They discussed specific user needs and preferences, the physical demands of the task concerned (i.e. spike-in procedure for 100 ml medical fluid bottle), essential product features and functions, and different design options. Based on the consensus of user opinions, consideration of technical feasibility and budgetary issues, the final version of the prototype was fabricated for trial in the real life situation by nurses of the focus group for a period of one month. Focus group members then discussed on the prototype design in order to make the final version of the device. The procedures involved were then adopted as part of the protocol for usability evaluation by prospective users of other work units.

II.   USABILITY EVALUATION BY PROSPECTIVE USERS

Subjects

Ten work units from different clinical specialties in one hospital voluntarily joined the study after attending a briefing session to introduce the rationale and potential benefits of the ergonomic device. At least three nursing staff from each of these units participated in the user testing process during their normal working hours and a total of 31 subjects were successfully recruited.

Procedure

The final version of prototype design was tested in each of the subjects' own work environments. The testing process was conducted on a one-to-one basis in an isolated area. The purpose of which was to prevent others from familiarizing themselves with the device prior to evaluation, because the first part of the evaluation involved estimating the guessability of this product. The second part of the evaluation demanded subjects to actually use the device once, and this was performed immediately after a standardized training session provided by the investigator. The subjects were then asked to fill in the survey in order to complete the entire process for this usability study.

Data collection

The first part of data collection focused on guessability, which included the time required to successfully guess the way in which the device was used and the ease with which the guessing occurred. After the trial involving the use of the prototype, the subjects were required to rate their estimate of the extent to which the device was an improvement using an 11-point numeric scale covering the aspects of learnability, possibility in the reduction of injury or accident, efficiency, convenience of performing the work procedure and comfort. The importance of the device for reducing musculoskeletal discomfort or overuse syndrome and the overall user satisfaction with the device were two additional attributes considered to be useful information to be collected for this study. Subjects were also asked to give specific opinions on the design of the prototype in order to make it more suitable for use in their particular context.

FINDINGS

A total of thirty-one subjects were successfully recruited for this study: 87.1% were female and 77.4% were aged above 30. The majority of participants (74.2%) were registered nurses. 80.6% worked in an acute care setting, while others worked in extended and ambulatory care units. 84% had more than 10 years of experience in performing the spike-in procedure.

User opinions were expressed through an 11-point numeric scale, with 0 representing the lowest possible score and 10 the greatest possible score. The rating was further categorized as,

- Score 1 to 3 = low degree of improvement
- Score 4 to 6 = moderate degree of improvement
- Score 7 to 10 = high degree of improvement

The mean scores on the seven parameters for overall user feedback indicated that the device was perceived to be highly learnable (8.68), moderately helpful in reducing the possibility of injury or accident (5.61), to give a moderate level of improvement in comfort (4.71), a low degree of improvement in terms of efficiency (3.19) and low convenience in performing the spike-in procedure (3.39). Most subjects considered that the device was important in the reduction of musculoskeletal discomfort or overuse syndrome (5.03), and was satisfactory to use (5.58).

Analysis of Variance (ANOVA) and Kruskal-Wallis test were conducted to examine whether there were any statistically significant differences in user opinions. In terms of guessability, no significant differences were found between subjects with different experience in performing the spike-in procedure, in terms of ease of guessing how to use the device and the time required to guess how to use the device. In other words, participants who were more experienced in the spike-in procedure did not have an advantage in guessing how to use the device.

For the other parameters concerning user opinions, female staff gave significantly higher ratings than male staff on perceived improvement in convenience in performing the spike-in procedure ($H = 5.867$, $df = 1$, $p < 0.05$). The results of Kruskal-Wallis also demonstrated significant differences among subjects working in different types of settings on perceived improvement in convenience in performing the spike-in procedure ($H = 8.392$, $df = 2$, $p < 0.05$), the extent of improvement in comfort ($H = 6.487$, $df = 2$, $p < 0.05$), the level of importance of the device in the reduction of musculoskeletal discomfort or overuse syndrome ($H = 6.503$, $df = 2$, $p < 0.05$), and overall satisfaction with the device ($H = 6.838$, $df = 2$, $p < 0.05$). However, no difference was found in terms of overall satisfaction with the device among subjects having different previous experience in performing the spike-in procedure. In comparing the opinions of subjects reported to have musculoskeletal discomfort and those without musculoskeletal discomfort or overuse syndrome (MSD) (Table 1), statistics showed a significant difference in the improvement of efficiency ($H = 5.412$, $df = 1$, $p < 0.05$), convenience in performing the spike-in procedure ($H = 8.316$, $df = 1$, $p < 0.05$) and comfort ($H = 5.96$, $df = 1$,

$p < 0.05$). Major differences were also found between these two groups of subjects in terms of overall satisfaction with the device (H = 6.8, df = 1, $p < 0.05$) and the importance of the device in reducing musculoskeletal discomfort or overuse syndrome (H = 6.484, df = 1, $p < 0.05$), but there were no significant differences ($p > 0.05$) for the other two parameters (i.e. ease of learning and possibility in the reduction of injury or accident).

Spearman Rank Correlation analysis was conducted to investigate the interrelationships among all the parameters for user feedback. Correlations were found between overall user satisfaction with the device and the following parameters (Table 2):

* Possibility in the reduction of injury or accident
* Improvement in work efficiency
* Improvement in convenience in performing the spike-in procedure
* Improvement in comfort
* Importance in the reduction of musculo-skeletal discomfort or overuse syndrome (MSD)

Table 1. Comparison of subject feedback between the group with MSD and the group without MSD

		Learnable	Injury reduction	Efficiency	Convenient	Comfort	Satisfaction	Important to reduce MSD
Subjects with MSD	Mean	8	8	7.33	9	8.33	8.33	8.33
	SD	1.00	0.00	1.155	1.00	0.577	0.577	0.5774
Subjects without MSD	Mean	8.75	5.36	2.75	2.79	4.32	5.29	4.679
	SD	1.602	3.021	2.888	2.558	3.007	1.977	2.1612

Table 2. Spearman Rank Correlation Coefficients of subject feedback after actual trial. (n = 31)

	Satisfaction	Injury reduction	Efficiency	Convenient	Comfort	Important to reduce MSD
Satisfaction	1	.634**	.713**	.656**	.814**	.760**
Injury / Accident		1	.303	.212	.528**	.455*
Efficiency			1	.794**	.558**	.481**
Convenience				1	.625**	.509**
Comfort					1	.652**
Importance to reduce MSD						1

** Correlation is significant at the 0.001 level (2-tailed)
* Correlation is significant at the 0.05 level (2-tailed)

DISCUSSION & CONCLUSION

Although the ergonomic device (prototype) was developed with active inputs from nurses with vast experiences in spike-in procedure, feedbacks from prospective users still indicated diversifying needs in the functionality and features of product. 32.3% of participants requested to have a device to handle medical fluid packages of different design, and 9.7% subjects preferred the device to be fixed on a firm surface instead of a free-standing design. This type of device clearly suggested the need to involve larger number of nurses, working in units of different job demands, to join in the focus group during the design process so as to generate a prototype or prototypes able to be used in most work situations.

Results of Kruskal-Wallis test showed a significant difference in user feedback between participants with or without musculoskeletal discomfort or overuse syndrome and between those working in different types of units, in the aspects of convenience in performing the work procedure, comfort, importance in the reduction of musculoskeletal discomfort or overuse syndrome, and overall satisfaction with the device. It may be useful to modify the sampling method in future usability evaluations of ergonomic devices by recruiting subjects with musculoskeletal discomfort or disorders for isolated study. Also, in future, detailed analysis should be conducted on the job demands of the spike-in procedures in various settings so as to explore the kind of determinants (such as work cycles, nature of work surface, work pace, etc.) that may affect user opinions on the functionality and features of the product. The investigators here had previously thought that subject experience in performing the spike-in procedure might correlate with the ratings on guessability because experienced users would have greater chance of encountering similar products beforehand, also that their well-formed work habits might have affected their overall satisfaction with the device, however these relationships were not established in this study. Overall user satisfaction with the device was found to correlate highly with the extent of improvement in comfort ($r = 0.814$, $p < 0.001$) and with the perceived importance in the reduction of musculoskeletal discomfort or overuse syndrome ($r = 0.760$, $p < 0.001$). The investigators were acutely aware of the small sample size here and hence the tentative nature of the results which can serve as a base for building up evidence for use in future usability study.

With regard to the arrangements for trial use of the prototype, some participants expressed views that it was difficult to comment on usability with such a limited trial period. The investigators will therefore explore the feasibility of allowing prospective users to try the prototype in real life environments for extended periods of time so as to obtain more realistic feedback in forthcoming product evaluations.

The investigators gained valuable experience in conducting this usability evaluation for an ergonomic device in a hospital environment, and this was considered to be a pioneering attempt in the health care industry of Hong Kong. The tentative findings indicated potential areas for refining the evaluation protocol. With the focus on developing a systematic framework for collecting essential information for product design, more user-centered devices can be provided to hospital workers with special needs (e.g. workers suffering from musculoskeletal disorders or discomfort). This approach adheres to strongly held belief of the investigators in the importance of maintaining wellness at work through appropriate ergonomics intervention strategies.

Acknowledgement

The authors would like to thank for the data analysis partly conducted by Carrie Chan, City University of Hong Kong.

REFERENCES

Buurman, R.D. (1997), "User-centered design of smart products." *Ergonomics,* 40 (10), 1159–1169.

Devereux J., Buckle P., and Haisman M. (1998), "The evaluation of a hand-handle interface tool (HHIT) for reducing musculoskeletal discomfort associated with the manual handling of gas cylinders." *International Journal of Industrial Ergonomics,* 21(1), 23-34.

Jordan P.W. (1994), "What is usability?" In: Roberson, S. (Ed.), *Contemporary Ergonomics.* Taylor & Francis, London, 454-458.

Jordan P.W. (2002), *An introduction to usability,* Taylor & Francis Ltd – T.J. International Ltd, Padstow, UK, 5-12.

Kwahk J. and Han H.S. (2002), "A methodology for evaluating the usability of audiovisual consumer electronic products." *Applied Ergonomics,* 33, 419 – 431.

Li K.W. (1998), "Ergonomic design and evaluation of wire-tying hand tools." *International Journal of Industrial Ergonomics*, 30, 149–161.

Sauer J., and Sonderegger A. (2009), "The influence of prototype fidelity and aesthetics of design in usability tests: effects on user behaviour, subjective evaluation and emotion." *Applied Ergonomics,* 40, 670-677.

Shackel, B. (1984), "The concept of usability." In: Bennet, J., Case, D., Sandelin, J., and Smith, M. (Eds.), *Visual Display Terminals.* Prentice-Hall, Englewood Cliffs, NJ, 45–87.

Chapter 41

Design of Hand Operated Trowel: An Ergonomic Approach

Mahendra Singh Khidiya[1], Awadhesh Bhardwaj[2]

[1] Department of Mechanical Engineering
Maharana Pratap University of Agriculture and Technology
Udaipur, India

[2]Department of Mechanical Engineering
Malaviya National Institute of Technology (MNIT)
Jaipur, India

ABSTRACT

Hand tools are commonly used on Indian farms. In Indian agriculture, hand tools, animal-drawn equipment and tractor/power operated machinery are extensively used for various operations. These equipments are either operated or controlled by human workers. Research and development in many parts of the world have focused on methods for improving worker comfort and productivity. Tool design may play an important role in the development of work-related problems in the hand and forearm. By improving the ergonomic properties of hand tools the health of users and their job satisfaction might be positively affected In order to present ergonomically well-designed and comfortable hand tools, CATIA software is used and FEA analysis is made with ABAQUS and evaluation studies have been conducted with Subjective experiences of the subjects (to measure comfort or discomfort Corlett and Bishop technique is adopted) subjective measurements are preferred when evaluating hand tools on comfort and discomfort, as comfort and discomfort are subjective feelings. To achieve goal, a designed hand tool evaluation study was conducted and Comfort Questionnaire for hand tools (CQH) is used and evaluated the modified trowel.

Keywords: Trowel, Ergonomics, Agriculture, Hand Tool, Productivity, CATIA, ABAQUS

INTRODUCTION

Ergonomics should be seen as an opportunity to improve productivity and quality while increasing employee safety and morale. Benefits of ergonomics are higher productivity, higher quality, reduced operator injury, increased morale, greater job satisfaction, lower medical & insurance costs, reduced lost time, and lower absenteeism, less employee turnover.

The hand trowel also known as a hand hoe is most commonly used hand tool for weeding. The tool is used in squatting position. The hand trowel consists of a sharp, straight-edged metallic blade with a tang embedded into a wooden handle. The blade and a tang are forged in single piece to a shape from medium or high carbon steel. In some cases alloy steel (nickel, chromium or molybdenum or manganese) is also used for the fabrication of blade. The cutting edge is hardened and sharpened. The tang is joined to the wooden handle with the help of rivets. The shape and design of the hand trowel are region or location specific depending upon the soil and Cultural practices.

Figure 1: Hand Trowel
Courtesy: C.T.A.E. agriculture farm, Udaipur

For operation the hand trowel is held in one hand and pushed into the soil for removal of weeds or unwanted plants. The cutting or uprooting of the weed or undesired plant takes place due to shear and impact action of the blade of the hand trowel. The hand trowel is used for removing weeds and unwanted plants from the crop. The tool is also used for breaking the surface layer, aeration and mulching of the soil.

METHODOLOGY

    I.   Testing the existing trowel in the field & evaluation of comfort questionnaire

    II.   Modeling in CATIA of existing hand trowel

    III.   Analysis of existing hand trowel in ABAQUS

IV. Modifying the existing tool ergonomically according ABAQUS and anthropometric data.
V. Testing again in the field with modified trowel

DESIGNING AND ANALYSIS

The design of hand trowel related tasks which are performed by modern CAD system can be grouped into three functional areas:

- Geometric modeling
- Engineering analysis
- Designing review and evaluation
-

Geometric Modeling: In CAD, geometric modeling involves computer compatible description of the geometry of an object. The mathematical allows the image of the object to be displayed and manipulated on a graphical terminal through signals from the central processing unit of the CAD system. In geometric modeling, the designer construct the image of the object on the CRT screen of the interactive computer graphics system, by inputting three types of command to the computer.

Engineering Analysis: In the formulation of designing part of this project, some of analysis is required. The analysis may be stress-strain calculations. The computer can be used to assist in this work. The most powerful analysis feature of a CAD system is Finite Element Method (FEM) is used in such type of analysis.

Design Review and Evaluation: Checking the accuracy of design can be accomplished conveniently on the graphic terminal. Semiautomatic dimensioning and tolerance routines which assign size specifications to surface indicated by the user help in reducing the possibility of dimensioning errors. Animation helps in checking kinematic performance of like mechanism without resorting to pin board experiment.

EXPERIMENTAL DESIGNS

The subjects were asked to dig a hole of 1 feet depth and area of 1 square feet. Three different types of trowels were provided to volunteers. The soil was made moist by sprinkling water about 3 or 4 hours before the start of experiment. After informing the subjects about the study and completing a written informed consent, the subjects were asked to dig holes in the soil as quickly as possible without any rest breaks. The order of trowels was systematically varied among the subjects to avoid fatigue effects. After the subjects finished digging holes, the descriptors of the Comfort Questionnaire for Hand tools (CQH) were rated and—if necessary—the meaning of the descriptors was explained. At last, the subjects rated overall comfort. After a rest break of at least 5 min, the next digging-task started. This procedure was repeated for both trowels.

OVERALL DISCOMFORT RATING

Overall discomfort rating in Table 1 gives the capacity of the worker or subject to work and it is defined on a scale with a sliding pointer with graduations marked. The rate is important because with its help a workers body's overall capacity can be known during working and proper intervals during working can be given for rest.

Table 1: The Comfort Questionnaire for Hand Tools

Comfort descriptors						
Hand tool	Unsatisfied		Satisfied somewhat		Fully satisfied	
Fits the hand	1	2	3	4	5	6
Is efficient	1	2	3	4	5	6
Is simple in use	1	2	3	4	5	6
admirable transmits applied Force	1	2	3	4	5	6
Has a pleasant feeling handle	1	2	3	4	5	6
Offer a high task concert	1	2	3	4	5	6
Needs low hand grip force supply	1	2	3	4	5	6
Needs good function between hand and handle	1	2	3	4	5	6
Cause in inflamed skin of hand	1	2	3	4	5	6
Cause peak pressure on the hand	1	2	3	4	5	6
Feels sweaty	1	2	3	4	5	6
Cause lack of sensation in hand	1	2	3	4	5	6
Cause muscles cramping	1	2	3	4	5	6
professional Looks	1	2	3	4	5	6
Overall comfort						
This hand tool is	uncomfortable		Average		comfortable	
	1	2	3	4	5	6

To grade posture, the body parts focused on is back, neck, shoulder, elbow, hand/fingers, hip joint, knee and ankle. While force evaluation, the users have a table for assistance where measurements in kilos and Newton are represented. The following aspects are assessed:

- weight, lifted in standing or sitting position;
- Assembly force, exerted by fingers or hand;
- Grip opportunities and other heavy load handling.

386

The underlying factors of the hand tools are identified using principal components analysis (PCA). The relationships between comfort descriptors (i.e. statements in end-users' own words that are related to comfort) and comfort factors (i.e. groups of comfort descriptors) with comfort experience are calculated. It is concluded that the same factors (functionality, physical interaction adverse effects on skin and in soft tissues) underlie comfort in different kinds of hand tools, however their relative importance differed. Functionality and physical interaction are the most important factors of comfort in using hand trowel.

TOOL DESIGN IN CATIA

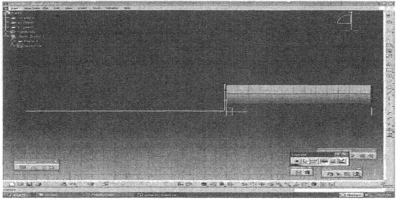

Figure 2
Solid part of tool generated in CATIAV5R18 in IGES
format for import in ABAQUS
Courtesy: MS Khidiya, Udaipur

ANALYSIS IN ABAQUS

Material properties are associated with part regions through the use of section properties. Property of solid section has defined that refers to the material created above and assign this section property to the part. Now the Mesh module in ABAQUS/CAE is used to generate the finite element mesh. The element shape and the element type values are inserted. ABAQUS/CAE offers a number of different meshing techniques. The default meshing technique assigned to the model is indicated by the color of the model when we enter the Mesh module; if ABAQUS/CAE displays the model in orange, it cannot be meshed without assistance from the user. ABAQUS/CAE opens the output database created by the job (**Job-1.odb**) and displays the un-deformed model shape. Visualization option is requested from dropdown menu to view the stressed part.

TESTING & EVALUATION

50 male agricultural subjects were randomly selected from different villages and it was assured that they had sound physical and mental health. For Testing of Existing Hand Operated Trowel, the Trowel was issued to the subjects and they had to work 35 minutes work followed by rest of 10 minutes and during rest they were asked to fill the CQH.

Figure 4: Existing trowel

The following procedure was used to test the existing hand operated trowel

MODIFICATIONS IN THE EXISTING TROWEL

The extra material was added at the bending of blade to remove stresses and to eliminate the vibration of tool due to repetitive work. Bend at blade in between blade has been provided according to anthropometric data and analysis in abaqus A firm grip designed according to local anthropometric data of local farmers for holding the tool is also provided. Bent handle are better than those with straight handles when the force is applied horizontally (in the same direction as workers straight forearm and wrist) (National Institute for Occupational Safety and Health) developed handle is Rounded, soft, and padded no sharp edges or deep grooves (reduces pressure points on fingers and hands) 55 mm in diameter and 90 mm long and High-friction surfaces or moldable substances may be added to handles to improve the grip

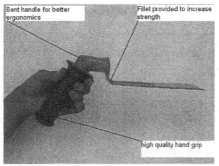

Figure 5: Improved Trowel   Courtesy. MS Khidiya, Udaipur

TESTING ON THE IMPROVED TROWEL

The following procedure was used to test the new trowel after this improved trowel was issued to the subjects and they had to work 35 minutes work followed by rest of 10 minutes. For measuring ODR they were asked to indicate a number on CQH. Determining Comfort Postural discomfort is the discomfort experienced by the subject because of muscular discomfort to maintain the body posture during the work. Discomfort is the body pain arising as a result of the working posture and/or the excessive stress on muscles due to the effort involved in the activity. In many situations, though the work may be well within the physiological limits, the body discomfort may restrict the duration of work depending upon the static load component involved in it and this is the case for most of the agricultural activities. For evaluating Comfort, CQH was used. In this technique the body was divided into 27 regions. The subject was asked to indicate the number on CQH for his work experience with modified tool and noted.

DETERMINING OVERALL COMFORT

Overall discomfort rating was taken on a ten point psychophysical scale (0=totally disagree with given CQH statement, 6= totally agree with given CQH statement) which is an adoption of Corlett and Bishop (1976) technique. A chart namely CQH was prepared including comfort descriptors. At the end of 35 minutes period the subject was asked to indicate the overall discomfort rating on this chart. The overall discomfort ratings given by each subject were added and scaled from one to six

	existing tool	modified tool	Percentage improvement
Sum	165	238	44.2%
Average	3.3	4.76	44.2%

Table 3: Analysis of existing and modified tool

Procedure:
The selected male subjects were asked to report to the field in the morning at 9 am. It was be ensured that they were in good health condition, have sound sleep in the previous night, and have a normal breakfast. They were free from stimulated beverages, cigarettes and recent exercise. Before the start of the operation subject was allowed to take rest for 15 minutes after that subject operated trowel.
Observations were taken for 35 minutes working and 10 minutes resting. After the operation, the subject was asked to indicate the rating on the CQH. This process was repeated for existing trowel and modified trowel.

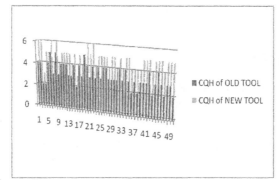

Figure 6 comparison of CQI I of an old and new tool

RESULTS AND DISCUSSION

From the above analysis it s observed that the maximum stress is developed at intersection of the planes of blades. Therefore to remove those stresses we have added fillet at intersection point in new improved tool. This imparts more strength & increases its durability. From the ergonomic point of view it is observed from field study that working with existing available hand tool is not comfortable. To provide comfort handle of hand trowel in bent in particular profile and a better quality of non slipping grip is used. Hand grip equipped with existing trowel was made of hard rubber which causes blisters on palm while working. This grip felt slippery on handle while working. But in improved tool the grip is provided it is made of durable & high quality rubber. This handle grip does not cause any blisters on palm while working. The grip surface is not slippery either. The handle bar of previous tool was straight as shown in figure 1 this was causing wrist strain due to bending of wrist. This reduced the workers comfort during long hour operations, therefore causing more fatigue. In improved trowel, handle bar was bent according to anthropometric data of local workers to ensure that wrist remains straight while working. This increased worker's comfort while working process.

CONCLUSIONS

The main objective of this paper has been to find a better way of working with less fatigue and better strength when using a trowel. For the better and easier utilization of the human power there is a need to develop a better working environment. In the process of making a better tool, the developed trowel has been shown to be better than the existing one in term of comfort level it is improved 44.2%. This paper indicates a way forward to improve the Trovel design and construction. Finally, is should be acknowledged that ergonomics does not offer an instantaneous process; rather it is a gradual process which must consider many factors. The present design as developed and explained in this paper may not be the ultimate solution but it is a step closer to a more satisfactory design which could offer improved working facility and fewer unwanted workplace musculoskeletal disorders and injuries.

REFERENCES

Amitabha De, Rabindranath Sen, R. Sen and A. De, 1992. A work measurement method for application in India. International Journal of Industrial Ergonomics. Vol.10 (4): 285- 292.

Ghugare, B.D, S.H. Adhaoo, L.P. Gite, A.C. Pandya, S.L Patel, 1991. Ergonomics evaluation of a lever-operated knapsack sprayer. Applied Ergonomics, Vol. 22(4): 241-250.

Gite, L. P. and D. Chatterjee, 2000. Proposed action plan on all Indian anthropometric survey of agricultural workers. AICRP on Human Engineering and Safety in Agricultural, CIAE Bhopal.

Kumar, V.J.F., C.D. Durairaj and V.M.Salokhe. 2000. Ergonomic evaluation of hand weeder operation using simulated actuary motion, Agricultural Engineering Journal. Vol. 9 (1): 41- 50.

Maegawa, H., H. Kiriyama, and T.Kurozumi, 2000. Ergonomical studies on redesign of working conditions in agriculture. Relationship between bed height, worker stature and working posture in strawberry culture. Bulletin of the Nara Agricultural Experiment Station (31): 1- 8.

Mohan D. and R.Patel 1992. Design of safer agricultural equipment: application of ergonomics and epidemiology. International Journal of Industrial Ergonomics Vol.10 (4): 301 – 309.

Shrawan K. and chengkung, C.1990. Spinal stress in simulated raking with various rake handles. Ergonomics. 33(1):1-11

Tewari, V.K. and S P. Geetha 2003. Occupational stress on Indian female agricultural workers. Proceeding 37th Convention of ISAE, FP – 10: 365-360.

Jafry and O'Neill, 2000 T. Jafry and D.H. O'Neill, The application of ergonomics in rural development: a review, *Applied Ergonomics* 31 (2000), pp. 263–268.

Khidiya M.S., Bhardwaj A., *"Study of Productivity and Ergonomic Application of Bent Hand Tools"* ,XXI National Convention of Agricultural Engineers & National Seminar on "Ergonomics and Safety management in Agricultural machinery and equipments" Organized by Institution of Engineers Udaipur,18-20 Jan 2008.

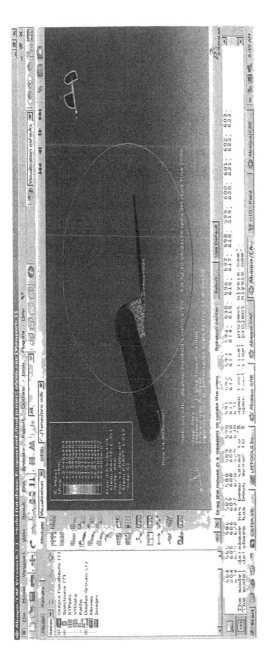

Figure 3
Deformed Shape with result
Courtesy: MS Khidiya, Udaipur

Table 2: CQH for old and new Hand

Subject no.	CQH of existing tool	CQH of modified tool	Subject no	CQH of existing tool	CQH of modified tool	Subject no	CQH of existing tool	CQH of modified tool
1	4	6	18	3	4	35	3	5
2	4	6	19	5	5	36	2	4
3	2	3	20	4	6	37	3	4
4	2	5	21	3	5	38	2	4
5	4	5	22	4	6	39	3	4
6	5	6	23	3	4	40	3	5
7	3	4	24	4	5	41	3	5
8	5	6	25	3	5	42	4	5
9	3	4	26	4	5	43	2	4
10	4	5	27	4	5	44	3	5
11	4	5	28	3	4	45	4	5
12	4	5	29	3	5	46	3	5
13	3	5	30	3	5	47	2	4
14	3	4	31	3	4	48	3	5
15	4	5	32	4	5	49	3	5
16	2	3	33	3	4	50	2	5
17	4	5	34	4	5			

Chapter 42

Modeling Precision Agriculture Device for Palm Oil Industry

Mohamad Fauzi Yahaya, Soo Li Choong

MIMOS Berhad,
Technology Park Malaysia
57000 Kuala Lumpur, Malaysia

ABSTRACT

The complexity of the palm oil industry, from planting, growth, harvest to the mill processing, has contributed significantly to its meager productivity. Weather, terrain, manual labor, palm oil species, age, pest and soil are some of the variables that large plantation companies have to endure, resulting in productivity fluctuations of about 22%. This paper reported the human factors considerations underlying the design and development of precision agriculture (PA) device for use in palm oil industry. The device is aimed to improve efficiency of triggering system at the pollination stage, whereby wireless sensor device and its decision support system can help to identify the most productive time for the pollination process. A fully pollinated flower will produce more fruits and a higher grade breed will fetch higher quality oil. A holistic user-centered design approach is employed in the design of the sensor, wireless transmission system as well as the data management system. The system is also designed so that they are operable indoor as well as outdoor and easily interpreted by professional agronomist and farmer. While the productivity improvement is not yet realized, the Human Factors approach of user-centric design guided by Ergonomics Quality in Design (EQUID) process has enabled a comprehensive design methodology to create a device and system for palm oil industry that is effective in application, efficient and easily operated.

Keywords: Palm Oil, Pollination, Wireless Sensor Network, User-centered Design, Data Management

INTRODUCTION

Malaysia is the biggest exporter of palm oil in the world. In 2009 the export value was worth US$19B. Currently, the plantation covers land area of 4.5 Million hectares and most of the activities on the plantation are carried out manually (Basiron, 2009). While there are some mechanizations already introduced in this agriculture industry, the challenges from the landscape alone is tremendous. The industry comprising of breeding, plantation and processing plant requires significant number of manual labor. The oil palm trees have height that ranges from 5 meters to above 10 meters and the pest range from a small bug to monkeys. The project has set the key success factors of achieving a higher pollination rate with improvement on the manual labor process of having to climb up on a regular basis. While the effect is minor, the impact is substantial in saving time of inspection through a more scientifically accurate prediction. Currently manual estimation by experienced climber determines the suitable timing for pollination. In this report, we present an initiative to implement wireless sensor network system on the breeding of oil palm plantation. The objective is to introduce semi automation in the pollination process through an intelligent triggering system. This system is important to ensure a good breed is produced with potential of high quality fruits. The system architecture is shown in Figure 1 below.

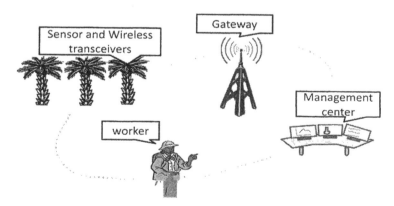

FIGURE 1 System Architecture of the wireless network device

In this system, the sensors which are placed on the trees will detect the suitable time for pollination process and relay the information through wireless communication via the gateway to the command center. An information management system will alert the worker to execute the pollination process within allocated time window. This intelligent system has saved significant time and resource requirement as compared to the regular repeated manual inspection that was carried out to determine optimum condition for pollination process. Since the products are designed for outdoor robustness, there are several Human Factor requirements that have to be considered to facilitate implementation. The systemic approach of Ergonomics Quality in Design (EQUID) process in the product development cycles was applied as demonstrated in the product evolution.

ERGONOMICS QUALITY IN DESIGN (EQUID) PROCESS - A SYSTEMS APPROACH

At the preliminary stage of EQUID process, initial technical specifications for the product were established by translating the user needs into technical requirements. The method of translating customer needs into technical requirements is out of the scope of this paper and therefore will not be discussed. Figure 2 illustrates the product development process in relation to EQUID.

Product Development Process and EQUID

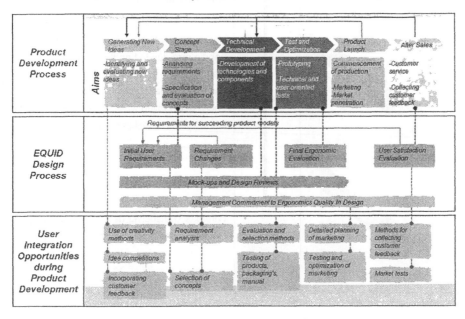

FIGURE 2 Product Development Process and EQUID. (Source: IEA, 2009; Khalid, 2009)

On the concept generation stage, several concepts were generated and some prototypes were evaluated and reviewed. At this point, using EQUID as a framework, the user experience considerations were checked based on a customized list of requirements. This checklist supplements conventional Design Review process for reviewing the form fit and function critical areas. Ergonomics reviews are conducted at the design stage. This design process is an iterative process until satisfactory results are obtained and all technical requirements are met (see Table 1). Final ergonomics evaluation was carried out during pilot testing at actual plantation site. The results of this usability testing are discussed at the end of this paper.

Table 1: Product Evolution of PA Device

Product Version	Design Direction	Analysis / Limitations
Version 1		Screw type interconnect constrained quick connectivity interconnection. Users spent longer time during maintenance. Smaller battery capacity shortened the product operating time. Thus increased the number of maintenance to be performed.
Version 2		Users were not comfortable with the bigger handle size due to handling discomfort during installation. It was not a pleasurable / affective design. The product failed water ingression protection test. Product redesign required.
Version 3		This is the final version of the Sensor and Wireless Transceiver device. The usability test was carried out on this design version.

HUMAN FACTORS APPLICATION ON THE PHYSICAL DEVICE

The product was designed with human factors considerations to meet the user needs in, handling, installation, robustness against pest animals and robustness in the outdoor tropical weather environment. The user is not the only factor because we also need to consider the animal factor as well. In this case we need to make sure that the animals will stay away from the device.

INSTALLATION AND MAINTENANCE

There are a few locations where the devices need to be installed.

- On top of the tree
- Within the fruit bunch
- On a transmitting tower

These are all high places that are not consistent because sometimes there are unexpected host on top of the tree such as snakes. In preparation for safety, the units need to be installed quickly. These user requirements are translated into lightweight and small size requirements, which have already been designed in the current device. For the installation we also need a special bracket that can improve the installation single handedly.

Easy maintenance is a common requirement for any electronic device. This includes battery replacement, faulty module repair and calibration. In order to address these issues, the device was separated into a few modules that are easily connected and disconnected. The design was changed from screwing type interconnect to latching type interconnect for quick connect and easy maintenance (Figure 3). Effective power management made the battery last longer.

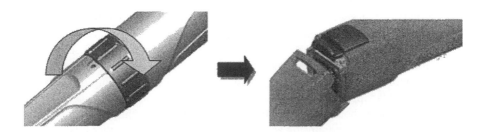

FIGURE 3 Design Changed From Screwing Type Interconnect to Latching Type Interconnect © 2009 MIMOS Berhad

PEST AND ANIMAL PROOF

Given that the oil palm plantation is a harsh environment with wild insects and animals abound, there is a high risk of animals such as monkeys that tend to damage the device while the devices are installed on the trees and left in the field. To overcome this problem, the plastic casings are molded in blood-red color. From our survey, blood-red acts as a visual deterrent that keeps monkeys away.

Besides monkeys, the device was designed to be dust proof which also prevented wild insects from making their way into the device and damage the internal components.

OUTDOOR ENVIRONMENT ROBUSTNESS

Oil palm plantation requires tropical climate with high humidity and heavy rain falls. Therefore the product will need to be able to survive high exposure to sunlight as well as heavy rain. In some cases, the sensor is used to measure soil properties. To ensure the product survives this condition, the device need to be chemical proof due to fertilization. Casing material selected requires strong resistant to this harsh environment.

In actual application, the device will be regularly moved to a new fruit bunch after a breeding cycle of one fruit bunch is completed because not all trees will pollinate concurrently. Since there is potential for slip and fall during installation on top of the trees, the entire system needs to survive stringent drop test requirement. This is one of the driving factors to ensure the product is lightweight and sturdy.

DATA MANAGEMENT CENTER GRAPHICAL USER INTERFACE (GUI)

The sensor collects data and sends to the management center for compilation. Through intelligent data processing, the command center alert workers on suitable time to climb up the tree and pollinate the flowers. Following Nielsen's Heuristics, the user interface design was created guided by the Ten Usability Guidelines (Nielsen, 1994). A well-designed screen keeps the window and icon designs clear and simple (Jansen, 1998). The details of each sensor node are stored in secondary levels. Information displays are designed using black text on white background to eliminate unnecessary reading difficulty even for people with viewing disabilities. (Web Quality-Readability, 1999).

Main navigation board is aligned on the left panel of the window, which grouped the major data monitoring activities into four groups. Another GUI design rule of thumb is that the range of options or choices should never be more than five or six (Jansen, 1998). For collapsible pull down menus, the group titles were highlighted in bold letters. This is in accordance with Nielsen's 'Recognition rather than recall' guideline to make objects, actions, and options visible. The user should not have to remember information from one part of the dialogue to another. Instructions for use of the system should be visible or easily retrievable whenever appropriate (Nielsen, 1994). GUI design is illustrated in Figure 4 and Figure 5 below.

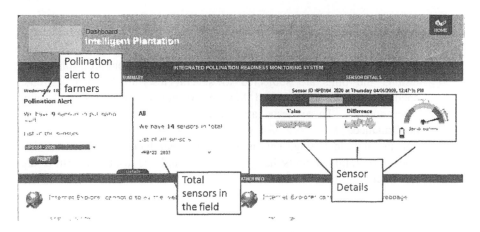

FIGURE 4 PA Device Pollination Alert Dashboard GUI Design © 2009 MIMOS Berhad

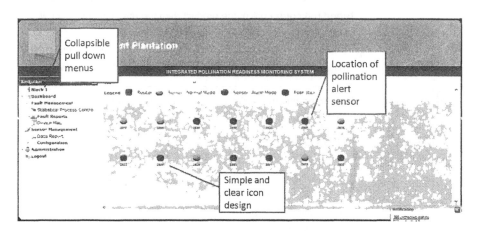

FIGURE 5 PA Device Location Tracking Dashboard GUI Design © 2009 MIMOS Berhad

USABILITY TESTING

We have conducted two types of usability testing:

- Extreme condition robustness testing. In the extreme condition robustness testing, the units are subjected to several tests to simulate extreme conditions such as high and low temperatures, vibrations, and salty environments. These tests are accelerated to predict product life performance.
- Functional field testing. This is a pilot test conducted at plantation site. There are total 14 units had been tested and the usability results are presented in Table 2.

Table 2: Usability Test Results

Usability Test	Results
Effectiveness	**Blood red color** - 0% of the product damage due to monkeys or wild animal pests was observed. **Installation** – Slow installation rate was observed. An average of 20 minutes was spent on one unit installation on the oil palm tree.
Efficiency	**Data transmission** – 50% of the products had intermittent issues during data transmission.
Satisfaction	**Easy to hold** – 52.63% of the survey (n=19) indicated they were very satisfied (score=4) on a 5 point-scale where 1 is dissatisfied, and 5 highly satisfied. 26.32% claimed they were highly satisfied, while 15.79% were moderately satisfied. **Product weight** – 42.11% of the survey (same sample size with the above survey item) claimed they were moderately satisfied (score=3) on a 5 point-scale where 1 is dissatisfied, and 5 highly satisfied. 31.58% claimed they were highly satisfied, while 21.05% were very satisfied. **Blood red color** – All users (100%) were highly satisfied on the product visual detection on the field.

KEY FINDINGS

Designing intelligent monitoring devices for agriculture requires greater understanding of the entire system. There is a need for more extensive surveys to complete the user centric requirements. Sometimes the users themselves have unknowns that can only be discovered after several pilot tests. Our pilot testing results have shown some challenges that need to be addressed in the future. This includes inherent properties of the plants which reduce wireless transmission capabilities, varieties of species requires different sets of constraint, age of trees affects the physical dimension of accessories, and so forth. However, the intelligent system has shown that the manual process of climbing the trees to verify optimum condition was greatly reduced because they now need to climb only once when the actual pollination process is required. This has significantly improved employee's productivity whereby their labor force can be channeled towards other activities such as fruit collections.

The ease of installation needs to be improved due to the mounting difficulty. There are various sizes of branches on palm oil trees and they also have thorns. Mounting of sensors also affects data accuracy and this consumed significant amount of setting time.

We also received feedback that the device looks fashionable even for agriculture. Our intention of incorporating affective design to attract users was slightly affected because the workers feel the device is quite fragile. They tend to associate the device with their personal devices such as cell phones. This could be beneficial to the company where there will be lesser breakdowns due to negligent handling.

Overall, there was high satisfaction on the product physical features such as easy to hold, product weight and product color. The project goal of reducing dependency on labor force can be achieved when the intelligent monitoring system is expended on other applications other than breeding.

CONCLUSION

Agriculture is a complex environment. There are so many variables which are difficult to control. In this project we have attempted to apply human factor and ergonomics in both the development process and product design.

In the development process, we adopted EQUID to complement existing development life cycle process. Basically, the user centric needs are identified very early during conceptual stage and audited as the project progressed.

In the product design, we incorporated critical features to satisfy the technical requirements identified.

- Handling, installation and maintenance.
- Design robustness to ensure product is animal and pest proof.
- Design robustness to withstand harsh climate.
- A good user interface design to manage the intelligent monitoring system.

The project success is on the consistency of the intelligent monitoring system to produce high quality breed as well labor reduction. There are several sequential advantages to be realized in the future. A well designed product will have to meet customer needs and Human Factors approach enables detailed requirements to be identified.

REFERENCES

About Palm Oil – Palm Oil Industry (website):
http://www.mpoc.org.my/Malaysian_Palm_Oil_Industry.aspx
Basiron, Y. (2009), *Trends and Potential of Malaysia's Plantation sector,* Malaysian Palm Oil Board.
IEA (2009), *EQUID. Communication at the EQUID Technical Committee Meeting,* IEA 2009 Congress, Beijing, China.
Jansen, B.J. (1998), *The Graphical User Interface: An Introduction,* Bernard J. Jansen Computer Science Program University of Maryland (Asian Division) Seoul.
Khalid, H.M. (2009), *EQUID. Communication at the Human Factors in Product Design course,* Damai Sciences, Kuala Lumpur.
Lim, C.K., and Foong, K.S. (Oct 2009), *Web-Based Management Software For Wireless Sensor Networks and Decision Support in Precision Agriculture,*Proceedings of the 3rd Asian Conference on Precision Agriculture (ACPA), Beijing, China.
Malaysia Weather – Malaysia Weather in General (website):
http://www.malaysiavacationguide.com/malaysiaweather.html
Nielsen, J. (1994), *Usability Engineering,* San Francisco, CA: Morgan Kaufmann
Noor, N.M. *Sustainable Agriculture System in Malaysia: The Role of Precision Agriculture in Sustaining Agricultural Practices among the Youth in Malaysia,* University Putra Malaysia.
Rasher, M. *The use of GPS and mobile mapping for decision-based precision agriculture,* National Cartography & Geospatial centre.
Wahid, M.B. (2010), *Overview of the Malaysian Oil Palm Industry 2009,* Malaysian Palm Oil Board.
Web Quality – Readability (1999) (website):
http://www.w3schools.com/quality/quality_readability.asp

CHAPTER 43

An Expert System for Risk Assessment of Work-Related Musculo-Skeletal Disorders

Sonja Pavlovic-Veselinovic[1], Alan Hedge[2], Miroljub Grozdanovic[1]

1 Faculty of Occupational Safety
University of Nis
Serbia

2 Cornell University
Dept. Design & Environmental Analysis
Ithaca, NY 14850, USA

ABSTRACT

This paper describes a menu-driven ergonomic expert system, SONEX, that is based on knowledge of work-related musculoskeletal risk factors and expert opinion. The software is easy to use, by experts and layman. Risk factors are divided into two main knowledge bases and four additional bases, and each of them is a separate expert system. The SONEX system has a large and diverse base of around 150 questions and knowledge base that contains over 230 factors, and around 500 possible answers. The SONEX system has been field validated by analyzing different jobs using traditional ergonomic methods and then comparing the outcomes with a SONEX analysis of the same jobs.

Keywords: Expert system, Work-related musculoskeletal disorders, Job evaluation, Ergonomic job design, Risk assessment

INTRODUCTION

During the past twenty years work-related musculo-skeletal disorders (WRMSDs) have become a major problem worldwide and both the economic and human costs are immense (Broberg, 1997; Morse et al. 1998; Pavlovic, 2001; Pavlovic et al., 2005). There are many different WRMSDs varying in the specific symptoms and their body locations (Kuorinka and Forcier, 1995). The National Institute for Occupational Safety and Health has defined WRMSDs as those diseases and injuries affecting the musculo-skeletal, peripheral nervous, and neurovascular systems that are caused or aggravated by occupational exposure to ergonomic hazards (NIOSH, 1997). Ergonomic hazards refer to physical stressors and workplace conditions that pose a risk of injury or illness to the musculo-skeletal system of the worker. Ergonomic injury risks include repetitive motions, forceful motions, vibration, temperature extremes (especially cold conditions), awkward work posture caused by the inadequate design of workstations, tools or other work equipment, and by improper work methods. The effects of ergonomic injury risks may be amplified by organizational factors such as shift work, work pace, imbalanced work-rest ratios, demanding work standards, lack of task variety, etc.

In the last 20 years considerable worldwide effort and research funding has been devoted to finding appropriate strategies for analyzing and preventing WRMSDs. There is general scientific agreement that ergonomics plays a decisive role in these efforts, and a failure to adhere to ergonomic principles of work design are believed to be the leading factors in the development of WRMSDs .

Ergonomic interventions are recognized as the preferred engineering approach and technique for preventing WRMSDs in the work place. Interventions must be based on a correct assessment of the ergonomic risks in the work. There are many and various risk factors that contribute to the development of WRMSDs (Keyserling et al., 1991; Winkel, 1992; Sauter and Swanson, 1996), and each of these must be thoroughly investigated. WRMSDs prevention and identification involves many different experts such as ergonomists, occupational physicians, safety engineers, occupational psychologists (Pavlovic-Veselinovic, 2007). It seems evident that no single person can master all of the knowledge needed to identify and prevent WRMSDs. Therefore there is a need for a tool that will systematically identify the ergonomic risk factors that might lead to development of WRMSDs, in their early stages of development, and that can easily be used by ergonomists or other health and safety professionals to analyze jobs.

Many paper-and-pencil ergonomic tools have been developed to help with the analysis of specific risk factors, for example, several of these tools focus on analyzing inappropriate work postures, such as the Ovacko Work Analysis System (OWAS), the Rapid Upper Limb Assessment (RULA) or Rapid Entire Body Assessment (REBA), (see Stanton et al, 2005), but these fail to incorporate other risk factors. Other tools focus on the symptoms being experienced by the worker, such as the NIOSH body discomfort survey, but they fail to analyze postural or other work components. Previous developments of ergonomic expert systems have often relied on mainframe computers or has been very limited in the coverage of ergonomic risks.

THE SONEX SYSTEM

A new expert system (SONEX), which is a menu-driven program that is based on the WRMSDs knowledge of many different experts, has been developed (Pavlovic-Veselinovic, 2007). The aim of SONEX is to assist both, occupational health and safety professionals, and employees who are at risk of WRMSDs, in the identification, assessment and control of ergonomic risks.

The SONEX system has been designed so that it is easy to use, both by experts and also by the layman-worker. The general principles of software architecture for expert systems were followed during software development. The result of this is that the system consists of two databases: a knowledge database and a database of decision-making rules, derived from analysis of the at-large-ergonomics and related literature, from the extraction of knowledge from numerous experts, and from the authors' own experiences in the practical assessment of ergonomic risks.

The selected method of programming operated as a decision tree, with the factors in the basis/bottom of the tree, which when placed inside a set of "rules", established by the expert/experts give some conclusion/result. The software has been designed so that the bases are open (of course, with the protected levels of access), so the user can change the predefined facts related to the rules of deduction (conclusions), without adding the changes to the software package, i,e. without changing any "source code". The software has a number of decision-making factors, presented from left to right on the screen, where a user (if they have permission to access) can optimize and customize the rules. The system allows the user to define which question will represents which factor, and which possible values may occur as a response to the question, and it then defines all possible outputs that could be a result of an analysis of the given answers. The reason for this combined approach is that the aim of this research was not the realization of a general level expert system which would be applicable to other areas too, but the development of the current software was focused on providing decision support in ergonomic risk assessment and the resulting software operates (and gives the results) as an expert system.

For the knowledge base a thorough review of all of the relevant research literature on the risk factors for WRMSDs was undertaken. Many risk factors were identified and these are grouped into two main knowledge base modules (DISCOMFORT and ERGONOMIC RISK FACTORS), along with four additional specific modules (WORKING ENVIRONMENT, WORK CHAIR, TOOL, ORGANIZATIONAL FACTORS), and each of these modules is presented separately in the expert system. Each knowledge base module can be accessed individually using SONEX, because each of these modules itself is a unique single expert system. The main screen of SONEX system is shown in figure1.

It is necessary to point out that in the current version all risk factors are treated as risk factors with the same importance/weight (W = 1.0). A block diagram of the SONEX system is shown in the figure 2.

Figure 1. The appearance of the initial screen of the SONEX system (Note that the software is written in Serbian)

The menu-based interface leads the user, with simple questions, through each of the 2 main modules, and then it recommends other knowledge base modules that may be helpful for more detailed analysis. The SONEX system has a large and diverse base of around 150 questions and knowledge base contains over 230 factors, and around 500 possible answers. The architecture of the system allows this to be expanded as new knowledge becomes available. The intention in developing this software package is not to diminish the importance of ergonomic experts nor to replace the appropriate diagnostic procedure where this is required, but to augment the decision making capabilities of the user.

If a certain workstation is being analyzed by some expert professional, or by the employee's who is feeling some discomfort in part of their body, it is useful for them to run the DISCOMFORT module (see figure 3).

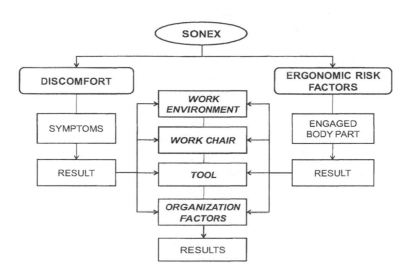

Figure 2. Block diagram of the SONEX expert system.

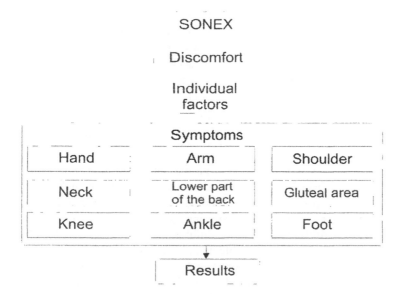

Figure 3. Block diagram of the submodule DISCOMFORT.

The rule-based approach that is used in the software is illustrated by the following example from the submodule "Gluteal area" (part of the DISCOMFORT module). This submodule comprises 5 questions, and the table 1 shows the different combinations and resulting conclusions which are defined in the following way:

0 - The question is answered with NO,
1 - The question is answered with YES,
X - The question does not affect the final conclusion

If, for example, a series of answers is 011X1 (combination No. 4), this means that on question No. 1 the answer is NO, and for questions 2, 3 and 5 have the answers that are YES, and question No. 4 does not affect on the final conclusion. This illustration of the coding principle for SONEX applies to all other factors and conclusions, so that the SONEX system has a knowledge base of over 230 factors, and over 500 possible answers/results.

Table 1. Possible combinations of answers to the questions and possible results

Number of Combination	Questions	Conclusion
0	00XXX	No problems in this part of the body
1	010X0	Muscle strain of lumbar spine
2	010X1	Osteoarthritis
3	011X0	Muscle strain of lumbar spine
4	011X1	Osteoarthritis
5	1X00X	Ischiogluteal bursitis
6	1X010	Muscle strain of lumbar spine
7	1X011	Ischiogluteal bursitis
8	1X10X	Ischiogluteal bursitis
9	1X110	Muscle strain of lumbar spine

After the selection of the body part with the most discomfort, the software asks the user a series of questions. The software allows the selection and analysis of all parts of the body (9 parts – hand, arm, shoulder, neck, back, gluteal area/hip, knee, ankle and foot). This module also contains questions related to individual risk factors. At the end of the analysis of a painful part of the body, the SONEX system will finally give the result as a possible diagnosis, with a description of symptoms, along with a list of possible ergonomic and organizational risk factors that may influence the development of a disorder (see the example shown in figure 4). The software also suggests additional knowledge base/modules for further analysis.

The final results from each module are appropriately color-coded in accordance with the principles of ergonomic design, given in the European standard EN 614 (EN614-1, 1995).

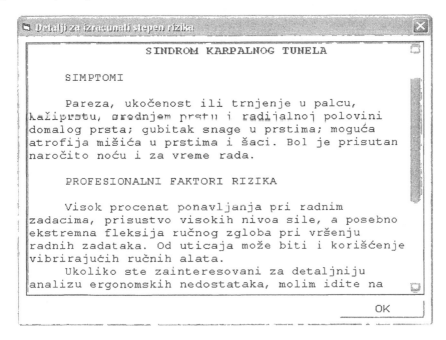

Figure 4. An example of the final result of the analysis of hand discomfort (Note that the software is written in Serbian).

The purpose of the ERGONOMIC RISK FACTORS knowledge base/module is the analysis of work tasks in order to assess the risk factors (ergonomic disadvantages) that these present. This is the first module that most technical personnel will start with for their analysis/assessment of WRMSDs risks, and it is recommended that the assessment includes input from an employee for this part of the analysis because worker awareness and participation are key success factors for the correct identification, reduction of potential risks and implementation of solutions.

SONEX can also be use as a tool to promote participatory ergonomics. In this module there are questions that allow the presence of risk factors such as repetition, force, vibration, static or awkward positions and low environmental temperature, to be identified.

The results are given in the form of work/ergonomic factors and the possible affected body part and WRMSDs which would be the result of exposure to these factors. As well as in the discomfort module, the results of the ergonomic risk factors module also suggest other extra modules which are useful to run for a more in depth analysis.

The architecture of the SONEX system allows it to easily be to include updated and expanded as new findings on ergonomic risk factors emerge. The system can also be translated into other languages. Future testing and development of the system for analyzing a wider variety of different types of work is planned.

VALIDATING THE SONEX SYSTEM

The validity of the SONEX system allows has been tested by comparing the results from conventional ergonomic analyses of a variety of jobs with those obtained from SONEX. The testing of SONEX capabilities to identify the ergonomic shortcomings that contribute to the development of the WRMSDs, as well as its ability to predict the occurrence of a musculo-skeletal disorder that may develop as a consequence of the observed shortcomings in the job, has been undertaken for several different jobs. The jobs were first analyzed with standard ergonomic methods (observational method, task analysis, anthropometrics and goniometrics measurements, Nordic questionnaires for analysis of musculoskeletal symptoms, interviews) to determine all of the problems that were present. The same jobs were also analyzed using SONEX and the results obtained from the software package demonstrated its ability to predict, with high accuracy, the nature and location of WRMSDs for different jobs, as well as the correct diagnosis of some of the WRMSDs, on the basis of symptoms which employees experienced and described. The results showed very good agreement between the two approaches; however, using the SONEX system was much quicker and easier for the analysis of these jobs.

CONCLUSIONS

The SONEX system has been developed as an easy to use and quick, yet comprehensive, ergonomics expert system. Testing of this system by comparing the results from the software with those from conventional ergonomic analysis methods has confirmed the accuracy of the system and the enormous speed advantage in conducting an expert evaluation.

In addition to analyzing the work situation SONEX also gives suggestions for ergonomic improvements, or corrections of the problems that have been identified. This means that SONEX can be used as a diagnostic tool, and for the ergonomic analysis of the workplace, and that the software can be offered to different users as a tool that will enable early detection and prevention of a number of different WRMSDs.

The architecture of the SONEX system allows it to be easily updated and expanded as new findings on ergonomic risk factors emerge. The software architecture also facilitates translation into different languages. Future testing and development of the system for analyzing a wider variety of different types of work is planned.

ACKNOWLEDGMENTS

This paper is written as part of project TR 21030, financed by the Ministry of Science and Technological Development of the Republic of Serbia.

REFERENCES

Broberg E. (1997), Official statistics of Sweden, National board of occupational safety and health, *Statistics Sweden.*

EN 614-1 (1995), *Safety of machinery. Ergonomic design principles. Terminology and general principles,* European Commitee for Standardisation, Brussels.

Keyserling W., Armstrong TJ., Punnett L., (1991), *Ergonomic job analysis: A structured approach for identifying risk factors associated with overexertion injuries and disorders*, Appl Occup Environ Hyg, 6(5), Taylor&Francis, pp. 353-363.

Kuorinka I., Forcier, L. (1995), *Work related musculoskeletal disorders: A reference book for prevention*, Taylor&Francis, London.

Morse T,. Dilon C,. Warren N., Levenstein C., Warren A., (1998), *The economic and social consequences of work-related musculo skeletal disorders; The Connecticut upper-extremity surveillance project*, Int J Occup Environ Health, vol. 4, pp. 209-216.

NIOSH (1997), *Musculoskeletal disorders and workplace factors. A critical review of epidemiologic evidence for work-related musculoskeletal disorders of the neck, upper extremity, and low back.* NIOSH Publication no. 97-141.

Pavlovic S. (2001), Work-related musculo-skeletal disorders, *SIE 2001, III International Symposium of Industrial Engineering*, Belgrade,149-151.

Pavlovic S., Grozdanović M., Grozdanović D. (2005), Prevalence of work related musculo-skeletal disorders, *IHCI 2005*, (on CD), Las Vegas, Nevada.

Pavlovic-Veselinovic S. (2007), An expert system for ergonomic risk assessment, *Proceeding "Ergonomics 2007"*, Belgrade, 105-110.

Pavlovic-Veselinovic S. (2007), Work-related musculo-skeletal disorders risk factors, *Proceeding "Ergonomics 2007"*, Belgrade, 94-98.

Sauter SL., Swanson NG, (1996), Psychological aspects of musculoskeletal disorders in office work, Psychosocial factors and musculoskeletal disorders, Taylor&Francis, London.

Stanton, N., Hedge, A., Brookhuis, K. Salas, E. and Hendrick, H. (2005), *Handbook of human factors and ergonomics methods,* Boca Raton, CRC Press.

Winkel, W. (1992), Occupational and individual risk factors for shoulder-neck complains, *Int J Ind Ergon*, vol. 10, pp. 79-83.

Chapter 44

Towards More Humane Calendar Applications

Stine Klit Engemand, Simeon Keates

IT University of Copenhagen
Rued Langgaards Vej 7
2300 Copenhagen S, Denmark

ABSTRACT

This paper describes a study about how people currently use computer-based calendar applications and examines which features do not meet their needs and expectations. Survey data was collected from 88 participants based in 6 organisations using 3 different calendar applications. Detailed follow-up interviews and focus groups were conducted with 13 users. Based on the survey and follow-up interviews, 3 new prototype designs for calendar applications were developed and evaluated. Based on findings in the empirical data, suggestions and recommendations that will help make calendar applications more acceptable, usable and satisfactory are being developed.

Keywords: Calendars, usability, survey, user-centred design

INTRODUCTION

Computer-based calendar applications are principally used by people wishing to manage their time efficiently and effectively. Consequently, those applications themselves should function efficiently and effectively as well, to be acceptable to the target users. However, their overall acceptability is not just based on the quality of the user interface (UI), but also the organisational policies on their use (Grudin, 1996).

This paper describes a study into whether users find their current calendar applications to be acceptable and usable. Based on the results of that study, 3 new calendar user interface (UI) prototypes were developed and evaluated. This paper focuses on the initial user survey and the user evaluations of the 3 prototypes.

RESEARCH QUESTIONS

The overall aim of this paper is to understand which elements in use of calendar applications contribute to making users manage their time more efficiently and in a socially acceptable way.

To reach this understanding the following research questions were considered:

- How do people currently use their calendar applications and what do they think of them?
    a. What are their preferences, what is satisfactory and what causes problems?
- How can the core of the calendar application be improved and which features can be added to extend the application in new and novel ways to make it more acceptable, usable and satisfactory than the current calendar?
- What design options exist?
    a. Of the design options: which ideas, features and elements work/do not work, and which need improvement?
- Is it possible to make design recommendations for one overall calendar application?
    a. If so, what recommendations can be given?

BACKGROUND

The role of calendars has become increasingly important as a tool to help not only individuals, but also groups, manage their time. There are two principal concepts underlying the use of calendars: time and scheduling.

The notion of what constitutes "time" is a source of philosophical debate. However, in the modern capitalist sense, time is now regarded as a limited resource and thus has an intrinsic value attached to it. This is commonly summarised by the notion that "time is money." So, a person's time needs to be optimised by whatever criteria they consider to be important. This process of optimisation leads directly to the concept of "scheduling", i.e. locating an activity in time.

However, there has been a very important shift in the ownership of an individual's calendar in recent years with the advent of the groupware calendar. Whereas earlier versions of calendar applications were largely standalone and private (and thus under the exclusive control of the person owning the calendar), more modern software allows multiple users to access any particular person's calendar. This allows colleagues and managers to view a person's time and even schedule meetings based on their interpretation of when would be a good time for the notional owner of the calendar, rather than what the owner would necessarily prefer.

It is easy to see how this change of ownership raises potential areas of conflict, especially in light of the idea that time is a limited, and thus valuable, resource. This paper strives to examine how the design of calendar applications can be modified to minimise the potential upset for users by making them feel more in control of both their calendar and consequently also their time, thus improving their efficiency in both time and task management.

METHOD

An online survey was conducted consisting of 31 questions and hosted on Survey Monkey. The survey was deliberately wide-ranging and explorative to uncover as many possible views and opinions on the use of calendars. For example, the survey addressed, the interaction with instant messaging (IM) systems as part of an integrated approach to collaborative work – although it turned out not many respondents used them together.

88 responses were received from 6 different organisations, e.g. IBM and MAN, around Copenhagen. These organisations consisted of both commercial companies and not-for-profit organisations. The companies also used 3 different calendar applications: GroupWise, MS Outlook (2003 and 2007 versions) and Lotus Notes.

The aim of the survey was to understand how users in different organisations and at different levels of seniority (i.e. number of years of service with their organisation and also position within the organisational hierarchy) used calendar applications and whether they were satisfied with the overall usability of the particular piece of software that they used. If they were not satisfied, the intention was to see if the cause of their dissatisfaction could be identified. The overall acceptability of the use of the software, though, is not solely a function of the design of the respective UI. It is also a function of the overall corporate-level strategy adopted (i.e. what the organisation mandates how its employees are supposed to use the software) and also the level of adoption by colleagues of the respondents. The survey was designed to examine all of these issues.

Following the results of the survey, 13 respondents were invited to participate in either focus groups or one-on-one interviews to provide a more in-depth understanding of their survey responses. The respondents were selected to represent a cross-section of all of the organisations invited to participate in the survey and also each of the different calendar applications. Further, the in-depth study also involved asking the users to describe order of creating events and sketch their ideal calendar UI.

Following this, 10 respondents were also asked the scheduling questions (i.e. put in events into a largely completed calendar). The aim was to understand how people like to schedule their time, for example by spreading meetings out across a day or by preserving contiguous blocks of time as meeting-free.

The results from the survey and follow-up interviews/focus groups were analysed to develop a design specification. From this specification, 3 prototype calendar UIs were developed using MS PowerPoint based on those results and a separate usability analysis of GroupWise, Outlook and Notes. Each prototype was designed with a unique interaction style:

- Prototype 1: A derivative of the existing applications' UIs – this was chosen to see whether comparatively simple modifications to the existing UIs could significantly alter the users' perceptions of the applications. Figure 1 shows an example screenshot.
- Prototype 2: A primarily graphical (i.e. icon-based) UI – this was chosen because graphical, icon-based style UIs are often considered to be more "user-friendly". Figure 2 shows an example screenshot.
- Prototype 3: A purely text-based UI, framed around "interviews" – this was chosen to simulate how a personal assistant might try to ask a manager about when to schedule a meeting and thus be closer to a more "human" interaction. Figure 3 shows an example screenshot.

The new UIs were shown to 6 users who each represented users of the 3 calendar applications.

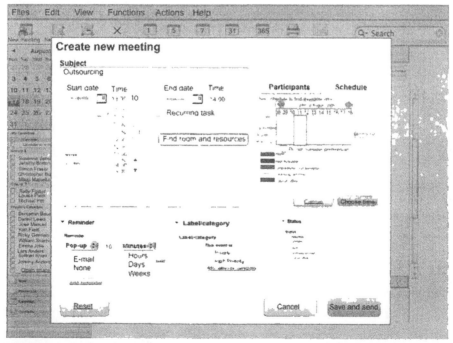

FIGURE 1. Setting the time of a new meeting for Prototype 1.

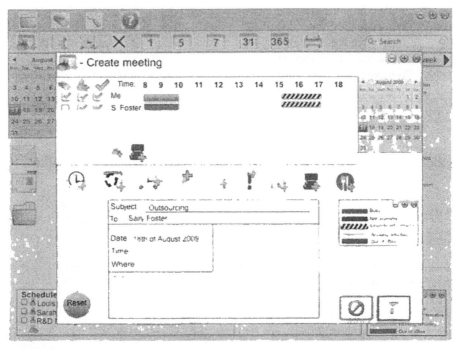

FIGURE 2. Setting the time of a new meeting for Prototype 2.

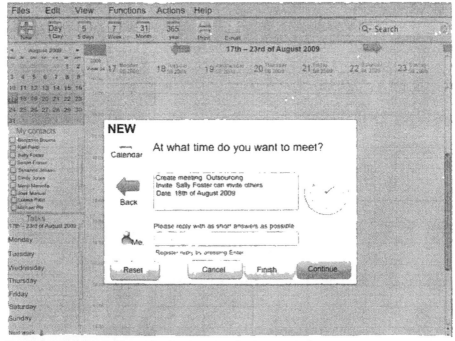

FIGURE 3. Setting the time of a new meeting for Prototype 3.

RESULTS

As discussed earlier in the paper, the results presented here will focus on the initial survey results and also the user evaluations of the three prototypes.

SURVEY RESULTS

Overall many of the respondents were generally satisfied with their current calendar applications. However, there were still a number of areas where improvement is possible. For example:

- The few users reporting as being "very satisfied" both typically planned and used more functions within the calendars, yet still experienced problems with the calendar.
- The areas of UI design, collaboration, hardware and organisational use proved to be of particular interest for possible improvement.
- Results from the survey showed confounding and interacting problems from multiple domains, which did not seem to be due to a simple cause and effect. For example, one organisation had issues with the adoption of a new calendar that were exacerbated because the calendar was also affected by a slow response time that coloured the perception of the system by the users. e.g. organisational issues with adoption and (in the same organisation) slow systems influenced the results for that particular organisation.
- The applications are often not suitable for long-term planning and overview of one's time.
- The users generally do not make a distinction between lack of differentiation between time management and task management, whereas most calendars enforce such a difference.
- The users have difficulty in understanding and learning the logic and structure of the calendar applications, especially Lotus Notes and GroupWise.
- There were technical issues with the hardware, e.g. it is too slow and it crashes for users at two of the organisations that have recently upgraded to Outlook 2007.
- The respondents considered that there were too many elements of the applications that were "predefined", but it was difficult to identify what they were specifically referring to. This may be symptomatic of a general lack of feeling of being in control.
- The users said that they could not plan everything and often they had their own methods of planning e.g. excel sheets and paper lists.
- There were specific problems with associated with creating recurring meetings and editing both meetings and recurring meetings.
- The users feel that they lack overview of other colleagues and their availability. However, this is due to the fact that not all users within their organisation use the calendar, some use it incorrectly and some do not

share their calendar. It is also an issue when scheduling a meeting with an external contact or client, since calendar availability is often restricted to within an organisation.

- The incorrect use of the calendar applications, etc., may be due to the fact that users have little training and most users do not know the corporate policies. This also results in users not knowing the functionalities of the calendar applications.

- The users regarded calendars as extremely valuable and important aids in their professional lives. They prefer being in control of their own calendar to be able to plan more efficiently. However, the users have often had little time to learn the systems. Most users had learned to use the calendar "sufficiently" and were basically "satisfied", yet they planned less and used fewer functions than might perhaps be expected by the developers of the applications.

- The layout and available functionality of the calendar applications change according to the context of use, e.g. home computer vs. office computer and/or on a PDA/smart-phone. This means that each new version must be learnt and there is a limitation of available features.

- Users have different preferences for meeting times and breaks between meetings. In addition, the users also have different strategies for scheduling their time.

- Users with a PDA schedule both their private and work time using the calendar applications, while others mostly plan work in their work calendar and personal life in, e.g., a paper calendar.

- Many respondents book their own time as "occupied" to ensure that no-one else tries to schedule a meeting in time that they wish to keep clear.

- The users wanted to control their own calendar and scheduled meetings just to block time. In addition, the users wanted to know the status of their colleagues and they did not have rights to edit in a scheduled meeting. This proves that users wanted more rights hence more power. This may challenge the organisational negotiation and also organisational power hierarchy, which might be there for a purpose i.e. managers vs. employees. This demonstrates the influence of IT on the organisational hierarchy and relations, which previous experimental studies seem to ignore.

- Monte Carlo simulations showed that there are many practical issues to consider if implementing the preferences for meetings and buffers in organisations (socially nice, but hard to implement).

Based on the results from the survey, it became clear that users are interested in setting up meetings based on more than simply whether their time is marked as free. One option for achieving this is the use of preferences to establish, for example, a minimum time between meetings.

USER EVALUATION RESULTS

- The users noticed the new functionalities but called attention to the fact that organisations must design policies so users know how to use the new functionalities e.g. setting preferences.
- The users did not read all options but instead suggested doing what they normally do.
- The order of testing of the prototypes does not play a significant role; the results were more influenced by the design of the prototypes and the tasks.
- The users took generally longer time at finishing tasks on prototype 3 and they seemed confused when doing the user trials. This might be due to the radical design, the new interaction style and the fact that users spend rather a lot of time on reading.
- The mean times on tasks on prototype 1 and 2 were almost the same. In addition, the T-test analysis proves that they are at no point statistically significantly different. This indicates that it is not possible to conclude that prototype 1 is faster than prototype 2. This is interesting in regards to prototype 2 being radically different from prototype 1, which is a derivative of the existing calendars.
- Most users said that they would prefer prototype 2 in the long term.
- Both qualitative and quantitative results were influenced by the fact that users were fastest and preferred what they were familiar with [as predicted by Nielsen, 1993]. However, this was somewhat limited by the many positive elements of prototype 2.
- The evaluation method for prototype 3 may not have been suitable, since the users seemed to have trouble with shifting their focus between the questions, the box with the overview and what to reply while at the same time getting instructions from the interviewer. Often the users did not read the questions and just clicked the text entry field to move forward.
- Prototype 3 was the least preferred, but 2 users found qualities in it. The positive elements of prototype 3 were that it appealed to users who preferred using the keyboard to navigate and write commands and users with repetitive strain injury. It also provided much more comprehensive guidance for novice users.
- When considering the new navigation on prototype 2 and the users' means value in performance in regards to the issue with familiarity, it is clear that it is possible to create a radical design, as long as elements that can satisfy the users in other ways are integrated e.g. large colourful targets, less text and fewer clicks that can make them more efficient in creating an event.
- The users were satisfied with functions that could speed up their use of the calendar e.g. a placement of elements that they were familiar with, "quick task", icons they knew instead of reading, larger targets, more reminders for one event, quick overview of elements in a created event with regards to invitees, resources, recurrence etc.
- The users understood the concept of user-defined meeting "preferences";

they saw the relevance and suggested improvements with working hours. Users comments during the evaluation and the Monte Carlo simulation called attention to corporate policies on the use of the preferences for buffer and schedule and rules in the algorithms to consider e.g. conflicting preferences. It also raises the issue of how users will react and make use of it in real situations.

- The designs may not just change the individual use, since the users have been given more freedom and control of their time. This may challenge the organisational negotiation and also organisational power hierarchy, which might be there for a purpose i.e. managers vs. employees. This shows the influence of IT on the organisational hierarchy and relations, but via these changes the users' needs have been considered and not other stakeholders.

CONCLUSIONS

Users are generally satisfied with the design of their calendar applications, but there is room for improvement. Following a detailed user study, 3 new UI prototypes were developed and evaluated. Overall, while the users accepted that both of the radically new UIs (text-based and graphical) offered distinct improvements in certain areas over the more "typical" design, Nielsen's observation that familiarity is an important factor in which design users are likely to prefer was shown to be correct (Nielsen, 1993). However, all 3 prototypes developed showed promising design features that could be incorporated into new versions of calendar applications.

It is also worth noting that the designs may not just change the individual use, since the users have been given more freedom and control of their time. This may challenge the organisational negotiation and also organisational power hierarchy, which might be there for a purpose i.e. managers vs. employees. This shows the influence of IT on the organisational hierarchy and relations, but via these changes the users' needs have been considered and not other stakeholders.

FURTHER WORK

The results presented in this paper are being used to develop design recommendations for the designers of calendar applications. These will be reported in a later paper.

Other further work that is suggested includes:.

- The design of the UI can facilitate the individual in working more efficiently but the full benefits will not be achieved unless the calendar use reaches the critical mass and it is used extensively as a social calendar of all employees. This fact proves the importance of more studies of users' reactions to these new technologies and to test it where the system will be used.

- Based on the evaluation; it must be studied how prototypes 1 and 2 can be combined, to improve both. Prototype 3 must be evaluated with a method suitable for its interaction style e.g. Wizard of Oz or instant messaging software.
- The outcome of the Monte Carlo simulations proved that organisational use and how to design the rules for the preferences for schedule and buffers must be considered. Thus it would be beneficial to discuss the 3 prototypes with various stakeholders of time and task management to get insight into needs of e.g. the organisations, the system, the developers, their agendas and learn how they think of it e.g. as a management tool and what this mean for the design. Furthermore aspects of the calendar that are not directly linked to the users' needs should be considered such as organisational needs and the interaction between e.g. e-mail, folder and calendar applications.
- The study and the results also generated ideas for other aspects of calendar studies. To consider the new patterns in calendar use and the problems that the users experienced it would be interesting to examine the areas of collaboration between different calendar applications and platforms e.g. internal between different calendar applications and with the calendar on a PDA. It would also be interesting to study the communicative practises in organisations regarding time and task management. Furthermore it would be interesting to examine the impact of organisational guidelines for use and introduction to the system to see if it improves the adoption of the system and reduces the issues concerning social use.

REFERENCES

Grudin, J. (1996). A case study of calendar use in an organization. SIGOIS Bull., 17(3), 49-51.

Lee, H. (2003). Your time and my time: a temporal approach to groupware calendar systems. Inf. Manage. 40(3), 159-164.

Nielsen, J. (1993) Usability Engineering. Morgan Kaufmann.

Palen, L. A. (1998). Calendars on the new frontier: challenges of groupware technology. University of California at Irvine. Retrieved July 24, 2009, from http://portal.acm.org/citation.cfm?id=335810.

Chapter 45

User Interface with Bird's Eye View Function

Takako Nonaka, Katsuya Oishi, Tomohiro Hase

Ryukoku University
Japan

ABSTRACT

This paper proposes an interface in which input is done by the user moving the equipment, offering bird's eye view like motions so that a 3D angular motion corresponds to the 3D indication on the display screen. An experimental prototype with an LCD was developed using on an embedded system. When the user moves the device, an acceleration sensor detects 3D angular motion normally. The parameters for data entry and the indication of rotation and direction of motion are selected in the opposite direction. The device achieved our proposed idea that the users can react naturally to the motion only by moving the information device and that the indication intuitively agrees with the operation.

Keywords: μT-Engine, acceleration sensor, image viewer

INTRODUCTION

As recent information equipment becomes smaller, it is difficult to place buttons, jog dials, and other input devices on its surface. It is also difficult to get a natural human feeling when handling buttons and jog dials to indicate scrolling, scaling, and rotation. Previously, the authors achieved the actions of movement, rotation, and scaling in a magnifying-glass type action when using a gyro sensor (Kokaji, Yonezawa, Kurinami, Nonaka, and Hase, 2009). Based on the previous prototype, this paper proposes an interface in which an input operation is carried out by the user moving the equipment, offering bird's eye view like motions so that a three dimensional (3D) angular motion corresponds to the 3D indication on the display screen.

PROPOSED SYSTEM

This paper describes the development of an image viewer system with an acceleration sensor. Figure 1 shows the proposed system.

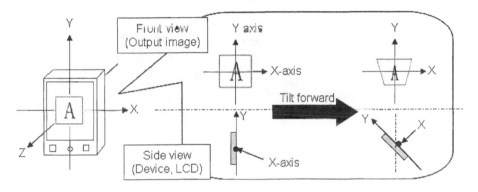

Figure 1 Relationship between input and output in the proposed system.

This system has an intuitive user interface (UI) in which the image viewer changes viewpoint according to the user's operation and the device's tilt. In addition, the system will allow the device to be small because it does not need special input devices such as a mouse or keyboard.

The motion and operation of the device by a user are detected as 3D angular motions using the three-axial acceleration sensor. The detected 3D angular motion and its intensity were used to calculate the 3D indications such as the movement, rotation and scaling parameters of affine transformations. In this device, the parameters for data entry and indication of rotation and direction of motion are selected in the opposite direction.

VERIFICATION EXPERIMENTS

Supposing that it will be used in consumer appliances, we built an experimental device for trial, using an embedded microcomputer and embedded Linux in order to verify our idea.

Figure 2 shows the appearance of the experimental prototype system. The acceleration sensor sends the gradient data to the standard development platform for embedded systems, μT-Engine, via the A/D conversion board.

The results are shown on the LCD display of the experimental device. Figure 3 shows the appearance of the verification experiments and the output image on the LCD screen. The device achieved our proposed idea that the users can react naturally to the motion only by moving the information device and that the indication intuitively agrees with the operation.

Figure 2 View of the experimental prototype

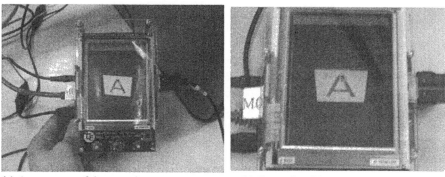

(a) Appearance of the experiment.   (b) Example of the output screen.

Figure 3 Verification experiment.

CONCLUSIONS

The idea proposed in this paper has demonstrated that the use of acceleration sensors can achieve motion that is similar to a bird's eye view. The combination of our proposed idea and the appearance motion of the magnifying-glass, which we previously proposed, realizes a more natural user interface.

REFERENCES

Kokaji R., Yonezawa, M., Kurinami, K., Nonaka, T. & Hase, T. (2009), *Magnifying glass-like user interface for game control*. Proc. 1st International IEEE Consumer Electronic Society Games Innovation Conference, 149-153.

Chapter 46

Ergonomic Assessment of Pronating-Reducing Mouse Designs: Anatomical Compliance in Three-Dimensional Space

David Feathers, Alan Hedge, Kimberly Rollings

Department of Design and Environmental Analysis
Cornell University
Ithaca, NY 14853, USA

ABSTRACT

The main objective of this study was to measure and model how the human hand interacts with examples of contemporary mouse designs and how this physical interaction may be associated with subjective aspects of comfort. The physical interaction between the mouse and the hand was captured for twenty-four healthy individuals using an electromechanical coordinate measurement machine, generating three-dimensional landmarks and surface features for the hand, mouse, and work surface. Four mouse designs, representing progression along the pronation-supination continuum were assessed. Results for three-dimensional hand-mouse measures of the metacarpophalangeal arch, dorsum of the hand, and vertical displacement of surface landmarks of digits I, IV, and V are reviewed and related to measures of perceived comfort. Negative perceptions of comfort were associated with an increase in the displacement of the hand from the work surfaces for landmarks associated with digits I, IV, and V.

Keywords: Anatomical compliance, Mouse design assessment, 3D measurements.

INTRODUCTION

Manipulation of a computer mouse is a blend of prehension and skilled movements. Current trends in mouse design have considered altering wrist position as potential strategy in reducing cumulative trauma associated with the use of computer mice. Promoting a decrease in pronation to favor a more 'neutral' wrist position mitigates upper extremity trauma (Aarås et al., 1997; Fagarasanu and Kumar, 2003; Gustafsson and Hagbert, 2003; Houwink et al., 2009; Lee et al., 2008; Rempel and Gordon, 1998). New "pronation-reducing" designs introduce sloping to the proximal portion of the hand.

A change in hand and wrist orientation may disrupt tactile and kinesthetic inputs (Jones and Lederman, 2006) and resultant hand position on mice across the continuum of pronating-reducing designs will likely change these inputs. Pronating-reducing designs will change hand and wrist orientation, and the location of contact areas of the hand to the mouse and work surface. Contact with the mouse can offer control through digit opposition and associated thenar and hypothenar eminence contact. Contact with the work surface offers kinesthetic input and may support active haptic processing as the hand feels mouse movement relative to the work surface. Training has been demonstrated to reduce postural issues and related musculoskeletal stress for pronating-reducing mouse designs (Houwink et al., 2009) and may mitigate some tactile and kinesthetic differences across the pronating-reducing mouse design continuum.

ANATOMICAL COMPLIANCE

Physical interaction of the human hand to the external environment has been studied in a variety of contexts such as postural variation and its relation to object shape (Liu, 2008; Santello and Soechting, 1998; Santello et al., 1998), to anthropometric variables characterizing hand movement (Buchholz et al., 1992). The concept of anatomical compliance (AC) is defined here as the congruency of morphological response of the human body to an external artifact/environment. This individual, adaptive response can be characterized as a static posture or as an active, dynamic set of behaviors. In the case of the hand molding to a variety of mouse designs, there will be an active component and a resting component. The morphological changes of the hand as it rests on the mouse in a 'poised-to-do-work' posture is of empirical interest in this work. To this end, there are several anatomical areas of interest for understanding AC across the pronating-reducing mouse design continuum, namely, MCP joints, dorsum of the hand, and vertical displacement of surface landmarks of digits I, IV, and V. The wrist, the MCP joints, and ultimately the fingers will be responding to these designs across the pronating-reducing continuum. Methods to classify AC across this continuum are based on posture and contact areas.

RESULTANT POSTURE

Pronating-reducing mouse designs offer a wide variety of asymmetry, typically in some combination of slopes and contouring that dictate asymmetric hand position. A mediolateral slope (from MCP II to MCP V) allows for a potential reduction in forearm pronation as it offers a surface contact that places the MCP II joint higher relative to the work surface than the MCP V joint. The hand resting on this surface will likely have less involvement of pronator teres and pronator quadratus, both innervated by the median nerve (Jenkins, 2009). A proximodistal slope results from the rise around the area of the MCP joints (especially MCP II and III), and descends to work surface height toward the wrist. This slope passively places the hand in a slight to mid-extension posture at the wrist, and may be contoured to both receive the palm and other ventral points of contact for hand control. The proximodistal slope also aids in positioning the proximal portion of the hand so that digits II-V may be slightly flexed and but remain non-weight-bearing so that flexor digitorum profundus can actuate the distal mouse buttons. Given the variation of contouring and sloping, the proximodistal slope may have other associated benefits.

AREAS OF CONTACT

Contact areas allow for active haptic sensing strategies (Jones and Lederman, 2006). Contact areas for the hand on the work surface during mousing tasks change across the pronation-supination continuum, increasingly becoming dependent on the hypothenar eminence and other more medial structures as mouse designs depart from fully pronated designs. Work surface contact areas have the immediate provision of musculoskeletal rest. Houwink et al. (2009) demonstrated that participants who received instructions on upper extremity positioning for a mouse design that departed from traditional, fully-pronated design, had more surface contact area with the work surface and therefore received the requisite musculoskeletal benefits. This research considers four mouse designs existing in the pronate-to-mid-prone/supine continuum. Anatomical compliance is investigated for mouse designs across this continuum. Understanding the match between an individual's anatomy and each mouse design offers an assessment of design efficacy. Effective design should support individual anthropometric variation across a multitude of tasks and should be perceived as subjectively favorable. This chapter details the physical parameters of anatomical compliance in mouse designs existing in a spectrum intended to support the hand. Individual hand variation is assessed along with the complexity of surface features for each mouse design in three-dimensional space to support CAD/CAE models. The impact of mouse design is assessed through positional behavior as individuals interpret subjective comfort of each mouse.

METHODOLOGY

DATA COLLECTION AND MODELING

Four computer mice representing a range of commonly-used mice across the pronate-to-mid-prone/supine designs were considered for the study, a fifth mouse was used but was excluded from this paper as it falls under a 'vertical' mouse design and therefore offers a radical departure of resultant posture and contact. The four mouse designs were (*coded in results*: see Hedge et al. 2010 *this book for further details of those mice*)·

- Conventional mouse design (HP, model M875U) (*triangle*)
- Anthropometrically-scaled- 4 different right-hand sizes (Contour Design, models CMO-BLK-S-R;M-R; L-R;XL-R) (*circle*)
- Switch Mouse (HumanScale, model SMUSB-adjustable length) (*diamond*)
- Wireless Laser Mouse (Microsoft, model 1083) (*rectangle*)

The mice and work surface were digitized using a six-axis electro-mechanical device (FaroArm Fusion; Faro Technologies, Lake Mary, Florida) to collect three-dimensional landmarks and surface features for the mouse and work surface for each individual. Each mouse was digitized and modeled as a three-dimensional solid NURBS model (Rhino 4.0; Robert McNeel & Associates, Seattle, Washington) (see figure 1). Three non-collinear reference points were created to resolve movement artifacts. The work surface (desk) was also digitized, modeled, and referenced in three-dimensional space.

FIGURE 1 Model of the *circle* mouse and work surface.

Participants were required to compete the Hedge et al. study (2010; *this volume*). After having become acquainted with the form and active use of each mouse design, participant hands were measured in a total of seven positions: palm-down (dorsum) on the desk; palm-up (ventral) on the desk, and hand in a static position representative of use across all five mice (the fifth was not reported here). Surface landmarks digitized included traditional osteological (bony) landmarks, non-bony surface landmarks (such as boundaries- e.g. finger tip), and other landmarks such as skin folds. A total of 58 landmarks were taken on the ventral aspect of the hand and 54 on the dorsal aspect of the hand (see figure 2).

FIGURE 2 Ventral and dorsal landmarks captured in three-dimensional space.

These landmarks formed a complete hand model created in Rhino 4.0. When each participant adopted their static, comfortable hand position for each mouse, a subset of dorsal landmarks (36) were digitized. This subset of measurements then informed the complete hand model as to positional changes in three-dimensional space.

ANALYSIS

Anatomical compliance was assessed for resultant posture and contact. For measures of posture, there were two methods of analysis: the dorsal (proximodistal) rise of MCP II (dorsal aspect) from the Radial Styloid in a constructed angle from the work surface and the dorsal slant (mediolateral) measured from the MCP II (dorsal aspect) to MCP V (dorsal aspect) in a constructed angle from the work surface. The proximodistal rise of the proximal portion of the hand was modeled as the resultant angle (a.) of MCP II (dorsal aspect) and the radial styloid to the work surface (see figure 3).

FIGURE 3 Dorsal Rise: Angle 'a' represents the angle created by the work surface, and a constructed line connecting surface landmarks MCP II (Dorsal Surface) to Radial Styloid.

A basic model of the mediolateral slant of the MCP arch was constructed using MCP II (dorsal aspect) and MCP V (dorsal aspect) landmarks, creating an angle (b.) to the work surface. Figure 4 shows the modeling of the dorsal (mediolateral) slant of the MCP arch for digits II-V.

FIGURE 4 Dorsal Slant (mediolateral) of MCP Arch

Two measures of contact were assessed: MCP displacement (ventral) parallel to the mouse surface and vertical displacement for landmarks of digits I, IV, and V.

The MCP displacement parallel to the mouse surface was assessed for the ventral aspects of MCP II-V, and the lateral (ventral-most) landmark for MCP I. Parallel displacement was measured relative to the surface contours of the mouse in an effort to normalize displacement parameters. Figure 5 shows the ventral portion of the hand and the associated MCP landmarks.

FIGURE 5 Metacarpophalangeal displacement (ventral) parallel to the mouse surface.

The vertical displacement from the work surface for digits I, IV and V classifies landmark-specific proximity. The landmarks included in this analysis are: finger tip, lateral portion of distal interphalangeal (DIP) joint, ventral portions of DIP, proximal interphalangeal (PIP) joint, and MCP. Figure 6 illustrates the vertical displacement of these landmarks from the work surface.

FIGURE 6 Vertical displacement from work surface for digits I, IV, and V.

Assessments of subjective comfort were analyzed and briefly reported in this chapter. These data were from the Hedge et al. (2010, *this book*). Participants completed a survey on comfort, ease of use, and control for each mouse, then rated all five mice on comfort, ease of use, visual attractiveness, and ranked overall preference in a final survey. Rated responses on all surveys were recorded by participants on analog, linear-analog scales (10 cm; 3.94 inches).

RESULTS

PARTICIPANTS

A total of twenty-four healthy, right-handed participants from Cornell University were enrolled in the study. All participants were to be familiar with computer and mouse use, and had just completed participating in the mouse study described by Hedge, et al., (2010, *this volume*). Experimental procedures were approved by Cornell University's Institutional Review Board for Human Participants and informed consent was given.

MEASURES OF RESULTANT POSTURE

Dorsal rise and slant was assessed for all twenty four individuals across all four mouse designs. Since there is no standardized way to measure the pronation-supination angle on a mouse, no attempt was made to investigate individual departure from the mouse design for either dorsal rise or dorsal slant, rather, angles 'a.' and 'b.' were reported from the desktop.

Table 1 Resultant posture for dorsal rise and dorsal slant in mean (S.D.) degrees across mouse designs

Resultant Posture	▲	●	◆	■
Dorsal Rise	24.9 (5.2)	24.8 (6.7)	36.7 (6.1)	24.6 (7.2)
Dorsal Slant	28.5 (9.2)	31.9 (8.5)	12.2 (5.0)	44.0 (9.2)

MEASURES OF CONTACT

The metacarpophalangeal arch was analyzed in table 2 with respect to perpendicular displacement from the mouse surface.

The vertical displacements from the work surface are explored in table 3 finger tip, lateral portion of distal interphalangeal (DIP) joint, ventral portions of DIP, proximal interphalangeal (PIP) joint, and MCP. Table 3 details the median displacement (mm) to the work surface for selected landmarks for digits I, IV and V across the four mouse types.

Table 2 Mean (S.D.) displacement (millimeters) for ventral aspects for MCP II-V across mouse design from the work surface

Displacement (mm) for MCP Joint	▲	●	◆	■
II	0.6 (0.2)	0.3 (0.2)	0.4 (0.2)	1.2 (0.9)
III	0.1 (0.1)	0.3 (0.3)	0.2 (0.2)	1.0 (0.3)
IV	8.9 (4.2)	0.5 (0.3)	0.2 (0.1)	2.6 (1.4)
V	15.8 (7.1)	1.2 (1.0)	2.9 (1.1)	4.2 (2.9)

Table 3 Select measures of median displacement (millimeters) for ventral aspects for MCP II-V across mouse type from the work surface (desk)

Displacement (mm) for Landmark	▲	●	◆	■
Digit I: Finger Tip	7	18	25	37
Digit I: DIP Lateral	8	26	32	45
Digit I: DIP Ventral	20	24	35	40
Digit I: Lateral MCP	9	23	48	40
Digit IV: Finger Tip	7	34	20	8
Digit IV: DIP Lateral	20	40	24	20
Digit IV: DIP Ventral	15	37	20	18
Digit IV: PIP Ventral	30	48	27	35
Digit IV: MCP Ventral	38	45	48	45
Digit V: Finger Tip	7	12	17	4
Digit V: DIP Lateral	15	15	22	11
Digit V: DIP Ventral	12	14	20	10
Digit V: PIP Ventral	18	20	25	20
Digit V: MCP Ventral	17	37	40	29

PHYSICAL INTERACTION AND MEASURES OF SUBJECTIVE COMFORT

The following measures of physical interaction were compared to subjective ratings of comfort: Dorsal Rise, Dorsal Slant, and Displacement of MCP joints to the mouse. Comfort takes into account physical, cognitive, and emotional elements which were not studied in their entirety here. In general, across mouse type, the higher the dorsal rise, the lower the subjective comfort rating score (see figure 6). This was especially the case for those individuals who had a greater MCP

434

displacement for MCP joints IV and V.

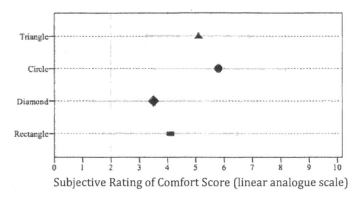

Subjective Rating of Comfort Score (linear analogue scale)

FIGURE 6 Subjective comfort ratings across mouse type (Mean values represented by basic shape, standard deviation in grey bars)

DISCUSSION

The concept of anatomical compliance is outlined and a novel approach to modeling the physical interaction of the hand to mouse is presented here. The postural modeling effort primarily emphasizes the proximal portions of the hand and the measures of contact show basic measures of hand contact for the work surface and mouse. The static modeling effort outlined in this study is by no means exhaustive or representative of the complexity of every aspect of the hand-mouse interaction.

There are, however, specific hand-mouse interactions important to note, such as the manner in which contact areas of the MCP IV and V joints, the hypothenar eminence, and other medial aspects of the hand increasingly become weight bearing as the pronation angle of the mouse decreases. This reduction in pronation places an emphasis on wrist abductors (radial deviation) to lift and move the mouse.

Houwink, et al., 2009 have investigated the impact training has on hand posture on pronating-reducing mice, and they found that post-training posture showed greater work surface contact for digits IV and V (p. 51). In the present research we were unable to explore the role of practice or training and this can be undertaken in future studies. When comfort ratings were examined as a function of anatomical compliance measures we found that negative perceptions of comfort (low ratings) were associated with an increase in the displacement of the hand from the work surface for landmarks associated with digits I, IV, and V.

ACKNOWLEDGEMENTS

This research was supported by funds from the College of Human Ecology and the

Human Factors and Ergonomics Laboratory, Cornell University.

REFERENCES

Aarås, A., Dainoff, M., Ro, O., Thoresen, M. (2001), Can a more neutral position of the forearm when operating a computer mouse reduce the pain level for visual display unit operators? A prospective epidemiological intervention study: Part II. *International Journal of Human-Computer Interaction*, 13(1), 13-40.

Buchholz, B., Armstrong, T., Goldstein, S. (1992), Anthropometric data for describing the kinematics of the human hand. *Ergonomics*, 35, 261-273.

Dennerlein, J., Johnson, P. (2006), Changes in upper extremity biomechanics across different mouse position in a computer workstation. *Ergonomics*, 49(14), 1456-1469.

Fagarasanu, M., Kumar, S. (2003), Carpal tunnel syndrome due to keyboarding and mouse tasks: a review. *International Journal of Industrial Ergonomics*, 31, 119-136.

Gustafsson, E., Hagbert, M. (2003), Computer mouse use in two different hand positions: exposure, comfort, exertion and productivity. *Applied Ergonomics*, 34, 107-113.

Hedge, A., Feathers, D., Rollings, K. (submitted), Ergonomic comparison of five mouse designs: Performance and posture differences. *Proceedings of the Third International Conference on Applied Human Factors and Ergonomics*. Miami, Fl. July, 2010.

Houwink, A., Oude Hengel, K., Odell, D., Dennerlein, J. (2009), Providing training enhances the biomechanical improvements of an alternative computer mouse design. *Human Factors*, 51(1), 46-55.

Jenkins, D. (2009), Hollinshead's Functional Anatomy of the Limbs and Back. Saunders Elsevier, St. Louis, Missouri.

Jones, L., Lederman, S. (2006), Human Hand Function. Oxford University Press, New York, New York.

Lee, D., McLoone, H., Dennerlein, J. (2008), Observed finger behaviour during computer mouse use. *Applied Ergonomics*, 39, 107-113.

Liu, K. (2008), Synthesis of interactive hand manipulation. *Eurographics/ACM SIGGRAPH Symposium on Computer Animation*, 163-171.

Rempel, D., Gordon, L. (1998), Effects of forearm pronation/supination on carpal tunnel pressure. *Journal of Hand Surgery*, 23A(1), 38-42.

Santello, M., Soechting, J. (1998), Gradual molding of the hand to object contours. *Journal of Neurophysiology*, 79, 1307-1320.

Santello, M., Flanders, M., Soechting, J. (1998), Postural hand synergies for tool use. *The Journal of Neuroscience*, 18(23), 10105-10115.

Chapter 47

Ergonomic Comparison of Five Mouse Designs: Performance and Posture Differences

Alan Hedge, David J. Feathers, Kimberly A. Rollings

Department of Design and Environmental Analysis
Cornell University
Ithaca, NY 14853-4401, USA

ABSTRACT

This study investigated task performance (point-and-click and dragging), wrist posture (extension/flexion and ulnar/radial deviation), when using a conventional optical computer mouse, a contoured mouse design, a slanted design and 2 vertical optical mouse designs. Twenty four participants (12 male, 12 female), between 18-35 years of age, were tested. Hand anthropometric measures were taken. Mouse design significantly affected performance for point-and-click (P=0.000) and dragging tasks (P=0.000). Performance was slowest with the conventional mouse and fastest for the vertical mice. Although ulnar deviation was slightly less for the vertical mice (P=0.000) wrist extension was substantially larger for these designs (P=0.000), which poses a greater risk of a musculoskeletal hand/wrist injury. Wrist posture was best for the slanted mouse design.

Keywords: computer mouse, wrist extension, Fitts' task performance, musculoskeletal injury risk

INTRODUCTION

A conventional mouse design is used with the hand in pronation. More recently, contoured, slanted and vertical mouse designs have become available that reduce wrist pronation. These alternative mouse designs have been developed supposedly to reduce the risks of musculoskeletal disorders, and to increase user comfort and performance (Aarås et al., 2001; Gustafsson and Hagberg, 2003; Houwink et al., 2009; Lee et al., 2008; Odell and Johnson, 2009).

Investigation of the effects of mouse design on performance typically have relied on a Fitts' task with performance improvements being indicated by faster movement time (Aarås et al., 2001; Po et al., 2005; Scarlett et al., 2005). A dynamic mouse movement taxonomy has been developed by Lee et al. (2008) and typical mouse tasks include moving the mouse (non-clicking or dragging), primarily using extrinsic hand muscles, and activating mouse buttons (left click, right click, scrolling), using a combination of intrinsic and extrinsic hand muscles. Gender affects how a mouse is used (Chang, 2006; Lee et al., 2008) and males appear to extend their fingers more and lift their mouse more than females (Lee et al., 2008). The upper extremity musculoskeletal demands of mouse use are affected by the relative positioning of the mouse (Dennerlein et al., 2006) and individual anthropometry (Hedge et al., 1999).

Fagarasanu and Kumar (2003) note that force, repetition, and the degree of wrist deviation are factors that elevate the risk of developing a musculoskeletal injury, such as carpal tunnel syndrome. Medical imaging has confirmed that non-neutral flexion and extension wrist postures change the morphology of the carpal tunnel which in turn detrimentally affects the function of the tendons and the median nerve; extreme wrist extension is widely regarded as the riskiest postural deviation of the wrist (Bower et al., 2006; Mogk and Keir, 2007). Changes in carpal tunnel pressure in response to computer mouse design have also been studied with comparable findings (Keir et al., 1999).

The conventional symmetrically-shaped computer mouse is used with a pronated hand. The earliest "ergonomic" mice also required a pronated hand but had a curved design to reduce some of the radial deviation associated with using the conventional mouse out to the side of an extended keyboard. Since then a variety of alternative mouse shapes and design styles have emerged. Today a user can choose from a plethora of so-called "ergonomic" mouse designs. Users have different hand sizes and some mice are available in different sizes while others have length adjustability built into the mouse. A recent trend in mouse design has been to present the product as a sloped design or even a vertical design. The present study is a detailed ergonomic evaluation of five current optical mouse designs, ranging from a conventional mouse to vertically-oriented mice. Effects of these designs on performance and posture were evaluated.

METHODS

Participants

Twelve men and 12 women between 18-35 years of age volunteered to participate in the study. All were right-handed, healthy with no history of musculoskeletal problems, and familiar with using a computer mouse. Participants were paid $25 for the study. The project was approved by Cornell University's Institutional Review Board for Human Participants.

Materials

Hand anthropometric dimensions were measured using calipers and also the detailed morphology was captured with a digitizing system (Faro arm). A twin axis electrogoniometer (Biometrics Ltd., model SG65) recording at 50Hz was used to measure wrist extension/flexion and radial/ulnar deviation movements. A laptop (Dell Latitude D620) was used to record wrist posture data.

A laptop (Dell Latitude D600) was used to present the performance tasks (Generalized Fitts' Law Model Builder software V. 1.1C; Sourkoreff & MacKenzie,1986). Software speed settings were adjusted to the same level for all mice. A small cross-hair starting target appeared on the computer screen followed by one of a circular array of 8 target circles (9.8mm diameter) positioned at 45° intervals and at one of two distances (80mm and 160mm). The sequence of targets was randomized for each trial. Each target was presented 4 times in each location and all sequences were randomized for each participant. Targets were presented singly and they disappeared after acquisition.

Two variants of the task were used: a point-and-click task, where participants clicked on a starting cursor and then moved the cursor to a click on the target, and a dragging task, where the participant dragged the cross-hair to the target circle. These performance tasks are comparable to those described as relevant mouse performance measures in ISO 9241-9 (ISO, 2000).

Five optical computer mice were selected for the study. For each mouse the manufacturer's name and any identification information was covered. Each mouse was randomly assigned a solid shape symbol as follows:

▲ (Triangle) - a conventional mouse design (HP, model M875U) operated with a pronated hand
● (Circle) - an ergonomic mouse design with 4 different right-hand sizes (Contour

Design, models CMO-BLK-S-R;M-R;L-R;XL-R). This mouse was contoured and operated with a semi-pronated hand

♦ (Diamond) - a Switch Mouse (HumanScale, model SMUSB). This is an adjustable length, sloped mouse design that is operated with a semi-pronated hand

■ (Square) - a vertical mouse design (Evoluent LLC, model VM3R2-RSB) that is operated with a mid-pronate/supinate hand.

■ (Rectangle) - a vertical mouse design (Microsoft Natural Wireless Laser Mouse model 1083) that is operated with a mid-pronate/supinate hand.

FIGURE 1. The five mice used in the study.

To standardize the extent of mouse movements for the different mice and different participants all tasks were performed on a horizontal circular-shaped mousing platform (26cm diameter: ~10 inches diameter) that was attached to a negative tilt keyboard tray (HumanScale 5G) that in turn was attached to the undersurface of an electric, height adjustable table (Workrite with Linak mechanism) that was set to the comfortable working height for each individual. Participants sat in an ergonomic chair that was adjusted to each person (Freedom Chair, HumanScale).

Procedure

Hand and wrist dimensions were measured with a caliper and ruler. Participants were sized, per manufacturer's instructions, for the Circle and Diamond mice. Then, using double-sided and medical tape, the electrogoniometer was attached to the dorsal side of the right hand and forearm to record right wrist flexion/extension and radial/ulnar deviation. The reference or zero position was set with the pronated forearm, wrist, and hand held at a neutral position with the palm facing down on the flat worksurface. Participants first practiced five repetitions of each of the multipoint standard Fitts' point-and-click and dragging tasks using the Triangle mouse. Two upright mice (Square and Rectangle), however, were intended to be held with the ulnar side of the hand and wrist facing the worksurface. In this position the "left click" button, located under the index finger, could be activated by a horizontal press of the finger towards the mouse. These vertical mice could be moved with hand/wrist movements. The last two mice (Circle and Diamond) positioned the hand mid-pronate/supinate and horizontal mouse movement could also engage more use of the arm to reduce wrist extension. The order of presentation of the mice was balanced and randomly assigned to participants, but the point-and-click task was always completed before the dragging task for each mouse. A stopwatch was used to record the rest time between the point-and-click and dragging tasks, and between each mouse while surveys were completed (60 seconds, minimum).

Data Analysis

All data were entered into a spreadsheet (MS Excel 2007). The first and last 20 seconds of electrogoniometer data was omitted to eliminate any extraneous movements related to starting or stopping the tasks. All data subsequently were analyzed with a multivariate statistical package (SPSSv18), using either Pearson correlations, independent t-tests or repeated-measures analysis of variance.

RESULTS

Hand Dimensions

There was a significant difference in the anthropometric dimensions of the hands of the men and women, and men's hands were longer and wider (Table 1.) Hand length was significantly correlated with hand width (P=0.001) and wrist width (P=0.005); hand width and wrist width were also significantly correlated (P=0.000).

Table 1 Hand anthropometric dimensions in cm for men and women in the sample (mean values; SD [ ], in cm).

	Men	Women	P
Hand length	19.5 [0.7]	17.7 [0.8]	0.000
Hand width	8.6 [0.5]	7.7 [0.5]	0.000
Wrist width	5.8 [0.4]	5.3 [0.5]	0.011

Task Performance

There was a significant difference between the 5 mice in the movement times for the point-and-click task (P=0.000). Movement times were faster for the Circle

(Contour), Square (Evoluent), and Rectangle (MS Vertical) mice and slower for the Triangle (HP) and Diamond (Switch) mice. Figure 1 shows the mean movement times for each mouse and also for the 80 mm and 160 mm movement amplitudes (note the profile is almost identical for each movement distance) for the point-and-click task.

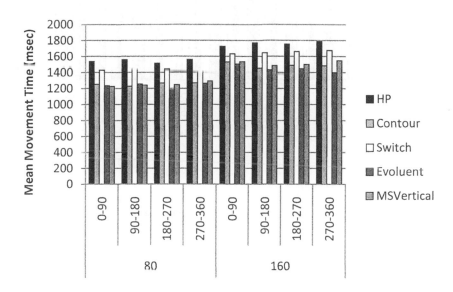

Figure 2. The mean movement times for the 5 mice for the two amplitudes (80mm and 160 mm) and grouped into angular quadrants for the point-and-click Fitts' task.

For the performance data that were gathered there was a significant difference in the movement times between the 5 mice for the cursor dragging task (P=0.000). For the dragging task the movement times were faster for the Square (Evoluent), and Rectangle (MS Vertical) mice than for the other 3 designs. Figure 2shows the mean movement times for each mouse and also for the 80 mm and 160 mm movement amplitudes (note the profile is almost identical for each movement distance) for the dragging task.

Wrist Posture

For the point-and-click task there was a significant difference (P=0.001) between men (23.8°) and women (34.1°) for wrist extension and between the mouse designs (P=0.000) but no significant interaction of gender and mouse design. For the point-and-click task there was no significant effect of gender on ulnar deviation but there was a significant difference among the mice (P=0.000).

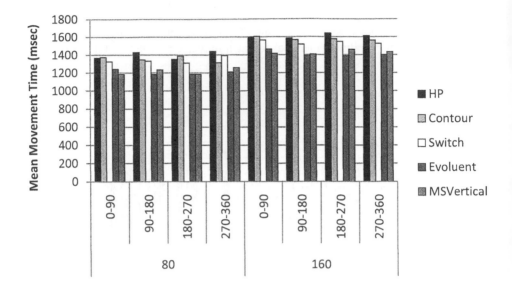

Figure 3. The mean movement times for the 5 mice for the two amplitudes (80mm and 160 mm) and grouped into angular quadrants for the dragging Fitts' task.

For the dragging task there was a significant difference (P=0.004) in wrist extension between men (24.1°) and women (35.7°) and between mouse designs (P=0.006) but no significant interaction of gender and mouse design for wrist extension. There was no significant effect of gender on ulnar deviation but there was a significant difference among the mice for the dragging task (P=0.000).

Figure 4 shows that the average wrist extension was highest, exceeding 30° wrist extension for the point-and-click task, for the Square (Evoluent) and Rectangle (Microsoft) vertical mouse designs, where the hand was mid-pronate/supinate. Wrist extension was least for the Diamond (Switch) mouse design where the hand was semi-pronated, and the average wrist extension was 12° for the point-and-click task. Wrist extension results were similar for the dragging task, and average wrist extension exceeded 30° for the Square (Evoluent) and Rectangle (Microsoft) vertical mouse designs, where the hand was mid-pronate /supinate, but was least for the Diamond (Switch) mouse design (13°).

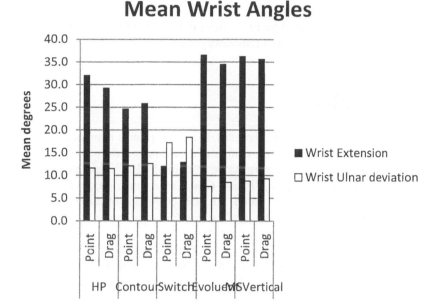

Figure 4. The mean wrist extension and ulnar deviation angles for the 5 mice for the point-and-click and dragging Fitts' tasks.

DISCUSSION

Current results compare well with previous work. Fitts' task testing of the Contour mouse (Scarlett et al., 2005) found an average movement time of ~1.04 seconds (averaged for 37.5mm and 127.5mm amplitude) compared to our result of 1.4 averaged for 80mm and 160mm amplitudes. For movement times the vertical designs performed better than the other designs and the conventional mouse was the slowest. Keir et al. (1999) found that wrist extension averaged 28.2° for dragging tasks and 25.6° for point-and-click tasks with a Contour mouse. Our data were 25.8° and 26.9° respectively. Gustafsson and Hagberg (2003) used a text editing task and found 23° wrist extension for a conventional mouse and18° for the Evoluent mouse: our results were 30.6° (conventional) and 35.5° (Evoluent) with the Fitts' tasks. Odell and Johnson (2009) used a Fitts' task and reported average wrist extension between 35.0° and 41.2° for different vertical mouse designs. Only the Switch mouse was used with wrist extension consistently below 15°. Movement times were slightly faster with the vertical mouse designs but wrist extension was substantially worse. The slanted mouse design may be the best balance for neutral posture and acceptable performance.

444

ACKNOWLEDGEMENTS

This research was supported by funds from the College of Human Ecology and the Human Factors and Ergonomics Laboratory, Cornell University.

REFERENCES

Aarås, A., Dainoff, M., Ro, O., Thoresen, M. (2001) Can a more neutral position of the forearm when operating a computer mouse reduce the pain level for visual display unit operators? A prospective epidemiological intervention study: Part II. *International Journal of Human-Computer Interaction*, 13(1), 13-40.

Bower, J., Stanisz, G., Keir, P. (2006) An MRI evaluation of carpal tunnel dimensions in healthy wrists: implications for carpal tunnel syndrome. *Clinical Biomechanics*, 21, 816-825.

Chang, C., Amick, B., 3rd, Menendez, C., Katz, J., Johnson, P., Robertson, M., Dennerlein, J. (2007) Daily computer usage correlated with undergraduate students' musculoskeletal symptoms. *American Journal of Industrial Medicine*, 50, 481-488.

Chang, C. (2006) Gender differences in the relationship between daily computer usage and musculoskeletal symptoms among undergraduate students. *Proceedings of the Human Factors and Ergonomics Society Annual Meeting*, 5, 1356-1360.

Dennerlein, J., Johnson, P. (2006) Changes in upper extremity biomechanics across different mouse positions in a computer workstation. *Ergonomics*, 49, 1456-1469.

Fagarasanu, M., Kumar, S. (2003) Carpal tunnel syndrome due to keyboarding and mouse tasks: a review. *International Journal of Industrial Ergonomics*, 31, 119-136.

Gustafsson, E., Hagberg, M. (2003) Computer mouse use in two different hand positions: exposure, comfort, exertion and productivity, *Applied Ergonomics*, 34, 107-113.

Hedge, A., Muss, T., Barrero, M. (1999) *Comparative study of two computer mouse designs*. Cornell Human Factors Laboratory Technical Report: RP7992, Ithaca, New York.

Houwink, A., Hengel, K., Odell, D., Dennerlein, J. (2009) Providing training enhances the biomechanical improvements of an alternative computer mouse design. *Human Factors*, 51(1), 46-55.

ISO, 2000. *ISO 9241-9 International standard: Ergonomic requirements for office work with visual display terminals (VDTs)--Part 9: Requirements for non-keyboard input devices*. International Organization for Standardization.

Karlqvist, L., Hagberg, M., Köster, M., Wenemark, M., Ånell, R. (1996) Musculoskeletal symptoms among computer-assisted design (CAD) operators and evaluation of a self-assessment questionnaire. *International Journal of*

Occupational Environment and Health, 2, 185-194.

Keir, P., Bach, J., Rempel, D. (1999) Effects of computer mouse design and task on carpal tunnel pressure. *Ergonomics*, 42(10), 1350-1360.

Mogk, J., Keir, P. (2007) Evaluation of the carpal tunnel based on 3-D reconstruction from MRI. *Journal of Biomechanics*, 40: 2222-2229.

Odell, D., Johnson, P. (2007) Evaluation of a mouse designed to improve posture and comfort. *Proceedings of the Work with Computing Systems Conference-International Ergonomics Association*, WWCS 2007, 115.

Po, B., Fisher, B., Booth, K. (2005) Comparing cursor orientations for mouse, pointer, and pen interaction. *CHI (Computer-Human Interaction): Smart Interaction Techniques*, 1, 291-300.

Scarlett, D., Bohan, M., Io, L., Jorgensen, M., Chaparro, A. (2005) Psychophysical comparison of five mouse designs. *HCI International 2005: The 11th International Conference on Human-Computer Interaction* Las Vegas, NV.

Sourkoreff, W. & MacKenzie, S. (1986). *Generalized Fitts' Law Model Builder (GFLMB), Version 1.1C*, University of Guelph, Canada.

CHAPTER 48

Five-Key Text Entry: An Empirical Comparison of Novice Performance with Three Keyboard Layouts

*Barbara Millet[1], Shihab S. Asfour[2],
James R. Lewis[3], Arzu Onar-Thomas[4]*

[1]Texas Tech University

[2]University of Miami

[3]International Business Machines

[4]St. Jude Children's Research Hospital

ABSTRACT

This paper describes an empirical comparison for five-key text entry of three keyboard layouts (Alphabetical, Predictive, and Hybrid). The Alphabetical layout positioned characters in alphabetical order. The Predictive layout employed Bellman and MacKenzie's (1998) Fluctuating Optimal Character Layout (FOCL), relying on letter-pair probabilities to dynamically rearrange letters to minimize selector movement. The Hybrid layout combined a fixed alphabetical and a reduced (seven-key) predictive character set. Twenty-four participants entered four text types (Words, Sentences, Addresses, and Internet URLs) with each of the three keyboards. Empirical results showed no overall difference in novice performance (speed and error rates) among the three keyboard layouts across the four text types. There were, however, substantial differences among the methods in efficiency and user preference, with the Hybrid layout significantly more efficient than and preferred to the Alphabetical and Predictive layouts.

Keywords: Text Entry, Keyboards, Predictive Keyboards, Five-key Text Entry, Selection-Based Methods

INTRODUCTION

An emerging area of research in human factors engineering is the evaluation of text entry for input-constrained devices using a small number of keys. These methods, often referred to as selection-based methods, use a virtual keyboard that has a cursor that moves over a matrix of characters in response to directional key presses. Although these techniques can employ as few as two keys, there are no commercial examples of two- or three-key methods. However, five-key text entry techniques are widely used on consumer products that do not have keyboards, such as in-car navigation systems, television remotes, mobile communication devices, and gaming controllers. The five-key method uses four directional keys (up, down, left, and right) to move the cursor and a fifth key for selection.

Selection-based methods are easy to learn (Wobbrock, Myers, & Rothrock, 2006) and can be operated with one hand. However, as indicated by the number of keystrokes required per character and high visual scan time, these methods tend to be inefficient. This paper focuses on alternative keyboard layouts for five-key text entry techniques, with the main goal to create and validate novel five-key text entry techniques that are more efficient than currently available methods for input-constrained devices.

METHODS

KEYBOARDS

The three keyboard layouts were Alphabetical, Predictive, and Hybrid (see Millet, Asfour, & Lewis, 2008, for the design factors considered and tools used to develop these layouts). The character set for each of the three keyboards occupied only two rows. These keyboards displayed a minimum of 51 characters (26 letters, 10 numbers, and 15 punctuation marks and special characters). Numbers, punctuations, and symbols were in logical groups. The top row was mostly letters and the bottom row mostly the extended character set. All three keyboard layouts included a subset of modifier and edit keys, used a snap-to-home cursor mode (whereby the cursor snapped to SPACE after each entry), and employed a wrap-around cursor.

The main advantage of an alphabetical keyboard (Figure 1) is user familiarity, which reduces visual search demand. Unfortunately, text entry with this layout requires a relatively large number of keystrokes.

FIGURE 1. Alphabetical Keyboard

The Predictive keyboard (Figure 2) used Bellman and MacKenzie's (1998) Fluctuating Optimal Character Layout strategy, which significantly reduces the keystrokes per character (KSPC) required to enter text. However, it is relatively difficult to learn because it exposes users to 27 different layouts, greatly increasing visual search demand.

FIGURE 2. Predictive Keyboard

The Hybrid keyboard (Figure 3) had a fixed alphabetical layout and a small set of seven predictive (next-most-likely) keys. The design goal was to allow users to take advantage of the reduced KSPC in the predictive portion of the layout while allowing easy access to a fixed alphabetic layout when the desired character was not one of the next-most-likely characters.

FIGURE 3. Hybrid Keyboard

TEST ENVIRONMENT

The experimental evaluations were conducted in a usability laboratory. The experimental station consisted of a laptop, an external monitor, and a customized external keyboard used for cursor control. Cursor control key mappings met the following criteria: (1) the Left/Right/Up/Down (navigation) keys formed an inverted-T shape and (2) the Enter key was beneath and to left of the navigation keys for easy acquisition by the thumb of the right hand (as in Bellman & MacKenzie, 1998). Participants used their index, middle, and ring fingers for navigation and the thumb for selection. Participants remained seated during the experiment. The default height of the desk was its standard height of 26 inches, raised or lowered as required to accommodate varying participant heights. The monitor was directly in front of the user at an 18-inch viewing distance.

TEST APPLICATION

An interactive, extensible research platform was developed to support this research (Millet, Asfour, & Lewis, 2009). The platform ran within a Microsoft Windows operating system and read the keyboard layouts and test phrases contained in two

XML files. The application implemented the three keyboard layouts. The user interface consisted of a large text field at the top of the screen where stimulus phrases appeared, an output field displaying the characters typed by the participant, and the character set of the input method under evaluation. The input keys controlled the cursor on the screen, allowing users to navigate the character set. Although the use of the keyboard and prototype application did not directly map to potential real world use, participants used the same apparatus for all input methods (in other words, all input methods received the same treatment).

DATA COLLECTION

The evaluation platform automatically recorded all participant interactions with the keyboards. Data collection began with the first keystroke for each phrase and ended with the last keystroke. Specifically, for each test phrase the software collected both the presented and transcribed text with a time stamp and key code for each keystroke. Errors underwent manual analysis. For the purpose of this research, the level of analysis was the overall error rate (no breakdown of errors by type).

PARTICIPANTS

The participants were 24 adults, all right-handed, fluent in English, and with normal or corrected-to-normal vision. By design, there was an equal mix of gender, age groups (<40 years old and ≥40 years old), and "texting" experience groups (Non-Expert and Expert). Modifying the criteria of Curran, Woods, and Riordan (2006), a non-expert was someone who sends fewer than 15 messages per week on average and an expert was someone who sends more. None of the participants had experience with any of the tested selection methods. All participants could type at least 22 words per minute as rated by an online typing test. Participants received monetary compensation of $15 for their participation in this study.

TEST TASKS

The experimental task was the input of phrases of text. In this study, there were four types of text, designed to simulate realistic text entry when using internet-enabled devices. The text phrases encompassed characters across the sets of the evaluated keyboard layouts. This allowed evaluation of the relative ease in entering any character in the sets. Additionally, the text types chosen allowed for comparison of the results against the literature. The text types were:

- **Words/Spaces:** This task involved entering only lower case letters, words, and spaces. For consistency with prior experiments, the MacKenzie and Soukoreff phrase sets (MacKenzie & Soukoreff, 2003, modified to use only American English spellings and only lower case letters) were the source for the test texts. Random selection from a base set of 500 phrases determined the specific phrases to use.

- **Sentences:** This task involved entering a sentence consistent with writing a text or short email message. As in prior work, the Brown Corpus was the source for a set of randomly selected sentences (Lewis, 1995). All of the selected sentences ranged in length from 25 to 35 characters.
- **Addresses:** This task was to enter a few addresses. The purpose of the task was to assess the relative ease of entering numbers with the different keyboards. As in prior work, the study used addresses randomly selected from the Human Factors and Ergonomics Society member directory (Lewis, 1995).
- **Web:** For this task participants entered relatively short URLs or email addresses for the purpose of assessing the relative ease of entering special characters with the different keyboards. This is the most rarely tested text type in current text entry research, mainly because most layouts do not have extended characters sets to allow the entry of these types of text. Therefore, it was necessary to create these texts for this study. However, the test texts of Sears and Zha (2003) used in their evaluation of QWERTY soft keyboards served as a model for this study's test texts for web addresses.

There were nine test phrases for each text type – a total of 36 phrases. This was the source for the formation of three data sets, each containing three test phrases of each text type. Each data set contained a total of 12 phrases. The order of presentation of the phrases within a phrase set was randomized, which also randomized the order of presentation of the text types.

PROCEDURE

At the start of the experiment, each participant received a brief tutorial on how to use the respective keyboard layout and time to familiarize themselves with the particular key mappings. Participants practiced entering text for two minutes using the following phrase (43 characters including spaces):

the quick brown fox jumps over the lazy dog

This is a well-known (over-learned) phrase that contains each letter of the English alphabet. The training data were not a part of any subsequent analysis. After the practice period, participants began working with the first data set.

Overall, participants entered 12 phrases using the respective keyboard layout, with instruction to enter the phrases as quickly and as accurately as possible. If an error occurred, participants could make corrections by selecting the delete key.

After completing the test tasks with all keyboard layouts, participants ranked the keyboards in order of preference. Participants were tested one at a time, with participation lasting approximately 90 minutes, including three short breaks.

EXPERIMENTAL DESIGN

The experiment used a 3x4 within-subjects factorial design. The two factors were Keyboard Layout (Alphabetical, Predictive, and Hybrid) and Text Type (Words/Spaces, Sentences, Addresses, and Web). The experimental design used a diagram-balanced Greco-Latin rectangle (Lewis, 1993) to simultaneously counterbalance the presentation of keyboard layout, the phrase set, and the pairing of the keyboard layout and phrase set. The dependent measures were entry rate (in corrected WPM (CWPM)), total error rates (%), and efficiency (in KSPC). After using the three layouts, participants ranked them in order of overall preference.

RESULTS AND DISCUSSION

TEXT ENTRY RATE

The overall results for average text entry speed (in CWPM) were: Alphabetical M— 4.980 (SD= 1.243), Predictive M= 5.030 (SD= 1.533), and Hybrid M= 5.087 (SD= 1.716). Overall results by text type for mean text entry speed were: Address M= 4.175 (SD= 0.923), Sentences M= 5.549 (SD= 1.387), Web M= 4.268 (SD= 1.079), and Words/Spaces M= 6.136 (SD= 1.542).

A repeated measures ANOVA was conducted to assess whether keyboard layout, text type or other factors were associated with CWPM. The final model for this design included keyboard, age, text type, an age by keyboard interaction, an age by text type interaction, and a text type by keyboard interaction. The results indicated that the mean CWPM was not associated with keyboard layout ($F(2,824)=$ 1.568, $p=$ .209). However, the age by keyboard interaction was significant $F(2,824)=$ 5.602, $p=$ .004, with participants less than 40 years old achieving greater speeds, on average, with the Hybrid keyboard than the Predictive ($p=$ .003) and the Alphabetical ($p=$ .005) keyboards. However, this text entry speed advantage was not evident in typing with the Hybrid keyboard rather than the Predictive ($p=$ .098) or the Alphabetical ($p=$ .737) keyboards for participants who were 40 or more years old.

Furthermore, the age by text type interaction effect was significant ($F(3,824)=$ 5.553, $p<$ .001). Participants less than 40 years old tended to type faster across all text types than participants who were 40 or more years old. Multiple comparisons tests showed that participants less than 40 years old typed Words/Spaces faster, on average, than Sentences ($p<$ .001), and typed Sentences faster than Addresses ($p<$ .001) and Web text ($p<$ .001). Similarly (but with somewhat different magnitudes), participants 40 or more years old typed Words/Spaces faster, on average, than Sentences ($p<$ .001), and typed Sentences faster than Addresses ($p<$ .001) and Web text ($p<$ .001).

Additionally, a text type by keyboard interaction effect was significant ($F(6, 824)=15.486, p<$ .001). Specifically, participants typed the Words/Spaces text type, on average, slower using the Alphabetical keyboard than with the Hybrid ($p<.001$) and Predictive ($p<.001$) keyboards. Similarly, participants also typed Sentences, on average, slower using the Alphabetical keyboard than with the Hybrid ($p<.001$) and Predictive ($p=.002$) keyboards. However, participants achieved greater mean text entry speeds using the Alphabetical than with the Predictive ($p<.001$) and the Hybrid ($p<.001$) keyboards when typing Web addresses. Furthermore, there was no significant difference in the typing speed of participants for Addresses between the Alphabetical keyboard and the Predictive ($p=.624$) as well as the Alphabetical and the Hybrid ($p=.018$, based on Bonferroni adjusted $\alpha=0.01$). It is reasonable for the Predictive and Hybrid keyboards to not surpass the Alphabetical keyboard when typing physical Addresses and Web addresses, given that these text types do not conform to the lexical modeling that defined the dynamic layouts.

ERROR RATE

The total error rate was the total number of error and corrective events divided by the total number of input events (both productive and corrective). The mean total error rates across keyboards were 7.6% for Alphabetical, 7.4% for Predictive, and 8.4% for the Hybrid keyboard. The mean total error rates across text types were 14.3% for Addresses, 8.1% for Sentences, 5% for Web addresses, and 3.8% for Words/Spaces.

Friedman tests were conducted given that the total error rate data were non-normal and resistant to successful transformation. The test results indicated that the total error rate was not associated with keyboard layouts ($X^2(2)=2.33$, $p=.311$). However, the total error rate was associated with text types ($X^2(3)=61.40$, $p<.001$). A post-hoc analysis based on Friedman rank-averages indicated that typing Addresses resulted in significantly greater error rates over Sentences ($p<.01$), Web ($p<.0001$), and Words/Spaces ($p<.0001$). Furthermore, typing Sentences resulted in significantly greater errors rates over Words/Spaces ($p<.001$). The remaining rank averages were not significantly different.

EFFICIENCY

KSPC is a metric for quantifying the efficiency of the three methods. The observed overall results for mean KSPC by keyboard were: Alphabetical M= 9.60 (SD= 1.521), Predictive M= 8.0 (SD= 1.998), and Hybrid M= 7.1 (SD= 1.799). The observed overall results by text type for mean KSPC were: Address M= 8.833 (SD= 1.056), Sentences M= 7.516 (SD= 1.922), Web M= 9.888 (SD= 1.529), and Words/Spaces M= 6.695 (SD= 1.963). Figure 4 shows both the computed and observed values. The observed values included typematic keystrokes (i.e., counted virtual key-presses during auto-repeat).

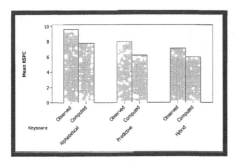

FIGURE 4. Computed and Observed KSPC by Keyboard

For the three keyboards, the observed KSPC (including typematic key-presses) was higher than the computed KSPC. Linguistic differences between the specific set of phrases entered and the language model could account for such differences. The effect of this difference depends on the statistical structure of each phrase. This effect, however, should be minor due to the high correlation between the letter frequencies in the phrase set and those in the reference corpus. The more likely cause of such differences is suboptimal entry due to inefficiency in keyboard usage. As depicted in Figure 4, participants entered more keystrokes per character than necessary: 23.2% more for Alphabetical, 27.8% more for Predictive, and 19.4% more for Hybrid. The percent difference was greatest for the Predictive keyboard. This implies that participants did not realize the intended benefits of language modeling with the Predictive keyboard to the same extent as the Hybrid. The most likely reason is that participants tended to overshoot and adjust more often with Predictive than with Hybrid, so participants did more work than necessary with the Predictive keyboard.

To determine whether keyboard layout, age, texting experience, or gender affected KSPC, a repeated measures ANOVA was conducted. The KSPC data were non-normal, but amenable to transformation by the natural log of KSPC plus 1. The final model for this design included keyboard, gender, text type and all associated interactions. There was a significant text type by keyboard interaction ($F(6,818)=$ 54.557, $p< .001$). Post-hoc comparisons revealed:

- The Alphabetical keyboard required significantly more keystrokes per character than Hybrid ($p< .001$) and Predictive ($p< .001$) when entering Addresses.
- Text entry with the Hybrid keyboard for Words/Spaces, Sentences and Web phrases was more efficient than Predictive ($p< .001$) and Alphabetical ($p< .001$).

PREFERENCE

The mean ranks for keyboard layout were 2.2 for Alphabetical, 2.5 for Predictive, and 1.4 for Hybrid (a lower mean rank is better- closer to first place). The Hybrid keyboard received the most first-place votes (16/24). A Friedman test showed a significant effect of keyboard ($X^2(2)$= 15.08, $p<$ .001). A post-hoc analysis based on Friedman rank-averages indicated that participants significantly preferred the Hybrid keyboard over the Predictive keyboard ($p<$ .001) and the Alphabetical keyboard ($p<$ .025). The remaining rank comparisons were not significantly different.

CONCLUSIONS

This experiment addressed novice five-key text entry with three alternative keyboard layouts across four text types. Results show no overall difference in novice performance (speed or errors) among the three keyboard layouts across four text types. However, the Alphabetical keyboard surpassed both the Predictive and Hybrid keyboards in text entry speed when typing Web addresses. The Predictive and Hybrid keyboards performed superior to the Alphabetical keyboards in typing Words/Spaces and Sentences, but performed no better in typing Address strings than the Alphabetical. This is most likely an effect of the optimization strategy, based on English digraphs, employed in designing the predictive layouts. Even so, it appears that performance did not attenuate as much, across text types, for the Alphabetical keyboard as with the Hybrid and Predictive keyboards. All keyboards followed the same basic pattern in performance across text types, but not to the same degree.

Results also suggest that using the Alphabetical keyboard at the novice level will not lead to an overall reduction in error rate in comparison to using the Predictive and Hybrid keyboards. Although accuracy was high overall, typing Addresses resulted in higher total error rates across keyboards. It appears that none of the predictive methods studied led to a significant gain in text entry rate or a reduction in error at this level of user training. Despite a lack of a strong association among the three keyboard layouts on the dependent measures for speed and error rates, there were substantial differences among the three methods in efficiency and user preference.

Keyboard designs were optimized to reduce KSPC. In this evaluation, of the three keyboard layouts, the Hybrid had the lowest difference in observed vs. computed KSPC. Further, the analysis indicated that the Hybrid keyboard layout minimized the required keystrokes for most text types.

When ranking layouts, the Hybrid keyboard received the most first-place votes. Participants' comments indicated that the advantages of the digram-based predictive layout are more accessible when typing with the Hybrid keyboard. Furthermore, participants' comments indicated that the Predictive keyboard was the most frustrating of the methods due to the high visual attention required. Overall, the Hybrid keyboard layout appeared to meet its design goals of increased efficiency and reduced demand on visual attention.

REFERENCES

Bellman, T., & MacKenzie, I. S. (1998). A probabilistic character layout strategy for mobile text entry. In *Proceedings of Graphics Interface '98* (pp. 168-176). Vancouver, Canada: Canadian Information Processing Society.

Curran, K., Woods, D., & Riordan, B. O. (2006). Investigating text input methods for mobile phones. *Telematics and Informatics*, 23, 1-21.

Lewis, J.R. (1993). Pairs of Latin squares that produce digram-balanced Greco-Latin designs: A Basic program. *Behavior Research Methods, Instruments, & Computers*, 25, 414-415.

Lewis, J. R. (1995). Input rates and user preference for three small-screen input methods: standard keyboard, predictive keyboard, and handwriting (Tech. Report 54.889). Boca Raton, FL: IBM Corp.

MacKenzie, I.S., & Soukoreff, R.W. (2003). Phrase sets for evaluating text entry techniques. Extended Abstracts of the *ACM Conference on Human Factors in Computing Systems* - CHI 2003, 754-755.

Millet, B., Asfour, S., & Lewis, J.R. (2008). Designing a hybrid layout for a five-key text entry technique. In the Proceedings of the *International Society for Occupational Ergonomics and Safety*, Chicago, IL, June 12-13, 2008, pp. 156-162.

Millet, B., Asfour, S., & Lewis, J.R. (2009). Selection-based virtual keyboard prototypes and data collection application. *Behavior Research Methods*, 41(3), 951-956.

Sears, A., & Zha, Y. (2003). Data entry for mobile devices using soft keyboards: Understanding the effects of keyboard size and user tasks. *International Journal of Human-Computer Interaction*, 16, 163-184.

Wobbrock, J. O., Myers, B. A., & Rothrock, B. (2006). Few-key text entry revisited: Mnemonic gestures on four keys. In *Proceedings of the ACM Conference on Human Factors in Computing Systems '06* (pp. 489-492). New York, New York: ACM Press.

Chapter 49

Adaptive Haptic Touchpad for Infotainment Interaction in Cars How Many Information is the Driver Able to feel?

Roland Spies, Werner Hamberger, Andreas Blattner,
Heiner Bubb, Klaus Bengler

Institute of Ergonomics
Technical University of Munich
85747 Garching, Boltzmannstr. 15, Germany

HMI Development
Audi AG
85045 Ingolstadt, Germany

ABSTRACT

This contribution deals with an innovative concept for in-vehicle infotainment control. Therefore a touchpad with an adaptive adjustable surface is suggested. To decrease the driver distraction from driving, the idea is to give additional information of the menu content via the haptic channel by elevating shapes on the touchpad surface. A test in a static driving simulator has been conducted to compare different shapes on a touchpad surface with each other.

Keywords: Automotive, HMI, Haptic, Touchpad, Infotainment

INTRODUCTION

To keep the increasing amount of information in modern vehicles easily controllable for the driver and also to minimize the mental workload, sophisticated presentation and interaction techniques are essential. This paper deals with the development and evaluation of an innovative input concept for menu-based automotive infotainment systems.

A challenge for car manufacturers today is the increasing amount of comfort functions in modern vehicles e.g. navigation, media and communication systems and soon internet services. To keep all these systems controllable while driving, car producers have integrated these functions in menu-based central infotainment systems which are mostly controlled by one multifunctional input device. Currently, many different solutions of such control devices are available on the market. These solutions can be divided into two different groups: First, an integrating approach represented by touchscreens and second an approach separating display and control element, e.g. turning knobs or joysticks in the center console. The latest concept with an additional touchpad for character input is delivered by Audi (Hamberger and Gößmann, 2009).

This contribution delivers a highly new approach for intuitive in-vehicle infotainment interaction via touchpad with an adaptive haptic adjustable surface. The idea is to control the whole menu content by just one single input device. The following chapter describes the control concept as well as the technical realization of a prototype.

THE HAPTIC TOUCHPAD

To make a touchpad useable for automotive applications, the idea is to give an additional haptic structure onto the surface for orientation. A prototype of a touchpad with such an adjustable haptic structured surface has been built up for evaluation purposes (Spics ct al., 2009a). In order to realize the haptic surface the technology of hyper braille is used (Hyperbraille, 2010). The touchpad surface can be adjusted to the displayed content on the middle screen, so that the user can feel and press every elevated element on the touchpad which is shown on the display. An additional visual feedback by highlighting the currently touched graphical widget on the screen shows the user the finger position on the touchpad. This guarantees an absolute compatible interface design between display and control device because of direct manipulation.

Moreover new interaction concepts, like it is suggested by Broy (2007) for example, are possible by using a two dimensional touch device. A touchpad enables a separation of display and control, what means that the position of the control element is independent from the display position. This means that infotainment information can also be provided in the Head Up display (Milicic, 2009) or further more innovative interaction concepts as it is proposed by Spies et al. (2009b) are thinkable.

Figure 1 shows the prototype of the haptic touchpad.

FIGURE 1: Prototype of the haptic touchpad

In case of menu operation while driving it is about a dual task situation which can cause interferences between the two parallel tasks. This means that menu control leads to driver distraction from the driving task. The major goal of an ergonomic interface design is to support the user in the phases of human information processing according to Bubb (1993). Especially the part of information detection via the visual channel from two different places while driving leads to a high number of glances away from the driving scene.

The idea of this contribution is to give the user redundant information via the haptic channel by elevating different kinds of shapes on the touchpad surface. Thus the haptic structure is not only for finding elements but also for getting the content of the element. A reduced number of glances away from the driving scene to the screen are expected. The next chapter describes the method of a usability test to clarify the following hypothesis:

1. Different haptic shapes enable transporting information via the haptic channel and reduce the visual effort for the menu control task
2. Giving information via the haptic channel by haptic shapes according to the menu content leads to safer driving performance
3. Sensing the menu content via the haptic shapes enables a faster menu operation

USABILITY TEST

A usability test in a static driving simulator with 36 test persons (25 M, 11 F, Ø= 39,36 years, S= 10,73 years) has been conducted by using a defined subject design. The driving task consisted of following a vehicle in a constant distance. The test persons had to adjust the brightness of the navigation map while driving. A slider and buttons as two different widgets are considered for such a menu task according to Spies and Bubb (2009).

For both kinds of widgets a standard version and a version with a shape according to the menu content are compared with each other (see figure 2).

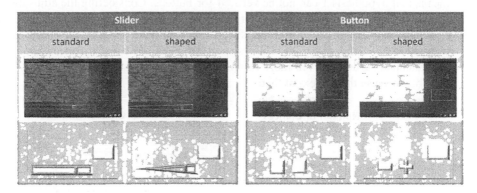

FIGURE 2: Variants of interaction widgets for adjusting the brightness of the navigation map

The Control speed, the gaze behavior as well as the driving performance were measured.

RESULTS

The statistic comparison is done by several t tests. The benefit of shaped haptic elements could not be shown in this simulator test.

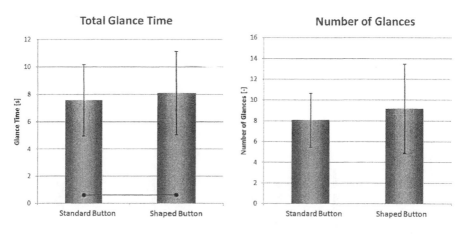

FIGURE 3: Gaze behavior of the test persons while adjusting the map brightness via the two different button versions

Comparing the data of gaze behavior of the test persons while adjusting the

460

brightness of the navigation map with the two different variants of buttons shows that there is no difference between the two variants as it is shown in figure 3. This means that hypothesis 1 can not be verified for buttons. Considering the adjustment task with the two different slider versions, figure 4 shows that there is even a significant higher number of glances as well as a significant longer total glance time with the shaped slider version. The test persons mentioned that the orientation on the touchpad surface is even more complex with different shapes and that it is not possible to sense the context of them.

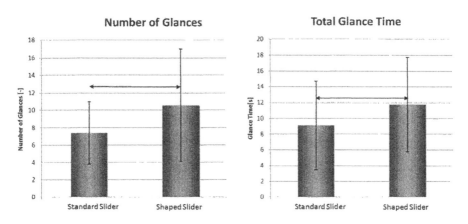

FIGURE 4: Gaze behavior of the test persons while adjusting the map brightness via the two different slider versions

Hypothesis 2 and 3 can not be verified either. Figure 5 shows that there is not any significant difference in control speed and lane deviation between the standard and the shaped button versions.

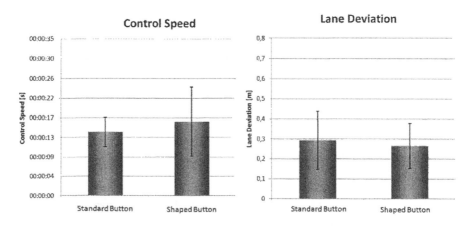

FIGURE 5: Control speed and driving performance of the test persons while

adjusting the map brightness via the two different button versions

Figure 6 even shows a significant slower control with a shaped slider than with a standard one.

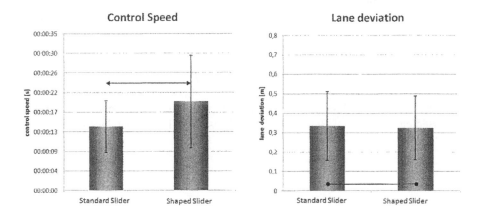

FIGURE 6: Control speed and driving performance of the test persons while adjusting the map brightness via the two different slider versions

As a conclusion it can be said that it is not possible to deliver context information via the haptic channel for menu operation in dual task situations. Different shapes make orientation even harder and lead to longer control times as well as worse driving performances. A reduced visual effort because of haptic information is not realizable. The haptic benefit is just given by simple rectangular shapes to help the user with orientating on the touchpad surface (Spies et al., 2009a).

REFERENCES

Bubb, H. (1993): *Informationswandel durch das System.* In: Schmidtke, H. (Hrsg.): Ergonomie. 3. Auflage. München u.a.: Carl Hanser

Broy, V. (2007): *Benutzerzentrierte, graphische Interaktionsmetaphern für Fahrerinformationssysteme.* Technischen Universität München, Dissertation

Hamberger, W., Gößmann, E. (2009): *Bedienkonzept Audi: Die nächste Generation.* In: VDI Wissensforum GmbH. (Hrsg.): Elektronik im Kraftfahrzeug. VDI-Berichte 2075, S. 677-686, VDI-Verlag, Düsseldorf

Hyperbraille (2010): http://www.hyperbraille.de/

Milicic, N., Platten F., Schwalm, M., Bengler, K. (2009): *Head-Up Display und das Situationsbewusstsein.* In: VDI Wissensforum GmbH. (Hrsg.): Der Fahrer im 21. Jahrhundert Fahrer, Fahrerunterstützung und Bedienbarkeit. VDI-Berichte 2085, S. 205-219, VDI-Verlag, Düsseldorf

462

Spies, R., Peters, A., Toussaint, C., Bubb, H. (2009 a): *Touchpad mit adaptiv haptisch veränderlicher Oberfläche zur Fahrzeuginfotainmentbedienung*. In: Brau, H., Diefenbach, S., Hassenzahl, M., Kohler, K., Koller, F., Peissner, M., Petrovic, K., Thielsch, M., Ullrich, D. & Zimmermann, D. (Hrsg.). Usability Professionals 2009. Fraunhofer Verlag, Stuttgart.

Spies, R., Ablaßmeier, M., Hamberger, W., Bubb, H. (2009 b): *Augmented interaction and visualization in the automotive domain*. In: J.A. Jacko (Ed.): Human-Computer Interaction, Part III, LNCS 5612, pp.211-220, Springer-Verlag Berlin HeidelbergDendrinos, D.S. (1994), "Traffic-flow dynamics: a search for chaos." *Chaos, Solitons and Fractals*, 4(4), 605–617.

Spies, R. and Bubb, H. (2009): *Infotainmentinteraktion der Zukunft - Touchpad mit adaptiv haptisch erfühlbarer Oberfläche*. In: VDI Wissensforum GmbH. (Hrsg.): Der Fahrer im 21. Jahrhundert Fahrer, Fahrerunterstützung und Bedienbarkeit. VDI-Berichte 2085, S. 247-258, VDI-Verlag, Düsseldorf.

Chapter 50

Design Approach of a Direct-Touch Input for Navigation Techniques Interface of Web Map

Hsuan Lin, Fong-Gong Wu

Department of Industrial Design
National Cheng Kung University
1 Ta-Hsueh Rood, Tainan, Taiwan70101, Taiwan

ABSTRACT

With the development and maturity of network technology and web map is change that the human use map. Through the computer can update the map information and display of suitable for human use. Nonetheless, the way a input from previous mouse into a touch-screen. The input device is changed, but the operating interface is still adopting previously operating interface. The purpose of this study is exploring the now widely way-finding tools – web map that user interface in the mouse and touch screen (Direct-Touch) to study better functions and interfaces of the Web map and to facilitate it usage. We collect and analyze the operating interfaces currently, and observe participants to operate the web map realistically. First, we through the world of famous search engines Google, Yahoo, etc. to collect 80 maps. After sifting carefully, we selected 8 representatives of web map that they can represent the type of web map operation function. Through the touch screen, 8 participants use the 8 maps performance actually. The result of operation: 1) when the fingertip press touch screen too much force or too less to effect touch and press accuracy of direction icon; 2) when the users who use right dominant hand across the screen, the right hand cover up sight line; 3) the back of middle finger, ring finger, and little finger easy to touch screen automatically; 4) Currently, sizes of direction icon of web map are too small to make the users operating troubled. According to the result of observation, the main of design not only resolve the area of fingertip too much but also improve the operating interface handing. As a result, we design Control Continuous-Enhanced Navigator. In the future, we will simulate testing on 4 different operating interfaces are conducted in the touch screen: 1) Combined-panning buttons or CPB; 2) Distributed-panning buttons or DPB; 3) Control Continuous-Enhanced Navigator or CCEN; and 4) Grab & Drag. We hope the result that would provide useful reference for us to design the new interface that suits for touch screen to operation. In conclusion, the results of this study in its analysis, experiments, and design all could be useful references for further research and practical applications of thematic web maps, PDA, GPS, Smart phone.

Keywords: Web Map, Input Devices, Operator Interfaces, Touch Screen

INTRODUCTION

With the new operating systems Win7 and iPhone OS introduced to the market, it is expected that the operation interface which previously relied on the mouse as the input device will gradually opt for the touch screen. As is confirmed by some studies, the touch screen is characterized by intuitive input, which makes it much easier for the user to learn and operate the device (Lu et al., 2005). Up to the present, the touch screen has been widely used in kiosks, ticket machines, automatic teller machines (ATM), and so forth. Thanks to the features of the touch screen, the public has made use of kiosks more and more frequently (Albinsson and Zhai, 2003). kiosks are mainly intended to provide convenient and instant services, such as web maps, cash withdrawals, museum sitemaps, and self-service gas stations (Kules and Kang, 2005; Cartwright et al., 2001). The general public will usually turn to an kiosk, accessing the web map to get familiar with the new environment. Therefore, the usability and functionality of the web map have been highlighted as an important issue.

As the web map is browsed, its navigation mode, or how it is presented, is a key factor influencing the user's viewing and operation (Gutwin and Fedak, 2004). A well-designed navigation technique can effectively lead the user to browse the information space of the webpage; furthermore, the user can explore its content by activating various functions (Neumann, 2005). As some researches have found, if the user is unfamiliar with the conceptual model, he/she will be inclined to commit operational errors. As a result, he/she will easily suffer a sense of frustration and take less interest in the web map (Norman, 1988). In view of the above, when the web map is being designed, the significant task facing the designer is whether the navigation mode will effectively provide the user with correct cognitive guidance and information feedback.

For most people, the mouse is the most common input device; consequently, nearly all operational interfaces, including the web map, base their navigation on the mouse and are designed as well as operated in the same way. If the mouse is replaced by other input devices, like the touch screen, the user may have some difficulty in operating and suffer lower efficiency. In the past, some researches were made to compare the functionality of different input devices (Card et al., 1978; Esenther and Ryall, 2006; Sears and Shneiderman, 1991; Forlines et al., 2007; Kin et al., 2009; Benko et al., 2006). Nevertheless, there were few studies which focused on the input devices and performed the in-field, simulated tests on the navigation techniques of the web map.

The purpose of this study is exploring the now widely way-finding tools – web map that user interface in the touch screen (Direct-Touch) to study better functions and interfaces of the Web map and to facilitate it usage.

CASE STUDY AND OBSERVATION ON MAP OPERATION INTERFACE

SURVEY OF WEB MAPS

In July, 2008, the authors employed the search engines, like Google and Yahoo, entered the keyword "web map", and searched both the Chinese and English map websites. From the searching result, the first eighty websites were arranged in descending order of relativity; thereafter, the web maps were singled out which possessed either continuous control or discrete control. Those websites which had been out of service or had an unstable connecting speed were excluded. Only one website was chosen from those which provided similar services. Moreover, to render it easier to make comparisons between web maps, the domestic maps were confined to those of Taiwan while the foreign maps were confined to those of the United States proper (excluding Alaska and Hawaii). In the end, eight web maps were chosen for the research under discussion.

Based on the eight experimental samples, we manipulated the functions of panning, zooming, and moving owned by the web map (Neumann, 2005; You et al., 2007), practiced searching for the targets, and identified navigation techniques currently adopted by the web maps. All functions were collected and classified as shown in Table 2 and 3. In the tables, we discovered that the navigation modes of the web map consist of two main functions, namely, control and display. The control function refers to the functionality of the press button, which is divided into continuous control and discrete control. The display function refers to the presentation mode when the image action is executed. Similarly, there are two modes of display: continuous display and discrete display.

Table 1: Zoom functions of web map operation interface

Functions	Zooming					
	Zoom-in (re-center)	Zoom-out (re-center)	Zoom-in (original center)	Zoom-out (original center)	Zoom-in by marquee	Zoom by fixed scales
Gi map						X
Go map	X+					X
M24 map					X	X+*
MM map	X+*		X+*	X+*		
P map						X
U map						X
Y map						X
MI map			X	X		X

Note: X denotes available function available, *denotes continuous control, and + denotes continuous display.

Table 2: Pan and move functions of web map operation interface

Functions	Panning		Moving	
	Panning (grouped buttons)	Panning (distributed buttons)	Move by re-center	Move by dragging
Gi map	X+		X+	X+
Go map	X+			X+
M24 map			X	
MM map	X+*			X+
P map		X+*	X+	X+
U map			X+	X+
Y map			X+	X+
MI map	X	X		

Note: X denotes available function available, *denotes continuous control, and + denotes continuous display.

OBSERVATION: USER OPERATING INTERFACE SURVEY AND OPINION

This research was targeted at the students of National Chengkung University, from whom volunteers were recruited. There were 8 participants in total, with males and females in equal numbers. The ages of the subjects ranged from 22 to 33 (Mean of age=26.57, SD=3.55), and they were all right-handers. Hardware included a desktop computer accompanied by a monitor or touch screen (manufactured by 3M and the model number being M170). The researchers tested map operation functions for the 8 selected sites with main focus on basic map operations, such as zooming and panning. The procedure used was as follows: (a) search the same places (Tainan City for Taiwan maps, Manhattan, New York for worldwide maps) in all 8web maps; (b) click the map operation buttons to navigate the area and learn the methods of use; and (c) record available functions, the methods of use , and the icon and layout design.

RESULTS OF OBSERVATION AND ANALYSIS

By making experimental observations as well as collecting and analyzing the user feedback, the researchers have obtained the following results, as is shown in figure 1.

- When operating the distributed arrow keys, the right-hander will move the right arm over the screen to reach the left-view image and press those keys on the left side, suspending the arm above the map and occluding much of it from view.
- When operating the combined arrow keys, the user is apt to press the wrong key because the desired key is too small in size.
- When operating the combined arrow keys, some users press the desired one so hard as to distort the finger. In such a case, the area covered by the fingertip is so large that the wrong key is also pressed. As a result, the system can not detect the pressing correctly and even cannot recognize it at all.
- When operating the distributed arrow keys, the user needs more time to move the hand onto the desired one. Since the keys are scattered on all the edges of the screen, the relative distance is made longer.

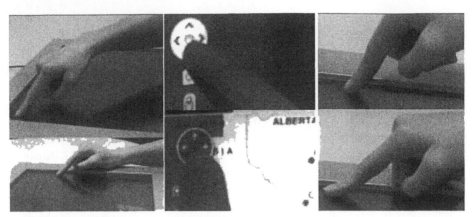

Figure 1. Handing in the   Figure 2. The control component   Figure 3. The area covered by
air and wrist local support.   occluded by the finger support.   the user's fingertip.

Based on the above observations and analyses, this study has identified the restrictions on the existing navigation tools adopted by the touch screen and chosen the course to take in developing the new navigation tool. The following aspects of design must be emphasized and enhanced:

- The area where the fingertip contacts the screen should be decreased to eliminate the wrong pressing.
- The cases where the hand spans the screen and occludes the map must be reduced; likewise, the pain of the muscles and bones arising from the suspended arm must be lessened.
- The arrow keys must be enlarged so that they may be pressed with ease.
- The relative distance between the hand and the arrow key must be shortened so that the time needed to move the arm may be decreased.
- The operational stability and controllability must be upgraded.

The fruits of relevant designs will be presented in the coming paragraphs.

DESIGN DEVELOPMENT

DECIDING ON OPERATIONAL MODES

As Fitt's Law indicates, with the distributed arrow keys scattered on the edges of the screen, the relative distance gets longer (Fitts, 1954). When operating such keys, the user will move the arm over a longer distance and thus spend more time in doing the pressing. Consequently, it will take the user more time to browse a certain map.

Moreover, by means of experimental observations, the researchers have discovered that the distributed arrow keys cause the right-hander to suspend the arm over the map involuntarily. In consequence, the suspended arm occludes the map from view and easily creates the pain of the arm muscles and bones (Lee et al., 2008). In view of the above, the navigation tool designed by the researchers adopts the centralized mode instead of the distributed mode.

In employing the navigation tool, the user has to click different arrow keys, panning or zooming the interface. However, this research finds out that some of the control components are not adequately designed, with problems including a tiny size, unacceptable proximity, and redundant clicking. Such problems lead to the mutual interference between the keys and lower the efficiency of the interface (Burigat et al., 2008). Also, it is discovered that the surface area where the fingertip contacts the screen exerts an effect on whether the arrow key will be clicked or pressed precisely (Clifton et al., 2007). To solve the abovementioned problems, the single control ball is adopted by us instead of the arrow keys; therefore, grabbing or dragging the ball is performed instead of clicking or pressing the key.

As for directional control, either the combined or distributed arrow keys have just eight directions available and set a limit on mobility. Therefore, we have decided on the control ball, which offers a 360-degree motion and facilitates smooth operation. Besides, the two-speed control has been added to the newly-designed navigation tool. The high or low speed is determined by the distance from the ball to the center of the circle. There are two zones, corresponding to the two kinds of speed. When the ball is within one zone, the speed remains constant and low; however, when it is in the other zone, the speed increases in proportion to the distance from the center.

PROTOTYPE OF THE CONTROL CONTINUOUS-ENHANCED NAVIGATOR

Starting from the concept of human factors design and making a series of researches, observations, and analyses, we have presented the design of the Control Continuous-Enhanced Navigator (CCEN), as is shown in figure 4. Its main design elements are described as follows: (1) Grabbing and dragging are the acts needed to operate the control ball (figure 5). (2) The distance between the ball and the center of the circle is directly connected to the speed of motion. (3) The high-speed or low-speed zone is marked in a different color. (4) The interface is operated through grabbing and dragging in the continuous control mode. Therefore, to perform the continuous action, the user is not required to repeat the same command; furthermore, the muscles and bones of the hand will not be overworked, with the burden on them reduced.

470

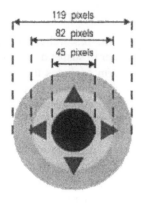

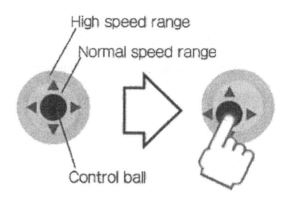

Figure 4. Prototype of CCEN.   Figure 5. Handing in the control ball.

CONCLUSIONS

In the future, we started to test navigation techniques functions, especially zooming and panning. Besides, we practiced searching for the targets and identified the navigation modes currently adopted by web maps. According to the result of observation, we will simulate testing on 4 different operating interfaces are conducted in the mouse and touch screen: 1) Combined-panning buttons or CPB; 2) Distributed-panning buttons or DPB; 3) Control Continuous-Enhanced Navigator or CCEN; and 4) Grab & Drag.

We wish that the research would provide useful reference for us to design the new interface that suits for touch screen to operation. In conclusion, the results of this study in its analysis, experiments, and design all could be useful references for further research and practical applications of thematic web maps, PDA, GPS, Smart phone.

REFERENCES

Albinsson,P., Zhai,S. (2003), High precision touch screen interaction, in, ACM New York, NY, USA, 105-112.

Benko, H., Wilson, A., Baudisch, P., (2006), Precise selection techniques for multi-touch screens, Google Patents.

Burigat, S., Chittaro, L., Gabrielli, S., (2008), Navigation techniques for small-screen devices: An evaluation on maps and web pages, International Journal of Human-Computer Studies, 66, 78-97.

Card, S., English, W., Burr, B., (1978), Evaluation of mouse, rate-controlled isometric joystick, step keys, and text keys for text selection on a CRT, Ergonomics, 21, 601-613.

Cartwright,W., Crampton, J., Gartner, G., Miller, S., Mitchell, K., Siekierska, E., Wood, J., (2001), Geospatial information visualization user interface issues, Cartography and Geographic Information Science, 28, 45-60.

Esenther, A., Ryall, K., (2006), Fluid DTMouse: better mouse support for touch-based interactions, ACM, 115.

Fitts, P., (1954), The information capacity of the human motor system in controlling the amplitude of movement, Journal of experimental psychology, 47, 381-391.

Forlines, C., Wigdor, D., Shen, C., Balakrishnan, R., (2007), Direct-touch vs. mouse input for tabletop displays, ACM, 656.

Gutwin, C., Fedak, C., (2004), Interacting with big interfaces on small screens: a comparison of fisheye, zoom, and panning techniques, Canadian Human-Computer Communications Society, 152.

Kin, K., Agrawala, M., DeRose, T., (2009), Determining the benefits of direct-touch, bimanual, and multifinger input on a multitouch workstation, Proceedings of Graphics Interface 2009, 119-124.

Kules, B., Kang, H., Plaisant, C., Rose, A., Shneiderman, B. (2005), Immediate usability: Kiosk design principles from the CHI 2001 photo library, citeseer. csail. mit. edu/571542. html. Last accessed, 22.

Lee, D., McLoone, H., Dennerlein, J., (2008), Observed finger behaviour during computer mouse use, Applied Ergonomics, 39, 107-113.

Lu, Y., Xiao, Y., Sears, A. Jacko, J., (2005), A review and a framework of handheld computer adoption in healthcare, International Journal of Medical Informatics, 74, 409-422.

Neumann, A., (2005), Navigation in space, time and topic, International Cartographic Conference, 11-16.

Norman, D., (1988), The psychology of everyday things, Basic books New York.

Sears, A., Shneiderman, B., (1991), High precision touchscreens: design strategies and comparisons with a mouse, International Journal of Man-Machine Studies, 34, 593-613.

You, M., Chen, C., Liu, H., Lin ,H., (2007), A usability evaluation of web map zoom and pan functions, International Journal of Design, 1, 15-25.

Chapter 51

Study on a Haptic Interaction with Touch Panel

*Daiji Kobayashi, **Sakae Yamamoto

*Department of Global System Design
Chitose Insitute of Science and Technology
756-65 Bibi Chitose Hokkaido, 066-8655 Japan

**Department of Management Science
Faculty of Engineering
Tokyo University of Science
1-3 Kagurazaka Shinjyuku-ku Tokyo, 162-8601 Japan

ABSTRACT

Recently, personal digital assistance (PDA) and PC with touch panel are in widespread use in Japan. Although touch panel operation including intuitive manipulation is low-skilled interface for young individuals, the user experience could be influenced by various effects. Therefore, the characteristics of the touch panel operation should be considered. In this study, the touch panel operation of evacuation route map on a PDA were compared with the handling a sheet of printed evacuation route map through experiments. The evacuation route map on the PDA included two semi-transparent buttons for zooming and the map position was moved by rubbing the touch panel with a finger. The regular evacuation route map was printed in black and white onto a sheet of A4 paper. Ten participants of each tried to walk from the specified location to the evacuation center using different types of evacuation route map; however, we focused on the participants' operation of the map. As the results, the participant's haptic interaction with the map was found, and it was revealed that the haptic interaction was useful for understanding the present location and planning the evacuation route; however, the map on the screen with touch panel prevented the user from haptic interaction, and the user experienced some kind of frustration.

Keywords: touch panel operation, haptic interaction, map

INTRODUCTION

The government of Japan orders the inhabitants to evacuate by self-help or by mutual assistance in urban disaster situation. When urban earthquake hazards are occurred, the bumps and obstacles on the street could block the evacuation behavior. More, the ever-changing blaze along the evacuation route jeopardizes the evacuee's life. Although the systems distributing information about the evacuation such as broadcasting network have existed, the appropriate information for each evacuee's situation should be available. The personal navigation system which informs the appropriate evacuation route in accordance with the user's physical and cognitive characteristics is a way to deal with the issues. In order to design the system's human interface, using the personal digital assistance (PDA) is a possible way. The touch-panel-equipped PDA is now in widespread use and the touch panel operation is more intuitive than before. The intuitive touch operation contains such as tracing a line with the fingers on the touch panel. Our previous study has addressed the usability issues of the evacuation route map on PDA and the evacuee's cognitive process was grasped to some extent (Kobayashi et al., 2009). Meanwhile, many to move touch panel operations were observed regardless of age or experience. The to move touch panel operations were assumed to be caused by various reasons, for example the sensitivity of touch panel included in the mechanical matter, the user's touch panel operation based on the user's cognitive and/or physical characteristics and so on. The mechanical matter should be the issues for PDA manufacture; therefore the factors relating to the aesthesis and cognition should be research. The user's aesthesis and cognition in touch panel operation can be relevant to the haptic interaction. Thus, we investigated the user's behavior from the viewpoint of haptic interaction.

METHOD

In order to compare the haptic interaction between two different types of evacuation route map, an observational study was conducted through experiment. The task for participant was moving from the specified location to an evacuation center in Shinjuku ward, Tokyo with the map in hand. The two types of map were the printed map onto a sheet of A4 paper and the map shown on screen with touch panel. Both of the maps indicated the same starting location, the goal and impassable spots.

The PDA was running Microsoft® Windows Mobile 6.1 Professional operating system, Japanese edition, and had a 4.1 inches wide VGA display with touch panel.

The custom software, displayed the map on screen of the PDA, was running on Adobe® Flash Player. The PDA showed the map and two semi-transparent buttons for zooming, and the map position was moved by rubbing the touch panel with a finger.

The participants were 19 students with ages ranging from 20 to 23 years and an elderly with ages 64 years. They were divided into 2 groups including 10 participants of each. The participants of group-A including 10 students with ages ranging from 20 to 21 years walked on foot seeing the printed evacuation route map. The group-B, consist 10 participants including 9 students with ages ranging from 20 to 23 years and an elderly with age 64 years, walked on foot using the map on the PDA. All participants were not inhabitant in the district and every participant of group-B did not have experience in using PDAs.

The participant of group-A was showed the present location (staring location) on the map, and walked to the evacuation centre depending on the sheet of map. The group-B were instructed on how to zoom and move the map on the PDA, and allowed to practice touch panel operation. After receiving the instructions, each participant walked from the start point to the evacuation center.

The participant's behavior including the operation of PDA or the handling of the sheet of map was recorded using a video camera and their thinking aloud protocol was recorded using a voice recorder. After arriving at the evacuation center (goal), the participant of group-B assessed the user experience in using the map on PDA based on a usability questionnaire. The questionnaire we made, consisted of 7 statements to which the user rates agreement on a 5-point scale of "Strongly Disagree" to "Strongly Agree," and the score of agreement was the followings:

1. Strongly disagree
2. Disagree
3. Neither agree nor disagree
4. Agree
5. Strongly agree

Seven statements represented the user experience on specific map operation were as follows:

1. I thought the semi-transparent buttons were easy to touch.
2. I thought the buttons were easy to find.
3. I thought the role of the buttons was understandable.
4. I thought the reaction of zooming was preferable.
5. I thought the magnification range of the map was preferable.
6. I thought the way to move the map was understandable.
7. I thought the map was easy to move.

RESULTS AND DISCUSSION

Although every participant made it to walk to the goal, the averaged score from the questionnaire result by group-B is as shown in Table 1. The questionnaire result indicates that the averaged score in 7th statements about the way to move map is 3.0. This is the lowest score with the maximum of standard deviation. However, the averaged score in 6th statement about the way to move map is 4.7 (S.D. =.5). Regarding the scores in 6th and 7th statement, the participants responded 'It was easy to zoom the map'; 'It was difficult to move the map by a finger', 'The sensitivity of touch panel was not understandable, and so button operation was preferable'.

The percentage of invalid operations in the all touch or rubbing operations observed was ranging from 8 to 27 (mean=17.7, S.D. =10.0) and it was difficult to say that the scores in 6th and 7th statement correlated strongly with the percentage of invalid operations.

Considering the above results of group-B, the experience in moving the map depended on the person. However, the assessment of the operation for moving the map was not preferable. Thus, it was assumed that the issue of haptic interaction existed. Therefore, we compared the haptic interaction with the map between group-A and group-B.

Table 1 Scores from the Questionnaire Result

Statement	Score	
	Mean	S.D.
1. I thought the semi-transparent buttons were easy to touch.	4.0	1.2
2. I thought the buttons were easy to find.	4.1	1.2
3. I thought the role of the buttons was understandable.	4.9	0.3
4. I thought the reaction of zooming was preferable.	3.5	1.6
5. I thought the magnification range of the map was preferable.	4.1	1.2
6. I thought the way to move the map was understandable.	4.7	0.5
7. I thought the map was easy to move.	3.0	1.6

OBSERVATION OF HAPTIC INTERACTION WITH THE MAP

In order to reveal the characteristics of haptic interaction with touch panel, we observed the all participants' behavior recorded in the video camera and voice recorder from the view of the way of handling or manipulating the map. From the results, every participant of group-A referred the map sheet when they planned the evacuation route or when they became disoriented, and touch the map actively, or traced the route on the map with a finger. Briefly, the participants seemed to access the map by actively-touching and actively-tracing route with fingers for two reasons; one is to get information from the map well and the other is to help their understanding present location on the map.

On the other hand, 5 of 10 participants of group-B were point the map on screen impulsively when they planned the evacuation route and/or became disoriented, and 3 of the 5 participants managed to touch the map and the map was moved in the opposite direction without their intent. The other 5 of 10 participants of group-B did not touch the map on screen with care and 4 participants rated the experience in moving map worse (the score was 1 or 2). Considering these results, map on screen with touch panel prevents the participants' haptic interactions with the map and influences on the user experience.

CONCLUSION

The observational results suggest that touching and tracing map with fingers is natural and active behavior when we use the map in hand. Although many PDA have screen with touch panel today, many users has to take care not to touch the map without intent. Consequently, there is a possibility that the user experiences frustration with the map on screen because the touch panel prevented user from haptic interaction with the map on screen. Therefore, the map on screen with touch panel should be made in view of haptic interactions between the user and the map in the future.

REFERENCES

Tullis, T. and Albert, B. (2008), *Measuring the user experience*. Morgan Kaufmann

Chapter 52

Universal Communication Through Touch Panel

Naotsune Hosono*, Hiromitsu Inoue,
Hiroyuki Miki, Michio Suzuki, Yutaka Tomita

*Oki Consulting Solutions Co. Ltd.
4-11-15 Shibaura, Minato-ku
Tokyo 108-8551
Japan

ABSTRACT

This paper discusses the effectiveness of haptics or tactile interaction on touch screens for universal communication during an emergency or while traveling. The original idea was proposed by deaf people for communicating in an emergency, such as a disaster or accident situation. The solution consists of simple menu-like metaphors implemented on cards or on remote terminal equipment display of a public terminal unit. Previous research results showed that using cards shortened the time needed to communicate symptoms by 30%. This Universal Communication Card can be then used not only by deaf people but also by universal users such as foreigners.Human Centred Design is the underlying concept behind this research. It compares cards and remote terminal equipment display like iPhone. Assessors evaluated from the perspective of tactile sense applying originally designed Marble method and Multivariate Analysis. The results showed the remote terminal equipment display to be more favorable compared to cards in that it is quicker to reach the target card and back to the homepage. This is particularly useful when traveling.

Keywords: Universal communication, haptics, tactile, mobile, disability, emergency, traveling

INTRODUCTION

Previous research discussed the usefulness of emergency communication with simple cards that were originally proposed by deaf people (Suzuki, Hosono et al., 2005). Although deaf people appear no different from others in daily life, communicating during unexpected situations such as disasters or accidents can be very problematic. For instance, a deaf patient being carried by an ambulance may have difficulty conveying his problem to the emergency rescue personnel. Simple cards will be one solution for interfacing between the patient and emergency rescue personnel. Since the cards are simple menus like picture cards, the patient just points to the simple metaphors of pain location or severity level on the cards.

The acquired simple metaphors are now implemented on Remote Terminal Equipment (RTE) displays such as iPhones. This paper discusses the effectiveness of haptics or tactile interaction for the universal communication during emergency or travel of people with hearing disabilities, hard of speaking as well as foreigners with underlying Human Centred Design (HCD) concept. This research compares effectiveness, efficiency and user's satisfaction between the cards and RTE display in order to evaluate from the perspective of tactile sense. The evaluation was conducted using an originally designed Marble method combined with Multivariate Analysis (MVA).

BACKGROUND

One of the authors of this paper is deaf, and Universal Communication Card (UCC) was originally his proposal (Hosono and Tomita, 2008). The idea for the proposal came after he experienced communication problems in an ambulance. He tried to convey where the severe pain was, but the emergency rescue personnel asked his name and address. He not only suffered physical pain, but mental frustration as well. There are similar requirements from emergency rescue personnel in ambulances, nurses and doctors in hospitals for tools to communicate with deaf people.

UCC was created to aid the communication between deaf people in distress and their helpers. The card is used to tell the location of the pain with a picture of the body just like a menu. The deaf patient simply points to the portion of body with the finger to tell how deep and severe.

REQUIREMENTS

Following the Human Centred Design (HCD) concept (Miki and Hosono, 2005), the survey began by asking people with hearing disabilities about their difficulties and experiences concerning sudden illnesses and problems they encountered due to their disability. To start, deaf people were asked about their hardship by way of inquiry.

Twelve deaf people participated in the inquiry. Eleven experienced communication problem in the hospital and five while in the ambulance.

In parallel, the data of patient complaints collected by Tokyo Fire Department (TFD, 290,471 patients) and Keio University hospital (2,421 patients) were analyzed. The collected data in this case are not of deaf people but all patients. According to the collected data, more than 30% of complaints were pain related.

FUNCTIONAL ITEMS IN UNIVERSAL COMMUNICATION CARD

Reviewing TFD and university data under the context of use, the requirements and needs of end users are clarified in order to create functional items for the menu-like metaphors on the UCC (Graf, 1992).

It was concluded that among the ten selected items, the card was most useful, effective and particularly efficient in determining pain/ache/grief complaints for hearing disabled patients. Aches and pains were isolated in the head, face, chest, back, stomach, waist, hands, and leg/foot. The depths of the aches were in the skin surface, viscerally or at the bone level. During emergencies, patients view easy-to-understand picture metaphors or icons depicting pain/ache/grief and external injury. A minimum of carefully selected keywords augments the Universal Communication Card icons (Fig.1). Ache areas and ache depth and severity are presented horizontally.

Considering context of use, there will be four similar but different in detail types of cards. The main differences are the places to be used. First, a disabled person may access a particular Internet home page and print out a small Universal Communication Card to carry or download it to a mobile phone as a safety measure. It can be used in public places, in the ambulance or at the hospital. Second, it is always made available in public places such as railway stations, airports and department stores for emergencies should a disabled person suddenly become ill. In this case, the ill person mainly requires calling an ambulance. Third, it is installed in ambulances for communication between the ill person and emergency rescue personnel. The ill person will point at his/her ache points and its severity (Fig.2).
The card is not only for deaf people but also for foreigners who do not speak the native language, elderly people with hearing problems and those with throat disease (Hosono and Tomita, 2008; Suzuki, Hosono et al., 2005).

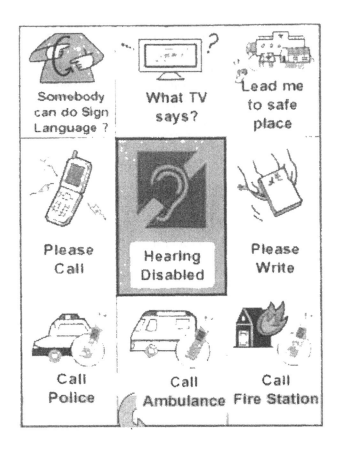

FIGURE 1 UNIVERSAL COMMUNICATION CARD (UCC) -1

SIMULATED EVALUATION OF THE UNIVERSAL COMMUNICATION CARD

Usability is measured by effectiveness, efficiency and satisfaction (ISO, 1998) underlying HCD. Effectiveness refers to how well the patient's condition is communicated to the emergency personnel. Efficiency is how quickly the information is conveyed compared to the situation without the card. Satisfaction is the mental relief the patient feels by carrying the card. Considering the nature of the card in such an urgent situation, efficiency is the most important factor. To measure usability, three simulated situations were prepared for the evaluations (Kurosu, 2003).

In usability testing, hearing disabled persons were asked to evaluate the UCC. Three tasks were conducted both with and without the card. In the first task, the disabled person asks the conductor at a railway station to call an ambulance. The second is letting the emergency rescue personnel in the ambulance know where the pain is located. The final task is telling the complaint to the nurses at the hospital. Four assessors joined the deaf disabled during the tasks. The railway station task took 22 seconds and 51 seconds with and without the card, respectively. The ambulance task was 152 seconds and 234 seconds with and without the card, respectively, for the first assessor, but 234 seconds and 321, respectively, for the second. On average, it was about a 30% improvement in the swiftness of the communication. The hospital task clocked in at 201 seconds and 269 seconds with and without the card, respectively. Overall, there was a 35% improvement in the communication time.

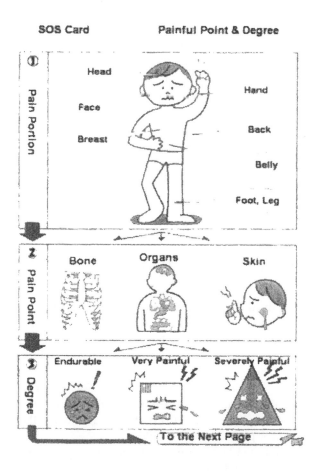

FIGURE 2 Universal Communication Card (UCC) -2

END USERS' EVALUATION BASED ON HCD

Human Centred design method (HCD) consists of five steps; 1. Plan the human center process, 2. Specify the context of use, 3. Specify user and organizational requirements, 4. Produce design solutions, 5. Evaluate designs against user requirements. During the first HCD design cycle, requirement surveys and evaluations were conducted using paper prototypes of the Universal Communication Card. Then the Universal Communication Card was placed on a Web site (http://www.aajd.org/) and made available for download by anyone who required it. Main users were hearing impaired who wanted to personally carry the card.

Observing the recent activities of the hearing impaired, more of them are carrying mobile phones for the email feature. Until now, phones were not practical for them. However, with the introduction of mobile mail, mobile phones have become an essential tool. Now the Universal Communication Card can be downloaded into not only personal computers (PCs) but also into mobile phones such as Apple's iPhone.

For the second HCD design cycle, UCC on mobile phones were evaluated by 34 assessors between the ages of 20 and 60 years old including 22 who were hearing impaired. The questions were about its usefulness and measure of relief. The results were produced by applying the Semantic Differential (SD) Method between 5 (Max.) and 1 (Min.). The usefulness result was 4.52 favorable with 0.51 standard deviation (SD). Measure of relief result was 4.55 M favorable with 0.62 SD.

TACTILE EFFECT EVALUATION

Through end user evaluation including many deaf people, it was found that many of them were interested in seeing UCC implemented on mobile phones particularly iPhones (Fig.3). The next step then is to compare human interface usability between cards and mobile phones with touch panels for sensory evaluation of the tactile following the HCD process.

For the experiment, two sample sets of UCC and RTE for iPhone with the metaphors were prepared with Persona Method (Cooper, 2007). For this evaluation 20 assessors, 16 males and 4 females, in their 20's participated. Three of them used iPhones daily. They are 3rd year grader of 20 years old.

The experiment was conducted between pairs of assessors. At first they were explained the purpose of the UCC and briefly instructed on the minimum operation of the iPhone jumping between the top page and object pages. Then the first assessor was requested to communicate just using samples of either the UCC or iPhone without speaking. A single experiment consisted of four cycles for a pair of assessors using either the card or iPhone. The task was that one assessor tries to communicate a scenario using either the card or iPhone and the receiving assessor scripts down the messages. The communication time was also measured with a stopwatch.

FIGURE 3 Universal Communication Card (UCC) metaphors on iPhone

Soon after this experiment, the assessors were required to measure their preferences on communication method with the tactile between UCC and iPhone through Marble Method. This method was originally created for sensory evaluation, a concept similar to the visceral level of emotion designed by D. Norman (Norman, 2005). In the preparation for this sensory evaluation, the required measurement and analysis were accomplished by utilizing Marble Method and Categorical Principle Component Analysis (CATPCA) of MVA (SPSS, 2009) of Statistical Package for Social Science (SPSS). The assessors are given fixed numbers of marbles (tokens) to distribute as votes between the two samples of UCC and iPhone. They were requested to place 7 (seven) marbles into 10 slots similar to Semantic Differential method (SD method) for three group evaluation items (Fig.4). First group item was the effectiveness with three factors of precise, easy to communicate and intuitive. The second group item was the efficiency with less practice, perspective feature, quick to search and easy to return. The third group item was satisfaction with comfort to use, emergency use, travel use and secure feeling. The outcome was analyzed by biplot chart where both items from above and assessors were plotted on the same plane (Fig.5).

Biplot result showed that iPhone with tactile function is better in efficiency with perspective, quick to search and easy to return. It is also favorable in travel use.

Female assessors tend to be positioned close to the weighting point of total average with slight bias toward precise, easy to communication and secure feeling. Among daily iPhone users, two are positioned to the weighting point whereas the third placed more emphasis on emergency use. All the assessors with overseas experience pointed out hard of communication. However, they are positioned separately.

FIGURE 4 An experiment view conducted by a pair of assessors with iPhone

Observing their evaluation manner, many assessors spread out all UCC on the table whereas iPhone is single screen. Judging from their behavior, they were uneasy with operating unfamiliar screens for the first time, even daily iPhone users. To reflect this, it must be necessary to ensure more preparation time to operate. When travelling, iPhone with a single touch screen must be preferable to use for its portable feature.

CONCLUSION

The concept of Universal Design must be readily achievable (FCC, 1996) and be an undue burden (DOJ, 2000) for not only disabled people but for manufactures as well. Universal Communication Card is one of the solutions to help overcome the difficulties the deaf people face without being overly expensive. Through evaluation of touch screen such as on iPhone, Biplot result showed that iPhone with tactile function is better in efficiency with perspective, quick to search and easy to return. It is also advantageous in travel use. During the process, it was found that the credibility to carry everyday brings much mental comfort and relief to the deaf people.

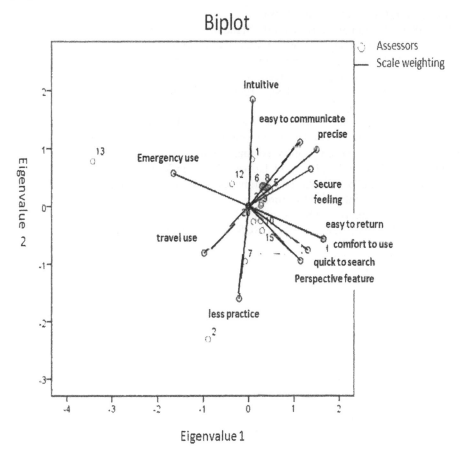

FIGURE 5 Biplot chart with items and assessors analyzed by MVA

ACKNOWLEDGMENTS

This research is supported by SCOPE (Strategic Information and Communications R&D Promotion Program) project organized by Ministry of International Affairs and Communication (MIC) of Japan.

Collected requirement data was provided by the Keio University Hospital Department of Emergency Medicine and the Tokyo Fire Department. The evaluations were done by in Kyushu/Japan, Suginami-ku Deaf Society in Tokyo and Prof. Dr. Sakae Yamamoto laboratory of Tokyo University of Science. This research was supported by the Oki Volunteer Fund.

REFERENCES

Cooper, A. (2007), About Face 3, Wiley.

DOJ (2000), Section 508 Accessibility Standards for E&IT.
Retrieved from http://www.section508.gov/index.cfm

FCC (1996), Section 255 Telecommunication access for people with disabilities.
Retrieved from http://www.fcc.gov/telecom.html

Hosono, N. and Tomita, Y. (2008), Urgent collaboration service for inclusive use, Vol.11, No.4, pp99-106.

ISO (1998), Ergonomic requirements for office work with visual display terminals (VDTs) -- Part 11: Guidance on usability 9241-11.

Kurosu, M. (Ed.) (2003), Usability Testing, Kyoritu, Tokyo.

Miki, H. and Hosono, N. (2005), IT Universal Design Maruzen, Tokyo.

Norman, D. A. (2005), Emotional Design: Why we love Everythings, Basic Books.

Graf, D. (1992), point it, Graf Edition.

SPSS (2009), Categories in Statistical Package for Social Science Ver.18, SPSS.

Suzuki, M., Hosono, N. et al. (2005), SOS card for deaf people, Proceedings of JES.

Chapter 53

AV Remote Controller with an Input Device to Recognize Handwritten Characters

Masahiro Yonezawa, Takako Nonaka , Tomohiro Hase

Ryukoku University

ABSTRACT

This paper proposes the recognition of characters directly written on the touch panel of an AV remote controller. First, we determined input on the touch panel was possible with an experimental system. Next, we confirmed recognition of characters directly written on the touch panel, and supported alpha-numeric characters and, symbols. Finally we confirmed the recognition rate of the system by the input experiment of ten users. The recognition rates were more than 95 percent when including a fifth characters candidate, and our proposed user interface was confirmed as effective in recognizing handwritten characters.

Keywords: Handwritten Character Recognition, Zinnia, Touch Panel, μT-Engine

INTRODUCTION

Recent audio visual (AV) devices are becoming more multi-functional, and AV systems consisting of these devices are becoming larger in scale. Consequently, an AV remote controller requires a number of buttons and icons. On the other hand, AV remote controller should be small enough to fit the palm for usage, and it is difficult to arrange many buttons, keyboard or other input devices on the surface. Furthermore very recently in Asian countries which use Chinese characters, it is becoming necessary to enter the relevant characters for the name of programs when using an electronic program guide (EPG).

This paper proposes the recognition of characters directly written on the touch panel of an AV remote controller by the user in order to solve the conventional problems. Figure 1 shows an image of the proposed system.

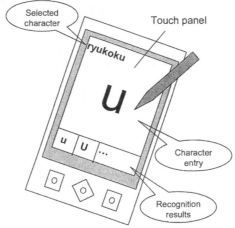

FIGURE 1   Image of proposed system

PROPOSED SYSTEM

For verification of this proposal, an experimental device was created for trial by using the embedded microcomputer and real-time OS, supposing that the device would be used in consumer appliances. Table 1 lists the specifications of the prototype system.

Table 1   Specifications of the Prototype System (T-Engine forum, 2003)

Item		Specification
µT-Engine	CPU	M32799 (M32700SAWG)
		Clock: 300 MHz
	Memory	Flash: 8 MB
		SDRAM: 16 MB
	OS	Linux/M32R, Kernel: 2.6.14.6
	Board Size	60 * 85mm
LCD	Quality	QVGA 320*240
	Screen Size	60 * 80 mm

Users can write characters directly on the touch panel mounted on the surface of the LCD in order to enter the data to the AV remote controller. Users can enter data by handwriting any of more than 100 characters, including letters, numerals, symbols and others that are available to the EPG function. To deal with cramped characters written by general users, the system is provided with a support vector machine for character recognition (Zinnia 2009), and further, with a function in the character database section so that it can register new characters one after another.

Figure 2 shows a flowchart of the experimental system.

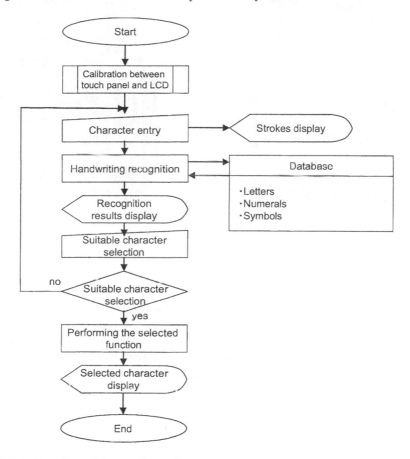

FIGURE 2   Flowchart of the experimental system

VERIFICATION WITH THE EXPERIMENTAL SYSTEM

Ten users were selected to enter 200 arbitrary characters, and as a result, the system could show a first candidate character with the right answer ratio being approximately 70 percent, and a fifth candidate character with the right answer ratio being approximately 95 percent or higher. Then, each user made registered data for

200 arbitrary characters in advance, and the system showed a 10 point improvement in the recognition rate for the first candidate character. The time of recognition was 0.3 seconds or less, allowing the users relaxed operation. Figure 3 shows the display screen image, and Figure 4 shows the operation of the prototype system.

FIGURE 3   Display screen image   FIGURE 4   Operation of the prototype system

CONCLUSION

This paper has proposed the recognition of characters directly written on the touch panel of an AV remote controller. In this system, the user can input to the AV remote controller by writing a character on the touch panel. As a result, our proposed user interface was confirmed as effective in recognizing characters handwritten by users directly on the touch panel of an AV remote controller.

REFERENCES

Zinnia: Online hand recognition system with machine learning, (May 31, 2009) Retrieved Nov 11, 2009, website: http://zinnia.sourceforge.net/index.html
T-Engine forum (2003), *T-Kernel standard handbook*, Personal media.

Chapter 54

AV Remote Controller Using Speech Recognition

Shuji Sugiyama, Takako Nonaka, Tomohiro Hase

Ryukoku University

ABSTRACT

This paper proposes a user interface that enables the user to operate an AV remote controller by speaking the name of the function that they want to use. Verification experiments are conducted using a prototype which operates with Linux on M32R. The proposed controller is considered a device that all family members can use. Ten users were selected to pronounce 200 arbitrary characters, and the rate of successful recognition was approximately 80 percent. The time for recognition was within one second, allowing the users relaxed operation.

Keywords: Speaker-independent recognition, Julius, Consumer appliances, User Interface

INTRODUCTION

Recent audio visual (AV) devices have become more multi-functional, and AV systems using these devices have become larger. Consequently, an AV remote controller requires a number of buttons and icons. On the other hand, it should be small enough to fit the operator's palm, but it is difficult to arrange so many buttons, jog dials, and other input devices on the surface of the unit. This paper proposes a user interface that enables the user to operate the AV remote controller by speaking the name of the function that they want to use.

AV REMOTE CONTROLLER USING PROPOSED SPEECH RECOGNITION TECHNOLOGY

The authors considered the following points when applying speech recognition technology to AV remote controllers. Speaker-independent, rather than speaker-dependent, speech recognition was adopted. This was because all family members should be able to use the same device. Word recognition was selected instead of sentence recognition to meet the type of usage. In order to reduce the load on the CPU, memory and other resources, only 120 words were selected including the function names that are necessary to operate AV equipment as well as numerals and Roman letters that are needed to make settings. Recognizable words were indicated on the GUI of LCD display. This was for two reasons: one is that user may only speak recognizable words, and the other is that the user should learn whether their spoken word has been successfully recognized or not by seeing a change in the GUI indication. Figure 1 shows a processing flow-chart of the proposed system.

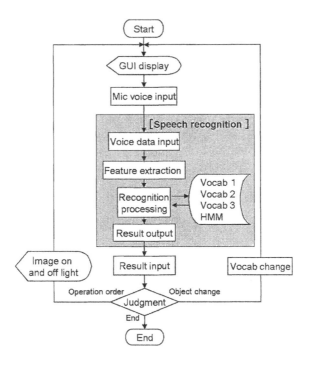

FIGURE 1 Flow-chart of the processing on the proposed system.

VERIFICATION BY PROTOTYPE

The authors used the experimental equipment described in the previous section to verify its effectiveness. For the experiment, an experimental evaluation device using an embedded micro processor unit (MPU) and an embedded operation system (OS) was created for trial, supposing that the device would be used in consumer appliances. Figure 2 shows an overview of the experimental equipment and Table 1 shows the hardware and software specifications of the experimental equipment. Considering household use, as shown in Table 1, the experimental equipment uses an embedded MPU and less expensive peripheral devices. Julius that is free speech recognition decoder software (Kawahara et al., 2000) was implemented to the prototype.

FIGURE 2 General view photograph of experimental equipment.

Table 1 The hardware and software specifications of the experimental equipment

ITEM	SPECIFICATION
CPU	OPERATION CLOCK: 300 MHz
MEMORY	ROM: 8 MB RAM: 16 MB
MOVEMENT VOLTAGE	3.3 V
OS	Linux / kernel 2.6.18
SIZE	125 mm x 85 mm x 45 mm
MIC	470Ω±30%, -44±4dB

Figure 3 shows the state of the operation check. In the verification experiment, the output GUIs on the prototype were checked and result messages were displayed on the monitoring PC display to confirm normal processing.

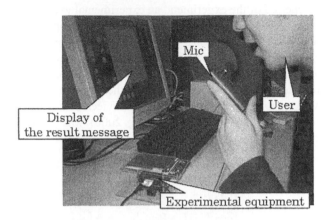

FIGURE 3 State of the operation check.

CONCLUSIONS

This paper has proposed a user interface that enables the user to operate an AV remote controller by speaking the name of the function that they want to use. Ten users were selected to pronounce 200 arbitrary characters, and the rate of successful recognition was approximately 80 percent. The time for recognition was within one second, allowing the users relaxed operation. The above results confirm that our proposed user interface is effective and that the user can operate the AV remote controller by speaking the name of function that they want to use.

REFERENCES

Kawahara, T., Lee, A., Kobayashi, T., Takeda, K., Minematsu, N., Sagayama, S., Itou, K., Ito, A., Yamamoto, M., Yamada, A., Utsuro, T., and Shikano, K. (2000), "Free software toolkit for Japanese large vocabulary continuous speech recognition." *International Conference on Spoken Language Processing*, 4, 476–479.

CHAPTER 55

Bridging the Disconnect Between Back Injuries and Prevention Using Interactive Computer-Based Systems

R. J. Banks, J. A. Hay

State Compensation Insurance Fund
San Francisco, CA 94103-1410, USA

ABSTRACT

The traditional method of back injury prevention has always been to train workers in how to lift properly when, in fact, the emphasis should be on educating them on risk identification and problem solving. A common safety poster illustrates a worker lifting a box from floor level, while being reminded to bend the knees and keep the back straight. This approach does not address the root cause of the hazard—the need to manually lift the box from floor level. While training has always been the first line of defense for employers because it is quick, inexpensive, and can be conducted in any environment, training alone is ineffective in back injury prevention (Daltroy, et al., 1997; Zwerling, et al., 1997). As injuries continue, the worker is blamed for not lifting properly and more training is prescribed. Given this, State Compensation Insurance Fund, set out to develop an interactive, computer-based DVD-ROM which guides the user through the process of planning, risk factor identification, implementation and effectiveness of controls (cost benefit), training, and follow-up.

Keywords: Back injury prevention, computer-based, employee participation, problem solving process

INTRODUCTION

Back pain and injuries have plagued general industry for years. A statistic often cited is that 85% of the population will experience low back pain in their working life, some of it reoccurring (Von Korff, et al., 1988). Any low back injury prevention process has to acknowledge that people may experience low back pain from both occupational and non-occupational or non-work-related factors (National Research Council, 2001). In the report by the National Research Council and the Institute of Medicine of the National Academies, musculoskeletal disorders of the lower back can be attributed to particular jobs and working conditions including heavy lifting, repetitive and forceful motions, and stressful work environments.

Intervertebral discs are most vulnerable to injury in the first 90 minutes of the day when subjected to torso flexion (Snook, 1998). This normally takes place in a non-occupational setting with such early morning activities as brushing teeth, picking up the newspaper, or putting on shoes. Material handling of loads in an occupational setting involves repetitive lifting, lowering, pushing, pulling, carrying, and twisting, exposes the soft tissues and discs of the lumbar spine to high levels of compressive forces and shear loading. The result is micro-fractures of the disc. The cumulative effect is considered an "illness" because of a failure to assign a specific incident or time of occurrence. Unless a traumatic injury occurs, such as a fall or being struck in the back, it is difficult to separate occupational from the non-occupational "causes" of a back problem, yet that is what the workers' compensation system asks treating physicians to do. It doesn't really matter if the injury was caused by the job or by outside activities. If an employee is unable to work because of a back problem, the costs are usually paid for by the employer with a loss of productivity on the part of the employee. Consequently, experts in this area agree that low back pain must be managed at work (and home), so it does not develop into a disability (Carter, 2000; Rowe, 1983; Fitch, 2004).

NEEDS ASSESSMENT

State Compensation Insurance Fund, a provider of workers' compensation insurance to California employers, recognized that back disorders are the number one cost driver. Early in the design process of the DVD-ROM titled State Fund's Back Connection®, phone surveys were conducted where State Fund policyholders (employers) were asked what their concerns were about back injury prevention. Each survey lasted approximately 18 minutes. The targeted audience consisted of policyholders from construction, healthcare, and manufacturing representing various levels of paid premium. The purpose was to access practices, attitudes, motivators, and constraints with respect to back injury prevention in their organization. Survey responses indicated that protecting employees is seen to be the leading benefit to workplace safety. Participants indicated that workplace safety has a positive impact on a company's financial health; one in three employers were willing to commit some time and resources for a 10% reduction in back injuries; and, help was needed in identifying risk, designing solutions, and tracking and measuring the results of their efforts.

Results of the survey allowed the developers to create a DVD-ROM featuring a toolbox (see Figure 1) where drawers open to reveal various tools for the development of a back injury prevention process. The toolbox concept allows the user the ability to maneuver through the process in a logical manner, and select tools for risk identification in which workers use pictures to identify postures or postures in their jobs, educational videos in English and Spanish, and methods to propose, implement, and evaluate interventions.

The DVD-ROM is fully functional via a Windows-based computer and has certain features that can be accessed by a DVD player.

Figure 1. Toolbox

PLANNING

Planning is the most important part of a successful process. Unlike programs which have a beginning and an end, a process is ongoing—it evolves; it can be built upon; and, it can be tailored to an organization's needs. All successful projects, programs, and processes begin with a plan and proper administration. A back injury prevention process is no different. Without a good plan, and the necessary tools and resources to accomplish that plan, a back injury prevention process will be doomed to failure from the start. In the top of the toolbox is the Site Plan, which offers a blueprint of the whole process, and lays out the steps using the analogy of building a house.

Building upon this, the Planning Tools drawer contains tools to help with the organization and implementation of a back injury prevention process (see Figure 2). Some of those tools include planning and tracking the progress of a back injury prevention process (depicted by the carpenter's pencil); managing the claims costs of back injuries (depicted by the first aid kit); and, calculating the cost savings of interventions or determining a return on investment (depicted by the tape measure).

Figure 2. Planning Tools Drawer

DETERMINING A COST BENEFIT

When the Dollars and Sense tape measure is launched, the user can choose the Simple Cost Calculator or the Cost Analysis Tool to calculate costs up front, before an actual injury or claim occurs. Much like a contractor estimates a job, knowing these costs up front can be extremely useful information when making critical business decisions regarding injury prevention.

The Simple Cost Calculator demonstrates the potential financial impact of a single claim and associated indirect costs on a company's profitability. It provides a concise, one-page report showing the estimated direct and indirect costs of an injury and the projected additional revenue that must be generated to pay for that injury.

The Cost Analysis Tool provides a concise, one-page report showing the estimated return on investment and payback period for purchasing equipment. A purchase can be analyzed using either expected injury prevention or estimated productivity improvements (see Figure 3).

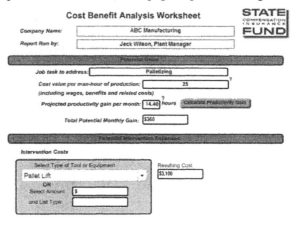

Figure 3. Worksheet from Cost Benefit Tool

PROBLEM SOLVING

To ensure worker buy-in and involvement in the injury prevention process, employers must invite their workers to actively participate in the process of risk identification and problem solving (Morkin, et al., 2002). In the Ergonomics Tools drawer, there are tools to educate the workers in identifying risk factors, understanding root cause analysis, and developing cost-effective interventions. This Risk Analysis Process (RAP) is a comprehensive approach to identify risk factors; brainstorm, select, and implement solutions; and, monitor and re-evaluate jobs after solutions have been implemented.

RISK IDENTIFICATION

The Risk Analysis Process, or the "card sort", is a unique way to get workers involved in identifying the risk factors which cause low back disorders. Workers are more aware of the tasks necessary to perform their job and should provide the most accurate identification of risk factors. In the card sort, workers sort through a series of 37 cards. Each card depicts a different posture or position found to be a risk factor to the development of low back disorders such as bending, lifting greater than 50 pounds, reaching beyond 12 inches, etc. The cards are in English and Spanish.

The worker is presented with a screen showing, in random order, a card depicting a posture or position and is asked if that posture or position is present in their job (see Figure 4). This is the first sort. The worker responds by selecting the Yes or No button. Their decision moves to the next screen until all 37 cards are sorted. The worker should pay particular attention to the posture or position illustrated and the written descriptions provided with each card, rather than to the specific job being performed in the illustration.

From the cards depicting risk factors determined to be present in a job, the worker then performs the second sort to determine the amount of effort expended while in that posture by selecting the appropriate button (see Figure 5)—light being less than 40% of strength; moderate being between 40% and 60% of strength; and, heavy being more than 60% of strength (Eastman Kodak Co., 1986; Rodgers, 1992).

For the third and final sort, the worker continues using the cards from the first and second sort to determine how often (frequency) they are in that posture by selecting the appropriate button (see Figure 6)—rarely is less than once a month; occasionally is once a month to six time a day; frequently is six times a day to six times an hours; and, constantly is more than six times an hour (Eastman Kodak Co., 1986; Rodgers, 1992).

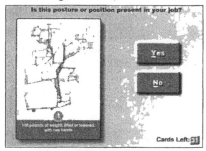

Figure 4. First card sort screen

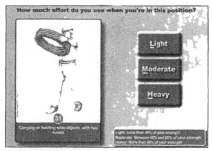

Figure 5. Second card sort screen

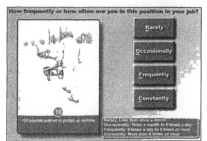

Figure 6. Third card sort screen

Based upon the responses, the final screen will show, in priority, up to five cards that represent the greatest risk factors based on effort and frequency (see Figure 7). If a risk factor identified by the worker does not appear, the program has determined that it does not fall into a high-risk category. This page can be printed out or the worker can press "Form" and the results of the card sort will auto-populate the Risk Analysis Process Worksheet, which will then lead the worker to the Just Ask Why® root cause analysis process. This process involves workers in developing, implementing, and evaluating solutions.

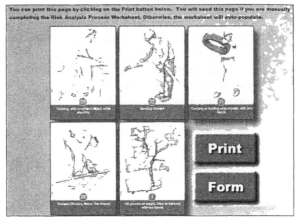

Figure 7. Results of card sort screen

TRAINING

Training is a critical component of any back injury prevention process. It is an opportunity to educate all levels of employees, from senior management to front-line workers. Though not the answer to a potential risk or hazard in itself, it is a key element in helping to ensure a work process or method is understood and undertaken properly.

In the Training Tools drawer are an awareness video and slideshows specifically targeted for workers and management, 24 "toolbox talks", and a series of five, 10-minute back safety videos, built around a late night talk show format, designed to facilitate short safety meetings. All these tools are available in both English and Spanish.

ADDITIONAL RESOURCES

The Resource Drawer provides tools to support all phases of a back injury prevention process—an equipment guide, web-based resources, a written ergonomics program, and the literature review referred to as "What Does the Research Tell Us". The literature review forms the backbone of State Fund's Back Connection®. More than one hundred articles on occupational low back disorders were reviewed to validate what works—employee involvement, risk factor identification, problem solving—and what doesn't work—back belts (McGill, 2004; NIOSH, 1994), training (Daltroy, et al., 1997; Zwerling, et al., 1997), another poster, video, or how-to-lift guide.

The last drawer, the Help Drawer, provides a glossary, contact information, acknowledgements, the licensing agreement, and instructions to open, edit, create, print, and save forms onto your hard drive.

CONCLUSIONS

The goal of this product was to develop a process of back injury prevention designed to work within a company's normal course of business—not compete with it. An additional benefit became evident for return to work. If the risk factors which caused the injury have been identified, the fear of re-injury is minimized. Various industries that have used State Fund's Back Connection®—for-profit as well as non-profit; from office, construction, healthcare, manufacturing, restaurants, and agriculture—find it to be an effective method of back injury prevention and return to work as it does what it was designed to do—involve the worker in the process.

Back Injuries   **Prevention**

ACKNOWLEDGEMENTS

The authors wish to acknowledge the following individuals for their contributions to this product: Doug Baker, Scott Boggess, Helen Chandler, Mary Fortune, Ryan Horton, Chase-Knoble-Hamilton (posthumous), Ira Janowitz, Joanette Lima, Michael Melnik, Suzanne Rodgers, Kristy Schultz, Chris Shulenberger, Jeff Tiedeman, Lou Vicario, and Vision Realm, Inc.

REFERENCES

Carter, JT and LN Birrell, ed. (2000). *Occupational Health Guidelines for the Management of Low Back Pain at Work—Principal Recommendations.* Faculty of Occupational Medicine, London, 54 pages.

Daltroy, LH, MD Iverson, MG Larson, R Lew, E Wright, J Ryan, C Zwerling, AH Fossel, and MH Liang. "A controlled trial of an educational program to prevent back injuries", *New England Journal of Medicine 337(5)*: 322-328, 1997.

Fitch, JL and BE Fitch, (2004). "Back Injury Prevention. Exploring Causative and Preventative Forces". *Occupational Health and Safety*, pages 119.

Kodak, Eastman Company (1986). *Ergonomic design for people at work*, Vol. 2. New York: Van Nostrand Reinhold Company.

McGill, S. M. Linking latest knowledge of injury mechanisms and spine function to the prevention of low back disorders. *Journal of Electromyography and Kinesiology*, 14(1):43–47, 2004.

Morken, T, B Meen, T Riisea, SHV Hauge, S Holien, A Langendrag, H-O Olson, S Pedersen, ILL Saue, GM Seljebo, and V Thoppil. "Effects of a training program to improve musculoskeletal health among industrial workers – Effects of supervisors' role in the intervention", Spine 25(1):137, 2002 (cited in Ergoweb 6/24/02).

National Research Council, 2001. Musculoskeletal Disorders and the Workplace. Washington, DC: National Academy Press, 492 pp.

National Institute for Occupational Safety and Health (NIOSH) (1994). *Workplace use of back belts review and recommendations* [DHHS (NIOSH) Publication No. 94-122].

Rodgers, S.H., A functional job evaluation technique, in Ergonomics, edited by J.S. Moore and A. Garg, *Occupational Medicine: State of the Art Reviews.* 7(4):679-711, 1992.

Rowe, M.L. *Backache at Work* (1983). Fairport, NY: Perinton Press.

Snook, S.H., B.S. Webster, R.W. McGorry, M.T. Fogleman, and K.B. McCann (1998). "The reduction of chronic non-specific low back pain through the control of early morning lumbar flexion, a randomized controlled trial." *Spine 23(23):* 2601-2607.

Von Korff M, Dworkin SF, Le Resche L, Kruger A. An epidemiologic comparison of pain complaints. *Pain.* 32(2):173–183, 1988.

Zwerling, C. et al. Design and Conduct of Occupational Injury Intervention Studies: A Review of Evaluation Strategies. *American Journal of Industrial Medicine.* Vol. 32, pp 164-179, 1997.

CHAPTER 56

Modeling of an Integrated Workstation Environment for Occupational Safety and Health Application

Abdul R. Omar[a], Isa Halim[b]

[a] Faculty of Mechanical Engineering
Universiti Teknologi MARA
40450 Shah Alam, Selangor, Malaysia

[b] Faculty of Manufacturing Engineering
Universiti Teknikal Malaysia Melaka
Locked Bag No. 1752, Pejabat Pos Durian Tunggal
76109 Durian Tunggal, Melaka, Malaysia

ABSTRACT

Ergonomics evaluation tools appear to be useful means to analyze occupational risk factors, however they are presented as an isolated tool. Hence ergonomics practitioners could not obtain total solution to solve the occupational safety and health problems. The objective of this study is to integrate selected ergonomics evaluation tools in a Knowledge-Base Decision Support System (KBDSS) to provide detail information to ergonomics practitioners in managing ergonomics hazards. The Model Oriented Simultaneous Engineering System (MOSES) was adopted to develop the information architecture. The value of the developed KBDSS is demonstrated through a case study in a manufacturing environment.

Keywords: Knowledge-Base Decision Support System, Ergonomics Evaluation Tools, Occupational Safety and Health

INTRODUCTION

In the new era of industrialization world, occupational safety and health plays an important role to preserve the competitiveness of industry. Poor occupational safety and health management in the workplaces contribute to occupational injuries and may reduce the performance of the industry such as less productivity, high medical costs and compensation claims, and may demoralize the workers.

Expert system has been recognized as a vital contributor to rapid revolution in information technology and generates numerous advantages to all industries.

Expert system is widely used in various fields; among them is occupational safety and health management.

CHALLENGES OF ERGONOMICS IMPROVEMENT IN WORKSTATION

Many ergonomics assessment tools are available to improve working condition in workplaces, for instance, NIOSH Lifting Equation 1991 has been developed to evaluate risk levels associated with manual materials handling (Dempsey, 2002). All tools have been validated and shown as potential tools in identifying and evaluating risks due to ergonomics hazards, however through observation the existing tools have shown few limitations that can be summarized as follow:

- Almost existing tools are pen and paper-based observation, hence need more sheets and time when assessing large number of subjects,
- Since data collection and interpretation are performed in a paper form, thus it is difficult to retrieve electronically when necessary,
- The existing tools are presented in a conventional form as pen and paper are mechanisms to execute the assessment. It requires an assessor to process the data manually. This practice may lead to error during calculation, especially when involving huge data, and
- The current assessment methods are seen as individual and isolated tools. It is difficult to perform multi assessments concurrently, for instance when assessing working posture, assessment on muscle fatigue could not be performed.

The present study is carried out to develop a Knowledge-Base Decision Support System (KBDSS) to record, process, and store the information systematically and integrated so that data processing and analysis can be performed efficiently.

MODELING THE INTEGRATED ERGONOMICS EVALUATION TOOLS

The major tasks involved in the development of KBDSS are model the information architecture, and establish the decision support system. The KBDSS is developed using JAVA language, and Object-Oriented Programming (OOP) was used as a programming method.

INTEGRATED ERGONOMICS EVALUATIONS SYSTEM DEVELOPMENT

An object-oriented approach has been used to model the information. Model Oriented Simultaneous Engineering System (MOSES) (Harding, J. A. et al., 1999) has been applied to construct the architecture. The MOSES architecture consists of two information models which are Product Model and Manufacturing Model that can be assessed by open set of application programming via integration environment. A Product Model is a representation of product in a computer and contains adequate information about the product to satisfy the product information needs. A product can be an actual product or services. On the other hand, Manufacturing Model provides information on the manufacturing resources and capability of manufacturing enterprises. Both Product Model and Manufacturing Model are instances of data model. In this study, the workstation constituents are modeled as part of Manufacturing Model.

The system is developed by imaging a worker in his/ her workplace. Usually, a workplace comprises company, department, workstation, and worker. In a workstation, it comprises at least a worker, machine, and working environment. Since ergonomics is a scientific discipline that concerning the interaction between human (worker), machine (equipment and tool), and working environment, the researchers identified that machine vibration and IAQ elements should be incorporated in the system. Hence, machine vibration and IAQ have been assigned to the class Machine and Environment respectively. The system constituted by several classes represented by the clouds as shown in FIGURE 1.1. The primary class of the system is COMPANY. In general, a department should be incorporated in a company, and DEPARTMENT has been assigned as a class under COMPANY. Similarly, in a department, there is at least one workstation, and a class WORKSTATION has been linked to DEPARTMENT. In a workstation, there are worker, a machine, and working environment. All elements are demonstrated by class WORKER, MACHINE, and ENVIRONMENT to represent worker (human), equipment and tools, and workplace environment respectively.

In WORKER class, it is divided into five classes namely Job Activity, Posture, Muscle, Standing Duration, and Holding Time. These five classes represent factors which are related to a worker and need to be assessed when he/she is performing a job. In addition, each class has different attributes, for instance in the Job Activity there are few attributes such as prolonged standing to determine activities performed by the workers in their workstations. There is no ergonomics evaluation carried out in the Job Activity, however the information is useful for further analysis. For example, if a worker is exposed to Job Activity associated with 'standing with forward bending', further analysis should be carried out is working posture assessment to ensure the worker could perform his/her job in safe working posture.

Another class that associated to WORKER is Posture. The main purpose of the Posture in the system is to assess working posture so that proactive actions could be taken to prevent workers from occupational injuries associated with awkward postures. In other words, it is useful to classify the posture of worker either it is safe or unsafe working posture. Similarly, posture also has few attributes (Hignett and McAtamney, 2000) that are critically examined to analyze working posture of a worker. The outputs of the Posture are action levels which are reflected to musculoskeletal loading associated with the worker's posture. Action level determines the level of intervention that is required to reduce the risks of injury due to physical loading on worker. "Action Level 0" indicates that the current posture is good and this condition has to be maintained. "Action Level 1" defines that the current posture has low risk, however further action may be required. "Action Level 2" classifies that the current posture has medium risk, hence further action is also necessary. "Action Level 3" represents that the current posture has high risk and further action is necessary soon. "Action Level 4" concludes that the current posture has very high risk, therefore further action is required immediately (Hignett and McAtamney, 2000).

Further ergonomics evaluation that is integrated in the system is muscles activity analysis. The analysis is represented by a class called Muscle. The purpose of muscles activity analysis is to analyze the risk for muscle fatigue during performing jobs. This method is suitable to evaluate the risk for fatigue accumulation in jobs that are performed for an hour or more plus ergonomics risk factors associated with awkward postures and frequent exertions (Suzanne, H. R, 1988).

In almost industrial workplaces, workers are required to perform jobs in prolonged standing. Since prolonged standing can lead to discomfort and fatigue, it is important to include an analysis tool in the system for assessing risk associated with prolonged standing. A guideline for standing at workplace (Paul and Hanneke, 2007) has been considered in the system and named Standing Duration class. In addition, standing duration attribute has been assigned to the Standing Duration class to enable the system to compute which risk level experienced by the worker.

In the last class of WORKER, Holding Time was associated to enable an assessor to identify recommended holding time with respect to posture. Holding Time class has attributes shoulder height and arm reach distance. The tool classifies postures on the basis of maximum holding time (MHT), and developed ergonomic solutions for the MHT of different postures categories. A comfort posture can be obtained when workers dealt with moderate working height of 50%, 75%, 100%, 125% with respect to shoulder height (SH) and small working distance of 25%, 50% corresponding to arm reach (AR), then the recommended MHT should be more that 10 minutes. In addition, moderate posture could be obtained if the work is performed at moderate working height of 50%, 75%, 100%, 125% from SH and large working distance of 75%, 100% from AR, then the ergonomic recommendation of MHT should be between 5 to 10 minutes. On the other hand, discomfort posture is recognized when the working height is too low or too high that is 25% and 150% of SH respectively. This condition requires ergonomic intervention of MHT of less than 5 minutes (Mathilde, et al., 1997).

In the Machine class, it has Vibration class. The aim of this class is to capture information about vibration and will enable vibration analysis to be conducted. The system will be able to compute recommended threshold limits corresponding to frequency and acceleration of vibration and the time during which worker are exposed (ISO 2631).

A workplace should maintain good indoor air quality for a healthy indoor environment. In contrast, poor indoor air quality can cause a variety of short-term and long-term occupational health problems (DOSH, 2005). In recognition the importance of indoor air quality in the workplace, the system includes a class called indoor air quality (IAQ). IAQ is linked to WORKSTATION through Environment class. The purpose of IAQ analysis is to determine the maximum limit of IAQ contaminants to ensure the workplace is healthy to workers. FIGURE 1.1 illustrates the information architecture that is applied in the system to represent a relationship between human-machine-environment in a workplace.

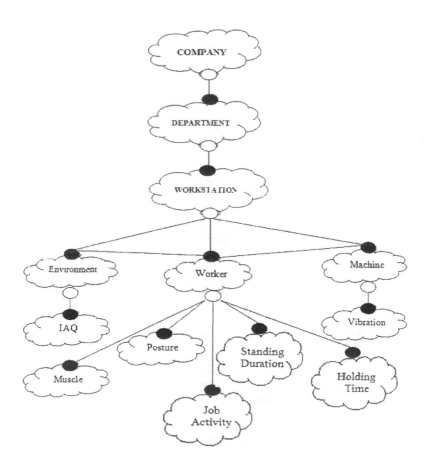

FIGURE 1.1 Information architecture of the system to represent the correlation between WORKER-MACHINE-ENVIRONMENT.

KNOWLEDGE-BASE DECISION SUPPORT SYSTEM APPLICATION

In the Knowledge-Base Decision Support System (KBDSS), each tool has their own KNOWLEDGE class and WORKING MEMORY class. The KNOWLEDGE consists of Rule Set and Rule that represented by *if ... then* rules. Each input data will be sent to the KNOWLEDGE class to process the data for analysis. The system will process the input data through information embedded in the KNOWLEDGE class, and then stored the output in the WORKING MEMORY class. KNOWLEDGE class has few objects that contain rules such as Posture Rule, Muscle Rule, Standing Rule, Holding Rule, Vibration Rule, and IAQ Rule. Each analysis data and results will be saved in the WORKING MEMORY class. The WORKING MEMORY class inherits classes namely Posture Memory – to store data and results from Posture Rule; Muscle Memory – to store data and results from Muscle Rule, Standing Memory – to store data and results from Standing Rule, Holding Rule – to store data and results from Holding Rule, Vibration Memory – to store data and results from Vibration Rule, and IAQ Memory – to store data and results from IAQ Rule. FIGURE 1.2 shows the relationship of each class in the KBDSS.

FIGURE 1.2 Correlation between WORKING MEMORY and KNOWLEDGE BASE and their objects.

The sequences of using the KBDSS can be summarized as follow:
- Identify the workplace profile,
- Identify the worker to be assessed,
- Determine the kind of analysis to be carried out and capture the data;
- Initialize and run the analysis, and
- Display the input data and results of analysis.

In the first stage, it starts from retrieving the workplace profiles of worker to be assessed from the database. It includes the name of company, department, and workstation that worker performed his/ her jobs. The main purpose of the information is to enable the assessors to retrieve the specific worker. For example, when large numbers of workers have been assessed, there are possibilities of having few similar workers' names in the database. Hence the required worker can be retrieved by examining through his/ her company, department, and workstation.

The second stage is performed once the workplace profile has been completely stored. At this stage, the KBDSS requires the assessor to record the profile of worker to be assessed. A Graphical User Interface (GUI) is used to record the personal details and job activity of worker. Personal details include worker's name, employee number, position, age, gender, body mass, height, working experience, job tenure, health status, working mode either shift or normal working hours, and smoking. Meanwhile, list of various job activities contain prolonged standing, standing with side bending, standing and walking frequently, prolonged sitting, prolonged bending, standing on the vibrated area, straight standing, static standing, standing and walking rarely, reaching goods, prolonged kneeling, sitting on the vibrated area, standing with forward bending, standing with body twisted, standing and sitting intermittently, lifting heavy goods, squatting, and handling imbalance goods. These activities are provided with check boxes so that the assessor could select any appropriate box (es). The information of workplace profile and worker profile will be stored in the Worker Memory and it will be retrieved when further analysis takes place.

In the third stage, the KBDSS requires the assessor to choose type of analysis to be carried out. The assessor can run KBDSS using single analysis or multi analysis (two or more analyses are carried out simultaneously). Each analysis is equipped with a GUI to capture their input data for further analysis. For instance, if an assessor wants to perform working posture analysis, he/ she is required to enter the working conditions of worker to be assessed such as posture and condition of body parts, mass of load handled by the worker and its condition, coupling condition of load, and posture activity. Once all necessary information have been keyed-in in the GUI, the assessor can proceed to the fourth stage, initialize and run the analysis.

Initialize and run the analysis stage requires the assessor to activate the rule of analysis and create the working memory.

In the last stage, the assessor can preview the data and a result of analysis has been carried out. The KBDSS will display the profile of the worker that has been assessed, input data, and results of analysis.

CASE STUDY

A case study was carried out in a metal stamping company situated in Shah Alam, Malaysia. The main operation of the company is metal stamping process. A workstation at the end of stamping production line was selected as a case study. Through video recording, this workstation required workers perform jobs in awkward working posture, thus further investigation should be made immediately. In the workstation, a worker was assigned to collect the stamped parts from an incoming conveyor and arrange them in a cage. To perform this job, four working postures have been adopted: 1) 20° to 40° torso flexion while reaching the products from the incoming conveyor, 2) ~180° body rotation to transfer the products to a cage, 3) 20° to 30° torso flexion while attempting to load the products into the cage, 4) more than 90° torso flexion to arrange the products in the cage, as illustrated in FIGURE 1.3. The worker has to perform the job manually with job cycle more than four times per minute.

Based on the working condition, the authors found that assessment on working posture should be carried out as it is potentially contributing to occupational injuries. Before the posture and muscles activity being analyzed, the workplace profile of worker such as company name, types of industry, category of industry, department name, workstation name, and profile of worker were recorded in the system. To analyze working posture, data on posture such as trunk posture and its condition, neck posture and its condition, leg posture and its condition, upper arm posture and its condition, lower arm posture, wrist posture and its condition, mass of load handled by the worker and its condition, coupling condition of load, and posture activity were captured. After all necessary data have been captured, the analysis of working posture is carried out by initializing the database and run the application.

FIGURE 1.3 Working postures of workers while they attempt to arrange the products in cages.

RESULTS AND DISCUSSION

FIGURE 1.4 shows data and result of working posture assessment while the worker performed jobs at his workstation. Based on the posture analysis, all the assessed activities contributed to unsafe postures and required immediate improvements. The unsafe postures were identified when the worker performed the jobs in torso flexion and body rotation. Further improvement could be implemented by providing an adjustable platform to the workstation so that it can accommodate worker's height. Furthermore, the cage for finished products is suggested to be equipped with a lifting mechanism and its location should be close to worker's side. The proposed lifting mechanism allows the workers to adjust the height of cage to an appropriate working level. This modification eliminated body rotation and extreme flexion while the worker transferring and arranging products in the cage.

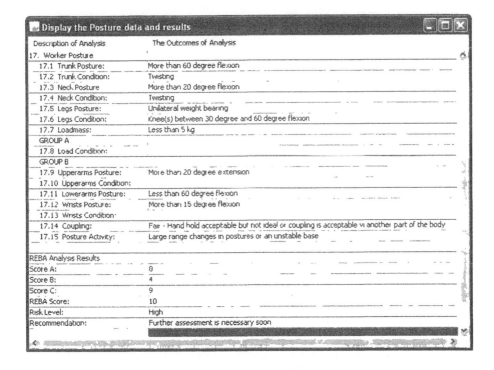

FIGURE 1.4 Display of data and result from working posture analysis.

CONCLUSION

Information model for integrated workstation constituents have been developed in a KBDSS to analyze working condition in a workplace. The value of the information model has been demonstrated through a case study. The carried out study concluded that:

- the developed system provides easy data processing and analysis, and time saving,
- the system has a database and could be retrieved and manipulated when required, and
- selected ergonomics evaluation tools have been integrated in a system to enable an assessor to perform analysis efficiently.

ACKNOWLEDGMENT

The authors would like to acknowledge the Ministry of Science, Technology and Innovation of Malaysia for funding this research under e-Science Research Grant, and the Universiti Teknologi MARA for providing facilities and assistance in carrying out this study. Special thank also goes to Miyazu (M) Sdn. Bhd. to facilitate case study. The authors would like to thank Mr. Hadi for data collection.

REFERENCES

Dempsey, P. G. (2002), "Usability of the Revised NIOSH Lifting Equation". *Ergonomics*, 45, 12, 817-828.

Department of Safety and Health of Malaysia (DOSH), Code of Practice on Indoor Air Quality. Ministry of Human Resources, Malaysia, 2005.

Harding, J. A., Omar, A. R., Popplewell, K. (1999), "Applications of QFD within a concurrent engineering environment". *International Journal of Agile Management Systems,* 1/2, 88-98.

Hignett, S. and McAtamney, L. (2000), "Rapid entire body assessment (REBA)". *Applied Ergonomics*, 31, 201-205.

ISO 2631 Mechanical vibration and shock – Evaluation of human exposure to whole-body vibration, 1997.

Mathilde, C. M., Marjolein, D., Jan, D. (1997), "Recommended maximum holding times for prevention of discomfort of static standing postures". *International Journal of Industrial Ergonomics*, 19, 9-18.

Paul, M. and Hanneke, J. J. K (2007), "Prolonged standing in the OR: a Dutch research study". *AORN Journal*, 86, 399-414.

Suzanne, H. R. (1988), "Job evaluation in worker fitness determination". *Occupational Medicine: State of the Art Reviews,* 3 (2), 219-239.

CHAPTER 57

Technology Acceptance and the Ageing Population: A Review

Ke Chen, Alan H.S. Chan

Department of Manufacturing Engineering and Engineering Management
City University of Hong Kong
Kowloon Tong, Hong Kong

ABSTRACT

As we move into the 21st century, we are confronted by the convergence of two major worldwide trends: Populations will have relatively more elderly people and technological innovation is likely to occur at an unprecedented rate. If elderly people are able and prepared to use new technologies there is the potential to greatly improve the quality of their lives. China has a significant percentage of the population of the world but, no previous study has investigated the acceptance of technology by elderly Chinese. Here, a review of the literature on technology acceptance models and theories was conducted and it was found that the existing technology acceptance models have not considered the unique characteristics, abilities, and limitations of the elderly. More research is needed to generate a better understanding of determinants of technology acceptance by the elderly.

Keywords: Technology acceptance model, elderly population

INTRODUCTION

This 21st century has witnessed the unprecedented, pervasive, and enduring demographic and technological changes on a worldwide basis. At the moment, there is no standard numerical criterion, but the United Nations has agreed that those of

514

60 years of age or more be referred to as the older population (WHO, 2001). In 2000, 10 percent of the world population was aged 60 or more. According to a report (United Nations, 2004), 'Population ageing is unprecedented, without parallel in human history - and the twenty-first century would witness even more rapid ageing than did the century just past'. In China, the number of people aged over 60 exceeded 160 million in 2009, accounting for about one half of Asia's over-60s and one fifth of the world's total. This number is expected to expand by three times to reach to 440 million by 2050 (United Nations, 2009). Figure 1 illustrates the predicted course of growth in the proportion of elderly people in the Chinese population[1]. This predicted shift in proportion has caused China to be characterized as a country 'becoming old before it gets rich' (UNFPA, 2006). Meeting age-related demands such as social security systems, health care services and community services with the present relatively low level of economic development will be a tough challenge for China.

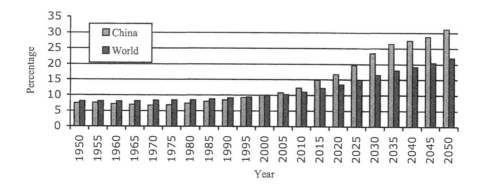

Figure 1. Proportion of population of 60 years or older to total population, China and World, 1950-2050 (United Nations, 2009)

In terms of technology, from October 1997 to 2009, the number of Internet users in China increases 545 times from 620,000 to 338 million (China Internet Network Information Center, 2009). The number of mobile phone subscribers in China has grown significantly from 43.4 million subscribers in 1999 to 547 million in 2007 (MIIT, 2008). Technology is rapidly becoming an integral part of Chinese and all societies. Application of technology can improve the living environment and quality of life for the elderly (Fozard, 1997). However, many studies have shown that most elderly people do not fully accept modern technology (Karahasanović, et al., 2009; Yu, et al., 2009).

The factors that contribute to the perception and acceptance of technology by elderly Chinese people have not yet been explored. The investigation of the key

---

[1] Elderly population refers to population aged 60 and over in this paper.

factors affecting technology acceptance and usage behavior of the elderly is important for both research and practice. This study reviews the state of knowledge about technology acceptance and considers whether there are any inadequacies in the current models as they apply to the elderly.

TECHNOLOGY ACCEPTANCE

TYPICAL THEORIES AND MODELS OF ACCEPTANCE

Technology acceptance has been described as 'the approval, favorable reception and ongoing use of newly introduced devices and systems' (Arning & Ziefle, 2007). Technology acceptance contains an attitude towards a certain behavior and the behavior itself. There are several models and theories that identify the major factors that affect a user's perception about a technology.

Both the theory of reasoned action (TRA) and theory of planned behavior (TPB) offer powerful explanations about human behavior in general. The TRA was proposed by Fishbein and Ajzen (1975) and suggests that a person's behavior is driven by his/her intention to perform the behavior and that this intention is, in turn, determined by attitude toward the behavior and his/her subjective norm. TPB extends TRA by adding a construct of perceived behavioral control, which is theorized to be an additional determinant of intention and behavior (Ajzen, 1985).

A technology acceptance model (TAM) was introduced by Davis et al (1989) to predict the information technology acceptance and usage behavior. According to the TAM, the two most important factors in explaining acceptance and usage of an information system are perceived usefulness (PU) and perceived ease of use (PEOU). Perceived usefulness was described as 'the degree to which a person believes that using the particular technology would enhance his/her job performance'. Perceived ease of use was defined as 'the extent to which a person believes that using a technology is free of effort'. PU and PEOU jointly determine the attitude towards using behavior (AT), i.e., an individual's positive or negative feelings about using the system. PU also mediates the effect of PEOU on AT. PU and AT predict the behavioral intention (BI), which directly affects actual usage behavior. TAM also assumes some 'external variables' such as user differences, and system characteristics, the effects of which are fully mediated by PU and PEOU. Numerous empirical studies have confirmed that the TAM is a robust and powerful model for explaining acceptance behavior (Cheng, et al., 2006; Teo, 2009; Lu, 2009).

CONTEMPORARY RESEARCH ON TAM

There is rapidly growing interest in TAM among researchers. Recently, some researchers have extended TAM by combining other theories like innovation diffusion theory (Wu & Wang, 2005; Mao & Palvia, 2006) and flow theory (Moon

& Kim, 2001; Lu et al., 2009); or by incorporating other constructs like social norms (Venkatesh et al., 2003; Lu et al., 2009; Zhang & Mao, 2008), trust (Gefen et al., 2003), perceived risk (Pavlou, 2003, Wu & Wang, 2005), cost (Wu & Wang, 2005), job relevance (Hu, 2003), and self-efficacy (Zhang & Mao, 2008). A number of studies provide evidence that TAM is also valid in the context of the world-wide-web (Lederer et al., 2000; Moon & Kim, 2001), electronic-commerce (Chen et al., 2002; Gefen et al., 2003), mobile commerce (Zhang & Mao, 2008), and telemedicine (Chau & Hu, 2002).

Venkatesh and Davis (2000) introduced TAM2 which included social influence processes (subjective norm, voluntariness, and image) and cognitive instrumental processes (job relevance, output quality, result demonstrability, and perceived ease of use) into TAM. Subjective norm, defined as 'a person's perception that people important to him/her think he/she should or should not perform the behavior', exerted a significant influence on perceived usefulness by internalization and identification, and also a direct effect on usage intentions for mandatory, but not voluntary usage contexts. This effect of subjective norm has also been found in other studies (Venkatesh et al., 2003; Lu et al., 2009; Zhang & Mao, 2008). Venkatesh et al. (2003) reviewed eight user acceptance models and formulated a unified theory of acceptance and use of technology (UTAUT). The UTAUT identifies three direct determinants of intention of usage (performance expectancy, effort expectancy, and social influence), two direct determinants of usage behavior (behavioral intention and facilitating conditions), and incorporated four moderators (gender, age, experience, and voluntariness of use).

AGE DIFFERENCES AND TECHNOLOGY ACCEPTANCE

There has been a lot of discussion about technology and technological devices and their use but much of it has been focused on use by younger adults, with little attention paid to acceptance and usage of technology and devices by elderly people. Most studies suggest that demographic characteristics are less important than the characteristics of the technology itself in determining acceptance and usage of specific technologies (Davis et al., 1989; Agarwal & Prasad, 1999). However, some studies provide preliminary evidence that different age groups may think and make different decisions when it comes to technology use and adoption (Venkatesh & Morris 2000; Morris et al., 2005; Arning & Ziefle, 2007).

In the study of Arning & Ziefle (2007), age-specific processes were found in the assessment of the PEOU and the judgment of PU. In the evaluation of the PEOU, younger adults referred to task efficiency (time on task) but older adults referred to task effectiveness (success in solving tasks). For older adults, PEOU played the main role in the assessment of the PU of a technical device, whereas the PEOU had a lower explanatory power for younger adults. The study of Morris et al. (2005) showed that, gender differences in technology perceptions became more pronounced among older workers. With increasing age, men placed greater emphasis on attitude, while women placed greater emphasis on subjective norm and perceived behavioral control.

Most elderly people are not necessarily interested in new technology and demonstrate a low level of technological self-efficacy than younger adults (Steele, 2009; Karahasanović et al., 2009; Ryu et al., 2009). They may feel they are not able to control what happens around them and this could lead to severe frustration. Previous studies showed that older people often encounter difficulties when attempting to adopt a new technological device. Barriers to use of a technology are largely associated with the design and usability of these devices and services (Czaja & Lee 2001; Mallenius et al. 2007). Another concern of the elderly is that using technology may reduce human contact and relations, and people will be alienated from each other (Niemela, 2007). From these findings, it can be concluded that it is not sufficient to examine only young users' technical acceptance and usage and then generalize the results to the whole user population.

EARLIER RESEARCH ON ELDERLY TECHNOLOGY ACCEPTANCE MODELS

A few studies have considered age-related factors in TAM. Ryu et al (2009) examined the elderly (aged above 50) people's adoption of video user-created content services in Korea. This study introduced elderly-specific constructs such as perceived physical condition (physical age), life course events (psycho-social age), perceived user resources, prior similar experience, and computer anxiety, each reflecting the complex ageing process. The results indicated that the effects of elderly-specific variables were mediated by internal beliefs (PU and PEOU), which implies that the elderly-specific variables can be seen as antecedents for conventional TAM constructs. Renaud & Biljon (2008) interviewed senior mobile phone users and presented a senior technology acceptance and adoption model for mobile technology (STAM). The STAM related technology acceptance factors (user context, perceived usefulness, facilitating conditions, ease of learning and use, etc.) to the adoption phase (objectification, incorporation and non-conversion phase). But the reliability and validity of STAM had not been verified. Cost, which was neglected as most previous research conducted in organizational settings, has been found to be the most critical determinant in determining an elderly person's acceptance of technology (Mallenius et al., 2007; Steele, 2009).

No previous study, however, has investigated Chinese older adults. It is not known to what extent the findings for other populations can be generalized to the elderly population in China. Also, more systematic research is needed to generate a better understanding of determinants of technology acceptance for the elderly in general.

CHARACTERISTICS OF THE ELDERLY

There are two categories of change when people grow old: biophysical and psychosocial change (Ryu et al., 2009; Steele, 2009).

BIOPHYSICAL CHARACTERISTICS

Biophysical change in ageing is associated with functional loss in many areas: visual and auditory perception, touch and movement, working memory and cognition, etc (Berkowitz& Casali, 1990; Chaffin, et al., 1994; Laux, 2001). Roger (1997) pointed out that, 'individuals over age 65 experience declines in sensory, perceptual, motor, and cognitive abilities that may interfere with their ability to interact with systems ranging from doorknobs to microwave ovens to computers'.

Difficulties of vision perception reported by elderly adults are mainly in spatial vision (acuity and contrast sensitivity), slowing of vision processing, seeing in poor light and near distance, processing color information, and visual search (Scialfa et al., 2004). Hand-held devices (mobile phone, PDA and auto GPS Navigation) rely primarily on traditional graphical user interface to present information. Age deficits in vision may influence the ease with which these technologies are used. Devices communicated by transmission of sound, like telephone rings and sound alarm, need to consider the hearing impairments of elderly adults when utilize auditory output. Use of new technologies, such as Instant Messengers and Web 2.0, require learning new skills and how to locate, access, manipulate, and use information sources. Given that the elderly experience declines in cognitive abilities such as working memory, they are slower to acquire computer skills than younger adults and require more help and hands-on practice when searching for information in electronic environments (Czaja & Lee, 2001).

PSYCHOSOCIAL CHARACTERISTICS

The psychosocial changes of ageing include status loss, loneliness, fear of illness and death, poverty, harmful life-styles and deterioration of the quality of life (Chen, 1987). The sense of personal control decreases with age because the elderly are likely to have experienced events beyond their control, such as loss of loved ones, loss of the work role, and loss of health. It has also been reported that social interaction declines in elderly adulthood (Erber, 2009). Social isolation may also be positively correlated with physical (hearing and vision) impairment (Weinstein, 2000). It was proposed by her that 'late life is a period of transition and adjustment to loss'. Transition includes retirement and relocation. China has strikingly lower retirement ages than other countries (Xie, 2007). Early retirement is associated with an abrupt change in lifestyle, e.g., loss of career identity, social detachment, and loss of sense of value from one's contribution to society (Xie, 2007).

Technology has been found to be a valuable tool for improving living conditions for elderly people, as well as providing benefits to family members, caregivers, and service providers. The general technology application areas for elderly people in China may be considered to fall into three categories: traditional technology, medical or assistive technology, and information and communication technology (ICT). Traditional technology includes fax machine, camera, tape recorder, microwave oven, television, automatic teller machines, etc. The medical/healthcare area is one of the major application fields of technology. Medical/assistive technology devices, such as electronic care surveillance devices, wireless sensor networks and smart home, enable the elderly to 'age in place', that is, to remain in their own homes for as long as confidently and comfortably possible (Pollack, 2005). Medical technologies can compensate for declining abilities due to ageing and help the elderly to live independently at home. ICT applications e.g., mobile phone, computer and Internet, can remotely connect elderly people with family and friends and help alleviate social isolation. Xie (2007) found that negative feelings and attitudes of the elderly towards post-retirement life can be improved by adopting ICT applications. For example, learning to use the Internet can make the lives of elderly Chinese's after retirement more meaningful and improve self-evaluation as well as other people's view on them. Through the Internet, the elderly also can seek out health information, purchase products, and engage in social networking.

Knowing the needs and capabilities of the elderly help us to better understand the factors which influence their acceptance and adoption of a new technology. Current research on technology acceptance, however, has not so far considered the unique characteristics, abilities, and limitations of the elderly. Understanding and creating the conditions under which technology can be incorporated into the lives of the elderly remains a high-priority research issue.

CONCLUSION

Population ageing impacts significantly on socio-economic and public health areas. Technology offers a challenge and an opportunity in providing support and in enhancing the daily lives of elderly people. Knowledge about which factors determine elderly adult usage of technology and how these factors operate is rather limited at present. In order to understand how the elderly can interact successfully with the software and hardware of technological devices and systems, it is essential to understand the biophysical and psychosocial characteristics, abilities and problems experienced by the elderly. Age-related factors have to be integrated into the current TAM model in order to generalize it to include elderly population. The present review is intended to assist towards an understanding of the current situation and problems regarding elderly technology acceptance, and to encourage further research on the development a unified elderly acceptance model.

REFERENCES

Agarwal, R., and Prasad, J. A. (1999), "Are individual differences germane to the acceptance of new information technologies?" *Decision Sciences*, 30(2): 361–391.

Arning, K., and Ziefle, M. (2007), "Understanding age differences in PDA acceptance and performance." *Computers in Human Behavior*, 23(6): 2904-2927.

Berkowitz, J. P., and Casali, S. P. (1990), "Influence of age on the ability to hear telephone ringers of different spectral content", in *Designing for an Aging Population: Ten Years of Human Factors/ Ergonomics Research*, Rogers, W. A. (Ed.). (1997). Santa Monica: Human Factors & Ergonomics Society.

Chaffin, D. B., *et al.* (1994), "Age effects in biomechanical modeling of static lifting strengths", in *Designing for an Aging Population: Ten Years of Human Factors/ Ergonomics Research*, Rogers, W. A. (Ed.). (1997). Santa Monica: Human Factors & Ergonomics Society.

Chau, P. Y., and Hu, P. J. (2002), "Examining a model of information technology acceptance by individual professionals: An exploratory study." *Journal of Management Information Systems*, 18(4): 191-229.

Chen, L. D., *et al.* (2002), "Enticing online consumers: An extended technology acceptance perspective." *Information & management,* 39(8): 705-719.

Chen, P. C. (1987), "Psychosocial factors and the health of the elderly Malaysian", *Annals of the Academy of Medicine*, Singapore, 16(1): 110-114.

Cheng, T. C. E., *et al.* (2006), "Adoption of internet banking: An empirical study in Hong Kong." *Decision support systems*, 42(3): 1558-1572.

China Internet Network Information Center. (July 16, 2009), *The 24th Statistical Report on Internet Development in China (Report).* Available at www.cnnic.org.cn.

Czaja, S. J., and Lee, C. C. (2001), "The internet and older adults: Design challenges and opportunities.", in *Communication, technology and aging: opportunities and challenges for the future*, Charness, N., Parks, D. C., and Sabel, B. A. (Ed.). Springer Publishing Company, New York, pp. 60.

Davis, F. D., Bagozzi, R. P., and Warshaw, P. R. (1989), "User acceptance of computer technology: A comparison of two theoretical models." *Management Science*, 35: 982-1003.

Erber, J. T. (2009), *Aging and older adulthood (2nd Ed).* Wiley-Blackwell.

Fozard, J. L. (1997), "Aging and technology: a developmental view", in *Designing for an Aging Population: Ten Years of Human Factors/ Ergonomics Research*, Rogers, W. A. (Ed.), (1997), Santa Monica: Human Factors & Ergonomics Society.

Gefen, D., Karahanna, E., and Straub, D. W. (2003), "Trust and TAM in online shopping: An integrated model." *MIS Quarterly*, 27(1): 51-90.

Hu, P. J. H., and Ma, W. W. (2003), "Examining technology acceptance by school teachers: a longitudinal study." *Information & management*, 41(2): 227-241.

Karahasanović, A., *et al.* (2009), "Co-creation and user-generated content-elderly people." *Computers in Human Behavior*, 25(3): 655-678.

Laux, L. F. (2001), "Aging, communication, and interface design", in *Communication, technology and aging: opportunities and challenges for the future*, Charness, N., Parks, D. C., and Sabel, B. A. (Ed.). Springer Publishing Company, New York, pp.153.

Lederer, A. L. *et al.* (2000), "Technology acceptance model and the World Wide Web." *Decision support systems*, 29(3): 269-282.

Lu, Y., Zhou, T., and Wang, B. (2009), "Exploring Chinese users' acceptance of instant messaging using the theory of planned behavior, the technology acceptance model, and the flow theory." *Computers in Human Behavior*, 25(1): 29-39.

Mallenius, S., Rossi, M., and Tuunainen, V. K. (2007), *Factors affecting the adoption and use of mobile devices and services by elderly people-results from a pilot study*. Proceeding of 6th Annual Global Mobility Roundtable, Los Angeles, available at: http://www.marshall.usc.edu/ctm/Research.

Mao, E., Palvia, P. (2006), "Testing an Extended Model of IT Acceptance in the Chinese Cultural Context." *Ergonomics Abstracts*, 37(2): 20-32.

Ministry of Industry and Information Technology of the People's Republic of China (MIIT). (2008), *Statistics of Communication Industry 2007 (in Chinese)*. Beijing, China.

Moon, J. W., and Kim, Y. G. (2001), "Extending the TAM for a World-Wide-Web context." *Information & management*, 38(4): 217-230.

Morris, M. G., *et al.* (2005), "Gender and age differences in employee decisions about new technology: An extension to the theory of planned behavior." *IEEE transactions on engineering management*, 52(1): 69-84.

Pavlou, P. A. (2003), "Consumer acceptance of electronic commerce: Integrating trust and risk with the technology acceptance model" *International journal of electronic commerce*, 7(3): 101-134.

Pollack, M. E. (2005), "Intelligent technology for an aging population: The use of AI to assist elders with cognitive impairment" *AI Magazine*, 26(2): 9-24.

Renaud, K., and Biljon, J. V. (2008), "Predicting technology acceptance and adoption by the elderly: A qualitative study" *ACM International Conference Proceeding Series*, 338: 210-219.

Rogers, W. A. (1997), *Designing for an Aging Population: Ten Years of Human Factors/ Ergonomics Research*. Human Factors & Ergonomics Society. Santa Monica.

Ryu, M. H., *et al.* (2009), "Understanding the factors affecting online elderly user's participation in video UCC services" *Computers in Human Behavior,* 25, (3): 619-632.

Scialfa, C. T., Ho, G., and Laberge, J. (2004), "Perception aspects of Gerotechnology", in *Gerotechnology: Research and Practice in Technology and Aging*, Burdick, D. C., and Kwon, S. (Ed.). Springer Publishing Company, New York, pp.19.

Steele, R. (2009), "Elderly persons' perception and acceptance of using wireless sensor networks to assist healthcare." *International journal of medical informatics.*

Teo, T., *et al.* (2009), "Assessing the intention to use technology among pre-service teachers in Singapore and Malaysia: A multigroup invariance analysis of the

Technology Acceptance Model (TAM)." *Computers & education,* 53(3): 1000-1009.

UNFPA China. (2006), *Population ageing in China-facts and figures.* UNFPA China Office, Available at http://www.un.org.cn/cms/p/resources/30/509/content.html.

United Nations, Department of Economic and Social Affairs. (2004),*World Population ageing 1950-2050.* New York, Available at http://www.un.org/esa/population/publications/worldageing19502050/.

United Nations, Population Division of the Department of Economic and Social Affairs of the United Nations Secretariat. (2009), *World Population Prospects: The 2008 Revision.* Available at http://esa.un.org/unpp.

Venkatesh, V., and Davis, F. (2000), "A theoretical extension of the Technology Acceptance Model: Four longitudinal field studies." *Management Science,* 46(2): 186-204.

Venkatesh, V., and Morris, M. (2000), "Why don't men ever stop to ask for directions? Gender, social influence, and their role in technology acceptance and usage behavior." *MIS Quarterly,* 24(1): 115-139.

Venkatesh, V., Morris, M., Davis, G., and Davis, F. (2003), "User acceptance of information technology: Toward a unified view." *MIS Quarterly,* 27(3): 425-478.

Weinstein, B. E. (2000), *Geriatric audiology,* Thieme Medical Pub, New York.

World Health Organization (WHO). (2001), *Definition of an Older or Elderly Person: Proposed Working Definition of an Older Person,* in Africa for the MDS Project, Health Statistics and Health Information Systems. WHO. Available at www.who.int/healthinfo/survey/ageingdefnolder

Wu, J. H., and Wang, S. C. (2005), "What drives mobile commerce? An empirical evaluation of the revised technology acceptance model." *Information & management,* 42(5): 719-729.

Xie, B. (2007), "Older Chinese, the internet, and well-being." *Care Management Journals,* 8(1): 33-38.

Yu, P., *et al.* (2009), "Health IT acceptance factors in long-term care facilities: A cross-sectional survey." *International journal of medical informatics,* 78(4): 219-229.

Zhang, J., and Mao, E. (2008), "Understanding the acceptance of mobile SMS advertising among young Chinese consumers." *Psychology & Marketing,* 25(8): 787-805.

Chapter 58

Evaluation of Dynamic Touch Using Moment of Joint Inertia

Takafumi Asao, Yuta Kumazaki,
Kota Kawanishi, Kentaro Kotani, and Ken Horii

Department of Mechanical Engineering
Faculty of Engineering Science,
Kansai University
3-3-35 Yamate-cho, Suita
Osaka 564-8680, JAPAN

ABSTRACT

The objective of this study is to clarify the perceptual mechanism for dynamic touch. In this study, the relation between the perceived length and moment of inertia of a rod about the elbow joint was analyzed for three arm posture. Ten university students participated in the experiment. The subjects vertically wielded nine kinds of rods using their elbows for a short distance; this was possible because their wrists and upper arms were fixed. For comparison, they held the rods statically. The perceived length of rod was not different for wielding and holding. However, the perceived length was greater for greater arm extension. Moreover, the simple regression of the perceived length on the static moment and moment of inertia of the rod about its center of mass (CM), wrist, and elbow was investigated. The contributing ratio of the static moment for holding was lower than that for wielding. The contributing ratios of the moments of inertia about the CM and about the wrist were higher than that about the elbow.

Keywords: Dynamic touch, haptic perception, moment of inertia

INTRODUCTION

Without the use of sight, a person can recognize the spatial properties of objects by wielding them or by even partial tactual contact. This haptic perception accompanied by joint movement is called dynamic touch. Through dynamic touch we can perceive the length (Solomon & Turvey, 1988), shape (Burton et al., 1990), holding position (Pagano et al., 1994), and direction of an object (Pagano & Turvey, 1995). The accuracy of a rod's length perception has been found to be better with wielding than with holding (Burton et al., 1990). Previous studies that obtained the above findings were directed at identifying the invariant in length perception. These studies proposed the moment of rotation, inertia tensor (Solomon & Turvey, 1988), and principal moment of inertia (Carello & Turvey, 2004) as the invariant for dynamic touch. Moreover, regression models of perceived length were offered, that involved the principal moment of inertia of the rod. It is appropriate for people to use the moment of inertia to perceive an object's mechanical and geometric information.

Muscular exertion is important for dynamic touch, which involves proprioceptive information from the muscle spindle and Golgi organ (Turvey, 1996). Even if the rod has invariable mechanical and geometric characteristics, the perceived length may be different due to arm posture, because muscle length varies according to the arm posture. In this study, the influence of arm posture on length perception was investigated with wielding and holding actions using elbow. Further, the perceived length was analyzed theoretically by using the moment of inertia at the wrist and elbow.

METHODS

ARM POSTURE AND APPARATUS

In a previous study on dynamic touch, experiments involving the wielding rods at the wrist, elbow, and shoulder were conducted (Pagano et al., 1993). When subjects move their arm by simultaneously using several joints, it is difficult to identify the invariant for a perceived length. In the present experiment, subjects can move their arm by using only their elbow. The wrist was fixed to the forearm by an athletic supporter (Daiya Industry, AWristGuard 000-3682) and the upper arm was immobilized on a slanted armrest, as shown in figure 1.

Figure 1 shows three arm postures adopted in the experiments. The upper arm was fixed to 62° from a horizontal plane. In figure 1, a) *verticality*, the forearm was perpendicular to the upper arm; b) *horizontality,* the forearm was horizontal to the ground; and c) *extension*, the elbow angle was 20° from the most extended position.

a) *verticality*

b) *horizontality*

c) *extension*

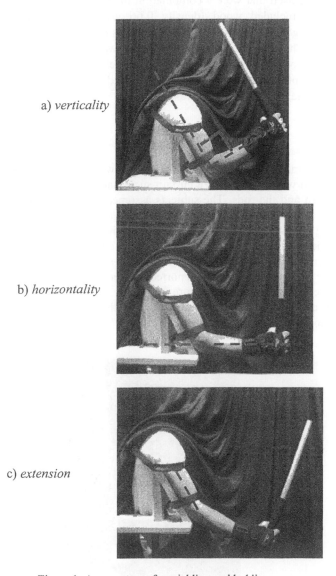

Figure 1. Arm postures for wielding and holding.

MATERIALS

Nine rods were used that were a combination of three materials and three lengths. The materials included wood (density 0.66 g/cm^2), plastic (1.4 g/cm^2), and aluminum (2.7 g/cm^2). The three different lengths of the rods used were 300, 400, and 500 cm. The external radius of each rod was 20 mm. Sponge grips were attached to the rods to avoid the affect of tactile feeling and specific heat of the rods on length perception. The internal and external radii of grips were 20 and 27 mm, respectively. The mass of the grips was 23 g.

PROCEDURE

Subjects were made to sit on a chair, and their wrist and upper arm were fixed. Their right arm was occluded by a curtain such that they were not able to see their own right arm and the rods. The subjects assumed one of the three arm postures, then one of the rods was placed in their right hand and they grasped it firmly. Under the wielding condition, subjects would wield the rod using the elbow such that the vertical position of their right hand was within 5 cm of the specific arm positions. On the other hand, under the holding condition, subjects would hold the rod and maintain the arm posture. Subjects reported the perceived length by clipping a stick using their left hand, as shown in figure 2.

Ten male university students participated in the experiments. Two trials were conducted for each condition.

Figure 2. Experimental conditions.

RESULTS

Figure 3 and 4 show the relation between the perceived and actual lengths. The dashed line indicates the position at which the perceived length and actual length are equal. The perceived length was greater for greater actual length. The perceived length for a rod of greater density was higher than that for one of lesser density, even if the actual rod length was equal. Some subjects reported that a large force had to be exerted for the *extension* arm posture. That is, they perceived a greater length as the arm was extended. Further, it should be noted that there was little difference in the perceived length between wielding and holding conditions. In this experiment, the difference in the conditions of wielding and holding was small since the subjects only wielded a rod within a short distance.

EVALUATION BY MOMENT OF INERTIA

In previous studies on dynamic touch, the perceived length of a rod was evaluated from the static moment and inertia tensor of the rod (Fitzpatrick et al., 1994; Turvey, 1996; Lenderman et al, 1996). In this study, the static moment and moment of inertia were adopted to evaluate the invariant of the perceived length.

Although there are nine elements in an inertia tensor, the products of inertia are zero if one axis of the coordinate system fixed on the rod with its origin at the center of mass of the rod is parallel to the lengthwise of the rod. Moreover, the arm movement in the experiment is only about the elbow joint axis. Therefore, only one moment of inertia about the rotating axis requires to be considered.

Table 1 shows the mean moments of inertia and mean static moment M_s calculated by using the rods' specification and subjects' anthropometric data. Variables I_g, I_w, and I_e are moments of inertia of the rod about the center of mass of the rod, about the wrist, and about the elbow, respectively.

Table 2 shows the contributing ratio r^2 of the simple regression of the logarithm of the mean perceived length to the logarithm of each mechanical parameter. The r^2 of static moment for wielding was higher than that for holding. The static moment is not appropriate to represent the perceived length for wielding (Burton & Turvey, 1990). In contrast, the moments of inertia of I_g and I_w expressed the perceived length well, since the r^2 values are high for wielding and holding. However, r^2 for I_e is not higher than that for I_g and I_w because the wrist joint was not constrained rigidly by the athletic supporter. Even if the arm posture varied, the moments of inertia I_g and I_w can represent the perceived length. However, the differences in the perceived length among the arm postures cannot be represented by the moment of inertia. Thus, it is necessary to investigate invariant to express the influence of the arm posture.

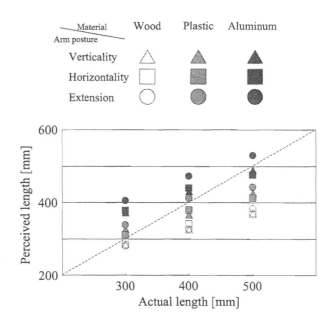

Figure 3. Perceived length and actual length by wielding.

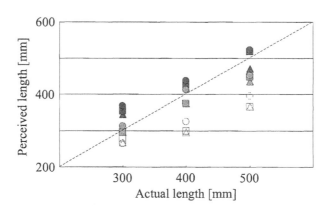

Figure 4. Perceived length and actual length by holding.

Table 1. Static moment and moment of inertia.

	Rod Material	Rod Length [mm]		
		300	400	500
M_s [gf m]	Wood	9.7	17.9	25.1
	Plastic	19.3	33.9	52.9
	Aluminum	35.1	63.1	99.6
I_g [g m^2]	Wood	0.6	1.4	2.6
	Plastic	1.1	2.6	5.1
	Aluminum	2.0	4.8	9.4
I_w [g m^2]	Wood	2.0	4.7	8.1
	Plastic	4.1	9.1	17.5
	Aluminum	7.5	17.1	33.2
I_e [g m^2]	Wood	10.2	15.1	18.7
	Plastic	19.5	27.7	38.3
	Aluminum	35.0	50.8	71.3

Table 2. Contribution ratio of simple regression for each invariance.

		r^2
Holding		
	M_s	0.95
	I_g	0.94
	I_w	0.95
	I_e	0.89
Wielding		
	M_s	0.91
	I_g	0.95
	I_w	0.95
	I_e	0.91

CONCLUSIONS

In this study, length perception was investigated for different arm posture by wielding and holding rods about the elbow. There were only small differences in the perceived length between wielding and holding conditions. In contrast, the perceived length was greater for greater arm extension. Moreover, the relationships between the perceived length and the static moment and moment of inertia were analyzed. The contributing ratio of the perceived length and the moment of inertia was high. In this paper, the invariant that expresses the difference in the perceived length because of the arm posture was not mentioned. If the arm posture varies, the joint torque should change. In future works, it will be necessary to investigate the relationship between joint torques and length perception. Moreover, the muscular state should be evaluated to clarify the perception mechanism for dynamic touch.

REFERENCES

Solomon, H.Y. and Turvey, M.T. (1988) Haptically perceiving the distances reachable with hand-held objects, Journal of Experimental Psychology: Human Perception and Performance, 14, pp.404-427.

Burton, G., Turvey, M.T., and Solomon, H.Y. (1990) Can shape be perceived by dynamic touch?, Perception & Psychophysics, 48, pp.477-487.

Pagano, C.C., Kinsella-Shaw, J., Cassidy, P., and Turvey, M.T. (1994) Role of the inertia tensor in haptically perceiving where an object is grasped, Journal of Experimental Psychology: Human Perception and Performance, 20, pp.276-284.

Pagano, C.C. and Turvey, M.T. (1995) The inertia tensor as a basis for the perception of limb orientation, Journal of Experimental Psychology: Human Perception and Performance, 21, pp.1070-1087.

Turvey, M.T. (1996) Dynamic Touch, American Psychologist, 51(11), pp.1134-1152.

Carello, C. and Turvey, M.T. (2004) Physics and psychology of the muscle sense, Current Directions in Psychological Science, 13, pp.25-28.

Fitzpatrick, P., Carello, C., and Turvey, M.T. (1994) Eigenvalues of the inertia tensor and exteroception by the "muscular sense", Neuroscience, pp.551-568.

Christopher, C.P., Fitzpatrick, P., and Turvey, M.T. (1993). Tensorial basis to the constancy of perceived object extent over variations of dynamic touch, Perception & Psychophysics, 54(1), pp.43-54.

Lenderman, S.J., Ganeshan, S.R., Ellis, R.E. (1996). Effortful touch with minimal movement: Revisited, Journal of Experimental Psychology: Human Perception and Performance, 22(4), pp.851-868.

Burton, G. and Turvey, M.T. (1990). Perceiving the Lengths of Rods That are Held But Not Wielded, Ecological Psychology, 2(4), pp.295-324.

Chapter 59

Improvement in Touch Operations by Tactile Feedback

T. Fukui[1], M. Nakanishi[1], Y. Okada[1], S. Yamamoto[2]

[1]Fac. of Sci. & Tech., Dept. of Administration Eng.
Keio University
3-14-1, Hiyoshi, Kohoku, Yokohama 223-8522, Japan

[2]Fac. of Eng., Dept. of Management Sci., Tokyo University of Science
1-3, Kagurazaka, Shinjuku, Tokyo 162-8601, Japan

ABSTRACT

Based on the background of eco-engineering, machinery investment is shifting from "replacement," which means introducing new machinery, to "rebuilding," which means changing software and adding human–machine interfaces such as touch screens to expand the functions of machinery used in many fields. Although conventionally the usability of such touch-screen-type control panels has mainly been improved by improving graphical user interface (GUI) designs, this method is not always effective, depending on work environment, task characteristics, and so on. In this study, we propose adding tactile feature to conventional visual and auditory interactions, as a new option for making touch operations in industrial fields more comfortable, and we examine their applicability. The results of the experiment, which involved operating the touch-screen-type control panel with tactile feedback from a standing position similar to that of the industrial fields, show the following: 1) the errors caused by insufficient pressure of a finger and deviation in pointing can be improved by adaptively changing the finger motions through tactile stimulus, and 2) such errors come to be easily noticed and corrected by changing tactile feedback according to the touched point.

Keywords: Touch operation, Tactile feedback, Industrial fields, Eco-engineering

INTRODUCTION

In recent years, the trend toward eco-engineering or sustainable engineering has been growing in different industries. Based on such background, machinery investment is shifting from "replacement," which means introducing new machinery, to "rebuilding," which means changing software and adding human–machine interfaces such as touch screens to expand the functions of machinery that have been used in many fields. Rebuilding is very effective for ecological management, but necessarily involves some restrictions because it often becomes difficult to install human–machine interfaces in ready-made machinery. For example, it is not always possible to fix touch screens in easily visible and easily touchable locations. Moreover, it is rather rare that large-sized screens can be fixed.

The usability of touch-screen-type control panels has been improved mainly by changing the GUI design, whose flexibility is high. Although not much, ergonomic knowledge of allocating command buttons or colour schemes exists, and it is also used in operating print machines, automatic teller machines, thicket machines and so on. On the other hand, as described above, touch screen-type control panels rebuilt in industrial fields cannot always been used in normal, ergonomically recommended environments, and it is often impossible to improve their usability by changing their GUI design.

Furthermore, various factors that change workers' activities are involved in industrial environments that are not involved in offices. For example, noise and lighting conditions may prevent industrial workers from getting information from systems. Moreover, in the case where workers have to control the touch screen while they sometimes check the condition of a machine or products, their attention is frequently required to be refocused.

Based on these factors, there are several situations in industrial fields that involve conditions that deviate from normal work environments. Thus, we propose a new way of improving the usability of touch-screen-type control panels, namely, adding tactile interactions as a new option that goes beyond conventional visual and auditory interactions, and examine their applicability. In this study, we first analyze difficulties in operating touch-screen-type control panels using a human-factor methodology, and conduct experiments to examine if and how tactile feedback from touch operations can improve these difficulties.

EXPERIMENT

As described in the previous section, the positions of industrial workers while touching control panels are often different from those that are ergonomically recommended. Thus, we first performed an experiment that simulated two situations: one in which the subjects operated a control panel located in an ergonomic work area from a sitting position, and the other in which they operated a control panel located out of an ergonomic work area from a standing position. We then compared the difficulties that were involved in the two tasks.

EXPERIMENTAL SYSTEM

A touch-screen-type control panel that is used as a human–machine interface in industrial fields is often installed in a machine, and in many cases, workers pay attention to it from a standing position. FIGURE 1 shows an example of metal-pressing machines that have touch screen-type control panels.

In this study, we constructed an experimental system that simulates the control panel on the metal-pressing machine. The system was composed of 15 types of images, command buttons to switch images, and buttons to enter numerical values related to metal processing. For some images, numerical command buttons popped up when certain blanks were selected. FIGURE 2-1 shows the screen on which a motion of the metal-pressing machine is selected, which was displayed at the start of the experimental task. FIGURE 2-2 (see Appendix) shows the screen of numerical command buttons, and FIGURE 2-3 (see Appendix) shows the screen on which numerical command buttons pop up. The GUI design of these images is a nearly identical copy of the original touch screens.

The touch screen used in the latter part of this experiment is the Tactile Touchscreen Demonstration Unit, manufactured by Immersion (FIGURE 3). Although this touch screen can give tactile feedback, its function was not used in this stage of the experiment, for which we used a touch screen that just involves a

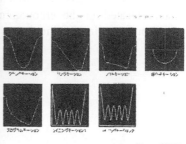

FIGURE 2-1. Touch screen in which motions of the metal pressing machine are selected.

FIGURE 1. An example of metal pressing machines with touch-screen-type control panels.
Courtesy of http://www.shinohara-press.co.jp

534

normal touch panel.

EXPERIMENTAL TASK

The subjects were 10 students of 21–23 years old. Every subject was right-handed. They repeatedly practiced controlling the experimental system to be familiar with it before starting the task. Their task was to control the experimental system according to the instructions displaying what values should be entered in which sections. The number of items to be filled in per task was 26. Table 1 shows an example of the instruction sheet.

EXPERIMENTAL CONDITIONS

We set the following two different patterns at the position taken by the subjects when controlling the experimental system as the experimental conditions: 1) the office pattern (FIGURE 4-1), where they touched the system located in the ergonomic work area from a sitting position, and 2) the

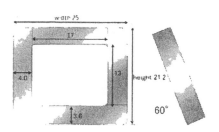

FIGURE 3. Touch screen used in this experiment. (unit: cm)

TABLE 1 An example of the instruction sheet

Moton 1	Pressure L m g	64 19
	Startng Pos	26 71
	Press ng Stroke	63 78

Moton 2	Pressure L m g	73 93
	Startng Pos	17 18
	Approach Stroke	72.37
	Press ng Stroke	25.28
	Press ng Startng Pont	92 95
	Pass Stroke	86 79
	Stop Duraton	38 91
	F nsh ng Pont	32.24

ON	CH1	Pos	93 75
		Pro No	60.52
	CH2	Pos	29 17
		Pro No	50.10
	CH3	Pos	51 47
		Pro No	46.06
	CH4	Pos	37 34
		Pro No	02.97
OFF	CH1	Pos	81 14
		Pro No	07 15
	CH2	Pos	83 62
		Pro No	09 19
	CH3	Pos	65 56
		Pro No	35.85
	CH4	Pos	69 01
		Pro No	

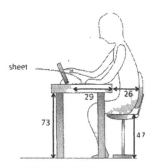

FIGURE 4-1. Subjects' position in the office pattern. (unit: cm)

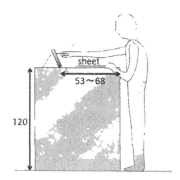

FIGURE 4-2. Subjects' position in the industrial pattern. (unit: cm)

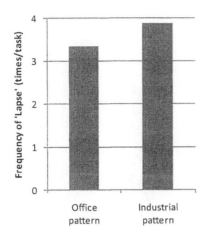

FIGURE 5. Frequency of "Lapses" (times/task) .

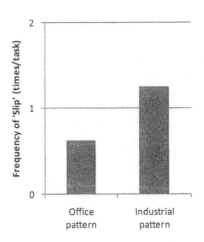

FIGURE 6. Frequency of "Slips" (times/task).

industrial pattern (FIGURE 4-2), where they touched a control panel located out of the ergonomic work area from a standing position.

The subjects experienced each condition once. Then, in order to eliminate the order effect, the order was switched.

MEASURED ITEMS

The subjects' operation logs were automatically recorded by the experimental system. We further recorded their actions during the task with two digital video cameras, fixing one camera at the back of the subject so that it could capture which command buttons he or she touched even if the experimental system could not sense it. We fixed the other camera at the top of the touch screen so that it could capture the subject's finger motions while touching the command buttons.

RESULTS

LAPSE, SLIP & MISTAKE

Next, we made three categories for why subjects could not correctly touch the command buttons indicated on the instruction sheet according to one of the most well-known human error categorization "lapse, slip, and mistake" (Norman; 1981, Reason; 1990): lapse, slip, and mistake. Specifically, we defined the cases where the appropriate command button was not touched as "lapse," the cases where another command button was touched unintentionally as "slip," and the cases where another command button was chosen wrongly as "mistake."

We however could not find any case that could be categorized as a mistake because the subjects were allowed to operate the system while looking at the instruction sheet. FIGURE 5 shows the average number of lapses per task, and FIGURE 6 shows the average number of slips per task. We can see that both lapse and slip occurred more frequently in the industrial pattern than in the office pattern. Moreover, from the video analysis, many more of the cases categorized as lapses than as slips, indicating that the members' touches were not sensed by the system even if they were sure that they touched the screen properly, with what was an imperfect touch.

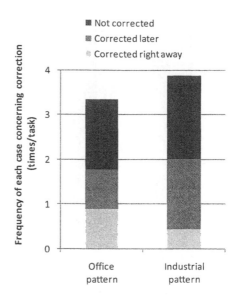

FIGURE 7. Frequency of each case concerning correction of "lapses" (times/task)

CORRECTION

Then, we focused on the subjects' actions just after the lapses occurred, and noted the following three cases: 1) those where the subjects noticed and corrected the lapses immediately, 2) those where they noticed and corrected them after they moved their finger to other locations, and 3) those where the subjects neither noticed nor corrected the lapses, and proceeded to shift to the next touch. FIGURE 7 shows the percentage of each case concerning the subjects' corrections of their lapses. From this result, we can understand that the touch operation in the industrial pattern has a tendency for the imperfect touch to be easy and noticing it to be hard.

ROOT-CAUSE ANALYSIS

We used root-cause analysis to determine factors causing the two problems of imperfect touch and not correcting it. FIGURE 8-1 and 8-2 show the result. It is difficult to remove the factors described in the painted boxes when the system is operated in the standing position such as in the industrial pattern. Thus, we examined the possibility of eliminating the factors that caused the above two problems, or cutting the connection by implementing the function of giving tactile feedback on each command button. In specific, we first expected that getting different tactile feedback from separate command buttons would enable subjects to easily notice an unintentional wrong touch; that is, we expected the feedback to eliminate the factor of "not noticing wrong input," which leads to the problem of

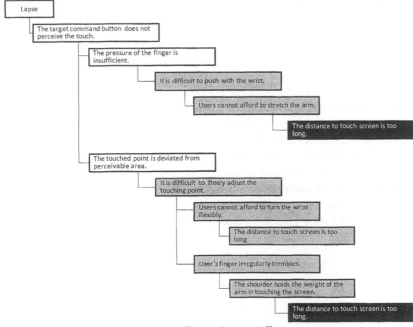

FIGURE 8-1. Root-cause analysis of "imperfect touch".

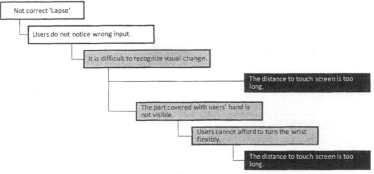

FIGURE 8-2. Root-cause analysis of "not correcting one's lapse".

not correcting imperfect touch. Second, we expected that tactile feedback would make subjects feel not like they were touching a flat command button that was a part of the screen, but like they were pushing an analogue switch button, and that their finger motion would adaptively change to improve the finger pressure and to adjust the touched point.

EFFECT OF TACTILE FEEDBACK

Our experiment investigates how implementation of tactile feedback changed the subjects' performance, particularly if it improved the two problems in the touch operation involved in the industrial pattern: imperfectly touching the screen and not correcting it.

METHOD

The experimental task was the same as the previous one described in the Chapter 3, and the same subjects participated. They performed the task once from the position of the industrial pattern shown in FIGURE 4-2. The measured items were also the same ones as described in Section 2.4.

When the command buttons sensed that the subject's finger had touched them, the surface of the touch screen vibrated differently according to the point at which it was touched and gave the subject's finger tactile feedback. FIGUREs 9-1 and 9-2 show the allocations of the tactile feedback to each command button, where the numbers corresponds to the ones on Table 2, which shows the characteristics of tactile stimulus.

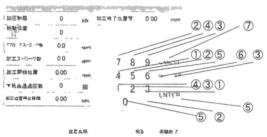

FIGURE 9-1. Allocation of tactile feedback to each numerical button.

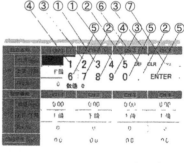

FIGURE 9-2. Allocation of tactile feedback to each button on pop-up.

TABLE 2 Characteristics of Tactile Stimulus

	Base effect	Relative magnitude	Duration(ms)	Repeat buffer(ms)
①	cnsp 1	7	11	90
②	cnsp 1	5	11	90
③	smooth 1	10	50	90
④	smooth 1	8	50	90
⑤	smooth 1	7	50	90
⑥	mag ramp up	various	320	0

RESULTS

FIGURE 10 compares the frequency of lapses when members operated the system with tactile feedback with the result, which was given from the previous experiment, and when they operated the system without tactile feedback from the same position such as in the industrial pattern. We can see that the implementation of tactile feedback in the industrial pattern reduced the cases of lapses to an even lower level than in the office pattern. Taking into consideration the root-cause

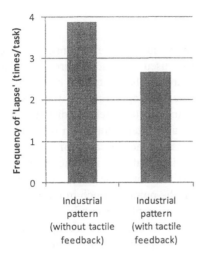

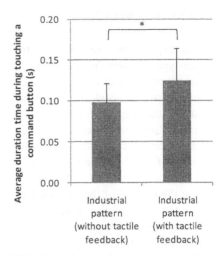

FIGURE 10 Frequency of "Lapses" (times/task)

FIGURE 11 Duration time during touching a command button. (*p < 0.05)

analysis discussed in the previous chapter, this result shows that insufficient pressure by the finger and deviation in pointing were improved. Moreover, we analyzed the duration time of touching a command button, and obtained the result shown in FIGURE 11. When we examine the cause of the reduction of lapses based on this result, we see that the subjects tended to apply more pressure in their touch when they used the system with tactile feedback, which seems to be a result of their finger motions adaptively changing due to tactile feedback.

FIGURE 12 compares the percentage of subjects correcting lapses when they operated the system with tactile feedback and when they operated the system without it, based on whether they noticed a lapse and how they corrected it. When they used the system with tactile feedback, they corrected almost 80% of lapses right away, while when they used the system without it, that percentage was only around 10%. From this result, we can confirm that tactile feedback enables subjects to easily notice their imperfect touches.

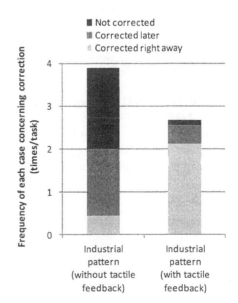

FIGURE 12. Frequency of each case concerning correction of "lapses" (times/task).

CONCLUSION

In this experiment, we focus on the fact that in industrial environments, touch screen-type control panels were not always used as they were in office environments, and discuss what the difficulties are and how they can be improved. In particular, we used an experiment to analyze how operating the touch screen from the standing position that is often adopted in industrial fields is different from operating the touch screen from the sitting position often adopted in offices. The results show that 1) the lapses tend to easily take place because of insufficient pressure of the finger and deviation in pointing, and that 2) it is difficult to notice and correct imperfect touch. Thus, in order to prevent these cases, we propose having the surface of the touch screen vibrate at the point of touch in order to give tactile feedback to the finger. Through our experiment, we confirm that 1) the subjects improved their finger motions adaptively in order to correctly touch the command button resulting in a reduction of lapses, and that 2) they came to easily notice imperfect and wrong touches resulting in more reliable correction.

Although conventionally the various difficulties in touch operations have been improved mainly by designing better GUI, this method is not always effective in industrial fields, depending on work environment, the characteristics of the task, and so on. In this study, as a measure to make touch screen-type control panels more useful for workers, particularly since they are expected to be increasingly introduced into industrial fields, we have suggested the applicability of tactile feedback. Depending on economic and social situations, industrial needs vary. Introducing a new human–machine interface offers a new option to meet such needs.

REFERENCES

Norman, D.A. (1981), "Categorization of action slips." *Psychological Review*, 88, 1–15.

Reason, J.T. (1990), *Human Error*. Cambridge University Press.

Shinohara Press Service. (2002), *Shinohara Press Service*. Retrieved, from http://www.shinohara-press.co.jp.

541

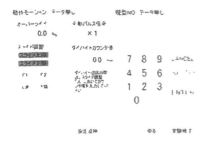

FIGURE 2-2. Touch screen that shows the numerical command buttons panel.

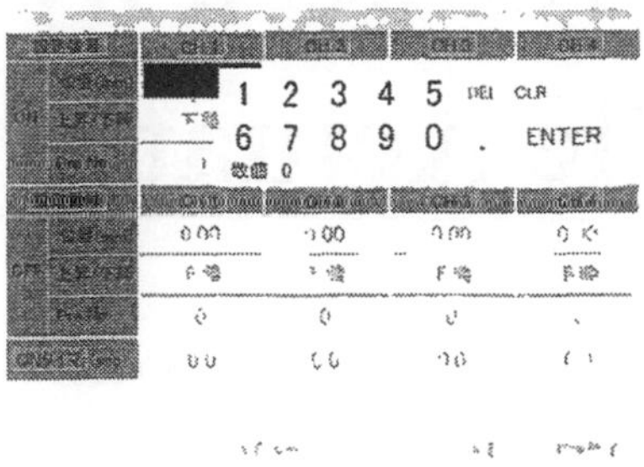

FIGURE 2-3. Touch screen in which numerical command buttons pop up.

Chapter 60

Effective Presentation for Velocity and Direction Information Produced by Using Tactile Actuator

Kentaro Kotani, Toru Yu, Takafumi Asao, Ken Horii

Department of Mechanical Engineering
Faculty of Engineering Science
Kansai University
Suita, Osaka 564-8680, JAPAN

ABSTRACT

The objective of this study is to clarify the effect of the posture of a recipient of tactile stimulus and the direction of the tactile stimulus on the perception of velocity generated by tactile apparent motion. Ten subjects participated in the experiment in which an air-jet tactile actuator was used to provide tactile stimuli. The stimuli were presented on the palm of each subject. The subjects perceived a set of stimuli in five postures and four directions of stimulus, and they estimated the velocity and the direction of tactile apparent motion. The percentage of correct responses regarding perceived velocity was analyzed for different postures and directions of stimulus sets. The results indicate that no specific posture had an advantage for velocity perception; however, there were significant differences in individual preference for accurate estimation of perceived tactile velocity. On the other hand, for the direction of stimulus, the accuracy of velocity perception was not affected by the direction of stimulus.

Keywords: tactile information, apparent motion, velocity perception

INTRODUCTION

Recently, tactile interfaces, which are user interfaces that employ tactile perception for input/output, have been the focus of an increasing number of studies (Hafez, 2007). However, tactile interface design guidelines to help optimize the characteristics of human tactile perception have not yet been fully developed. Hayashi et al. (2006) indicated that the fundamental characteristics of tactile interface design are not properly understood. Moreover, very few studies have been performed on the mechanics of velocity perception in order to establish design guidelines.

With regard to research on tactile displays, there have been very few studies on velocity perception compared to those on pattern representation and texture perception. Moreover, studies dealing with velocity perception have focused on the fundamental characteristics of the ability to discriminate velocity in response to tactile stimuli. Thus, advanced studies have not been conducted on the manner in which people perceive velocity in response to a discrete set of stimuli to the skin.

In this study, we focus on the role of posture for perceiving accurate velocity information, as generated by a tactile actuator. Clarification of the relationship between the posture employed when receiving tactual information and the accuracy of the perceived velocity information can possibly lead to the proposal of a methodology for designing a tactile interface with accurate and error-free tactile information transmission. In addition, a relationship may exist between the direction of velocity and the accuracy of information transmission. Therefore, the objective of this study is to determine an effective technique for presenting velocity information by evaluating the accuracy of velocity identification at different postures and for different directions of stimulus.

METHODS

A total of 10 subjects consented to participate in the study. All of them were recruited from the university community and were right-handed. An air-jet tactile actuator was used to provide tactile stimuli. Figure 1 shows the set of nozzles of the tactile actuators used in our study. The five nozzles could produce tactile stimuli in four directions by individual delayed air-jet emissions. The subject's palm was positioned on the actuator to receive stimuli. Pressurized air was generated by an air compressor and delivered using solenoid valves. The air pressure was regulated using a series of electropneumatic regulators. The intensity of the air-jet stimuli was controlled at 150 kPa, and stimuli were applied to the center of the left palm. The subjects verbally indicated the level of the perceived velocity for a given set of stimuli and the direction in which the set of stimuli was given. Three levels of velocity were presented, and each level was repeated five times for each direction. A total of five postures were tested to evaluate the accuracy of the perceived velocity of tactile apparent motion. Figure 2 illustrates the tested postures. These postures were selected under the assumption that each of them can be considered as being common and relaxed postures adopted by people in the course of daily life.

In this study, the stimulus set represented movement through apparent motion. Tactile apparent motion is defined as a perceptual phenomenon. It refers to the phenomenon wherein a person receives a pair of stimuli separated by a certain time duration but perceives the pair of stimuli as the motion of a single stimulus from the location of the first stimulus to the location of the subsequent stimulus presentation (Higashiyama et al. 2000).

The velocity of the apparent motion can be experimentally determined by using Kolers's equation (Kolers 1972), where the perceived velocity due to a pair of continuously applied tactile stimuli is calculated as the physical distance between the locations of the first and subsequent stimuli along the motion path divided by the interstimulus onset interval (ISOI). In this study, three levels of velocity were presented to the subjects for identifying the tactually perceived velocity. Table 1 lists the three levels of velocity along with the two corresponding parameters—duration and ISOI.

RESULTS AND DISCUSSION

Figure 3 shows the number of incorrect responses of all the subjects for velocity perception for each posture. Incorrect responses are those responses for which the subject responded erroneously to different velocity levels. Figure 4 shows the number of incorrect responses for the direction of motion for all subjects.

As shown in figure 3, the number of incorrect responses corresponding to posture 3 was the highest. However, the results of ANOVA showed that there were no significant differences between the postures ($F(4, 180) = 1.652$, $p = .17$). Interestingly, trends in the subjective responses showed that the subjects reported that there was a particular posture in which it was relatively easy to perceive velocity compared to other postures; this posture was, however, different for different subjects. Therefore, each subject may have had a particular posture at which they perceived velocity easily. However, these postures did not assure greater accuracy in responses to stimulus transmission because they just felt "easier" than other postures and were therefore, not "more accurate." In the studies focusing on postures receiving tactile-form information (Corcoran, 1977, Oldfield & Phillips, 1983), researchers found posture-related accuracy in recognizing characters, whereas Higashiyama et al. (2000) suggested that the effect of proprioception from the body may bias subjects' velocity perception. This finding was based on their observations regarding individual variations attributed to the relationship between postures and the accuracy of information transmission.

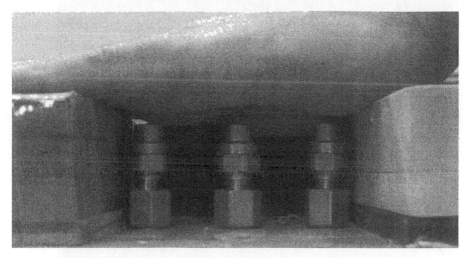

(a) Side view

546

(b) Top view

Figure 1. Air-jet-based tactile actuators

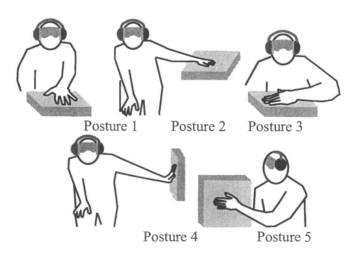

Posture 1   Posture 2   Posture 3

Posture 4   Posture 5

Figure 2. Postures for perceiving tactile stimuli

Table 1: Levels of velocity of apparent motion with corresponding durations and ISOIs

	Duration [ms]	ISOI [ms]
High velocity	50	30
Medium velocity	100	50
Low velocity	170	150

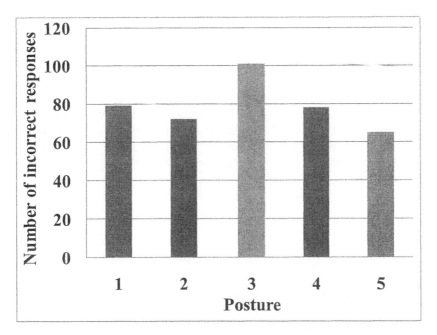

Figure 3. Number of incorrect responses for velocity according to posture (Total number of presentations: 600)

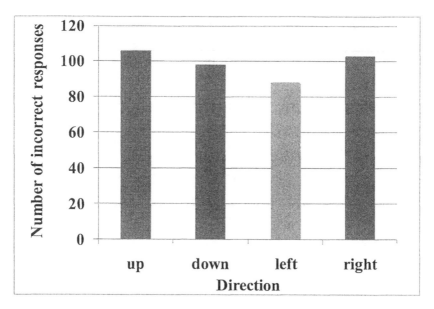

Figure 4. Number of incorrect responses for velocity by direction of motion (Total number of presentations: 750)

In figure 4, the number of incorrect responses was not significantly changed by the direction of apparent movement ($F(3,180) = 0.444$, $p > 0.05$). This result indicated that the accuracy of velocity perception was not influenced by the direction of tactile apparent motion. However, this conclusion will have to be verified by a systematic experiment to evaluate the origin of spatial coordinates in the human tactile localization system—the tactile egocenter. The range of velocities, i.e., high, medium, and low, may also effect the accuracy of perceived velocity. The relationship between the direction of tactile apparent motion and the accuracy of perceived velocity can be clarified by further analysis of the experimental results.

CONCLUSIONS

In the present study, we conducted an experiment to clarify the effect of posture and direction of tactile stimulus on the perception of velocity generated by tactile apparent motion. The result showed that there were no significant differences in the accuracy of perceived tactile velocity for different postures. On the other hand, the subjects reported that there was a particular posture in which it was easier to perceive velocity; however, this posture was different for each subject. A further study may include an increased number of postures while taking into account the factor of tactile egocenter. In addition, other parts of the body such as fingers and faces should also be considered to verify the relationship between the direction of tactile motion presentation and the accuracy of velocity perception since human mechanoreceptors are differently distributed in location and density on different parts of the body.

ACKNOWLEDGMENTS

This study was supported by the Kansai University Research Grants: Grant-in-Aid for Encouragement of Scientists, 2009, "Mechanisms of perception through human tactile communication" and JSPS KAKENHI (20370101).

REFERENCES

Corcoran, D. (1977). The phenomena of the disembodied eye or is it a matter of personal geography? Perception, 6, 247-256.

Hafez, M. (2007). Tactile interfaces, applications and challenges, The Visual Computer, 23(4), 267-272.

Hayashi, S., Watanabe, J., Kajimoto, H., Tachi, S. (2006). Study on motion after effect in tactile sensation, Trans. VR Soc. Japan, 11(1), 69-75 (In Japanese with English abstract).

Higashiyama, A., Miyaoka, T., Taniguchi, S., Sato, A. (2000). Tactile and Pain, Brain publication (In Japanese).

Kolers, P. (1972). Aspects of Motion Perception, Pergamon Press.

Oldfield, R., Phillips, R. (1983). The spatial characteristics of tactile form perception, Perception, 12, 615-626.

Chapter 61

Driver's Natural Anticipation Horizon in Deceleration Situations

Daria Popiv, [1]Klaus Bengler,[1] Mariana Rakic[2]

[1]Institute of Ergonomics
Technische Universität München
85747 Garching, Germany

[2]BMW Group Forschung und Technik
80992 Munich, Germany

ABSTRACT

The introduced investigation deals with the description and analysis of the driver's natural anticipation horizon and resulting driving behavior in deceleration situations. The evolution of the anticipation horizon is reflected by the change in the driver's situation and resulting actions through a course of time. Therefore, multiple points in time span during a deceleration situation are accounted for in the analysis. These points correspond to a change of driving situation and/or driver's situation, and are attributed by characteristic measurements. These are the metric distances to the situation, Time-to-Collision or Time-to-Contact (TTC) with the situation relevant objects, as well as the executed driving actions. The observed time and spatial span of the driving behavior is evaluated with respect to the safety, comfort, and efficiency in fuel consumption. This explores the potential of an anticipation improvement via the advanced driver assistance systems.

Keywords: anticipation, deceleration, efficiency

INTRODUCTION

This paper deals with the investigation of natural driving behavior in a classical controlled experiment. The presented analysis includes six deceleration situations. These situations are suitable for presentation in the fixed-base simulator regarding their comparability with real-life conditions in terms of the used analysis measurements. The goal of investigation is to describe the natural anticipation horizon of drivers during deceleration phases and their resulting actions. Based on the results, it is discussed in which situations the expansion of driver's anticipation with Advanced Driver Assistance System (ADAS) can be beneficial. The benefit is thereby defined by the gain in safety, comfort, and efficiency.

EXPERIMENTAL DESIGN

In the following the description of participants, hardware and software tools used for the implementation of experiment driving course and the analysis of the drives, as well as the analysis approach is provided.

SUBJECTS

26 participants (17 male and 9 female) took part in the experiment. All of them hold valid category B European driver's licenses. The average age of the test subjects was 34 years (standard deviation, SD = 13,6 years) at the time the experiment took place. The driving experience varies: seven participants drive less than 10.000 km per year, eleven – between 10.000 and 20.000, and eight – more than 20.000 km per year.

HARDWARE AND SOFTWARE TOOLS

The experiment was performed at the fixed-base driving simulator located at the Institute of Ergonomics, Technische Universität München. The field of the driver's front view is 180°.

The landscape and driving course are simulated using SILAB® software (SILAB website), which allows flexible and precise creation of the driving situations including the control over simulated traffic. The driving data of the subject vehicle as well as relevant situational data, e.g. distance to the relevant traffic participants, are recorded at 60Hz within SILAB framework. The visual data are collected at 25Hz using DIKABLIS® software (Ergoneers website).

The descriptive analysis of driving data is done with the help of MATLAB® and Excel.

EXPERIMENT DRIVING COURSE

The presented investigation deals with the description and analysis of driver's natural anticipation and driving behavior in six situations from the baseline experiment drive (drive without any driver's assistance). Three deceleration situations occur on the two-lane rural road, on which the allowed speed is 100km/h if not explicitly changed by the traffic signs. Two situations include a construction site on the road throughout the forest ("Construction site behind a right curve" and "Construction site in a left curve"). In front of both construction sites drivers have to decelerate before overtaking because of the oncoming traffic present on the opposite lane. The curve in "Construction site behind a right curve" has the radius of 700m, and it ends 200m before the construction site. In "Construction site in a left curve", the radius of the curve is 500m and the construction site is located directly in the curve. The third situation is the "Town entrance", where drivers have to reduce their speed to 50km/h because of the urban area entrance. In the town, drivers encounter the "Parking car" situation. It includes a parking car, which could not be overtaken immediately because of the oncoming traffic.

Two other situations occur on the highway road, where there is no speed limit if not otherwise explicitly set by traffic regulations. In the first highway situation, drivers have to decelerate from imposed allowed speed of 100km/h on this segment of the highway to 60km/h because of the traffic congestion (*"Stagnant traffic"*). In the second situation, they have to come to a full stop from deliberately driven speed in front of an idle traffic jam located behind a curve (*"Highway jam"*). This is the situation, where the safety criterion in terms of the collision avoidance with the idle vehicles in front plays the most important role.

ANALYSIS APPROACH

At first, following operational terms are defined: "traffic situation", "driving situation", "driver's situation", (adopted from Reichart, 2001), and "driving behavior" (adopted from Fastenmeier, 1995).

Traffic situation ("Verkehrssituation") is an objectively present consolation of traffic participants and relevant influencing parameters of the traffic surroundings. In this work, if not indicated otherwise, the term "situation" refers to the traffic situation. *Driving situation* ("Fahrsituation") is that part of traffic situation, which can be objectively seen/perceived by the driver. *Driver's situation* ("Fahrersituation") is the subjective representation of the situation established by the driver after perception. *Driving behavior* ("Fahrverhalten") means the driving actions, which are exercised in order to complete the current driving task.

The evolution of the driver's natural anticipation horizon is described and evaluated in terms of driving situation, driver's situation, and driving behavior. These are the points in time considered for the following analysis:
- Optimal coasting (CoastOpt) – point in time, when the start of vehicle coasting is optimal in terms of an efficiency criterion (coasting is releasing

an accelerator pedal and decelerating with the help of the motor drag torque). The needed distance and the proper point in time for CoastOpt depending on driven and goal speeds are calculated for the vehicle dynamics model used in the fixed-base simulator at the Institute of Ergonomics, which resembles one of the BMW 3i motors with automatic transition.

- Real coasting (CoastReal) – point in time, when the driver actually releases the accelerator and starts coasting.
- Situation visible (SitVis) – situation becomes physically visible.
- Situation seen (SitSeen) – situation is actually seen by the driver, defined by the first fixation on the situation which lasts longer than 100ms.
- Driver's reaction on the situation (React) – change of the driving action after the situation is seen.

Additional point is introduced for the safety critical situation "Highway jam":

- Optimal braking (BrakeOpt) – point in time, when the driver should start braking mildly (not exceeding $-3m/s^2$) in order to be able to stop in front of upcoming jam without experiencing uncomfortable decelerations.

The limit for a subjectively comfortable deceleration is chosen $-3m/s^2$ in this work (adopted from Farid et al., 2006). The distance required to brake from the driven speed to the desired/set speed with this deceleration is introduced as distance needed to brake on the border of comfort. In the situation analysis, it is compared with the distance at which drivers actually start actively braking. Distance needed to brake on the border of comfort is obtained using constant acceleration formulas, which describe the simple motion in one dimension.

For all of the analyzed points, metric and time measurements and their descriptive values (mean Ø and SD) are provided:

- Metric measurement – distance (DIST), in m, to the deceleration situation.
- Time measurement – Time To Collision, also known as Time To Contact (TTC), in s, generally calculated as [distance to the situation with the slower moving object, in m]/[driven speed, in m/s – speed of the slower moving object, in m/s]. In case of situations with static objects, it is [distance to the situation with the static object, in m]/[driven speed, in m/s]. In the later case TTC can be considered as the time headway (Cherri et al., 2005), or as time distance to the situation. More TTC-related information can be found in (van der Horst, 1990).

RESULTS

In the following, analysis results are presented for the six situations from the baseline drive. Further results regarding other investigated questions can be found in the works of (Popiv et al., 2010).

OVERVIEW

After the descriptive values were determined for every analysis point, the normal distribution of these values was proved and the relationship between them was examined.

As for efficiency criterion evaluation, the difference between CoastOpt and CoastReal values is presented later for every described situation. It can be mentioned that for the most of situations the time point of CoastOpt lies earlier than that of SitVis and CoastReal, therefore proving the ADAS potential for increasing the efficiency benefit by informing the driver about the upcoming situation.

Characteristic values of SitSeen and React depend on the situation. The observed tendency is that the vast majority of the drivers see non-critical situations involving static objects on time distances TTC greater than 6s (95% of the cases), and rarely – further than 12s even if the situation is visible before. After the situation is seen, the time of following reaction depends on the driving style and driver's experience, certainty in future development of the situation, availability of time to plan and decide upon the action, etc.

CONSTRUCTION SITE BEHIND A RIGHT CURVE

In this situation, it is possible to reduce estimated fuel consumption by providing information to the driver early enough about the deceleration situation. Without this information the drivers start coasting a vehicle at ca 160m farther from when it would have been optimal. Table 1 presents the description of evolution of natural anticipation horizon of drivers and their resulting behavior. The descriptive values of DIST and TTC are also provided for every analyzed point.

Table 1 Construction site behind a right curve - evolution of driving and driver's situations with resulting driving behavior

ANALYSIS POINT	DIST/ TTC (Ø±SD)	DRIVING SIT.	DRIVER SIT.	PERFORMED DRIVING BEHAVIOR
Coast Opt	544±50m 18,8±1,2s	Constr. site is not visible	Driver sees a light right curve in front	Driver keeps the foot on the accelerator
Sit Vis	450m 14,8±1,7s		Driver approaches the curve	Driver keeps the foot on the accelerator
Coast Real	384±93m 13,0±2,7s	Constr. site is visible	Driver concentrates on driving through the curve	Driver releases the accelerator pedal and starts coasting
Sit Seen	273±78m 10,2±3,0s		Driver sees the construction site	Driver keeps coasting
React	266±49m 9,4±1,8s		Driver realizes that the oncoming traffic will not allow overtaking	Driver starts braking

Drivers perceive the situation in general almost 200m after it becomes visible. The reason for that is that the drivers are concentrating on negotiating the curve. Nevertheless, the TTC at SitSeen point provides drivers with enough time to decide on the needed action and to perform it in a comfortable manner. This is confirmed by the beginning of the braking sequence which is initiated between 300m and 200m before the construction site is reached, which is more than enough for comfortable deceleration way (distance needed to brake on the border of comfort is 160m when decelerating from 100 to 0km/h).

CONSTRUCTION SITE IN A LEFT CURVE

This situation differs from the previously described by the sharper curve, and the construction site is located in the curve itself. Therefore, the situation becomes visible just at a distance of 280m (TTC 10,1±1,2s). The drivers see it at a distance of 210±37m with TTC of 8,0±1,9s, which is about 2,5-2,8s later for average driven speeds of 100-90km/h, which were recorded during the experiment for this part of the road. The braking action is performed at a DIST 165±89m, TTC 5,7±2,8s, and the observed braking distance is already on the border of comfort. Longer times and traveled distances compared to the previous situation between SitSeen and React are explained by the fact, that in the curve drivers do not easily see the oncoming traffic, and therefore delay the decision. In this situation, earlier information from ADAS could increase the comfort of the drivers.

Also the benefit in efficiency can be gained by expansion of the anticipation horizon of the driver. Because of the curve, drivers start coasting 200m after the CoastOpt point.

TOWN ENTRANCE

Overall, this is an easy situation to anticipate and react in a comfortable manner by the drivers themselves. It is well-visible due to a surrounding field landscape and well-recognizable because of the houses near the town entrance. It also does not involve any other driving vehicles and there is little uncertainty in the development of the situation.

Table 2 Town entrance - evolution of driving and driver's situations with resulting driving behavior

ANALYSIS POINT	DIST/ TTC (Ø±SD)	DRIVING SIT.	DRIVER SIT.	PERFORMED DRIVING BEHAVIOR
Coast Opt	ca 650±50m ca 21,4±2,7s	Entr-ce is not visible	Driver sees the straight part of the road	Driver keeps the foot on the accelerator
Sit Vis	500m 16,7±5,1s	Entr-ce is visible	Driver sees the light curve in front	Driver keeps the foot on the accelerator

Coast Real	448±135m 15,8±4,1s		Driver concentrates on driving through the curve	Driver releases the accelerator pedal and starts coasting
Sit Seen	332±132m 11,8±4,1s		Driver sees the town entrance but decides not to change a driving action	Driver does not change the behavior, i.e. keeps coasting
React	205±110m 6,5±3,4s·		Driver decides to decelerate stronger	Driver starts braking (minimum decelerations reached by the participants are usually > -3m/s^2)

The drivers see this situation significantly earlier than other described situations. Not only is it visible earlier compared to the two previous situations, but there are no demanding driving tasks to perform on this part of the road unlike in "Construction site in a left curve". However, there is still a difference of 200m between CoastOpt and CoastReal. It has to be proved by further investigation, if the driver with the help of ADAS would have preferred to coast earlier.

PARKING CAR

In general, drivers see the situation early enough and react fast, which can be explained by relatively short distances on urban roads to make a maneuvers and particular alertness to the numerous obstacles requiring deceleration.

The optimal coasting phase should have been started at approximately 150m away from the situation. However, in complex situations which are common on the urban roads, the optimal coasting usually cannot be started even if the driver is informed in advance about particular incoming deceleration situation. More imminent situations can be of the outmost importance at this point of time, e.g. driving through the intersection, making a turn, etc.

Table 3 Parking car - evolution of driving and driver's situations with resulting driving behavior

ANALYSIS POINT	DIST/ TTC (Ø±SD)	DRIVING SIT.	DRIVER SIT.	PERFORMED DRIVING BEHAVIOR
CoastOpt/ SitVis	150m 10,6±1,2s		Driver looks at the upcoming intersection, where the driven road is the main road	Driver keeps the foot on the accelerator
Coast Real	126±33 9,1±2,4s	Park. car is visible	Driver sees the parking car and oncoming traffic, which makes immediate overtaking impossible	Driver releases the accelerator pedal and starts coasting
Sit Seen	125±23 9,0±1,8s			
React	69±19m 4,9±1,2s			Driver starts braking*

*Distance needed to brake from 50 to 0km/h on the border of comfort is 32m

STAGNANT TRAFFIC

It is an easy situation to anticipate by drivers themselves. The imposed 100km/h speed limit on this part of the road implicitly suggests that slower moving vehicles are to be expected. During the experiments, the drivers perform not only comfortable braking action, but also start coasting almost at the CoastOpt point.

Table 4 Stagnant traffic - evolution of driving and driver's situations with resulting driving behavior

ANALYSIS POINT	DIST/ TTC (Ø±SD)	DRIVING SIT.	DRIVER SIT.	PERFORMED DRIVING BEHAVIOR
Coast Opt/Sit Vis	159±56m 24,5±12,6s		Driver sees the slower moving cars in front of him	Driver keeps the foot on the accelerator
Coast Real	152±50m 18,5±3,2s	Cars are visible		
Sit Seen	122±33m 14,7±4,6s		Driver approaches the slower moving cars	Driver releases the accelerator pedal and starts coasting
React	80±34m 9,6±4,0s		19 drivers decide to decelerate stronger	Drivers start braking mildly (less than -3m/s^2)

TRAFFIC JAM

In this situation, the safety criterion is crucial. The idle jam is located behind a curve with 700m radius. The driven speed is deliberately chosen by the drivers. During the experimental drives, the average speed driven at this segment is around 150km/h. In reality, such an idle jam located exactly behind a curve happens quite seldom, and can be caused by e.g. the accident. In such situations ADAS help can fundamentally increase safety.

Table 5 Traffic jam - evolution of driving and driver's situations with resulting driving behavior

ANALYSIS POINT	DIST/ TTC (Ø±SD)	DRIVING SIT.	DRIVER SIT.	PERFORMED DRIVING BEHAVIOR
Brake Opt	1000m/-	Jam is not visible	Driver sees straight part of the highway road and a following curve	Driver keeps the foot on the accelerator
Coast Real	494±259m/ 14,8±6,4s		Driver concentrates on driving through the curve	Driver releases the accelerator pedal and starts coasting
Sit Vis	260m/ 7,8±1,8s	Jam is		

		visible	Driver sees the jam	Driver does not yet change the behavior, i.e. keeps coasting (reaction time)
Sit Seen	247±10m/ 6,7±1,8s		Driver sees the jam	
React	225±50m/ 6,4±1,7s		Driver rapidly decelerates	Driver starts braking hard (average decelerations reached -8,4m/s^2)

Efficiency optimized action is not reasonable here – coasting from the driven speed of 150km/h would last for more than 1.5km. Therefore, the focus here is put on when and where the drivers should start moderately braking. The exact values are presented in the Table 5.

In-Depth Analysis of Accidents

During this situation, four accidents happened. The subjects that caused the accidents are all experienced drivers. In three of the cases (subjects 10, 11, 19), drivers perceived the situation at the time point with critically small TTCs < 5s (van der Horst et al., 1994). Their decelerations did not suffice to come to a full stop. In the beginning of emergence braking, non-trained drivers rarely are able to reach necessary maximum decelerations (Breuer, 2009).

Subject 16 saw the situation when it first became visible. The evaluation of the situation was wrong – the driver did not realize that the vehicles are idle, and as a result did not try to perform the emergency braking.

Table 6 Characteristic values experienced during accidents

Subject	Subject's Experience (km/year)	SitSeen Driven speed	SitSeen DIST/TTC	React DIST/TTC
10	15.000-20.000	180km/h (50m/s)	220m 4,3s	200m 4s
11	>20.000	184km/h (51m/s)	219m 4,2s	190m 3,7s
16	>20.000	168km/h (47m/s)	260m 5,5s	200m 4,3s
19	15.000-20.000	162km/h (45m/s)	209m 4,6s	170m 3,8s

CONCLUSION

The results of the analysis show, that not in every deceleration situation the driver's natural anticipation horizon suffices for the most efficient, comfortable, and safe action. The additional assisting information provided to the driver at the proper point of time can lead to the better comprehension of the evolving driving situation and increase gain in the efficiency and comfort. It can also help to avoid collisions in safety critical situations. The representation of information should not lead to the driver's distraction. In the following, the overview regarding the potential gain in efficiency, comfort, and safety for each of the six described situations is given.

Town entrance. This is a well-anticipated situation by the drivers.TTC at the point of when the situation seen is 12s, which definitely provides enough time for comfortable action. However, also in this situation the efficiency can be increased.

Parking car. Drivers see and react by braking on this situation with time distances of 9s and 5s. This suffices for a comfortable deceleration, but not maximal efficiency gain. Early information by ADAS about the upcoming deceleration situation in the urban area might not be taken into driver's account due to current driving situation.

Stagnant traffic. Driver's anticipation horizon conforms to safety, comfort, and efficiency criteria.

Traffic jam. This is a safety critical situation, in which the time distance of 6-7s of when the situation is generally seen demands an emergency braking action.

Table 7 Potential demand on ADAS in analyzed situations

SITUATION	Efficiency	Comfort	Safety
Construction site behind a right curve	+		
Construction site in a left curve	+	+	
Town entrance	+		
Parking car	?		
Stagnant traffic			
Highway jam		+	+

REFERENCES

Breuer, J., 2009, Bewertung von Fahrerassistenzsystemen. Handbuch Fahrerassistenzsysteme, GWV Fachverlage GmbH, Wiesbaden, pp. A55-A68

Cherri, C., Nodari, E., Toffetti, A., 2004, Review of existing Tools and Methods, Deliverable 2.1.1 of the Integrated Project AIDE, Brussels, AIDE Consortium

Ergoneers website: http://www.ergoneers.com, accessed January 2010

Farid, M., Kopf, M., Bubb, H., Essaili, A. (2006). Methods to develop a driver observation system used in an active safety system. In Proceedings of 22. Internationale VDI/VW-Gemeinschaftstagung, Wolfsburg.

Fastenmeier, W. (Hrsg.) (1995). Autofahrer und Verkehrssituation - Neue Wege zur Bewertung von Sicherheit und Zuverlässigkeit moderner Straßenverkehrssysteme (Mensch-Fahrzeug-Umwelt, Bd. 33). Köln: Verlag TÜV Rheinland.

Horst, van der, R., 1990, A time-based analysis of road user behaviour in normal and critical encounters, PhD Thesis, Delft University of Technology, Delft.

Horst, van der, R., Hogema, J., 1994, Time-to-Collision and Collision Avoidance Systems. Proceedings of the 6th ICTCT Workshop: Safety Evaluation of Traffic Systems, Salzburg, s. 109-121

Popiv, D., Rommerskirchen, C., Rakic, M., Duschl, M., Bengler, K. (2010). Effects of assistance of anticipatory driving on driver's behaviour during deceleration phases. Submitted for European Conference on Human Centered Design for Intelligent Transport Systems 29.-30.04.2010, Berlin

Reichart, G., 2001, Menschliche Zuverlässigkeit beim Führen von Kraftfahrzeugen, Forschritt-Berichte VDI, Reiche 22, Nr.7, Düsseldorf: VDI Verlag.

SILAB: http://www.wivw.de/ProdukteDienstleistungen/SILAB/index.php.de, accessed January 2010

Chapter 62

Haptic Gear Shifting Indication: Naturalistic Driving Study for Parametrization, Selection of Variants and to Determine The Potential for Fuel Consumption Reduction

[1]*Christian Lange,* [2]*Antonio Arcati,* [3]*Heiner Bubb,* [3]*Klaus Bengler*

[1]Ergoneers GmbH,
85077 Manching, Mozartstraße 8 ½
Germany

[2]Continental Automotive GmbH
Siemensstraße 12, 93055 Regensburg
Germany

[3]Institute of Ergonomics
Technical University of Munich
Boltzmannstraße 15, 85748 Garching
Germany

ABSTRACT

The present work deals with the parametrization of a haptic gear shifting indication as well as with the determination of the potential for fuel consumption reduction of this discrete indication at the gaspedal. With this kind of gear shifting indication the driver is getting told whenever he should change to gear upwards or downwards. The work is subdivided into two steps. In the first step the most intuitive and most liked haptical gear shifting indication is selected in a preliminary test from 8 possible indications. In the second step, this selected variant is evaluated in the main experiment concerning its potential for fuel consumption reduction in detail in a naturalistic driving study.

Keywords: Driver assistance, naturalistic driving study, ecology, fuel consumption, gear shifting indication

INTRODUCTION

Besides increasing active safety, future driver assistance systems shall also be able to assist the driver in a fuel saving manner of driving and thus reduce CO_2 emission. The present work presents the concept of a discrete gear shifting indication at the active gaspedal and proves its potential for fuel consumption reduction.

PRELIMINARY TEST

The preliminary test is to determine how intuitive and accepted 8 different haptical gear shifting indications are. These 8 variants are shown in the following Table 1. These 8 variants were previously parameterized by experts concerning the parameters "counterforce", "frequency" and "amplitude". Therefore, the experts were driving with a BMW 530i, which was equipped with the active accelerator, and the experimental leader changed the parameters online until the experts were comfortable with the parametrization. The parametrization for the selection of the most intuitive and accepted variant was the mean over all expert parametrizations.

In the preliminary test, again the BMW 530i is used, which is equipped with the active accelerator. The preliminary test is to select the most intuitive and accepted of the 8 variants for a haptic gear shifting indication shown in Figure 1. The experimental leader can easily switch between the 8 variants with the help of a control software. The behavior of the active accelerator changes according to the selection of the experimental leader.

Within the preliminary test, the 8 variants are evaluated by 12 participants. The participants drive with each of the variants for about 15 minutes. After the 15 minutes the subjects evaluate the different variants on a questionnaire concerning their likeability and acceptance and concerning their adequacy for a gear shifting indication. The results show that the "double tick" in both characteristics is evaluated best with great advance to all other variants. Afterwards follow the vibration (2s and permanent), the counterforce (2s and permanent) and at the end the combination of counterforce and a superimposed vibration (2s and permanent). Thus, the double tick is the variant that will be used in the main experiment.

Table 1 Overview of the 8 different hatical gear shifting indications of the preliminary test

Vibration for 2s	
Permanent vibration	
„Double tick" 3 times repeated, with a distance of 3s between the double ticks	
Permanent double ticks	
Counterforce for 2s	
Permanent counterforce	
Counterforce with a superimposed vibration for 2s	
Permanent counterforce with a superimposed vibration	

MAIN TEST

Within the main experiment to estimate the influence of a haptic gear shifting indication on fuel consumption, three different experimental conditions are compared to each other. The three experimental conditions are shown in Figure 1.

The variant "reference condition" represents the normal driving without gear shifting indication. When driving with the condition "optical gear shifting indication" an optical output tells the driver when to change the gear. The content of the optical output is the direction, up or down, and the target gear. This variant is equivalent to the series production status of a BMW E60. Within the "optical and haptical gear shifting indication" the driver gets in addition to the optical output a haptical output at the active accelerator when it's necessary to change the gear. Therefore the double tick is used, which was evaluated best in the preliminary test.

These three variants are compared to each other within a naturalistic driving study with 24 participants. The experimental car is again the BMW 530i. The experimental leader can switch between the three variants easily with the help of a control software. In addition, the control software records all relevant driving dynamics data like current fuel consumption, current gear, speed and so on. During the main experiment, the subjects drive three times the same test track, which lasts for about one hour. The test track is composed of parts of a highway, of rural roads and of urban roads. During the three drives, the subjects experience the three different experimental conditions.

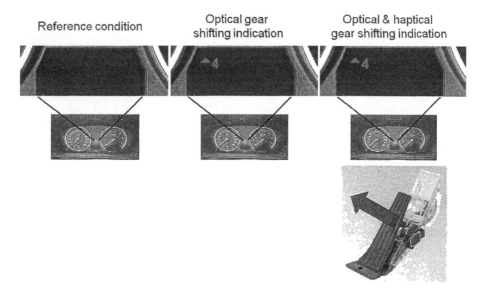

FIGURE 1 The three experimental conditions „reference condition", „optical gear shifting indication" and „optical and haptical gear shifting indication "

The analysis of the recorded data shows, as exemplarily depicted in Figure 2, a very positive influence of the „optical and haptical gear shifting indication" on the mean fuel consumption per 100km. The fuel consumption can be reduced by approx. 8% compared to the reference condition and to the sole optical indication. In contrast, the sole optical gear shifting indication has no influence on the fuel consumption of the participants. With the help of a detailed analysis of the recorded data can be clarified why the fuel consumption reduction takes place

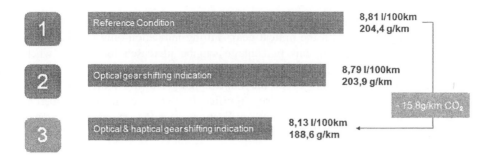

FIGURE 2 Mean fuel consumption per 100km depending on the three experimental conditions

As shown in figure 3, subjects follow the advice at the active gaspedal to shift upwards in 93% of all cases and the advice to shift downwards in 57%. This results in an overall shifting behavior of more than 83%. When driving the reference task, the subjects do only shift in the right time interval in 38,5%, what is significantly less. It has to be noted that the subjects didn´t get neither an optical nor a haptical advice when driving with the reference task. It was logged when the advice would have come, if the subject would have been driving with the optical or the optical & haptical condition. When the subject was driving with the reference task and the subject made a gear shift within 1s before and 10s after the advice would have come, it was counted as if the subject would have followed the shifting indication. Considering the optical gear shifting indication, there is no significant influence on the gear shifting behavior compared to driving the reference task

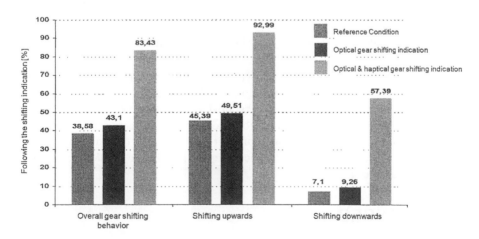

FIGURE 3 Mean following of the shifting indication depending on the three variants

The following of the gear shifting indication when driving with the optical & haptical gear shifting indication has a clear influence on the utilization of the upper gears (see figure 4). With the reference task subjects drive only in 7,59% of the time in the highest gear and with the optical gear shifting indication only in 15,22% of the time. Compared to that, this percentage can be increased to 37,81% when driving with the optical & haptical gear shifting indication.

Besides the clear advantages of the haptical gear shifting indication in the objective measurements there is also significant advantage in the subjective measures. The haptical gear shifting indication is rated significantly more comfortable, practical, attractive, sportive, motivating, easier and elegant than the optical gear shifting indication.

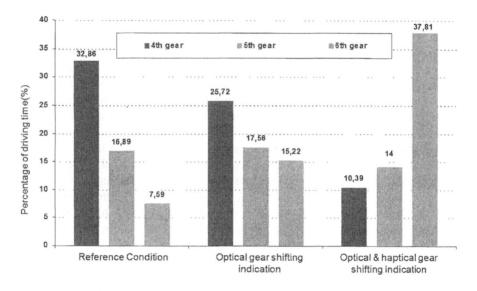

FIGURE 4 Utilization of the gears 4, 5 and 6 depending on the three variants

The present work shows the positive contribution of a haptical gear shifting indication for fuel consumption reduction and acceptance. Furthermore, future development potential to optimize the haptical gear shifting indication as well as further possibilities for fuel consumption reduction by an active accelerator are demonstrated. Fuel consumption can moreover be reduced by other measures like a constant assistance in speed- and distance-keeping (see Lange, 2008a; Lange et al. 2008b and Lange et al. 2006).

REFERENCES

Hassenzahl M., Burmester M., Koller F. (2003) *AttrakDiff: Ein Fragebogen zur Messung wahrgenommener hedonischer und pragmatischer Qualität*. In: J. Ziegler & G. Szwillus (Eds.), Mensch & Computer 2003. Interaktion in Bewegung (pp. 187-196). Stuttgart, Leipzig: B.G. Teubner.

Lange, C. (2008a) *Wirkung von Fahrerassistenz auf der Führungsebene in*

Abhängigkeit der Modalität und des Automatisierungsgrades, Dissertation, Technische Universität München, München 2008.

Lange C., Bubb H., Tönnis M., Klinker G. (2008b) *Sicherheitspotential und Verbrauchsreduzierung durch ein intelligent geregeltes aktives Gaspedal*, In: Tagungsband der 3. Tagung Aktive Sicherheit durch Fahrerassistenz, 7./8. April 2008, Garching

Lange C., Tönnis M., Bubb H., Klinker G. (2006) *Einfluss eines aktiven Gaspedals auf Akzeptanz, Blickverhalten und Fahrperformance*, In: Proceedings 22. Internationale VDI/VW Gemeinschaftstagung Integrierte Sicherheit und Fahrerassistenzsysteme, Wolfsburg 2006

Chapter 63

Knowledge Based Shaping of Working Conditions in Industry and Health Care

Teodor Winkler

Faculty of Organization and Management
Silesian University of Technology
41-800 Zabrze, Roosevelt Street 26, Poland

ABSTRACT

An overview of knowledge resources and their classification regarding the working conditions shaping is given. Knowledge Engineering methods supporting shaping of working conditions are presented. Applying of video recording, Motion Capture, RFID technology for knowledge acquisition, knowledge gathering and knowledge dissemination purposes are discussed. Limitations for this methods within industry and health care areas are described. Two case studies from the mining industry and from the hip surgery are presented.

Keywords: Knowledge Engineering, working conditions, mining industry, hip surgery.

INTRODUCTION

In the Institute of Mining Technology KOMAG for many years there have been carried out works that allow to determine common approach to shaping working conditions in seemingly distant areas which are mining and healthcare. (ENHIP, 2005, MINTOS, 2007, Knowledge based...2009).

Implementation of processes both in the industry and in the healthcare requires resources. Apart from the material, human and financial resources, the resources of knowledge play a vital role. Thank to these resources these processes become gradually knowledge-based.

Working conditions depend on the applied technical means: machinery, devices and tools, including manual tools. The life cycle of technical means comprises such stages as: designing, manufacturing, exploitation (usage), liquidation. The working conditions may be discussed in the stage of exploitation, however, they are determined by the usability features of the applied technical means which are defined at the designing stage. These stages are related to two types of resources: designing knowledge and exploitation knowledge. These resources are shared among the participants of both stages according to the model presented in Figure 1.

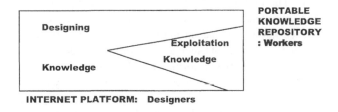

Figure 1. Model of knowledge sharing.

In the designing process there are used virtual designing methods which are based on scenarios acquired from the processes implemented in the exploitation stage. Proportions between the resources of designing and exploitation knowledge made available in the designing stage presents a "vertical" stream addressed to the designers. In the exploitation stage the stream of knowledge addressed to the users relates to maintenance, repairing and servicing, but also includes elements of designing knowledge which take form of drawings, schemes and models. In the stipulated "horizontal" stream there emerge relations, due to which the elements of designing and exploitation knowledge may be used in the situational context in which the user currently is found.

METHODS OF KNOWLEDGE ENGINEERING AND ICT TECHNOLOGIES SUPPORTING SHAPING WORKING CONDITIONS

The resources of knowledge on which Knowledge Engineering operates are not homogenous. Classification of types of knowledge is subject to continuous discussions. (Thagart, 2007). For the needs of this article there will be considered Explicit Knowledge, Tacit Knowledge, Descriptive Knowledge and Procedural Knowledge. Explicit knowledge is a type of knowledge coming out of human consciousness and presented in a determined form. It may be presented by the following means: words, figures, signs and symbols. Due to this fact its dissemination is possible with the use of commonly understandable (paper and electronic) forms of transfer, such as: reports, drawings, tables, graphs, etc. Transferring explicit knowledge between entities is relatively simple (Grudzewski and Hejduk, 2004). The author of the term "tacit knowledge" is M. Polanyi who described it in the following words: "we know more than we are able to express". (Polanyi, 1967). Tacit knowledge has an individual character – it exists only human consciousness. It comprises elements such as: individual abilities, experience, beliefs, intuition, non formalized practical information. Because it is created by personal experience and activity, it has rather practical dimension. This type of knowledge is difficult to articulate and its conversion into explicit knowledge is difficult or impossible. Therefore, transfer of tacit knowledge is a hard task. Both types of knowledge are mutually complimentary. Declarative knowledge is a set of facts and rules specific to a particular considered area. Declarative knowledge is subject to formal notation (Moczulski and Ciupke 2008). Procedural knowledge relates to the ways *how* particular activities should be done. It comprises a set of procedures, the action of which represents knowledge in a particular area. Some selected aspects of procedural knowledge are subject to formal notations, for instance by verbal description, check list, block diagram, (Moczulski and Ciupke 2008). Relating the aforementioned remarks to the model of knowledge sharing in fig. 1 it may be stated that the designing knowledge is majorly of the declarative type, but the exploitation knowledge bases mainly on the procedural type. Both types occur in the explicit and tacit form.

Knowledge Engineering uses particular methods of acquisition, gathering and dissemination of knowledge. From the operational point of view, for the needs of shaping the working conditions it is required to use knowledge in the explicit form. It is particularly important in dissemination of knowledge during trainings. Therefore, during the acquisition process there are preferred methods favoring conversion of tacit knowledge into explicit one. When acquiring the procedural knowledge necessary for safe and healthy performing the working activities Video

Recording and Motion Capture are useful. They are used in *in situ* acquisition of knowledge. There are however subject to limitations occurring in many workplaces, particularly in non-stationary ones. In mining the main reason is the methane threat. In many underground mines application of electronic devices (e.g. video cameras) is forbidden. There are only permitted device in special non-explosive cases conforming to the requirements of ATEX Directive (ATEX, 1994). The requirements of sterility which are in force in the hospitals exclude presence in the proximity of operational field of any other people but medical staff. This limits the space for observation of working activities.

The resources of knowledge are stored in the knowledge repositories; stationary and portable. Knowledge is stored in the form of Reusable Knowledge Elements represented by verbal descriptions, 2D drawings, 3D models, simulations, computer animations and films. Among them there are defined relations which facilitate using browsing mechanisms and searching.

Knowledge based shaping of working conditions requires access to the knowledge resources both by designers and the users. Designers use explicit knowledge stored in the repositories embedded in the Internet platform (MINTOS, 2007). The resources of knowledge necessary to the user for safe performing of the working activities directly in their workplace are stored in the portable repositories accessed with the use of PDAs (Personal Device Assistant) or UMPCs (Ultra Mobile Personal Computer). This type of knowledge is made accessible in the context of particular situation in which the user is found. This context is created with the use of ICT Technologies (Information and Communication Technologics), e.g. RFID (Radio Frequency Identification). RFID tags placed on objects or carried by people inform about the context of situation.

CASE STUDY - SHAPING WORKING CONDITIONS IN THE MINING INDUSTRY

This case study was carried out within the RFCS MINTOS Project, (MINTOS, 2007). A lot of manual work, which causes overloads in musculo-skeletal system, is still realized in the mining industry. Video recording of selected manual loading/unloading operations was realized in *in situ* conditions. Reflective markers were fixed in anthropometrical points (among others in elbow joints and knee joints) of miners bodies. Part of the movie presenting realization of given loading/unloading operations was selected, and then it was entered to 3D Studio MAX software. Superposition of recorded postures and anthropometrical models was realized by superposing the square markers (dummies), which are available in the software, on reflective markers. Due to inverse kinematics mechanism, the segments of anthropometrical models were adjusted to body segments of recorded postures, due to what the segments of models "took over" movement also from the recorded postures. This procedure carried out in the laboratory conditions is presented on Figure 2.

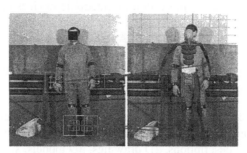

Figure 2. Superposition of recorded postures and anthropometrical models

File .BIP, comprised animated anthropometrical models associated with other recorded postures, was created. Continuous ergonomics analysis of animated models was conducted with the use of Anthropos-ErgoMAX software (plug-in to the 3D Studio MAX), (Anthropos-ErgoMAX, 1999). Figure 3 presents a sequence of models and diagrams of coefficient of static discomfort of spine vertebrae. Models of worst postures as regards loads in L5/S1 vertebrae are selected.

Files .JPG are created for these models, and then they were exported to 3D SSPP software for biomechanical analyses (3D SSPP 4..3, 2001). Selected animations are collected in the knowledge repository. Manual superposition of models recorded in .JPG format with models of 3D SSPP software is realized in 3D SSPP software, what enables determining joints angles. Results of biomechanical calculations for such defined posture are presented in. These results will be also included in the repository.

Some components of mining equipment are marked with RFID tags in mines due to logistic reasons. There are components of big size and weight among them. Stability of these components during transportation is crucial for people safety. Loss of load stability can not occur in any phase of transportation. Due to that during preparation of load for transportation the miner uses a check list, which is displayed on PDA screen. PDA computer is integrated with RFID reader.

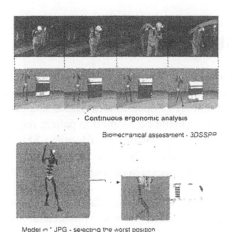

Figure 3. Continuous ergonomics analysis and biomechanical assessment of postures

On the basis of code, which is recorded in RFID located on transported component, a proper check list is taken from the repository with the use of decision tables.

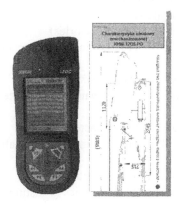

Figure 4. Check list supporting safely underground transport.

Diagram containing informations enabling stable transport of the load is presented on Figure 4. As the preparation of load for transportation requires manual activities, results of ergonomics and biomechanical assessments associated with the load, which were mentioned before, can also be taken from the repository.

CASE STUDY - SHAPING WORKING CONDITIONS IN THE HIP SURGERY

Surgical instruments are one of the factors that shape surgeon work conditions. Surgeon's work requires use of complicated surgical instrumentation that should meet criteria both of feasibility of operation activities and economic criteria. From one side the instrumentation should be adapted to patient requirements and from the other side to the surgeon psychosomatic features. This case study presents investigations carried out within the FP6 Project ENHIP, (ENHIP, 2005) coordinated by Instituto de Biomecánica de Valencia (IBV), Valencia, Spain.

Surgical tools of improper design, not adapted to the surgeon psychosomatic features may cause health hazards to musculo-skeletal system. Questionnaire inquires of 11 surgeons making hip alloplasty operations can testify for that. Questionnaire results enabled indication tools, use of which can cause pain in specified parts of musculo-skeletal system (ENHIP, 2006). Besides, questionnaire results included suggestions of changes that could improve both the design features of tools and the method of operation. The team consisting of following members decided to develop new surgical instruments:
- orthopedic surgeons (2 hospitals),
- surgical instruments designers (from 5 factories),
- Production Engineering and Material Engineering specialists (from the same factories),
- Mechanical Engineering specialists (1 institute),
- specialists from ergonomics and biomechanics (1 institute).
- orthopedists (2 hospitals),
- surgical instruments designers (from 5 factories),
- Production Engineering and Material Engineering specialists (from the same factories),
- Mechanical Engineering specialists (1 institute),
- specialists from ergonomics and biomechanics (1 institute).

So it was an interdisciplinary team that was geographically dispersed and having dispersed knowledge resources. The factories belonged to SMEs group and had not their own knowledge resources from advanced Mechanical Engineering, biomechanics and ergonomic technologies. Under such circumstances only virtual environment consisting of computer models of material objects (rooms, equipment, tools) as well as of computer models of anthropometrical features was the element that enabled joint work. Virtual prototyping of tools, according to accepted criteria, was carried out in that environment. Criteria were formulated by the mentioned above specialists. Virtual prototyping included simulation and animation of phenomena that occur during surgeons operation.

To gain about surgical hip operation, video recording of three such operations were made. Using the mentioned above "superposition" method, the surgeons bodies were replaced by their models having their anthropometrical features. In Figure 5a a fragment of operation in standing posture with use of Rasp Impactor was presented, and in Figure 5b with use of Rongeur. Models enabled analysis of surgeons postures during operation, access to operation field, method and direction of force exertion (hitting a hammer to Rasp Impactor), tool grasping.

Complementary analysis of video records enabled determining speed of movements, which determine the value of kinetic energy e.g. in cooperation with a hammer - Rasp Impactor. In the case of Rasp Impactor the criterion of weight reduction was considered as well as a criterion of reduction of value of vibrations caused by hammer impacts and improvement of tool grasping.

a)   b)

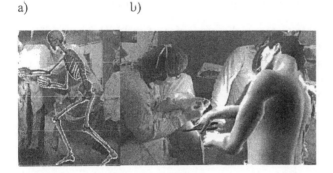

Figure 5. Video recorded postures of surgeons during the hip operation.

Using Multi-Body Analysis and Finite Elements Method software the Mechanical Engineering specialists made static and dynamic analyses, in a result of which reduction of tool weight and vibration level were achieved, Figure 6a. Optimized geometrical form of toll with grip plate, made of non metallic material is shown in Figure 6b.

a)   b)

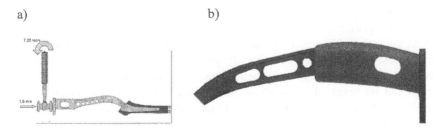

Figure 6. Dynamic analyses of existing tool (a) and modified tool with reduced mass (b).

Design features of tools from one factory were input data to those analyses. Computer models of modified tools were transformed into material models in one factory using Rapid Prototyping technology, Figure 7. In such a way surgeons could assess the tool in the light of ergonomics criterion.

Figure 7. Material model of the tool made by Rapid Prototyping technology.

Tool prototype does not enable to assess the properties of material from which a tool will be made as well as to assess the behavior of the material during operation. However it is possible to assess the functionality of modified tools. That part of research work was jointly realized by the surgeons and specialists from IBV. Results of virtual prototyping and tests of material model were included in a production process.

CONCLUSIONS

Presented case studies were focused on two lifecycle stages: designing and operation. They show how the knowledge, gained during operation, can help to improve work conditions. They confirm correctness of model of knowledge division given in Figure 1. By the methods of Knowledge Engineering a tacit knowledge is transformed into explicit knowledge. Increased resources of explicit knowledge, supported by ICT technologies, can help to create *on site* work conditions. In designing process the criteria for assessment of design solutions that influence work conditions are formulated on the basis of experts' knowledge. The second case study explains how complicated the environment of participants of technical mean life cycle can be. Then some limitations can appear in sharing, transferring and getting access to knowledge. In the research work, carried out within international projects, there are some legal aspects resulting from intellectual property issues that are core business of project participants. Terms explicit knowledge and tacit knowledge have their range. There is a question if all explicit knowledge can be revealed also outside the company where it was gained and developed? To what extend the knowledge gained in the joint project can be used in solutions that compete among each other? From the other hand the work conditions have their social aspect. Improper work conditions can lead to health damage, injuries and even death of many people. Ageing societies of developed countries put higher demands for work. At the same time free transfer of knowledge resources (the same as free transfer of goods) will enable to joint developing countries to information society. These are open questions and that problems can be solved on the base of other paradigms of society development that the current ones.

REFERENCES

ATEX, (1994). Directive 94/9/EC of the European Parliament and the Council of 23 March 1994 on the approximation of the laws of the Member States concerning equipment and protective systems intended for use in potentially explosive atmospheres.

Anthropos-ErgoMAX, (1999). User Guide Version3.0. IST GmbH Kaiserslautern.

ENHIP, (2005). Ergonomic Instruments Development for Hip Surgery an Innovative Approach on Orthopaedic Implants Design. Contract No. 017806-ENHIP.

ENHIP, (2006). Deliverable 1_3 Clinical Assessment Group Four-Monthly Evaluation Report Wp5_031_Tec_Ibv_V00_Deliverable1_3.

Grudzewski, W.M., Hejduk I.K. (2004). Knowledge Management in Enterprise (In Polish). Difin, Warszawa.

Knowledge based ... (2009). Knowledge based improvement working conditions in Health Care. National Centre for Research and Development (Poland). Contract No.: N R11 0026 06/2009

MINTOS, (2007). Improving Mining Transport Reliability. RFCS Coal RTD Programme, Contract No. RFCR-CT-2007-00003.

Moczulski W., Ciupke K. eds. (2008): Knowledge Acquisition for Hybrid Systems of Risk Assessment and Critical Machinery Diagnosis. Institute for Sustainable Technologies, Gliwice-Gdańsk.

Polanyi, M. (1967) The Tacit Dimension, Routlage & Kegan Paul Ltd., London

Thagard, P. (April 12, 2007), How to Collaborate: Procedural Knowledge in the Cooperative Development of Science, http://cogsci.uwaterloo.ca/Articles/how-to-collaborate.pdf.

3D SSPP 4.3, (2001). Three – Dimensional Static Strength Prediction Program Version 4.3. The University of Michigan's Center of Ergonomics.

Chapter 64

Innovation-Based Enhancing Work Conditions in Healthcare Organizations

Joanna Bartnicka, Teodor Winkler

Institute for Engineering of Production
Silesian University of Technology
41-800 Zabrze, ul. Roosevelta 26, Poland

ABSTRACT

The aim of the article is to present the results of research and the conception of solutions in the area of creation of work conditions in the hospitals with application of innovative technologies and methods. Particularly, the paper describes employment of such methods and technologies as: human body modeling, computer modeling, RFID (Radio Frequency Identification), GIS (Geographical Information Systems), CMS (Content Management Systems). The consideration undertaken in the article shows the method of integration of above-mentioned methods and technologies. This integration will allow to build the virtual copy of real hospital and create the complex system called Virtual Hospital. The essence of virtual hospital is creation of relation between the graphic interface, which is built in GIS system, and knowledge recourses concerning enhancement of work conditions in health care organizations. These resources are stored in knowledge repository and create the network of context relate information sharing by Internet Platform. The recourses included in the knowledge repository are divided into main three groups: essential knowledge (work methods, ergonomics), operational knowledge (the knowledge in range of maintenance management) and organizational knowledge (organization of work, organization of work space). The recipients of Virtual Hospital system are medical staff, nursing staff, administrative staff, as well as patients, who will learn about conditions of treatment and care in the hospital.

Keywords: healthcare, virtual work environment, virtual hospital, RFID, GIS, CMS

INTRODUCTION

The term 'work environment' should be understood as all the factors in an organization which are connected with the nature of the work being done and with the environment in which the work takes place.

With this definition in mind, it can be stated that the hospital work environment is highly complex and changeable in nature. This makes the process of shaping it difficult and all the organizational units and all the hospital staff must be involved in it. The complexity of the work environment is connected with a large number of factors influencing it. These factors can be divided into the following groups:

- material work environment, including: physical, chemical and biological factors,
- technical factors stemming from the structural and technological properties of the work means (technical means),
- organizational factors connected with work management and processes, including the work methods used, the way in which individual works are performed, work ergonomics,
- psychosocial factors resulting from the interaction between work requirements and the competences and individual characteristics of the workers.

The variability of a work environment is influenced by the following:

- the diversity of cases of disease,
- a wide range of medical equipment, including the equipment in hospital wards, surgeries, treatment rooms, specialist equipment, instruments, disposable equipment etc.,
- a wide range of patients in terms of their anthropometric features,
- a wide range of patients in terms of mobility and self-sufficiency,
- a variable number of medical personnel over 24 hours,
- varied work methods,
- varied scopes of patient treatment and care,
- no uniform and clear qualitative criteria for choosing medical equipment.

The health care service sector is a special area of activity where the work environment affects not only the staff working there but also the patient. The work environment may therefore help or hinder the provision of health care services. Thus it can also be said that the conditions of patient treatment and care are part of the work environment in health care organizations.

Negative impact of work environment on the operation of health care organizations has been identified. Disregard of or being unaware of the principles of ergonomics, bad work organization, bad workspace organization (e.g. room layout), no training in new work methods, including instruction in how to operate specialist equipment, an excessive number of administrative activities (such as completing documents with the same data), insufficient information, bad choice of medical equipment are only some of the elements affecting the following: The work being excessively burdensome, the process of treating and nursing the patient, a lower quality of health care services and consequently higher costs of running the hotel.

VIRTUAL HOSPITAL AS A TOOL SUPPORTING WORK ENVIRONMENT ENHANCEMENT

Considering the problem outlined in the Introduction the following question arises: How should the work environment in health care organizations be enhanced and how to reduce or eliminate the negative effects they bring about?

The answer to this question is supporting work environment management using innovative methods as well as information and information and communication technologies (ICT): human body modeling, computer modeling, RFID (Radio Frequency Identification), GIS (Geographical Information Systems), CMS (Content Management Systems). At present works are being carried out at the Institute for Engineering of Production at the Silesian University of Technology aiming to create a work environment enhancement system for health care organizations called the Virtual Hospital, which is based, among others, on the technologies listed above.

Fig. 1 shows the connections between the elements representing the respective areas of the hospital organization with the methods and technologies used to build the Virtual Hospital.

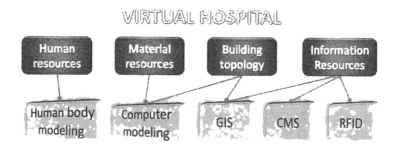

Figure 1. The connections between the component elements of the Virtual Hospital and the innovative technologies

The Virtual Hospital is a computerized 'enhanced copy' of a real-life organization. This copy represents, among others, the following elements of the organization:

- material resources,
- human resources and patients,
- information resources,
- building and surrounding area topology.

The main features of the Virtual Hospital are the context defined connections between the resources listed. These connections are in essence a computerized representation of real-life processes in a hospital. What is more, these processes are performed on the basis of the access to appropriate information resources.

As a result the Virtual Hospital can meet the requirements of a training and support system to enhance the work environment and help manage the hospital organization as a whole.

USING COMPUTER MODELING METHODS IN THE *VIRTUAL HOSPITAL* SYSTEM

This chapter describes how the respective methods and modern information and ICT technologies are used in the context of building and operating the Virtual Hospital.

The operation and the graphic interface of the Virtual Hospital are mainly based on the **GIS system**. The GIS system (Gotlib et al. 2008) is complex computer software used to record and present spatial graphic data. In the traditional view this data represent a generalized image of the earth's or its fragment's surface. However, this view can be extended by rendering a closed architectural space (building topology). Another important feature of the GIS system is that the graphic data are connected with the attributes stored in the GIS system's internal database or an external one, which can take the form of the already mentioned knowledge repository.

The GIS system can help enhance the work environment in health care organizations in particular by the following:

- providing support in planning the layout of patient rooms, other rooms and items of equipment, including medical equipment within the hospital's space,
- providing support in planning the layout of the hospital buildings in the open space,
- providing support in planning where to place patients on a ward,
- providing support in the organization of the following zones: White zone (clean zone) and dirty zone as well as planning efficient transport routes for people (respecting the patients' privacy) and equipment,
- providing support in planning escape routes and in the organization of the evacuation of people and equipment,
- identifying medical equipment and drugs storage areas (including small items of equipment) and providing support in keeping the record of the equipment,

- providing support for facility management, including operation management and the building's installations maintenance,
- providing employees and patients with assistance in orientation and navigation in the building as well as between buildings thanks to the visualization of the building's and the transport routes' topology,
- facilitating the mobility of people with a disability through notices giving information about the fact that the building is adapted to the needs of persons with different types of disability,
- allowing patients and their families to become acquainted with the appearance of the rooms and the whole hospital building prior to the planned visit.

All these functions can be fulfilled thanks to connecting the graphic elements, e.g. the plan of the building, with suitable information resources stored in the knowledge repository.

Both the graphic resources stored in the GIS system and the resources stored in the knowledge repository are ordered and linked according to a given situational context.

In particular the graphic data in the GIS system can be ordered thanks to the layer structure of the system. Thematically different graphic resources are placed on separate layers. This layer structure also makes it easier to manage the graphic resources through such actions carried out on layers as: switching on, switching off, filtering, searching, sorting etc.

Fig. 2 shows the structure of the Virtual Hospital, including the GIS system's layer structure and the connection with the knowledge repository. The choice of layers and graphic data recorded on them was based on the defined specifications of the Virtual Hospital.

The following thematic layers were identified in the GIS system:

- the topography of the area together with the hospital buildings,
- plans of the respective floors of the building,
- characteristic places within the building, such as the following zones: clean and dirty, registration, admissions, toilets, shops etc.,
- places with facilities for people with a disability,
- layers with a graphic record of equipment items (the respective types of equipment are placed on different layers),
- transport routes divided into users moving along the routes and the end location (the respective routes are placed on separate layers),
- escape routes,
- installations (the respective installation types are placed on separate layers).

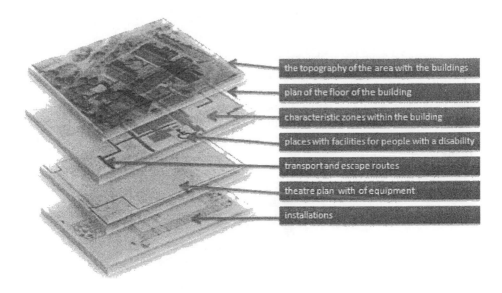

Figure 2. A drawing showing the structure of the Virtual Hospital

Figure 2 shows how the elements of the building plan are linked with the information in the knowledge repository, which is a context extension of the information shown on the plans. This linking enables the GIS layers to also take the form of a user's graphic interface, which is the basic level from which the Virtual Hospital is run.

The following solution is an example of how data from the knowledge repository are added to the graphic information on the GIS maps: operating theatre plan (GIS) - 3D computer visualization of the theatre - the types of surgical procedures performed in the operating theatre – descriptions of how the procedures are performed – work methods – equipment in the operating theatre – computer simulations of how to operate the equipment, health and safety at work, hazards connected with carrying out the activities (see fig. 3). As the example shows the data from the repository are presented in various forms. One form is 3D models and computer simulations which give a virtual representation of hospital work environment. The methods used to create such forms of presenting information are **computer modeling** and **human body modeling**.

Considering the aims and premises of the Virtual Hospital system these methods are used to provide a computer representation of the following material resources: the equipment, rooms and the following human resources: the employees' and the patients' anthropometric features. Therefore, they help create a virtual work environment and enhance work conditions through the following:

- computer simulations of work methods, including the activities carried out on the patient,
- computer simulations of equipment operation,

- providing support in the organization of work space,
- providing support in shaping the space according to the needs of people with a disability.

The method of creating a virtual work environment together with the methodology for its computer-based assessment are described in detail in the article (Winkler, Bartnicka 2009). The content of the presentation applies to the analysis of the work of the nursing staff.

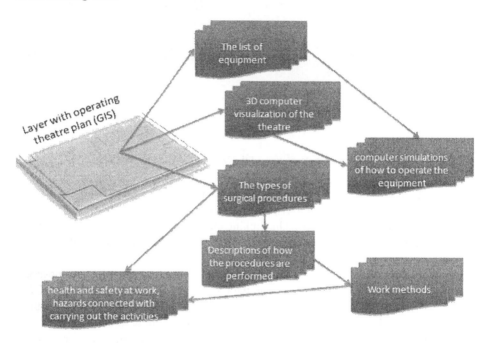

Figure 3. An example of how data from the knowledge repository are added to the graphic information on the GIS maps

ICT TECHNOLOGIES IN MAKING THE *VIRTUAL HOSPITAL* SYSTEM AVAILABLE

The ordered and formalized data resources stored in the Virtual Hospital system, which provide support in enhancing work conditions are distributed by means of Internet technologies. Such a solution enables the dispersed users (hospital staff, patients) to access the data possible at any time and in any place. With this in mind, **Content Management Systems (CMS)** were selected and tested with regard to the functionality of the Virtual Hospital.

The system in question is conducive to enhancing work conditions through the following:

- collecting data which support processes carried out in health care organizations,
- defining links between respective information resources,
- integrating the GIS graphic interface with the Internet browser's interface,
- making information resources available in text, graphic and multimedia forms,
- making the information resources available to users no matter where they are.

The support for the works connected with enhancing work organization can only be effective on the condition that quick and unrestricted (mobile) access to the relevant data resources stored in the Virtual Hospital is available.

The RFID technology provides automatic access to these resources and displays them on a palmtop screen. The **Radio Frequency Identification (RFID)** technology provides an automatic, touch-free identification and reading of data by means of radio waves of various frequencies. This technology is a development of the bar code technology. The following are the basic elements of the RFID: a label (also called a tag or a transponder), a reader, an antenna, a transmitter/receiver, software. The labels are made up of an electronic circuit which stores the data and an antenna which transmits them via radio waves. The RFID reader communicates with the labels in order to obtain the recorded data; the data can be recorded in an external base.

An important feature of this technology is its ability to define any object, animal and even a human by placing on them unambiguous identifying data by means of an electronic identifier and then read them by means of a wireless device (Dziennik Urzędowy, 2007)).

Work conditions are enhanced by the RFID technology in that context information, i.e. information connected with a specific problem situation at a given moment is provided.

The RFID technology operation was simulated in a selected hotel with the participation of the medical staff and patients.

Fig. 4 presents three situations which show how to use the RFID technology in making available the information resources obtained from the CMS database relevant for the given situational context (see Bartnicka, Smolorz 2010).

Figure 4. Examples of applying the RFID technology in making available context information resources

The first situational context: the data readout takes place at the moment of entering the patients' ward – the plan of the room is displayed on the screen of a PDA type computer.
The second situational context: the data readout takes place when the RFID reader is placed close to the band with the label on a patient's hand – a set of data about the patient is displayed on the screen of a PDA type computer.
The third situational context: the data readout takes place when the RFID reader is placed close to the label on an equipment item – an electronic operating manual for the item of equipment is displayed on the screen of a PDA type computer.
As the planned functionality of the Virtual Hospital requires it is possible to

connect to other information resources stored in the repository from the level of the screens displayed after using the RFID labels.

The simulation of the operation of the RFID technology in a medical organization showed to what degree the RFID technology supports work conditions management in a health care organization. Simultaneously, it points out a need to provide information about promoting innovative and tested for safety technologies supporting the functioning of health care.

CONCLUSIONS

The article was elaborated on the basis of experience in analyzing and assessing the work conditions in health care organizations as well as experience in creating knowledge management models and applying innovative information and ICT technologies. It was shown in particular that the technologies which, in the traditional view, support activities in industry (computer modeling), transport and environment engineering (GIS), logistics (RFID) can successfully support activities in health care. These activities include among others creating conditions for the treatment and care of the patients which on the one hand guarantee high quality of health care services and on the other ensure comfortable work conditions for the staff and create the right conditions for the development of their competences and skills.

Innovation-based enhancement of the work conditions in health care organizations according to the premises described in the articles and the results of the research carried out so far to verify its effectiveness prove the validity of the chosen research subject matter.

The article was elaborated as part of the development project (Agreement No.: N R11 0026 06/2009) called 'Knowledge-based enhancement of the work conditions in health care organizations' financed by the National Centre for Research and Development.

REFERENCES

Bartnicka J., Smolorz M. (2010): Zastosowanie technologii RFID w zarządzaniu zasobami w placówkach opieki zdrowotnej, w: (red.) R. Knosala: Komputerowo zintegrowane zarządzanie. Tom I, Oficyna Wydawnicza Polskiego Towarzystwa Zarządzania Produkcją, Opole

Dziennik Urzędowy Unii Europejskiej (2007): Opinia Europejskiego Komitetu Ekonomiczno-Społecznego w sprawie identyfikacji radiowej (RFID), (2007/C 256/13)

Gotlib D., Iwaniak A., Olszewski R. (2007): GIS. Obszary zastosowania. Wydawnictwo Naukowe PWN, Warszawa

Winkler T., Bartnicka J. (2008): Virtual working environment of nurses. 2nd International Conference on Applied Human Factors and Ergonomice, 2008 AHFE International Conference, 14-17 July 2008, Caesars Palace; Las Vegas, Nevada USA

Chapter 65

Heat Transfer Dynamics in Clothing Exposed to Infrared Radiation

Uwe Reischl[1], Budimir Mijovic[2], Zenun Skenderi[2], Ivana Salopek Cubric[2]

[1]College of Health Sciences
Boise State University, USA

[2]Faculty of Textile Technology
University of Zagreb, Croatia

ABSTRACT

Thermal mannequin heat loss (heat gain) during exposure to heat radiation was evaluated. Semi-nude and clothed configurations were tested. The relative effects of dry skin and sweating with increasing air velocities were also documented. The study was based on the use of a newly developed inflatable thermal mannequin placed inside a negative pressure wind tunnel. The experiments documented the IR attenuation effects of sweating and clothing. Additionally, the heat radiation "cross-over" point for clothing was identified were protective clothing can be beneficial initially while later becoming an impediment to metabolic thermal equilibrium. The studies showed that the new thermal mannequin technology platform can serve as an effective instrument for evaluating convective, evaporative, and radiative heat exchange properties of protective clothing.

Keywords: Heat Radiation, Protective Clothing, Thermal Mannequin

INTRODUCTION

Evaluating heat transfer characteristics of clothing systems is important in an attempt to provide thermal comfort and safety in the design of protective garments. Values for clothing thermal insulation can not only be used to estimate the potential heat stress experienced by persons exposed to hot environments but also the potential for cold stress when exposed to cold environments. Thermal insulation values also serve as important measures of garment effectiveness, functionality, and suitability in work and leisure.

Heat transfer takes place in the form of conduction, convection, radiation, and sweat evaporation. The insulation provided by clothing is impacted by the design and the materials used in the manufacturing and assembly of the garment, the body surface area covered, the patterns of fabric layers used, and the tightness of fit. While the measurement of the resistance to dry heat loss can be used to estimate the thermal comfort in cold and relatively moderate environments, the moisture permeability and ventilation characteristics play a more significant role in hot conditions where evaporation of sweat provides a large component of the overall metabolic heat dissipation.

The impact of radiant heat on clothing systems has always been difficult to assess especially in relationship to sweating and moisture build-up in garments. However, with the availability of a new thermal mannequin technology platform, it is now possible to evaluate the performance of garments during exposure to various levels of radiant heat.

METHODS

APPARATUS

An inflatable thermal mannequin in the shape of a standing adult male heated at a constant temperature and inflated with a constant pressure was used in all tests. The mannequin's dimensions corresponded to a "Medium" normally used in describing standard garments. A semi-nude standing configuration served as the reference (control) for all measurements. Arm and leg movements could be activated and controlled with a pulley and push rod system that allowed the mannequin to perform walking and running movements. Heat input to the mannequin was controlled through a variable transformer and maintained at a constant temperature of 42.0^0C. Input and output air temperatures were monitored using digital thermister temperature probes with a measurement accuracy of 0.1^0C. Total mannequin heat loss (watts) was calculated based on the measured difference between the temperature of the mannequin input air and the mannequin output air. Clothing and sweating were independent variables in the testing protocol. The difference between input and output air temperatures was used as the dependent

590

variable. A negative pressure laminar air flow wind tunnel was used for generating controlled air flow conditions. The functional components of the thermal mannequin system are illustrated schematically in Figure 1. A photograph of the mannequin in the semi-nude (control) configuration is shown in Figure 2. The gravity-flow irrigation system used for simulating sweating is illustrated schematically in Figure 3. The negative pressure wind tunnel used for generating controlled laminar air flow is illustrated in Figure 4. Average surface (skin) temperatures always maintained for the semi-nude (control) configuration are shown in Figure 4.

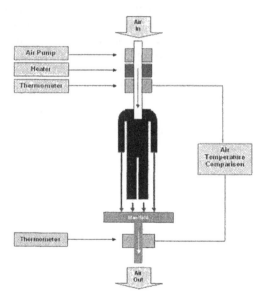

Figure 1. Schematic illustration of the inflatable thermal mannequin technology platform.

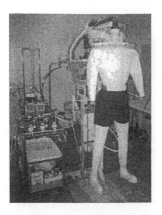

Figure 2. Photograph of the Inflatable thermal mannequin system in the semi-nude (Control) configuration.

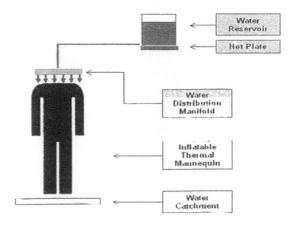

Figure 3. Gravity-flow irrigation system used for simulating mannequin whole-body sweating.

592

Figure 4. Negative pressure wind tunnel used for generating controlled laminar air flow conditions.

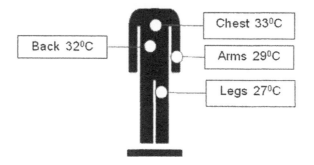

Figure 5. Average (dry) skin temperatures for the thermal mannequin in the semi-nude (control) configuration for an ambient air temperature of 42^0C at 30% relative humidity with no air flow.

PROCEDURES

All tests were performed inside a temperature controlled room. Ambient air was maintained at 22^0C (+/- 1^0C). Relative humidity ranged from 25% to 35%. The mannequin input air temperature was maintained at 42.0^0C (+/- 0.1^0C). The mannequin inflation pressure was constant at 4.7 mmHg. Volume air flow through the system was constant at $.01m^3$/sec. Output air temperature was monitored continuously. System thermal equilibrium was achieved within 8-10 minutes. Total heat loss from the thermal mannequin was calculated using Equation 1.

Equation 1. $\quad Q_{watts} = 1190 \, (V_{.01m3}) \times (T_{out} - T_{in})$

All tests were preceded by a "control" measurement where the inflatable thermal mannequin was operated for 15 minutes in a semi-nude configuration. The mannequin was then outfitted with the garment to be tested. To simulate sweating, the mannequin was saturated with warm water using a continuous flow gravity irrigation system mounted on the shoulders of the mannequin as illustrated in Figure 3. The temperature of the irrigation water at the point of contact with the mannequin was 45^0C. Excessive water flow was maintained for semi-nude conditions and clothed conditions to maintain skin saturation. Saturation was defined when water dripped from the feet down into catchment container on the floor.

Exposure to increasing air flow conditions was achieved by placing the mannequin inside a 2- meter long, 1- meter wide, and 2- meter high wind tunnel as illustrated in Figure 4. The wind tunnel was also located in the temperature controlled room. Air flow was generated by three variable speed axial fans installed at the down-wind position of the wind tunnel.

Controlled heat radiation exposures were achieved using a cluster of four 250 watt IR reflector lamps positioned symmetrically 1.25 meters in front of the mannequin. The four lamps were directed onto the mannequin's chest. Heat output was controlled by a variable transformer.

TEST CONFIGURATIONS

The following mannequin configurations were tested:

- Semi-nude: Dry
- Semi-nude: Dry + Air Flow
- Semi-nude: Dry + Jacket
- Semi-nude: Dry + IR Exposure
- Semi-nude: Wet
- Semi-nude: Wet + Air Flow
- Semi-nude: Wet + IR Exposure

RESULTS

Figures 6, and 7 summarize the heat transfer values measured for the thermal mannequin in a semi-nude (SN) configuration. This includes dry skin, wet (sweating) skin, exposure to increasing air flow conditions, and increasing heat radiation. The semi-nude configuration served as a "control" or "reference" for all experiments. Figure 6 also illustrates the changes observed in the total mannequin heat loss for both dry skin and wet skin as the ambient air speed was increased. With increasing air flow, the total heat loss from the wet (saturated) skin increased more quickly than the heat loss from the dry skin. Figure 7 illustrates the changes in the mannequin heat loss as the IR heat radiation intensity levels were increased. The dry skin condition exhibited a greater reduction in heat loss than the wet skin. Figure 8 illustrates the impact of a jacket on dry skin heat exchange as the IR heat radiation exposure intensity levels were increased.

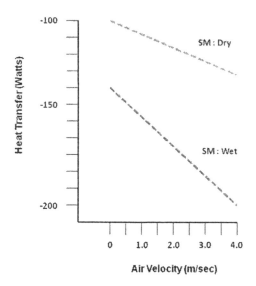

Figure 6. Mannequin heat transfer for dry and wet skin conditions exposed to increasing ambient air flow.

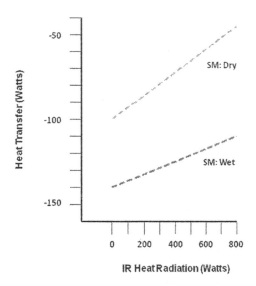

Figure 7. Mannequin heat transfer for dry and wet skin during exposure to increasing heat radiation intensity levels.

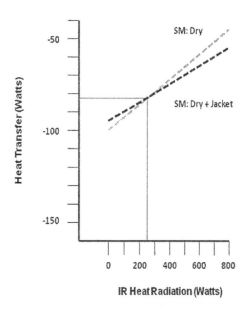

Figure 8. Mannequin heat transfer for dry skin with and without a jacket during exposure to increasing heat radiation intensity levels. A thermal "cross-over" occurs at the 250 watts IR Radiation level.

ANALYSIS

The heat transfer characteristics observed in the thermal mannequin does not simulate human physiological responses. The technology represents an ergonomic instrument useful in assessing the fundamental convective, evaporative, and radiative heat transfer processes. However, the shape, form, and dimensions of the mannequin, including the surface temperature distribution patterns, correlate to the basic characteristics of a human subject. The heat transfer characteristics exhibited by the semi-nude configuration ("control"), are comparable to measurements obtained on human subjects exposed to similar environmental conditions (Chen, 2004, Mijovic, 2009, Parsons, 1999). As seen in Figures 6, 7, and 8, the baseline for the mannequin heat loss of 100 watts is comparable to the heat loss measured for resting human subjects (Havenith, 1999, Brode, 2008).

The impact of air flow on convective and evaporative heat loss from the skin is shown in Figure 6. The dry skin (convective heat loss) increased by 30 watts when the air velocity was increased from 0 to 4 meters per second. The evaporative heat loss, however, increased by 60 watts for the same increase in air velocity. The magnitude of this difference is consistent with observations made on human subjects (Chen, 2004, Hodder, 2007, Zhang, 2001).

The effect of IR radiation exposure on body heat transfer summarized in Figure 7 shows that a reduction in body heat loss is most pronounced for the dry skin condition. Under this condition, the total mannequin heat loss without IR exposure is 100 watts but decreases to 45 watts when exposed to 800 watts of heat radiation. Such a reduction is not seen for the sweating condition. A net reduction of only 30 watts is observed for the sweating condition. This may be due to a "reflection" or "scattering" of the heat radiation or is the result of IR absorption by the moisture which then accelerates the evaporation process of the sweat. Regardless of the mechanism, the sweat on the skin appears to provide IR "attenuation".

Figure 8 summarizes the changes in mannequin heat loss for both the non-sweating semi-nude and the clothed configurations as heat radiation exposure is increased. The semi-nude configuration shows a more rapid change in heat loss than the clothed condition. At the "0" IR condition, the semi-nude configuration exhibits 100 watts of heat loss and 40 watts of heat loss at the 800 watts IR exposure intensity level. In contrast, the "clothed" condition shows 95 watts of heat loss at "0" exposure and 55 watts at the 800 watts IR exposure level. A "cross-over" point for the two configurations is seen at the 250 watts IR exposure intensity level.

The "cross-over" phenomenon highlights the potential role of clothing in providing protection against heat radiation. Below the cross-over point, the clothing imposes insulation on the mannequin in <u>excess</u> of the IR heat radiation blocked by the garment. Above the cross-over point, the clothing provides actual "protection" against the IR radiation by reducing the overall heat load on the body. The implications are the following: For relatively low IR exposure levels, protective clothing may impose an unnecessary (excess) heat stress factor while at higher levels, the garments can be beneficial. The "cross-over" point must be determined for each garment system separately.

CONCLUSIONS

The inflatable thermal mannequin offered a simple and practical technology platform for evaluating garment systems with regards to their potential impact on convective, evaporative, and radiative heat exchange. While the mannequin did not simulate human physiological responses to heat or cold, the system offered a realistic thermal profile relevant to ergonomic applications. Measurement of the impact of IR radiation on body thermal equilibrium for both dry and wet skin conditions showed that convective, evaporative, and radiative heat exchange can be considered separately. While one form of heat exchange may increase metabolic heat loss from the body, another may decrease it. The interaction can be manipulated by proper testing, design, and use of materials to assure comfort and safety.

REFERENCES

Chen YS, Fan J, Qian X (2004) *Effect of garment fit on thermal insulation and evaporative resistance.* Textile Res. J. 74: 742-748

Havenith, G. (1999) *Heat balance when wearing protective clothing.* Annals of Occupational Hygiene. 43(3): 289-296

Hodder, S. and Parsons, K.C. (2007) *The effects of solar radiation on thermal comfort.* International Journal of Biometeorology, 53(3): 233-250

Mijovic B, Skenderi Z, Salopek I (2009) *Comparison of subjective and objective measurement of sweat transfer rate.* Coll. Antropol. 2: 509-514

Parsons, K, Havenith, G., Holmer, et.al. (1999) *The effects of wind and human movement on the heat and vapour transfer properties of clothing.* Annals of Occupational Hygiene, 43: 347-352,

Brode, P, Kuklane, K, Candas, et.al. (2008) *Heat transfer through protective clothing under symmetric and asymmetric long wave thermal radiation.* Zeitschrift fuer Arbeitswissenschaft, 62: 267-276.

Zhang P, Watanabe Y, Kim SH, Tokura H, Gong RH (2001) *Thermoregulatory responses to different moisture-transfer rates of clothing materials during exercise.* J. Text. Inst. 92: 372-378

Chapter 66

Innovative Solutions for Adjustment of City Area to Disabled Persons and the Aged With Using Computer Techniques

Agnieszka Kowalska Styczeń[1], Joanna Bartnicka[1], Christophe Bevilacqua[2]

[1]Institute for Engineering of Production
Silesian University of Technology
41-800 Zabrze, ul. Roosevelta 26, Poland

[2]CITYTAK
121 rue Chanzy, Ruche Technologique du Nord
59 260 Hellemmes Lille

ABSTRACT

The article presents an innovative system supporting adjustment of city space to all inhabitants, including disabled and elderly persons. The system consists of technical and organizational solutions designed for helping disabled persons. The aim of the solutions is to enable all persons, including the disabled, access to information and improvement of mobility. The proposed system, in particular, consists of components of infrastructure designations, as well as recommendations for the design of city space, comprising the needs of amblyopic persons, blind persons, persons with physical dysfunctions and elderly persons. Above mentional designations take forms of maps, tables, information mats, which include information in the form of inter alia: text, graphics, colors and layers, as well as we study the possibility to improve the comprehension using fluorescent properties. Through the access to readable information and communicators, the proposed

600

system of designation aims at helping the people stay in the city space, move both in open and closed space, as well as to facilitate access to sphere of public life for all persons, but especially for ones, who so far have stayed in social isolation. The whole issues within the proposed project aims at contributing to a change in the approach to the development of social and personal life from the particular approach which is taking into account the needs of selected social groups, to non-discriminatory approach, aimed at equalizing opportunities for all people.

Keywords: disabled persons, the Aged, city space, accessibility, mobility, innovation, sphere of public, design for all

INTRODUCTION

Each person has a right to live in dignity and therefore to fully participate in social life free from discrimination. This right is granted in the Universal Declaration of Human Rights, adopted by the General Assembly of the UN on 10th December 1948.This is also included in Polish legal acts such as: the Constitution of the Republic of Poland (article 67, paragraph 1, article 68, paragraph 3, article 69), which specifies the rights of disabled people and the duties the authorities have towards them and stresses that 'all the people are born free and equal in their rights' and the Charter for Disabled People (Monitor Polski [Official Gazette of the Republic of Poland] No 50, item 475 of 13 August 1997), and also an anti-discrimination bill 'Equal chances for people with disabilities' (this bill is to go to the Sejm of the Republic of Poland in September 2009).It is therefore appropriate to look for innovative non-discriminatory solutions to increase mobility and improve access to information for everyone, including a disabled and an elderly person. Such solutions should be e.g. a system of marking urban space and the buildings in it appropriately and architectural solutions eliminating barriers on traffic routes. By innovative solutions we understand an integrated system helping to move around and providing access to information to everyone, including a disabled or elderly person rather than solutions and suggestions designed exclusively for these people. In our opinion, the answer to the legal regulations concerning discrimination as well as the answer to the challenges of creating a modern barrier-free urban space is the CITYTAK project described below. The originator and creator of this project is Christophe Bevilacqua. Creation of solutions and methods were preceded by studies in such domains as: haptic, ergonomic and psychology (see all references). At the stage of creating and testing the solutions Mr Bevilacqua entered into a cooperation with the THIM laboratory (Laboratoire Technologies, Handicaps, Interfaces et Multimodalités) at the University of Paris VIII, the E.R.D.V. de LOOS (Ecole Régionale pour Déficients Visuels) and with many associations helping disabled people. The works on the innovative solutions were preceded by a close observation of people in their everyday life (including disabled and elderly people), which helped identify the needs of people with various disabilities. All the concepts and solutions were verified and tested at each development phases by disabled and elderly people. This was possible thanks to the cooperation with numerous organisations and associations dealing with the issues of disabled people e.g. Valentin Haüy de Lille (www.avh.asso.fr), Handifac Lille (http://handifac.free.fr), Aventure et partage (www.aventure-partage.org).The CITYTAK project aims to develop the solutions suggested for the present moment as well as to introduce other non-discriminatory products. We hope that the Faculty of Organisation and Management at the Silesian University of Technology will play a significant part in this process because it has started cooperation with Christophe Bevilacqua and with the TIIIM laboratory at the University of Paris VIII.

PRESENTATION OF THE CITYTAK PROJECT – ADAPTING THE CITY AND THE BUILDINGS TO THE NEEDS OF DISABLED AND ELDERLY PEOPLE

The CITYTAK project is an idea for ethical, non-discriminatory products available for all. It is a 3D geographic information system designed for urban space and insides of buildings also containing numerous technical and architectural solutions. The objectives of CITYTAK are to help people throughout their progression across streets and buildings (from their departure to their arrival).

The suggested solutions for urban space are the following (see fig. 1):
- A situation plan or a street map covering an area of the city
- A street plate available and touchable for pedestrians
- A street plate
- Information plate (with the name of the institutions inside public buildings and their business hours)

Tactile paving systems signaling the location of maps, entrances into buildings, obstacles such as stairs etc. (the tactile paving are textured and contrasted in order to make it easier for blind people, people with weak vision and other people to be able to find the objects, especially in the evening and at night).

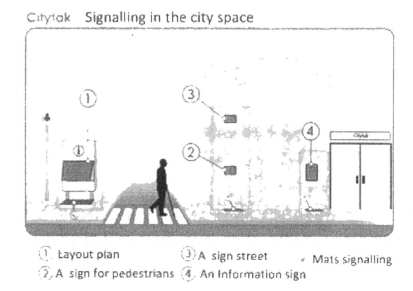

Figure 1 Signposting and information system

The starting point for the urban solutions presented here is the situation plan of a city area covering a maximum of 26 streets (see fig. 2).

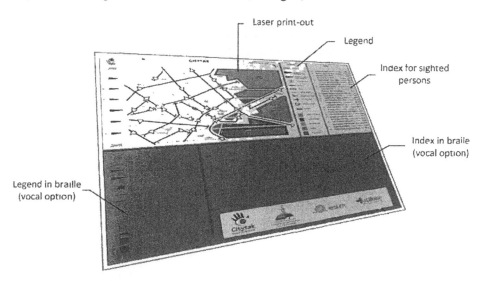

Figure 2 Situation plan showing a part of city area (in this case the centre of Lille, France)

In the system presented here the city is divided into 5 zones: downtown (marked white), north (marked orange), south (marked blue), east (marked green) and west (marked yellow). The same colour is used on all signposts of a city area. If a person leaves this area, the colour will change. This colour decomposition helps person with impaired cognition (in aged person as example) to know if they are in the right place.

Each city street map consists of a "sighted part" (clear sharp colours for people with good and weak vision, also suitable for colour-blind people) and a top transparent and raised part made of Plexiglas. The top transparent part contains a raised street map with a legend in braille. This map is different from regular ones also in that it is not north-oriented. The place where the map is supposed to be situated street is marked in its centre, therefore what is in front of us on the map is also in front of us in reality, what is on the left side of the map is also on the left side in reality and so on. This makes map reading and moving around the city much easier because it decreases the mental load.

The same principle applies to other solutions shown in fig. 1 - namely street plates for pedestrians (see fig. 3) and information plates (see fig. 4).

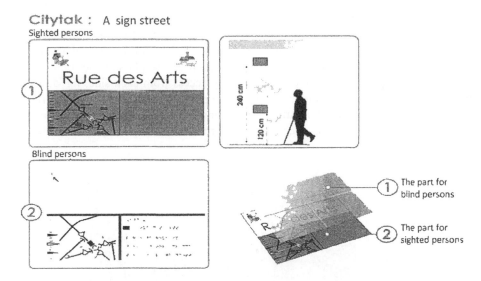

Figure 3 Street plate for pedestrians

604

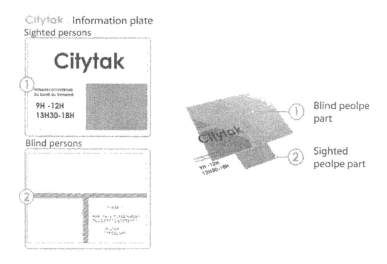

Figure 4 Information plate on a building.

The suggestion of tactile paving surfaces completes the signposting system. These mats use a different texture of the raised surface in order to inform blind people and people with weak vision where the situation plan is, where the street sign is as well as where stairs, pedestrian crossings and entrances to buildings are.

The CITYTAK project also offers innovative solutions for the very process of installing the signalling system (maps and information) within urban architecture. The suggested signposting are mounted at an appropriate height suitable for all the users, including people in wheelchairs with easy access from each side, fig. 5, and fig. 6.Also, the stands are equipped with electric power supply from solar batteries which lights the maps and information signs during the night. Therefore, an external source of electric energy is not necessary in order to install the stands.

Figure 5 Map stands

Figure 6 The information system during the day and the night

The CITYTAK products also include an information system inside public buildings. The following solutions are suggested in order to adapt the buildings (fig. 7):
- The situation plan of the building on each floor
- Information plates on the doors in the building (or next to them)
- Tactile paving surfaces informing e.g. about stairs, lifts or exits (like external tactile paving surfaces they have a different texture).

Citytak : Signalling in the building space

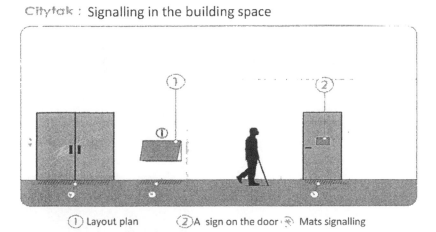

① Layout plan   ②A sign on the door Mats signalling

Figure 7 Signposting and information system inside the building

Like the solutions for the outside, the information system inside the building is based on a situation plan situated on each floor of the building. Like the city map, the map containing such a plan consists of two layers. The bottom part of the map contains a colour picture of the given floor and the top part, made of transparent Plexiglas, contains information for blind people.

USING COMPUTER TECHNIQUES IN CITYTAK PROJECT

The CITYTAK products are mainly composed by a 3 dimensions plate which include 2 maps : the first one is composed by layers and represents the environment for blind people, the second one is a usual map for sighted people.

To make the maps designing easier, CITYTAK has developed a software called CITYPOINT which permits to generated automatically some tasks. The fig. 8 shows an example of how the graphic data from the digital map are conversed to situation plan according to Citytak requirement in CITYPOINT software.

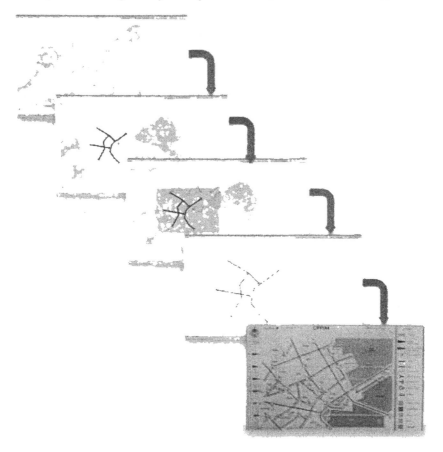

Figure 8 An example of how graphic data from the digital map are conversed to situation plan according to Citytak requirement

The principle of operation is relatively simple. After the opening the software, you can charge a building or a street map. The interface proposes tools to represent the environment and to fills information (street names for a street map, offices for a building map) in the fields devoted.

Consider the adding of a street as an illustrative example. You draw it on the blind map, the software generates the following actions : it draws the street on the sighted people plan, places it in the sighted people legend, translates automatically the name in Braille and puts it in the blind people legend.

The positions and specifications of the legends are defined in the software in such a way that when you have drawn all the streets, CITYPOINT has already automatically generated the blind and sighted people legends. This procedure allows to avoid some mistakes, especially during the translating in Braille.

CONCLUSIONS

The technical solutions presented here are non-discriminatory solutions, which makes them innovative. These solutions make navigation in urban space and public buildings easier for all the users, including disabled and elderly people. The signposting and maps are put at a height suitable for people in wheelchairs. The street and information plates are visible, but do not take up too much space and are integrated with the urban space or with the inside of the building. The project can have many applications, including in urban space:

- City street maps
- Industrial zone maps
- City transport network maps
- Maps of industrial, tourist facilities, parks, gardens etc.

Inside buildings:

- Plans of buildings
- Audio and touch labels e.g. in museums

Analysing the solutions which provide availability for everyone (non-discriminatory solutions) no other solution which would be similar and equally versatile could be found. In our opinion, the project presented here is therefore unique.

Currently many cities in France are interested in implementing the presented solutions, e.g. Lille (schools), Tourcoing (streets and public buildings), Roubaix (parks and gardens), Douai (a modern eco city area) and the AMIEN and LOUVRE Museums in the field of culture. As a result of the cooperation between the Faculty of Organization and Management at the Silesian University of Technology and the city authorities in Zabrze and Mr Bevilacqua the CITYTAK project solutions will also be introduced in Zabrze, Poland. Zabrze has the status of a pilot city in the project and will be the first one to introduce the innovative CITYTAK solution

608

system. The works will begin with introducing maps and information plates in the City Hall building, which is currently being renovated.

The authors thank Geoffrey Lepoutre from CITYTAK for his comments.

REFERENCES

Ballesteros, S., Manga, D., & Reales J.M (1997). Haptic discrimination of bilateral symmetry in 2-dimensional unfamiliar displays. Perseption & Psychophysics, 59 (1), 37-50.

Berger, C, & Hatwell, Y. (1996). Development trends in haptic and visual free classification: Influence of stimulus structure and exploration on decisional processes. Journal of experimental child Psychology, 63(3), 447-465.

Bertelson, P., Mousty, P., & D'Alimonte, G.(1985). A study of braille readind : 2. Patterns of hand activity in one –handed and two-handed reading. The quarterly journal of experimental psychology,37A, 235-256.

Bris, M.,& Morice, J.C.(1995). Conception du dessin en relief pour les personnes non-voyantes. Le courrier de Suresnes, 63,5-16.

Fagot, J., Lacreuse, A., & Vauclair, J. (1994) ; Hand-mouvement profiles in a tactual-tactual matching task: Effect of spatial factors and laterality. Perception & Psychophysics, 56(3), 347-355.

Fagot, J., Vauclair, J. (1993) . La téralisation chez les singes. La Recherche, 24(252), 298-304.

De Volder, A. (1997). Réorganisation corticale chez les déficients visuels jeunes. In Actes du Colloque du groupement de Recherche et d'information Consacré à la cécité et l'Amblyopie, (pp.121-129). Bruxelles, Belgique.

De Volder, A., Bol, A., Robert, A., Arno, P., Grandin, C., Michel, C., & Veraart, C. (1997). Brain energy metabolism in early blind subjects : neutral activity in the visual cortex. Brain research, 750, 235-244.

Chapter 67

Ambient Intelligence in the Classroom: an Augmented School Desk

Margherita Antona[1], George Margetis[1], Stavroula Ntoa[1], Asterios Leonidis[1], Maria Korozi[1], George Paparoulis[1] and Constantine Stephanidis[1,2]

[1]Foundation for Research and Technology – Hellas (FORTH),
Institute of Computer Science, Heraklion
Crete, Greece
E-mail: cs@ics.forth.gr

[2]University of Crete
Department of Computer Science
Greece

ABSTRACT

This paper discusses the opportunities and challenges of Ambient Intelligence (AmI) technologies in the context of classroom education, and presents the methodology and preliminary results of the development of an augmented school desk which integrates various AmI educational applications. The overall objective is to assess how AmI technologies can contribute to support common learning activities and enhance the learner's experience in the classroom. Young learners were involved from the first phases of the design of the desk and its applications using scenario-based techniques.

Keywords: Ambient Intelligence, educational applications, eLearning, English as a Foreign Language, learner-centered design

INTRODUCTION

Ambient Intelligence (AmI) is an emerging field of research and development that is rapidly gaining wide attention by an increasing number of researchers and practitioners worldwide [13]. As a consequence, the notion of AmI is becoming a de facto key dimension of the emerging Information Society, since many of the new generation industrial digital products and services are clearly shifted towards an overall intelligent computing environment.

AmI will have profound consequences on the type, content and functionality of the emerging products and services, as well as on the way people will interact with them, bringing about multiple new requirements for the development of the Information Society (e.g., [9]). While a wide variety of different technologies is involved, the goal of AmI is to either hide the presence of technology from users, or to smoothly integrate it within the surrounding context as enhanced environment artifacts. This way, the computing-oriented connotation of technology essentially fades-out or disappears in the environment, providing seamless and unobtrusive interaction paradigms. Therefore, people and their social situation, ranging from individuals to groups, and their corresponding environments (office buildings, homes, public spaces, etc), are at the centre of the design considerations.

AmI is often claimed to bring a significant potential in the domain of education [1]. Information and Communication Technologies already permeate the classroom environment in many ways. They can play an important role in education by increasing students' access to information, enriching the learning environment, allowing students' active learning and collaboration and enhancing their motivation to learn [5]. This paper presents a preliminary attempt to develop an augmented school desk and the related applications, which aim at integrating AmI technology in the learning process. Following a discussion of current issues in technology integration in the classroom, the paper describes the overall concept and the hardware and software characteristics of the prototype desk. Then, a scenario of use, and the related software applications currently under development are introduced. Finally, a formative evaluation experiment with a small group of young learners of English as a Foreign Language is reported and the results are discussed.

BACKGROUND

In the recent past, learning with the use of ICT was strongly related to concepts such as distance learning [2], educational games [11], intelligent tutoring systems and e-learning applications [3]. Overall, two major trends characterize recent efforts:

- Learner-centered design: learner-centered design approaches adopt and enhance traditional user-centered design practices in the educational context. Learner-centered design may be intended as design with the needs of the learner at the forefront, and possibly involving learners at various stages of the design process. Related work has addressed on the one hand the specific characteristics of the very young population as users of interactive technologies [8, 15], and on the other hand pedagogical and learning-related requirements [16].

- Adaptive and personalized eLearning: these approaches aim at exploiting intelligent tutoring and adaptive hypermedia techniques in eLearning in order to provide educational content matching the individual learning requirements of different learners, and to support content reusability [4]. In particular, in the domain of foreign language learning, which is of relevance to the work reported here, intelligent techniques are targeted to personalized support for learning strategies and error correction [12].

Furthermore, the notion of smart classrooms has become prevalent in the past decade [19]. Smart classroom is used as an umbrella term, implicating that classroom activities are enhanced with the use of pervasive and mobile computing, sensor networks, artificial intelligence, robotics, multimedia computing, middleware and agent-based software [6]. Following the rationale of augmented technology in the educational environments, new means of interaction - such as interactive whiteboards, touch screens and tablet PCs - have gained popularity and have become a major tool in the educational process, allowing more natural interaction. Smart classrooms, for example, may support one or more of the following capabilities: video and audio capturing in classroom [18], automatic environment adaptation according to the context of use, such as lowering the lights for a presentation [7], lecture capturing enhanced with the instructor's annotations, or information sharing between class members.

The work reported in this paper, which is conducted in the context of the AmI Programme of ICS-FORTH [10], constitutes an initial step towards investigating the role of AmI technologies in the educational context and in the classroom environment. The underlying approach is learner-centered, involving a small group of young learners from the very first steps of the design, and targeting to provide intuitive and seamless tools to improve the learning and classroom experience. On the other hand, adaptation and personalization techniques, as well as techniques in the domain of intelligent computer-supported language learning [12] are revisited here in an AmI perspective, aiming at exploiting the interaction possibilities offered by AmI technologies towards facilitating learning of English as a foreign language. An augmented school desk has been selected as a first AmI artifact to be developed in the context of the project, along with a series of educational applications which integrate ambient interaction as well as digital augmentation of physical paper (e.g., [17]).

DESIGNING AN AUGMENTED SCHOOL DESK

PHYSICAL ARTIFACT DESIGN

In the context of AmI, the classroom is a challenging environment. In practice, there are severe space and layout limits to the introduction of AmI equipment, which should be unobtrusive, hidden or embedded in traditional classroom equipment and furniture. It is very important that such equipment can be installed smoothly and easily moved around in the environment, and that space requirements are as limited as possible. This implies several constraints on how the AmI classroom environment can be developed. To address this issue, the AmI Classroom project has adopted an artifact-oriented approach, by stepwise introducing independent AmI augmented artifacts in the environment.

As already mentioned, the first such artifact is an augmented school desk (Figure 1), where an additional piece of furniture has been designed to fit typical school desks of standard dimensions according to EU normative[1]. Such an 'add-on' provides a custom plexiglass 27 inches diagonal wide screen whose inclination can range from 30° (with respect to desk surface) to completely horizontal. It embeds almost invisibly all the devices required for the operation of the AmI applications, and has a width of 40 cm, thus requiring relatively limited additional space with respect to the standard desk. A wooden prototype of the desk has been built. The green color has been chosen to facilitate vision processing.

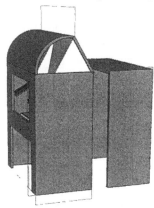

Figure 1. The design of the augmented school desk

HARDWARE AND SOFTWARE SET UP

An important consideration in the design of the augmented school desk was that AmI in the classroom should be compatible with the school of today, as the anticipated transition to the paperless classroom does not appear so imminent. Therefore, as a first step, the augmented school desk should smoothly support paper-based learning materials and the use of handwriting. This requires the adoption of vision techniques on the front part of the desk, as well as smart pen

---

[1] EN 1729-2:2006 Furniture. Chairs and tables for educational institutions. Safety requirements and test methods.

integration. On the other hand, it was considered important to embed in the enhanced desk vision-based back projection multi-touch interaction, in order to avoid ceiling mounted or hanging projectors and cameras and ensure gesture interaction quality under variable lighting conditions. This option had several important implications, as it was not an easy task to realize a back-projection multi-touch set-up within the available space. The resulting hardware set up includes (Figure 2):

- An Intel Core 2 Quad Core PC
- 2 DLP mini projectors located behind the screen
- 1 mirror for reducing the projection distance
- 2 cameras located behind the screen
- 4 infrared projectors located behind the screen
- 1 camera located on top of the screen and capturing images of the conventional desks
- 1 smart pen and its transmitting device[2]

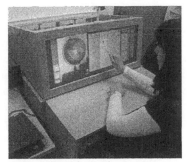

Figure 2. Hardware set up   Figure 3. Multi-touch screen

The PC runs MS Windows 7. To better exploit the custom screen dimensions, a horizontal window manager was developed which includes two application and two menu areas. In order to support multi-touch interaction, the software reported in [14] is used. The smart pen input is captured through the Pegasus SDK[3]. The necessary front vision software is currently under development. The multi-touch display embedded in the desk (Figure 3) allows augmenting physical learning materials with additional context-dependent information.

USAGE SCENARIO AND EDUCATIONAL APPLICATIONS

An important objective of the AmI Classroom project is to ensure that the developed technological artifacts go at the heart of the learning process. To this purpose, learning of English as a foreign language was selected as a testing domain, and current practices were examined to identify useful support which can be offered

---

[2] http://www.gandc.gr

[3] http://www.pegatech.com/_Uploads/Downloads/DevelopersWebSite/index.html

through AmI technologies, focusing on preparation for the first and advanced certificates (thus addressing an adolescent student population). Based on such an analysis, extensive usage scenarios were compiled. The scenarios mainly address activities which take place in the classroom; however, an important consideration is that the software to be developed should be general enough to support also learning at home or elsewhere, independently from the classroom infrastructure and the augmented desk itself. In particular, the scenarios address the seamless context-dependent provision of useful additional information during language-learning activities. Interaction with the provided facilities is based on gestures, either through the desk screen or directly on paper resources. For example, the learner can indicate a word on the page to view additional information about it, or the answer to a fill-the-gap exercise to receive feedback or hints. Text entry during learning activities is based mainly on handwriting using a smart pen. However, an on-screen keyboard is also currently under design to support small text entry tasks on the touch screen. The desk should also be able to identify each user through a personal object (e.g., a school diary, pen, etc).

The developed scenarios led to the identification of a set of software applications for enhancing English learning experience through the augmented desk. These include the login screen, the welcome screen, an individual personal area summarizing the current delivery status of all assignments, a dashboard where students can temporarily save material, the assigned exercises in electronic form with the supported hints and help, a dictionary-thesaurus application, a personal dictionary application, a note-taking applications, an application for viewing course related multimedia, and language-learning games (e.g., hangman). The dictionary-thesaurus presents a short definition of the word, its pronunciation, a button allowing the student to hear the pronunciation, some synonyms and a few examples of use in a sentence. In addition, the thesaurus offers options for extended descriptions, grammar information, complete list of synonyms, antonyms and several examples. Figures from 4 to 7 show some mock-ups of the designed applications.

Although the enhanced desk is designed for individual use, collaboration is foreseen for some tasks. For example, the students can exchange materials and the teacher can send materials (e.g., assignments) to the personal areas of all students through simple gestures. The smart environment automatically restricts actions that can be carried out by the students according to the current context. For example, when a reading task is active, the students cannot use some functionality in their system (e.g., multimedia, saving, printing, etc.), and when an exercise task or a test is being carried out, students are not allowed to send material to other student's personal areas. The system will also produce statistics regarding the hints that have been requested, per student and for the whole class, so the teacher will be able to monitor the student's progress. The system will also monitor the success rate of each exercise and provide statistics. Teacher can then combine this with the hint statistics to review results and adjust the difficulty (remove/alter exercise) in order to improve the learning curve. The system can also measure the time needed by each person to complete a specific task.

FORMATIVE EVALUATION

A formative evaluation experiment was conducted involving 5 young learners of English as a foreign language (1 male and 4 females, within the age range from 11 to 16, all studying for a first or advanced certificate in English, and all familiar with PCs and mobile phones, but not with AmI environments). The experiment was targeted to collect users' opinion regarding:

- The desk itself.
- The overall idea of interactive student desk in the AmI class and how it can assist learning.
- The usefulness of each application regarding the English course, and in particular the thesaurus, the multimedia application, the personal area for assignments and homework delivery, the myVocabulary area, and the hints and confirmations during exercises.
- The User Interface (UI) layout and aspects of the supported gesture interaction.

The experiment took place individually for each learner. After a very brief introduction, the learners were driven through a simplified scenario illustrating the main aspects of desk and of the related applications.

During the execution of the experiment, the children were asked questions about various aspects of the scenario, and notes were taken with all their comments.

Figure 4. The Personal area with assignments and their delivery status

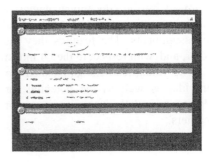

Figure 5. Gradual hints for solving fill-the-gap exercises

616

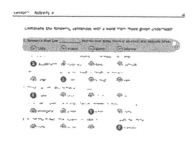

Figure 7. The Thesaurus application

Figure 6. An exercise in electronic form

After the completion of the scenario, they were asked to answer a questionnaire composed of 17 questions. Of the questions, formulated in an informal style to appeal the young learners, 15 used a Likert scale from 1 (sure!) to 5 (no way!). The remaining two questions concerned listing aspects of the scenario that the children particularly liked or disliked. The results of the questionnaire are depicted in figure 8, where lower scores correspond to positive and higher scores to negative answers.

Overall, the results are very positive, as all the children involved in the experiment were very interested about the desk and its applications, and some where enthusiastic about having such an artifact available in their classroom.

The preferred features of the desk were the personal area and the dictionary, followed by the educational games, the dashboard, the hints and confirmations, touch interaction, pointing at things and viewing info, and the electronic submission of assignments. The features they disliked most were mainly the desk size and color, and, with respect to the UI mockups, again the colors and the fonts.

The young learners appreciated the educational support which the desk aims to provide, as well as the potential for better organization of work between the classroom and the home environment and collaboration with teachers and other learners.

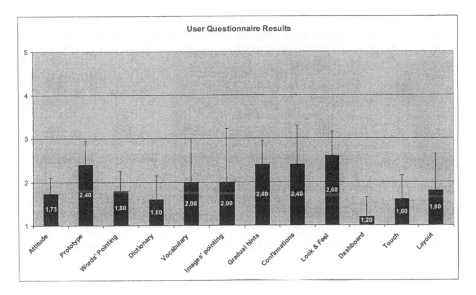

Figure 8. Questionnaire results

Some of the children also proposed to include a grammar application similar to the thesaurus application, displaying grammar rules, verb tables, etc, related to the task at hand. On the other hand, the young learners appeared to view the desk as a 'trendy gadget' and demonstrated to be very sensitive to aesthetic issues, asking for the possibility of personalizing the desk color, selecting fonts, colors, background images and avatar images. Regarding interaction, they appreciated the gesture-based applications, but they also asked for more traditional interaction means such as the keyboard.

CONCLUSIONS

The results of the conducted study confirmed that AmI technologies have the potential to enhance the classroom learning experience and to be positively viewed by young learners, provided that they are carefully designed and tested. Adolescents seem to be aware of the opportunities offered by novel technologies and willing to embrace them, but also face classroom technologies as fancy personal gadgets which should be colorful, aesthetically pleasant, and fun. They also appear aware of potential technological difficulties, as well as space and practical limitations in the classroom environment. On the other hand, young learners easily grasp the potential of novel interaction techniques, such as gestures, but do not ignore more traditional means with which they are familiar already.

Currently, the prototype desk has been assembled and all the necessary software building blocks have been installed. Fully functional applications are being developed, along with the vision software required for supporting page and word

618

recognition on paper material. Following full implementation, a larger evaluation experiment is being planned, involving also, besides children, teachers and parents. On the other hand, new collaborative games scenarios are being elaborated in order to fully exploit multi-touch interaction.

The designed enhanced desk is intended to constitute a significant part of a complete AmI classroom environment which will be realized in an AmI Research Facility currently under construction at ICS-FORTH.

Acknowledgements. This work is supported by the ICS-FORTH internal RTD Programme 'Ambient Intelligence and Smart Environments'.

REFERENCES

1. Abrami P. C., Bernard R. M., Wade C. A., Schmid R. F., Borokhovski E., Tamim R., Surkes M., Lowerison G., Zhang D., Nicolaidou I., Newman S., Wozney L. & Peretiatkowicz, A. (2006), *A Review of E-learning in Canada: A Rough Sketch of the Evidence, Gaps and Promising Directions*. http://ccl-cca.ca/NR/rdonlyres/FE77E704-D207-4511-8F74-E3FFE9A75E7E/0/SFRElearningConcordiaApr06.pdf
2. Bates A. W. (2005), Emerging trends: convergence and specialization in distance education, In: Technology, e-learning and distance education (pp. 6-9), Routledge Taylor & Francis Group (ISBN: 0-415-28437-6)
3. Brooks C., Greer J., Melis E., and Ullrich C. (2006), Combining ITS and eLearning Technologies: Opportunities and Challenges, In: Proceedings of the 8th International Conference on Intelligent Tutoring Systems (ITS '06), pp. 278-287
4. Brusilovsky, P. 2004. KnowledgeTree: a distributed architecture for adaptive e-learning. In Proceedings of the 13th international World Wide Web Conference on Alternate Track Papers &Amp; Posters (New York, NY, USA, May 19 - 21, 2004). WWW Alt. '04. ACM, New York, NY, 104-113.
5. Cook, D. J., Augusto, J. C., Jakkula, V. R. (2009). Ambient intelligence: Technologies, applications, and opportunities. *Pervasive and Mobile Computing*, Volume 5, Issue 4, August 2009, Pages 277-298.
6. Cook, D. J., Das, S.K. (2007). How smart are our environments? An updated look at the state of the art, *Journal of Pervasive and Mobile Computing* 3 (2) (2007), pp. 53–73.
7. Cooperstock, J. (2001): Classroom of the Future: Enhancing Education through Augmented Reality. In: *Proc. Conf. Human-Computer Interaction* (HCI Int'l 2001), pp. 688–692. Lawrence Erlbaum Assoc, Mahwah
8. Druin, A. (2002). The Role of Children in the Design of New Technology. *Behaviour and Information Technology*, 21(1) 1-25.
9. Emiliani, P.L., Stephanidis, C. (2005). Universal Access to Ambient Intelligence Environments: Opportunities and Challenges for People with Disabilities. *IBM Systems Journal*, Special Issue on Accessibility, 44 (3), 605-619.

10. Grammenos, D., Zabulis, X., Argyros, A.A., Stephanidis, C. (2009). FORTH-ICS internal RTD Programme 'Ambient Intelligence and Smart Environments, In the *Proceedings of the 3rd European Conference on Ambient Intelligence,* Salzburg, Austria, November 18-21, 2009.

11. Gros B. (2007) Digital games in education: the design of game-based learning environments. *Journal of Research on Technology in Education* 40, 23–38.

12. Heift, T., Schultze, M. (2007). *Errors and intelligence in computer-assisted language learning: parsers and pedagogues.* Routledge, Taylor&Francis, NY.

13. IST Advisory Group (2003). *Ambient Intelligence: from vision to reality.* Electronically available at: tp://ftp.cordis.lu/pub/ist/docs/istag-ist2003 consolidated report.pdf

14. Michel, D., Argyros, A. A., Grammenos, D., Zabulis, X., Sarmis, T. (2009). Building a multi-touch display based on computer vision techniques. In *Proceedings of the IAPR Conference on Machine Vision and Applications* (MVA'09), pp. 74-77, Keio University, Japan, May 20-22, 2009.

15. Nousiainen, T., Kankaanranta, M. (2008). Exploring Children's Requirements for Game-Based Learning Environments. *Advances in Human-Computer Interaction.* Volume 2008, Article ID 284056, 7 pages.

16. Quintana, C., Soloway, E., Norris, C. (2001). Learner-Centered Design: Developing Software That Scaffolds Learning, Advanced Learning Technologies, *Second IEEE International Conference on Advanced Learning Technologies* (ICALT'01), 2001, pp. 0499.

17. Robinson, J., A. Robertson, C. (2001). The LivePaper system: augmenting paper on an enhanced tabletop, *Computers & Graphics*, Volume 25, Issue 5, October 2001, Pages 731-743.

18. Shi, Y., Xie, E.A. (2003): The smart classroom: Merging technologies for seamless tele-education. *IEEE Pervasive Computing Magazine.*

19. Xu P., Han G., Li W., Wu Z., Zhou M. (2009). Towards Intelligent Interaction in Classroom, In: *Universal Access in Human-Computer Interaction. Applications and Services* (pp 150-156), Berlin: Springer.

Chapter 68

Cultural Modelling of Remote Communities

Alvin W. Yeo[1], Edwin Mit[1], Po-Chan Chiu[2],
Jane Labadin[2], Ping-Ping Tan[2]

[1]Centre of Excellence for Rural Informatic
[2]Faculty of Computer Science and Information Technolog
Universiti Malaysia Sarawa
94300 Kota Samarahan, Sarawak, Malaysia

ABSTRACT

There is an increasing number of initiatives in bridging the digital divide. This is reflected in a UNESCAP estimate, that half a million telecentres will be required in the Asia Pacific region. As such, there would be an increase in demand of software applications to serve these rural communities. However, existing software methodologies predominantly are targeted at organisations and communities in the urban areas. These methodologies generally do not accommodate target audiences such as those living in the rural areas. In this paper, we propose a generic cultural model which accommodates the different communities and explains the model using two remote communities in Borneo, East Malaysia. The resulting model can then be used to develop prototypes of software for these rural communities.

Keywords: User modelling, rural and remote communities, community-centred design methodologies, digital divide

INTRODUCTION

It is generally acknowledged today that there is a growing number of initiatives to bridge the digital divide. This growth is reflected in the increasing number of telecentres being built and the increasing amount of funding invested by both governments and non-government organisations to address the issues brought about by the digital divide. A telecentre is one way to provide access to ICTs for the underserved communities, such as the remote and rural communities. According to the United Nations Economic and Social Commission for the Asia Pacific (UNESCAP), approximately 476,896 telecentres are needed in providing access to the people in the rural area in the Asia Pacific region (UNESCAP, 2008). If the (conservative) cost of USD50,000 is required to build a telecentre in the remote area, the telecentres for just the Asia Pacific alone would cost USD5 billion.

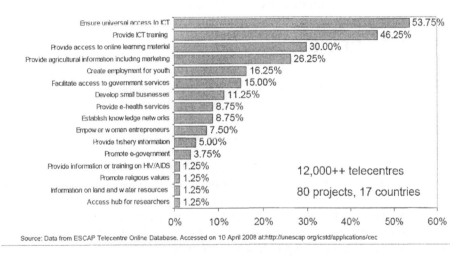

Source: Data from ESCAP Telecentre Online Database. Accessed on 10 April 2008 at:http://unescap.org/icstd/applications/cec

Figure 1: Usage of Telecentres in the Asia Pacific Region

Also, in a survey conducted by UNESCAP, it was reported that out of 12,000 telecentres in 17 countries (see Figure 1), more than half (54%) of the telecentres provide access to ICTs; and another 46% provide ICT training (UNESCAP, 2008). The other activities that are conducted at the telecentres suggest that there are numerous software applications that have been developed and provided. An interesting observation of these figures is that while much research has been conducted and knowledge generated from the implementation of telecentres all over the world, there is little work being carried out to understand the target audience of the telecentres, i.e. remote communities in rural areas. In addition, there is little reported work which encompasses the study and development of software for these remote communities. Much of the existing research are in the context of the urban dwellers, and includes work related to project management. Since the software is now to be deployed in the rural areas, there is a need to accommodate this new audience and context into the current software development process. In addition, this research would encompasses the need to study how these remote communities work, how they interact with each other, and also their way of life.

In this paper, we attempt to extend existing user models employed in software development to incorporate cultural components, given the different target audience of our software applications, i.e. the rural communities of different ethnicities. Existing software construction applications such as IBM Rational Rose, provides user modelling tools which typically cover components such as skills, roles, goals, tasks, skill sets, and user domains. We aim to include cultural components which pertain to the different ethnic groups, and to leverage on these differences in creating, a "richer" user profile. With these additions, it is hoped that software applications that meet the needs and capabilities of rural communities can be achieved. In the next section, we cover the existing tools and relevant research pertaining to development of software, specifically on user modelling.

LITERATURE REVIEW

The software development fraternity uses the software development lifecycle which basically depicts the significant phases or activities of a software development project from its conception until the completion of the product.

There are many methodologies of the software development lifecycle. The Agile methodology provides many opportunities to assess the direction of a project throughout the development lifecycle (Agile methodology, 2010). It promotes teamwork, collaboration and process adaptability throughout the lifecycle of the project. According to Su Kong (2007), organisational culture affects the organisational adoption of agile methodology. Su Kong (2007) also hypothesised that if an organisation's culture, reflected by its organisation members' language and values, and many layers of bureaucracy, then the organisation's culture is not compatible with agile culture.

Another methodology is the structured systems analysis and design method (SSADM). SSADM uses a combination of text and diagrams throughout the whole lifecycle of a system design, from the initial design idea to the actual physical design of the application (Webopedia.com, 2010). However, the SSADM approach does not emphasise on user-centered design throughout development process.

Unlike SSADM that builds the systems based on functions, object oriented (OO) system development takes objects as the basis all through the analysis and design stages where data structures and methods are both considered. These objects can be easily replaced, modified, and reused (Bahrami 1999). Ochimizu (n.d.) stated that the superiority of OO approach is the localisation of change effect (data abstraction or information hiding). This is crucial when cultural aspects are involved, as cultures evolve with time and very much depend on the local communities. The OO approach in software development was first introduced without standard notations. Only when the Object Management Group

(OMG) formally accepted the Unified Modelling Language (UML), did the OO approach have a standard modelling method. Due to the similarity of the OO approach in representing real life object, UML would be considered a very suitable modelling method to represent the cultural needs of the rural communities. Another advantage of the OO approach (Ochimizu n.d.) is the reuse of sub-classes. By using UML to model a culture for a specific community, and through OO localisation approach, the notation can be reused to model other cultures from other community.

The methodologies as established above are based on software which are developed for urbanites that have systems problems and where software developers will provide solutions for them. Unlike the urban dwellers, rural communities may not have experience using any software or system, and thus may have difficulties visualising its usage. In addition, the existing methodologies are also targeted typically at organisations, with well-defined structures and processes; both of which may not exist in the remote communities.

Interestingly, Preece (2000) has introduced a community-centered development approach which is a methodology that involves the members of the community in a participatory design process with the developers. However, the methodology introduced by Preece (2000) concerns only for online information system for the target community which comprises urbanites.

In addition, the existing software development methodologies do not incorporate cultural components into software development lifecycle which encompasses a study of how the remote communities think, how they interact with each other, how they use ICTs and how they live their lives. Thus, how we develop software for the rural communities will be different from that of conventional methodologies. There is a need to create "richer" user profiles and incorporate cultural components that involves rural communities into existing software development lifecycle.

The next section provides a model which was developed using the OO approach, based on an ethnic group, an instance of a remote and rural community.

CULTURAL MODELLING

Urbanisation and globalisation have drastically changed the lifestyle of the people in many communities. Remote villages are still generally rich in culture, and still retain the traditions that distinguish the different communities from one another. Preservation efforts of this diversity are necessary to ensure the continued existence of these cultures. For example, these cultures should be modeled, and reused in the development of software for other remote communities. To do this, we propose the cultural model.

We define a *Cultural Model* as a model that describes the cultures, such as taboos which are reflected in the activities of the remote and rural communities. A study has been carried out on these communities, in trying to model these rituals, activities and taboos, which are associated to every stage of their lives. It encompasses aspects such as the marriage process, having their own family and children, a growing child, beliefs, leader selection, song or chants, selection of place to stay, festival, related song and dance, economic activities, legends, and death. To obtain more accurate information, a field trip to the remote community is necessary to gather the necessary information about the community, in this case, the Penans.

REMOTE COMMUNITY SELECTED

We selected Long Lamai, a remote Penan rural village which the Centre of Excellence for Rural Informatics (COERI), Universiti Malaysia Sarawak, is working with. This community is located in Sarawak, a State in East Malaysia. The community is geographically isolated. To reach Long Lamai from a major town, one has to take an eight-hour journey on the logging road (untarred) and another hour by long boat upriver. There are 300 Penans living in the village, and they have been settled there more than 40 years. Unlike their nomadic counterparts, they have farms, grow hill padi, plant vegetables, and still carry the older tradition of hunting. Long Lamai at present has no access to power, and was only provided access to ICTs in 2009 by COERI.

METHODOLOGY: DATA COLLECTION

The most powerful tool used, to capture the information is open-ended interview with pre-defined questions. The interviews were conducted with the community leader, religion leader and the old folks. The researchers stayed with the community between 5 to 7 days for each visit; this follows a similar approach used by Mukerji (2009). However, Mukerji (2009) group the households into same socio-economic parameters (e.g., class, occupation, caste, religion, education, age, gender, institutional and political affiliations). Such grouping would result in loss of much information, which is required in the development of the software. In addition, the 'think-aloud', observation and ethnography approaches complement the process.

In the section, the information is analysed and modelled using UML notation. The prototype will be developed to show the remote communities' cultures and activities. The UML model in Figure 2 (see Appendix) shows the static model of a community based on their daily life and culture. This model includes how a member of the community looks for a partner in life, and the marriage process; they believe in fate that determines the success and happiness of a marriage. Fortune is determined by looking at a bird, for example, an eagle, and how it is flying during the marriage ceremony. This only can be done by the leader. The model also shows the birth process, naming of children, the growing child and what they should learn, for example, hunting and working at farm, how they select their leader and place to stay. The tradition of selecting a place to stay was only practised when they were still nomadic. Their beliefs, religion, songs, chants, and dances that are related to a particular festival is also modeled, as well as the rituals and taboos related to death.

BENEFITS OF CULTURAL MODELLING

After the cultural models of remote communities have been created, the software application can be developed to show the cultures and related activities. Every ethnic group has its own unique cultures, and therefore web applications developed for this community will be rich in their cultures. This will definitely boosts their visibility in the tourism industry. At the same time this will help to promote their unique artefacts to the tourist. There are number of future activities planned in this remote community such as home stay, jungle trekking, boat cruising or rafting, fishing, picnic, and cultural night. These may be organised in their unique traditional ways. For example, the Penan community at Long Lamai has built homestays based on their very traditional hut, where the tourist can really experience how they live their lives, i.e. staying in the open hut in the middle of the jungle.

EXPLANATION AND APPLICATION OF THE CULTURAL MODEL

In order to show that the cultural model in Figure 1 can be easily applied to another remote community, an explanation of the model will be provided using the Dusun community from Buayan village, another remote village situated in the north of Borneo island. Figure 2 shows People as the main class, depicting the Dusun people. Each individual (or object) has attributes such as name, identity card number and religion. Universally, a baby inherits features from the parents (represented as People class). Baby is produced through marriage.

Similar to the Penan communities, Dusun has their own language and their everyday activities include fishing, hunting, and farming. The model depicts these activities as separate classes because the Dusun community will have their own methods of performing those activities compared to the Penan communities where this model is derived from. Similarities and differences of the communities can be represented using the object behaviors and methods. In order to perform those activities, the community will have their own equipment.

In the proposed cultural model, the *a-part-of* relationship classes are provided as examples to better understand the model. When this model is applied, these classes can be added or removed according the communities' needs. In the Penan communities, blow pipe is used as weapons but in the Dusun community, blowpipe is not their weapon of choice. Instead, they use the spear as their weapon.

We can further observe this model of the Dusun community through their ceremonies. Both Penan and Dusun communities have their respective ceremonies. The model represented the ceremonies as a class of its own. This shows that both communities take part in ceremonies but they will have different methods to perform the ceremonies. The model again is valid as the Dusun community have their own marriage and death ceremonies, their own songs, music instruments and dances. Polymorphism in OO approach provides the capability to model these cultural needs of the communities. The same applies to taboos.

The cultural model proposed in this paper highlights that the cultural aspect is indeed a big factor for consideration during software development, which is absent in existing software development methodologies.

DISCUSSION AND CONCLUSION

As earlier mentioned, remote communities most likely have little experience in ICT or none at all. Thus, they will not have the need for software unless it will improve their livelihood. Farmers, for instance, will benefit from accessing information on agricultural techniques and innovations, as well as crop prices, if such information is available in their local language. Therefore, farmers can increase their productivity and sell their products at a better price. Apart from this, the remote communities can promote their culture via the Internet by selling their handicrafts and other ethnic products. This means that the software applications that one can develop for the remote communities need to improve the livelihood of the community. Thus, understanding the way they think, their way of life is crucial during the software development.

We have presented a generic cultural model with the objective to model the cultural aspects of the community. Having such model incorporated into our software development lifecycle, will ensure that their cultural aspects are captured and taken into account when developing such software. In addition, we also recommend local participation and an incremental approach in content creation and software development. Using this approach will ensure the development of quality software for the once neglected audience, the remote and rural communities.

ACKNOWLEDGMENT

The research was partially supported by the Ministry of Science Technology and Innovation, Malaysia, Demonstrators Application Grant Scheme: eBario Roll-Out.

REFERENCES

Agile methodology (2010). *Understanding Agile Methodology*. Retrieve February 24, 2010 from http://agilemethodology.org/

Bahrami, A. (1999). *Object Oriented Systems Development – Using the Unified Modeling Language*. The McGraw-Hill Companies. Singapore. Pg. 4.

Dennis, A., Wixom, B.H. and Tegarden, D. (2002). *Systems Analysis & Design – An Object Oriented Approach with UML*. USA: John Willey & Sons

Endsley, M.R. (2004). "Situation awareness: Progress and directions." In S. Banbury and S. Tremblay (Eds.), A Cognitive Approach to Situation Awareness: Theory and Application. Aldershot:Ashgate Publishing, 317-341.

Lewis, P. (2003). *The Object Oriented Approach*. Available online: http://users.ecs.soton.ac.uk/phl/ctit/oo/node3.html

Mukerji, M. (2009). ICTs and Development: A Study of Telecentres in Rural India. *10th International Conference on Social Implications of Computers in Developing Countries*. Dubai, United Arab Emirates.

Ochimizu, K. (n.d.). *Superiority of Object Oriented Approach (History and Perspectives)*. School of Information Science.

Preece, J. (2000). *Online Communities: Designing Usability, Supporting Sociability*. Chichester, UK: John Wiley & Sons

Su Kong (2007). *Agile Software Development Methodology: Effects on Perceived Software Quality and the Cultural Context for Organisational Adoption*. Dissertation. The State University of New Jersey, New Jersey.

Webopedia.com (2010). Retrieved February 24, 2010 from http://www.webopedia.com/TERM/S/SSADM.html

628

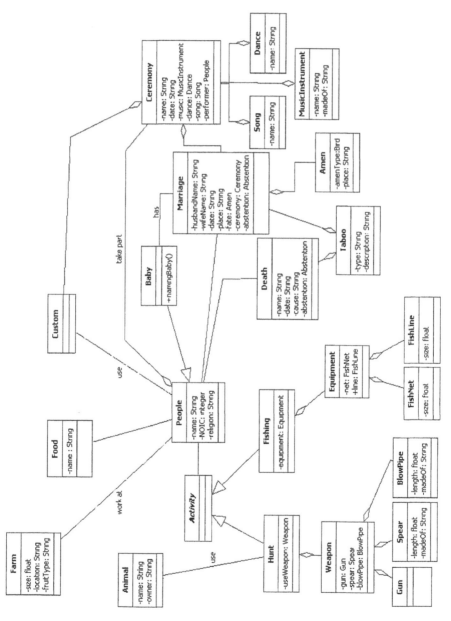

Figure 2 Unified Modelling Language (UML) Cultural Model

Chapter 69

Beyond Usability: Ordering e-Service Quality Factors

Sandrine Prom Tep[1,] Aude Dufresne[2]

[1]RBC Financial Group e-commerce Chair of HEC Montréal
3000 Chemin Côte-Ste-Catherine
Montréal, QC H3T 2A7, Canada
sandrine.prom-tep@hec.ca

[2]Département de Communication
Université de Montréal
Case Postale 6128, Station Centre-ville
Montréal, QC H3C 3J7, Canada
aude.dufresne@umontreal.ca

ABSTRACT

As consumers increasingly use the Internet to shop or search for product-related information, website effectiveness has become critical for companies offering online transactions or information that could lead to a purchase. The HCI literature on usability has mainly focused on either the intrinsic qualities of a website interface or the level of task performance it enables. More recently, several studies have used a combination of both approaches, and some have incorporated the marketing-specific notion of usability as part of service quality. This study extends this research, examining various other aspects of website quality in addition to usability (quality and quantity of information, visual design, information security and personalization of interactive features). This study confirms the validity of the Netqu@l scale for evaluation of online service quality by a factorial analysis conducted on a sample of over 4,000 subjects evaluating 21 websites in four industry sectors. In addition, this study examines the relative contributions of five components of service quality to website effectiveness (task completion) and user satisfaction. Discriminant and regression analyses were used to rank the factors promoting online service quality, clearly establishing the key importance of usability, both in terms of task performance and consumer perceptions.

Keywords: Quality of online services, usability, website effectiveness

INTRODUCTION

The media never cease to remind us that e-commerce has become indispensable. According to eMarketer projections for 2008, 63% of Canadian Internet users seek information online before buying a product (eMarketer, 2009) and 51% complete the purchase online in the US (eMarketer, 2010). But although this new medium is more essential every day, the quality of the services it provides remains highly variable. As pointed out by Agarwal and Venkatesh (2002), companies need to "understand the barriers and facilitators of online purchasing and to construct websites that can turn visitors into paying customers." This is especially true as consumers increasingly search for product information online, even if the final purchase is not completed by Internet.

Websites should therefore strive for quality in terms of both the transactions offered and the information provided. Although the importance of website quality and the multi-dimensionality of the "quality" construct are well-recognized, there is less agreement on proposed conceptual definitions of online service quality and corresponding assessment instruments (Bressoles, 2004; Parasuraman et al., 2005). However, one factor that has consistently been included in these definitions of quality is "usability" (or "ease of use" or "navigability"). The objective of this article is not to debate the emerging definition of online service quality, but rather to focus specifically on the various factors that contribute to online service quality and attempt to rank them in order of importance.

THE QUALITY OF ONLINE SERVICES

THE CONSUMER PERSPECTIVE

To define what constitutes the quality of an online service, it is important to clarify whether we are considering the question from an organizational (expert) perspective or individual (consumer) perspective. As noted by Basch (1992), "Quality has a lot to do with perceived value, with what the consumer considers 'quality goods'... Given this emphasis on user satisfaction, it's only natural for the online industry's customers to help define performance measures for the products they buy and use." In this study, online service quality was measured directly by consumers and corresponds to the website users' perception of its quality. As Internet services are delivered by a system, not a person (Hoffman and Novak, 1996), we must examine the users' interaction with that system to measure the quality of service provided.

USABILITY

The ISO 9241-11 quality standard provides the most classic and widely recognized definition of usability: "The extent to which a product can be used by specified users to achieve specified goals with effectiveness, efficiency and satisfaction in a specified context of use" (International Standard Organization, 1998). The HCI literature approaches usability issues in various ways, but has mainly focused on the intrinsic qualities of the interface (Ravden and Johnson, 1989; Bastien and Scapin, 1993; Shahizan and Li, 2007) or the level of task performance it enables (Nielsen, 1993; Bevan, 1995). More recently, pushed by increasing interest in usability from new fields of research, several studies have used a combination of both approaches (Palmer, 2002; Seffah et al., 2006; Venkatesh and Agarwal, 2006), and some incorporate the marketing-specific notion of usability as part of service quality (Bressoles, 2004; Parasuraman et al., 2005; Nantel et al., 2005). This trend toward increasing interest for the user-consumer within usability research has broadened to include research in the social sciences such as communications, anthropology and marketing (particularly consumer research). These fields now include the user-consumer in a wider conceptual framework incorporating symbolic dimensions related to culture, identity and emotions (Zhang and Li, 2005).

THE ONLINE EXPERIENCE AND WEBSITE EFFECTIVENESS

It has been recognized for some time that "creating a compelling online experience for cyber customers is critical to creating competitive advantage on the Internet" (Novak et al., 2000). It was to examine the user perception of the experiential quality of the interaction between consumers and websites that this study was undertaken. Indeed, although quality is in itself a clearly desirable objective for any online service, we sought to determine to what extent perceived quality is the combined product of usability, effectiveness and other more experiential factors.

Whereas analytical studies of website effectiveness rely on expert analysis, this study is empirical in nature, based on the observation of test users interacting with the website to complete a specific task. The effectiveness of each website was evaluated on the objective measurement of success or failure in task completion, namely finding a particular piece of information about a representative product. For example, study participants randomly assigned websites in the renovation sector were asked to find the price of the least expensive jigsaw offered on that site.

This study draws on data collected as part of a larger study on website effectiveness designed to identify the most effective sites in major industry sectors (Nantel et al., 2005). Here, we focus on the components of service quality and how they combine to foster website effectiveness and user satisfaction, to derive general conclusions as a contribution to our understanding of online service quality.

THE NETQU@L CONCEPTUAL MODEL

The Netqu@l scale was used to evaluate user perception of the quality of online services offered on each website (Bressoles, 2004). Five factors were measured: ease of use, quality and quantity of information, visual design, trust in the security of information entrusted to the site and personalization of the site's interactive features. Our empirical approach included two attitudinal dimensions (evaluation of perceived quality factors such as usability and overall satisfaction) and one behavioural dimension (task completion) as a measure of effectiveness.

STUDY VARIABLES

Three types of variables were examined:
1. **online service quality,** as measured by the five factors in the Netqu@l scale, including ease of use (usability);
2. **overall satisfaction** with the website, based on a reliable and validated five-item scale to measure attitudes toward a website (Novak et al., 2000).
3. **website effectiveness** based on task completion. Classification as "success" or "failure" was based on user answers after being asked to search for specific information on the website. "Failure" applied for responses such as, "I don't know" and all unmistakable errors, such as indicating "$0" as the price of the least expensive jigsaw.

ANALYTICAL FRAMEWORK

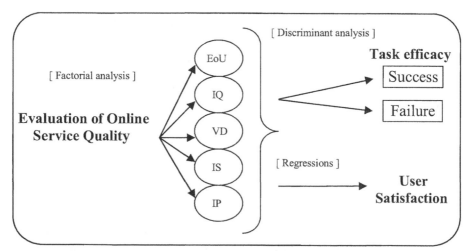

FIGURE 1 Research model. EoU, ease of use; IQ, information quality; VD, visual design; IS, information security; IP, interaction personalization.

Figure 1 illustrates our research model in conceptual and methodological terms.

Factorial analysis was first used to examine the relative contributions of the five factors of online service quality with our study sample. Next, used descriminant analysis o determine the extent to which each factor in the online service quality scale correlated with task outcome. Finally, multiple regression analysis was used to examine to what extent perceived website quality can explain overall user satisfaction, and which quality factors were most highly correlated with user satisfaction. Our hypothesis was that online service quality, particularly usability, is a good indicator of user satisfaction.

METHODS

STUDY SAMPLE AND DATA COLLECTION

The data for this study were collected as part of a larger study conducted jointly by the RBC Groupe Financier E-commerce Chair at the École des Hautes Études Commerciales of Montréal and the Léger Marketing firm (Bressoles and Nantel, 2005). A total of 4144 test website users were recruited at random from the Léger Marketing consumer panel, representing a wide range of experience with Internet navigation in general and with the websites used in the study. Each month from September 2004 through January 2005, approximately 1000 test users participated in a survey, split almost equally among four industry sectors. The B2C (business to consumer) websites of 21 major Canadian companies representing the electronics, travel, renovation and financial industries were examined (Table 1).

Table 1 Company Websites by Industry Sector

Financial Institutions	Electronics	Renovation	Travel
Desjardins	Future Shop	Home Depot	Destina.ca
National Bank of Canada	Best Buy	Rona	Exitravel.com
ING Direct	La Source	Canadian Tire	Expedia.ca
Toronto Dominion	Brick	Home Hardware	Travelocity.ca
Scotia Bank	Dumoulin		
CIBC			
BMO			
RBC Royal Bank			

For each website representing a particular industry, the study participants were asked to perform a plausible and similar search task. Specifically, for the electronics industry, they were to find the lowest priced 4-megapixel digital camera; for the travel sites, the best possible price for a trip to Cancun from May 1st to the 8th, 2005; for the renovation sites, the least expensive jigsaw; and for the financial institutions, the interest rate for a one-year fixed-rate mortgage.

After having succeeded (or failed) in accomplishing their task, each test user was asked to complete an online questionnaire to assess the quality of the website's online service (Table 2), indicate their overall satisfaction with the site, and answer a few socio-demographic questions. The test users participated from their homes and were remunerated for their participation.

Table 2 Online Service Quality and User Satisfaction Questionnaires

NetQu@l Scale: 7-point scale from "strongly disagree" to "strongly agree"	
Ease of Use (EoU)	
1.	This website is easy to use.
2.	It is easy to find information on this website.
3.	It is easy to browse and find what you're looking for on this website.
4.	This website's structure and layout make finding information easy.
5.	This website's layout is clear and simple.
Information quality and quantity (IQ)	
6.	This website offers detailed information on products or services.
7.	The information on this website is pertinent.
8.	The information on this website is accurate.
Visual design (VD)	
9.	This website is attractive.
10.	This website shows creativity.
11.	This website is visually appealing.
Trust in information security (IS)	
12.	Overall, I trust this website's security measures.
13.	This website guarantees that I can surf safely.
14.	I think that my private life is protected on this website.
15.	I trust this website not to use my personal information indiscriminately.
Interaction personalization (IP)	
16.	I can interact with this website and receive personalized information.
17.	This website helps me fulfill my specific needs.
18.	This website has interactive features that help me with navigating.
19.	This site saves my preferences and offers me additional services or information based on these preferences.
Chen and Wells Scale (1999): 5-point scale from "definitely disagree" to "definitely agree"	
1.	This website makes it is easy for me to build a relationship with this company.
2.	I would like to visit this site again in the future.
3.	I am satisfied with the service provided by this website.
4.	I am comfortable surfing on this website.
5.	I feel surfing the website is a good way for me to spend my time.
6.	Compared with other websites, I would rate this site as: (5-point scale from "one of the worst" to "one of the best")

ANALYSIS AND RESULTS

FACTORIAL ANALYSIS OF ONLINE SERVICE QUALITY FACTORS

The survey results were first analyzed to determine whether the five factors of the scale indeed helped determine online service quality. Principal component factorial analysis, with varimax orthogonal rotation, of the 19 items of the Netsqu@l scale clearly indicated that the five quality factors together accounted for 86.5% of the total variance in the following order: ease of use accounted for 25% of total variance; information security, 23%; visual design, 15%; information quality, 12%; and interaction personalization, 11.5%. Although ease of use was of only slightly greater importance to the test users than trust in information security, this finding confirms published reports of its dominant role in determining the quality of online services.

FACTORS CONTRIBUTING TO TASK COMPLETION

Discriminant analysis was then conducted to determine whether the five quality factors contributed to discrimination between success or failure in task completion. This analysis revealed a very significant ($p = 0.000$) discriminant function and allowed the online service quality factors to be ranked in order of their contribution to website effectiveness. As can be seen in the structure matrix (Table 3), ease of use had the highest correlation with the discriminant function (0.898). In comparison, trust in information security and visual design contributed the least to discrimination between success and failure in task completion.

Table 3 Discriminant Analysis Results

Service quality factor	Function Correlation
Ease of use	0.898
Information quality and quantity	0.838
Interaction personalization	0.655
Trust in information security	0.362
Visual design	0.308

The classification of results showed 74.6% successful classification by cross-validation (the most conservative computational method). This result is much greater than the chance ratio (50%) or the proportional chance criterion (70%), confirming the dominant role of usability in the online service quality construct. This result also indicates that, beyond ease of use, it is very important to provide good information in sufficient quantities, because the information factor often impacts the usefulness or pertinence of the site and contributes greatly to user perception of service quality.

FACTORS CONTRIBUTING TO USER SATISFACTION

Finally, we examined how closely the various factors of the online service quality construct were related to overall user satisfaction. First, using multiple regression analysis, we were able to confirm that the five factors of perceived quality (overall model) were closely correlated with satisfaction ($R^2 = 76.4\%$), with highly significant coefficients ($p = 0.000$). Simple regression analysis for each factor demonstrated that personalization of interaction ($R^2 = 0.634$) was most closely correlated with user satisfaction, just ahead of ease of use ($R^2 = 0.618$).

Table 4 Linear Regression Results

Service quality factors	Adjusted R^2
Ease of Use	0.618
Information quality and quantity	0.497
Visual design	0.458
Trust in information security	0.322
Personalization of interaction	0.634
Overall model of online service quality	0.764

CONCLUSIONS AND IMPLICATIONS

Although the quality of online services may be defined in various ways, certain factors such as usability stand out as central to the quality construct. These factors definitely influence consumer perception of quality, which in turn positively influences their overall satisfaction with a website and correlates with website effectiveness — the ability to inform users and allow them to complete tasks related to the site's intended purpose. This study also shows that among the five components of service quality we measured, ease of use showed the highest correlation with website effectiveness and was also one of the factors most closely associated with user satisfaction, slightly ahead of personalization of interaction.

These findings are of particular interest because they are based on a large (n = 4144) representative sample of Internet users drawn from the population at large, rather than relying on university student participation for the sake of convenience. In addition, this study used a large sample of websites (n = 21), each highly representative of their industry sector, being among the largest and most popular sites, which strengthens the external validity of the results.

Designing a quality website is crucial to the success of companies that offer an online storefront. As creating large commercial websites necessarily involves trade-offs among the available resources (time, personnel, budget, existing technology, etc.), managers need to know which factors are most important in designing a high quality site that is effective in building customer loyalty, both online and off. This study delivers a clear recommendation: all Internet storefronts should prioritize ease of use, information quality and personalization.

LIMITATIONS AND FUTURE RESEARCH

This study is not an exception to the rule and some limitations of its methodological and conceptual approach bear mentioning. In terms of methods, it is important to emphasize that the test users evaluated the website after having visited it and searched for certain information; the user satisfaction scores could thus be influenced by the users' success or failure in completing the requested task. Although the results could consequently be considered an interpretation of their success or failure, they nonetheless represent the aspects of service quality to which the user attributed their success or failure and which they would probably like to see improved.

In conceptual terms, this study confirms the positive correlation between perceived usability, as a major factor in website quality, and the website's capacity to satisfy users and enable them to accomplish the intended purpose of their visit. This study also highlights the great importance of the quality and quantity of information as a contributing factor to website effectiveness and of personalization of the interactive features as the primary contributor to overall user satisfaction. However, as previously noted, the task used here was an information search, that is, a specific goal-oriented task. Future research should examine whether the results are the same for more experiential browsing, less oriented toward a specific goal. Such research could determine the extent to which the nature of the task might influence the relative importance of visual design, which in other circumstances may well constitute a quality factor which contributes more to user satisfaction, if not to website effectiveness.

ACKNOWLEGEMENTS

The authors would like to thank the RBC Groupe Financier E-commerce Chair at the École des Hautes Études Commerciales of Montréal for permission to use the data from their study.

REFERENCES

Agarwal, R. and Venkatesh, V. (2002), "Assessing a Firm's Web Presence: A Heuristic Evaluation Procedure for the Measurement of Usability." *Information Systems Research,* 13(2), 168–186.

Basch, R. (1992), "The Seven Deadly Sins of Online Services." *Online*, July, p. 22–25.

Bastien, J.M.C. and Scapin, D.L. (1993), *Ergonomic criteria for the evaluation of human-computer interfaces*, Rapport technique No. 156, Institute Nationale de Recherche en Informatique et en Automatique, Le Chesnay, France.

638

Bevan, N. (1995), "Measuring usability as quality of use." *Software Quality Journal*, 4 (11), 115–130.

Bressoles, G. (2004), *La qualité de service électronique, Netqu@l: Mesure, conséquences et variables modératrices* (Doctoral dissertation), Université de Toulouse, France.

Bressoles, G. and Nantel, J. (2005), *Vers une typologie des sites Webs destinés aux consommateurs*. Cahier de recherche No. 05-03, Chaire de commerce électronique RBC Groupe Financier, HEC Montréal.

Chen Q. and Wells W.D. (1999), "Attitude Toward the Site." *Journal of Advertising Research*, 39 (5), 27–37.

eMarketer (2009) US Online Activities of US Internet Users, Q3 2009 (% of respondents), retrieved from emarketer.com.

eMarketer (2010) Online Buyers 2008 & 2009, retrieved from emarketer.com.

Hoffman D.L. and Novak T.P. (1996), "Marketing in Hypermedia Computer-Mediated Environments: Conceptual Foundations." *Journal of Marketing*, 60 (3), 50–68.

International Standard Organization (1998), *ISO 9241-11*. The ISO Website: http://www.iso.org/

Nantel, J., Mekki Berrada, A. and Bressolles, G. (2005), "L'efficacité des sites Web : quand les consommateurs s'en mêlent." *Revue Gestion*, 30 (1), 16–23.

Nielsen, J. (1993), *Usability engineering*. AP Professional, Boston.

Novak, T.P., Hoffman, D.L., Yung, Y., Measuring the Customer Experience in Online Environments: A Structural Modeling Approach, *Marketing Science*, Vol 19, no 1, 2000, pp.22.

Palmer, J.W. (2002), "Web site usability, design, and performance metrics." *Information Systems Research*, 13 (2), 151–167.

Parasuraman, A., Zeithaml, V.A., Malhotra, A. (2005), "E-S-QUAL: A Multiple-Item Scale for Assessing Electronic Service Quality." *Journal of Service Research*, 49(4), 406–13.

Ravden, S.J. and Johnson, G.E. (1989), *Evaluating Usability of Human Computer Interfaces: A Paractical Method*. Ellis Horwood, Chichester, UK.

Seffah, A., Donyaee, M., Kline, R.B., and Padda, H.K. (2006), "Usability measurement and metrics: A consolidated model." *Software Quality Journal*, 14 (2), 159–178.

Shahizan, H. and Li, F. (2007), *"Benchmarking the Usability and Content Usefulness of Web Sites: Developing a Structured Evaluation Framework"*, in: Social Implications and Challenges of e-Business, Li, Feng (Ed.). IGI Global Publishing, Hershey, Pennsylvania.

Venkatesh, V., and Agarwal, R. (2006), "Turning Visitors into Customers: A Usability-Centric Perspective on Purchase Behavior in Electronic Channels." *Management Science*, 52 (3), 367–382.

Zhang, P. and Li, N. (2005) "The Intellectual Development of Human-Computer Interaction Research: A Critical Assessment of the MIS Literature (1990-2002)." *Journal of the Association for Information Systems*, 6 (11), 227–292.

Chapter 70

A Curriculum for Usability Professionals (and Why We Need One)

Randolph G. Bias

School of Information
The University of Texas at Austin

ABSTRACT

There is no universally-accepted certification for usability professionals. In the US, one can pass the test of the Board Certification of Professional Ergonomists and become a "Certified Human Factors Professional." But few employers demand such a certification for someone they hire. Jakob Nielsen (2005), says "we will need to scale up by a factor of 100 in terms of available [usability] personnel" (p. 3). Elsewhere (Bias, 2003) I've written about the dangers of amateur usability engineering. How shall we train those new usability professionals? How shall we help avoid amateurism in the field? Here I offer my thoughts on usability curriculum for usability professionals and the real-world context that motivates the need for such a curriculum.

Keywords: Usability, human-computer interaction, user-centered design, curriculum

INTRODUCTION

Elsewhere (Bias, 2003) I've written about the dangers of amateur usability engineering. In short, if one does a poor job of programming, he/she gets discovered by system test time. But if one does a poor job of usability engineering, this doesn't get discovered until the Web site cuts live and visitors start leaving in droves, or the software application ships and the customer support phones start ringing.

Usability engineering *grand pere*, Jakob Nielsen (2005), says "we will need to scale up by a factor of 100 in terms of available [usability] personnel" (p. 3). How shall we train those new usability professionals? How shall we help avoid amateurism in the field, and promote the acceptance of usability as a professional discipline?

THE CONTEXT

Steve Krug (2006), the author of the justifiably popular and influential *Don't make me think!*, offers on his Web site the following "Tip of the month," and says that it also has been "Last Month's Tip, since June 1997": "If you really want to know if your Web site works, ask your next door neighbor to try using it, while you watch. (You bring the beer.)" (Advanced Common Sense, 2010). To this I say, if your target user audience is "tipsy neighbors," this is a great idea. If it is anyone else, such an extreme discount approach to usability engineering is dangerous.

There is no universally-accepted certification for usability professionals. In the US, one can pass the test of the Board Certification of Professional Ergonomists and become a "Certified Human Factors Professional." But few employers demand such a certification for someone they hire. Too, it is commonly the case that a software developer can get a bachelor's or master's degree in computer science or electrical engineering and take zero classes in usability or human-computer interaction. In my perhaps-should-be-humbler opinion, this has several repercussions:

- Anyone who wishes can claim to be a "usability professional,"
- Therefore, there is an unevenness in the excellence of usability support,
- Therefore, we suffer from a relative lack of respect in the Web or other software design/development world,
- And so, usability does not enjoy a routine seat at that design/development table,
- And when an organization DOES decide to invest in usability, hiring managers have a relatively hard time in finding and selecting usability professionals.

DANGERS OF AMATEUR USABILITY ENGINEERING

The absence of any widely-accepted certification for usability professionals, combined with the fact that SOME of usability engineering is "not rocket survery" (Krug, 2010), plus the advent of tools that allow even tyros to generate Web sites, has conspired to motivate more and more people to do some sort of perhaps rudimentary usability engineering. Of course every usability professional has to have a first day on the job. The danger comes when that day should be also the first day he or she conducts some usability evaluation.

Of course, there are short courses and other resources to help the would-be usability professional "ramp up," courses designed to help sensitize people to the importance of usability engineering, and tool them up so they can carry out certain methods. But with no overseeing board, and lack of widespread understanding of what constitutes excellent usability engineering practice, this state of affairs often produces a poor usability engineering product.

CREDENTIALS VS. TALENT

The issue is more than credentialing. It's about excellent, systematic usability engineering. It's about our field being perceived as a true, professional discipline. It's about enabling software development managers to distinguish whom to hire. Sure, it is also about enabling as many people as possible to do as good a job as possible usability engineering their designs. And I am NOT some credentials bigot who believes that some particular college degree is a prerequisite for conducting excellent usability engineering support.

ARE WE TO BE LIKE DOCTORS OR CIVIL ENGINEERS OR DECK BUILDERS?

Imagine some vice president of engineering finally springing for a little usability support and receiving BAD support from a usability amateur claiming to be a professional. My fear (and I have seen this happen) is that the VP will NOT think, "Hmm, I received poor usability support." Rather, he or she will tend to think, "Usability isn't worth it," and decide never, ever to spend one dollar on usability again.

Consider the last time you had surgery, did you have it performed by an AMA-licensed surgeon? What if I had offered to perform the surgery for you for half the cost? I'll bet you wouldn't've taken me up on my offer.

But, you say, perhaps, usability is a practice, a craft. A better analogy would be civil engineering. Now, if I wanted a bridge built, I'd STILL go to a licensed or somehow-certified engineer. But that doesn't mean that every person who wants to build a deck in his or her back yard should have to go earn an engineering certification before digging the first post hole.

PERCEPTION OF OUR FIELD

Are we doomed to having people think our field is more analogous to building a deck in their back yard than to building a bridge? I think there's a big difference between my building a deck in my back yard and my paying someone to build me a deck. Certainly my expectations are different. (And I know my wife has expectations that are way different – she would expect the deck we paid for to be not terrible.)

It is frustrating to me that we have to spend so much of our professional time trying to insinuate ourselves onto projects because there is no systemic acceptance of the value of usability. That's why Deborah Mayhew and I edited two books (1994, 2005) on *Cost-Justifying Usability*; it was the best way we found to gain credibility and a seat at the software development table. And as a usability consultant I find it hasn't gotten much better. I take it as a personal failing that I haven't convinced the world that usability is "worth it." What other service provider must spend so much of his/her time convincing would-be clients that they have a problem? When you need a plumber the evidence is sloshing around your feet. So often, when a team needs a usability professional they think, "Aw, heck, we're users. We can probably trust our own intuitions as to what is usable." And they fail to consider that their target audience may not have degrees in electrical engineering, and may not have spent the last 18 months of their lives working on this very product interface.

One-time great University of Texas football coach Darrell Royal once said, in support of having a run-oriented offense (as opposed to a pass-oriented one), "When you pass the ball, three things can happen and two of 'em are bad." Well, when Krug offers his "tip" to people, I think three things can happen, and two of 'em are bad. Now, people MIGHT get some good, quick, valuable feedback on their Web design. But they MIGHT get some BAD guidance, if the neighbor is unrepresentative of the target audience in a meaningful way (let's say the neighbor is a Web navigation genius). And the second possible bad thing is it gives people (and by people I mean "software development directors who might otherwise pay for real, professional, systematic usability evaluation") the impression that you can get all the usability you need for the investment of one hour and one or two long-necks.

A CURRICULUM

And so, we at The University of Texas at Austin School of Information choose to educate the next generation of usability practitioners, systematically and professionally. We teach usability and related classes in our master's program. We have launched students into industry for seven years with an MSIS (Master's of Science in Information Studies) and a focus on information architecture, design, and usability. And we are always looking to improve our pedagogy. We seek feedback from the employers who have hired our graduates, and seek proactive advice about what skills potential employers will need from our current and future students. And washing across it all is a fundamental belief in the steeping of our students in the theory of human information processing and human-computer interaction, a systematic coverage of usability engineering methods, an emphasis on the importance of the ability to conduct research and interpret others' research, and a focus on the vital nature of advocacy and an understanding of business process.

A person coming out of the University of Texas at Austin School of Information with a master's degree who intends to get a job as a usability professional will likely have taken an intro and an advanced class in information architecture, an intro and an advanced class in digital media design, and an intro and an advanced class in usability, each one of them 15 weeks. Plus he or she will have either written a thesis in the area or had some capstone experience wherein he or she conducted an industrial-strength piece of design work and/or usability evaluation for some company or agency. Plus, just as all master's students coming out of our school, he or she will have taken a Research Methods class covering descriptive and inferential statistics, experimental design, and qualitative research methods.

CLASSES

Our master's program entails a set of core courses covering organizing and managing information, an introduction to research methods, and user studies. In addition to this a student interested in usability can take courses in usability, information architecture, design, and human-information processing. In general the usability topics that are covered include:
- the scientific underpinnings of usability, including:
  - human perception
  - human cognition
  - memory
  - mental models
  - decision making
  - human-computer interaction
- organization of information to fit the task
- user profiles
- user requirements gathering
- principles of visual design
- principles of interaction design
- usability evaluation
  - inspection methods
  - testing methods
- business process, including cost-justifying usability

In addition to the usability training, the students with master's degrees who come out of our program, via the aforementioned research methods class, have a scientist's awareness of good research practice. For instance, they know that it is bad science to accept the null hypothesis. And thus often it is dangerous practice to make conclusions based on "no difference." I assert that the average (or even above average) computer programmer or tech writer does NOT know this. It is often the case that I see usability "professionals" find no difference – say between performance by one user audience on a new prototype design and performance by another user audience – and claim that the design serves both audiences equally well. When I'm "out there" in the world, serving as a consultant, I never worry when I compete for a contract with someone who is not savvy in the ways of science. Well, I might not get the contract, but I know my work will be superior, because I won't make mistakes such as this one. And so will the work of our students whose education is steeped in research design and practice.

EXPERIENCE

Beginning with class assignments and continuing through their "capstone" experience, usability-interested students develop a portfolio of design and evaluation projects. Indeed, the usability classes entail guest presentations by current usability professionals working in large software companies, small consultancies, state agencies, and other arenas. These people provide valuable, current, actionable advice about how best to implement the design and evaluation "book learning" our students receive.

Plus, all usability classes have a project orientation. In my Introduction to Usability class I have students pair up to conduct usability evaluations of the Web site of some nonprofit agency. They are required to write the test plan as though they were going to do it "right," but then they are free to test a convenience sample. In Advanced Usability they carry out another usability evaluation project, but this one is different in that a) it is an individual project and b) it must be "industrial strength." That is, they must test a sample representative of the target audience.

In these hands-on exercises, the students get a chance to try out usability evaluation methods under the guidance of a usability professional, to fine-tune their technique, and to get experience advocating for their data, in front of the product or site stakeholders.

MENTORING

As a long-time practitioner, my own teaching includes a good bit of focus on a master-apprentice approach. Yes, people can learn usability "on the job." But how much better is it if they are able to try out methods in the less stressful environment of a class or independent study exercise, with a usability professional offering advice along the way? In our program we provide real-time mentoring of our students, as they try out their new methods. Plus we serve as mentors for them as they enter industry.

PORTFOLIO

One explicit component of our program is our encouragement of our students to build a portfolio of their design and evaluation efforts throughout their information architecture, design, and usability classes, and throughout their independent studies and their capstone experiences.

One of the problems fostered by amateur usability engineering is the difficulty hiring managers have figuring our whom to hire. One might argue that if some Web- or other user-interface-development manager is inspired enough to want to hire someone with usability expertise, that he or she has the wherewithal to do a good job of that. The problem is, that hiring manager has seen a boat load of programmers – probably used to be one. But he/she may or may not have any experience with usability. So, in a field such as ours, where there is no widely acknowledged certification, only a fairly young history of actual degrees in the field, and a perhaps spotty work-place history it is doubly important to offer crisp, reliable distinguishers for who is and is not a professional. It is our hope that our structured program, and our students' portfolios, enable even the hiring manager without much experience with usability with the opportunity to make wise choices about new hires.

DISCERNMENT

One important thing for any budding usability professional to acquire is the DISCERNMENT to know what methods to employ when, and the discernment to know what methods the newly trained professional should try on his or her own, and for which ones should we "wait 'til the doctor arrives." As Abraham Maslow (1966) said, "I suppose it is tempting, if the only tool you have is a hammer, to treat everything as if it were a nail" (p. 15). If the only usability engineering method you know is heuristic evaluation, then that's the method you'll choose to apply to any design problem. The usability professional needs to know not only how to conduct usability methods, in gathering user data to inform and validate user interface designs, but also know which methods to use when.

A CODA: USABILITY FOR EVERYONE

Nielsen, when suffering over the lack of usability expertise, goes on to recommend a

second solution . . .: to expand usability beyond the usability professionals.

If everybody needs usability, then everybody should do usability. We may need 50 million people who know some usability, but [that] is easier to achieve than getting one million people who are full-time usability experts. (p. 3)

In addition to careers in usability, information architecture, and design, some School of Information students become school librarians, academic librarians, public librarians, digital librarians, archivists, conservators, and the like. Each will at some point design some interface, be it physical or virtual, to information resources. How much better might those interfaces be, how much more successful, efficient, and pleased will the information retrievers be, if these information professionals have had at least one course in applying usability methods to that design?

REFERENCES

Advanced common sense. (2010). http://www.sensible.com/. Accessed February 20, 2010.

Bias, R. G. (2003). The dangers of amateur usability engineering. In S. Hirsch (chair), Usability in practice: Avoiding pitfalls and seizing opportunities. Annual meeting of the American Society of Information Science and Technology, October, Long Beach.

Bias, R. G., & Mayhew, D. J. (Eds.) (1994). *Cost-Justifying Usability*. Boston: Academic Press.

Bias, R. G., & Mayhew, D. J. (Eds.) (2005). *Cost-Justifying Usability, 2nd Edition: Update for the Internet Age*. San Francisco: Morgan Kaufmann.

Krug, S. (2006). *Don't make me think!: A common sense approach to Web usability, 2nd edition.* Berkeley, CA: New Riders.

Krug, S. (2010). *Rocket surgery made easy: The do-it-yourself guide to finding and fixing usability problems*. Berkeley, CA: New Riders.

Maslow, A. H. (1966). The psychology of science: A reconnaissance. New York: Harper & Row.

Nielsen, J. (2005). Usability for the masses. *Journal of Usability Studies, 1*, 1, pp. 2-3